Biological Invasions in the Tropics

Monitoring, Mechanism, Management

Biological Invasions in the Tropics

Monitoring, Mechanism, Management

Edited by

Akshay Kumar Chakravarthy
Former Head, Division of Entomology & Nematology,
Indian Institute of Horticultural Research, Bangalore, India
Former Professor and Head, Department of Entomology,
University of Agricultural Sciences, GKVK, Bangalore, India

Pathour Rajendra Shashank
Division of Entomology, ICAR-Indian Agriculture Research Institute,
New Delhi, India

CABI

CABI is a trading name of CAB International

CABI
Nosworthy Way
Wallingford
Oxfordshire OX10 8DE
UK

Tel: +44 (0)1491 832111
E-mail: info@cabi.org
Website: www.cabi.org

CABI
200 Portland Street
Boston
MA 02114
USA

T: +1 (617)682-9015
E-mail: cabi-nao@cabi.org

A catalogue record for this book is available from the British Library, London, UK.

ISBN-13: 9781836991953 (hardback)
9781836991960 (ePDF)
9781836991977 (ePub)

DOI: 10.1079/9781836991977.0000

Commissioning Editor: Ward Cooper
Editorial Assistant: Emma McCann
Production Editor: James Bishop

Typeset by Exeter Premedia Services Pvt Ltd, Chennai, India
Printed and bound by CPI Group (UK) Ltd, Croydon, CR0 4YY

Contents

Foreword

One of the most significant and alarming global phenomena is the introduction, colonization, and establishment of invasive species, and their impact on biodiversity and economies. Often, an entire ecosystem is at a risk due to the introduction of invasive alien species (IAS), resulting in irreversible loss of endemic species. All the major taxonomic groups of plants and animals have been recognized for their invasion potential. Severe reduction in population or near extinction of native species due to biological invasions has been documented. Invasive species disrupt ecosystem functioning, causing direct or indirect cascades of collapse. Additionally, their adverse impacts on public health and agricultural and natural ecosystems are well recognized.

Records of eradication or successful management of invasive species are scarcer in the tropics than in temperate regions. This is due to lesser understanding of biological invasions and limited financial support in developing countries of the tropics. Degree of successful invasion varies among the species depending upon climatic and environmental factors. For instance, among arthropods, ants better respond to changing food and environmental conditions. Reproduction by budding, formation of super colonies, polygyny, polydomy and omnivory are some of the features that facilitate their adaptation to changing situations. The need for research on the ecology, spread and management of invasive species among developing countries of the tropics requires the utmost attention.

This book, *Biological Invasions in the Tropics: Monitoring, Mechanism, Management*, summarizes the need for research on invasive species in the tropics in general, and in South Asia in particular. It contains 18 chapters in three sections, with the first section containing two chapters on invasions and invasives. The first chapter explores evolution of invasives and the second chapter on invasives in South Asia. The second section focuses on invasives in ecosystems, with three chapters addressing tropical forests, tea ecosystem and the Himalayan region. The third section covers insects as invasives with 13 chapters on thrips, whiteflies, tomato pin worm, fall armyworm, beetles, scale insects and mealybugs. These chapters provide a 'window' into the diverse modes of spread and management options.

To tackle invasive alien species, countries must have an effective coordinating and networking mechanism with adequate funding support. Many South and South-East Asian countries lack policies, programmes and plans for management of invasive species. In limited cases, mostly supported by external funding for implementation, follow-up actions are often absent. This book emphasizes the need for national and international cooperation and coordination and execution of research and implementation plans. Hopefully, this book will trigger awareness on prevention, spread, eradication and sustainable management of invasive species.

Rangaswamy Muniappan
Director, IPM Program
Center for International Research, Education, and Development
Virginia Tech
526 Prices Fork Road
Blacksburg, VA 24061 USA

Contributors

Aarthi Helen P., Division of Entomology, ICAR- Indian Agricultural Research Institute, New Delhi, India.

Abhay Kumar Pandey, Department of Mycology and Microbiology, Tea Research Association, North Bengal Regional R&D Centre, Nagrakata, Jalpaiguri, West Bengal, India.

Abhinaya, Department of Entomology, College of Agriculture, Raichur, Karnataka, India.

Akshay Kumar Chakravarthy, Division of Crop Protection, ICAR – Indian Institute of Horticultural Research, Bengaluru, Karnataka, India (Retired).

Anbuchezhiyan Sneka, Department of Entomology, Faculty of Agriculture, Annamalai University, Chidambaram, Cuddalore, India.

Anil Kumar G., Department of Entomology, Agricultural College, Bapatla, ANGRAU, Andhra Pradesh, India.

Arindam Kumar, Division of Crop Protection, ICAR – Indian Institute of Horticultural Research, Bengaluru, India.

Arunkumara C.G., ICAR – National Rice Research Institute, Central Rainfed Upland Rice Research Station, Hazaribagh, Jharkhand, India.

Athulya R., Forest Protection Division, Institute of Wood Science and Technology, Bengaluru, India.

Atiraj Rathi, Forest Protection Division, Arid Forest Research Institute, Jodhpur, India.

Basavaraj K., Department of Entomology, College of Agriculture, Gangavathi, Karnataka, India.

Bhimashankar Gotyal, Division of Germplasm Conservation and Utilisation, ICAR – National Bureau of Agricultural Insect Resources, Bengaluru, Karnataka, India.

Chaitanya B., University of Agricultural Sciences, Raichur, Karnataka, India.

Chaithra G.N., Mahatma Gandhi Botanical Garden, University of Agricultural Sciences, GKVK, Bengaluru, India.

Chandra Ghosh, Department of Tea Science, University of North Bengal, Darjeeling, West Bengal, India.

Deepak Kumar Mahanta, Forest Entomology Discipline, Forest Protection Division, Indian Council of Forestry Research and Education (ICFRE) – Forest Research Institute (ICFRE–FRI), Dehradun, Uttarakhand, India.

Dhyan Chowdary D., Department of Entomology, Agricultural College, Bapatla, ANGRAU, Andhra Pradesh, India.

Elango Santhoshinii, Department of Plant Pathology, University of Agricultural Sciences, Dharwad, Karnataka, India.

Eurekha S., University of Agricultural Sciences, Raichur, Karnataka, India.

Eurekha Savadatti, Department of Entomology, College of Agriculture, Dharwad, Karnataka, India.

Godavari, Division of Entomology, ICAR- Indian Agricultural Research Institute, New Delhi, India.

Gundappa B., Division of Germplasm Conservation and Utilisation, ICAR – National Bureau of Agricultural Insect Resources, Bengaluru, Karnataka, India.

Ifrah Khan, Forest Protection Division, Arid Forest Research Institute, Jodhpur, India.

Ipsita Samal, ICAR-National Research Centre on Litchi, Mushahari, Ramna, Muzaffarpur, Bihar, India.

Kariyanna B., Division of Fluoro-Agrochemicals, CSIR – Indian Institute of Chemical Technology, Tarnaka, Hyderabad, India.

Keerthi M.C., Division of Entomology and Nematology, ICAR – Indian Institute of Horticultural Research, Bangalore, India.

Kishore Chandra Sahoo, Division of Entomology, Graduate School, ICAR – Indian Agricultural Research Institute, Assam, India.

Komal J., Basic Seed Multiplication and Training Centre, Central Silk Board, Pendari, Bilaspur, Chhattisgarh, India.

Kumara A.D.N.T., Department of Biosystems Technology, Faculty of Technology, South Eastern University of Sri Lanka, University Park, Oluvil, Sri Lanka.

Manish Bharadwaj, Department of Zoology, Satyendra Narayan Sinha College (Magadh University), Warisaliganj, India.

M.N. Rudra Gouda, Division of Entomology, ICAR – Indian Agricultural Research Institute, New Delhi, India.

Mogili Ramaiah, Division of Entomology, ICAR – Indian Agricultural Research Institute, New Delhi, India.

Mukesh Kumar Dhillon, Division of Entomology, ICAR – Indian Agricultural Research Institute, New Delhi, India.

Nadur Lakkanna Naveena, Department of Entomology, College of Agricultural Sciences, Keladi Shivappa Nayaka University of Agriculture and Horticultural Sciences, Iruvakki, Shivamogga, Karnataka, India.

Nagalapura Ramakrishnappa Prasannakumar, Division of Crop Protection, ICAR – Indian Institute of Horticultural Research, Bengaluru, India.

Nagaraju T., Division of Fluoro-Agrochemicals, CSIR – Indian Institute of Chemical Technology, Tarnaka, Hyderabad, India.

Nandhini D., Division of Entomology, ICAR- Indian Agricultural Research Institute, New Delhi, India.

Palanisamy Dhanapriya, Department of Entomology, Faculty of Agriculture, Annamalai University, Chidambaram, Cuddalore, India.

Pathour Rajendra Shashank, Division of Entomology, ICAR – Indian Agricultural Research Institute, New Delhi, India.

Peketi Krishna Reddy, Department of Entomology, College of Agricultural Sciences, Keladi Shivappa Nayaka University of Agriculture and Horticultural Sciences, Iruvakki, Shivamogga, Karnataka, India.

Poiyamozhi Harikrishnan, Department of Biotechnology, University of Agricultural Sciences, Dharwad, Karnataka, India.

Prabhulinga T., ICAR – National Bureau of Agricultural Insect Resources, Bengaluru, Karnataka, India.

Pramod Kumar M., Division of Fluoro-Agrochemicals, CSIR – Indian Institute of Chemical Technology, Tarnaka, Hyderabad, India.

Preeti Sharma, Division of Entomology, ICAR – Indian Agricultural Research Institute, New Delhi, India.

Priyanka Nayak, Institute of Agricultural Sciences, Siksha 'O' Anusandhan, Bhubaneswar, Odisha, India.

Rahul Kumar, Department of Zoology, Sheodeni Sao College (Magadh University), Kaler, India; University Department of Zoology, Vinoba Bhave University, Hazaribagh, India.

Ramya N., ICAR – Indian Institute of Agricultural Biotechnology, Ranchi, Jharkhand, India.

Sainath G., Division of Fluoro-Agrochemicals, CSIR – Indian Institute of Chemical Technology, Tarnaka, Hyderabad, India.

Saipooja Nelluri, Division of Fluoro-Agrochemicals, CSIR – Indian Institute of Chemical Technology, Tarnaka, Hyderabad, India.

Saleemali Kannihalli, Department of Entomology, College of Agriculture, Gangavathi, Karnataka, India.

Sampath Reddy, Division of Fluoro-Agrochemicals, CSIR – Indian Institute of Chemical Technology, Tarnaka, Hyderabad, India.

Samreen, Division of Entomology, ICAR – Indian Agricultural Research Institute, New Delhi, India.

Sangeeta Singh, Forest Protection Division, Arid Forest Research Institute, Jodhpur, India.

Santhosh Naik, Division of Entomology, ICAR – Indian Agricultural Research Institute, New Delhi, India.

Sarbani Roy, Tocklai Tea Research Institute, Tea Research Association, Jorhat, Assam, India.

Satyapriya Singh, Central Horticultural Experiment Station (ICAR – Indian Institute of Horticultural Research), Bhubaneswar, Odisha, India.

Shivakumara K.T., Division of Germplasm Conservation and Utilisation, ICAR – National Bureau of Agricultural Insect Resources, Bengaluru, Karnataka, India.

Somnath Roy, Tocklai Tea Research Institute, Tea Research Association, Jorhat, Assam, India.

Sourajit Bayen, Tocklai Tea Research Institute, Tea Research Association, Jorhat, Assam, India.

Sourav Chakrabarty, National Pusa Collection, Division of Entomology, ICAR –Indian Agricultural Research Institute, New Delhi, India.

Sringeswara A.N., Mahatma Gandhi Botanical Garden, University of Agricultural Sciences, GKVK, Bengaluru, India.

Stuti, Department of Zoology, Banaras Hindu University, Varanasi, India.

Subham Gambhir, Division of Entomology, ICAR – Indian Agricultural Research Institute, New Delhi, India.

Subrata Goswami, Department of Entomology and Agricultural Zoology, Institute of Agricultural Sciences, Banaras Hindu University, Varanasi, Uttar Pradesh, India.

Sudipa Das, National Pusa Collection, Division of Entomology, ICAR – Indian Agricultural Research Institute, New Delhi, India.

Suman Barman, Division of Entomology, ICAR – Indian Agricultural Research Institute, New Delhi, India.

Sunil Kumaraswamy, Department of Agricultural Entomology, University of Agricultural Sciences, Bengaluru, India; Department of Biology, Oklahoma State University, Stillwater, OK 74075, United States.

Sunitha N.D., Department of Entomology, College of Agriculture, Vijayapura, UAS, Dharwad, India.

Suresh M. Nebapure, Division of Entomology, ICAR- Indian Agricultural Research Institute, New Delhi, India.

Tambat B., Department of Forestry and Environmental Science, UAS, GKVK, Bengaluru, India.

Tanmaya Kumar Bhoi, Forest Protection Division, Arid Forest Research Institute, Jodhpur, Rajasthan, India.

Tulasi B., Division of Entomology, ICAR – Indian Agricultural Research Institute, New Delhi, India.

Usha Nandini A., Society for Science and Technology Applications (SSTA), Bengaluru, India.

Varsha Giri, Silviculture and Forest Management Division, Arid Forest Research Institute, Jodhpur, India.

Velmurugan Anithaa, Division of Germplasm Conservation and Utilisation, ICAR – National Bureau of Agricultural Insect Resources, Bengaluru, Karnataka, India.

Velusamy Naveen, Department of Entomology, University of Agricultural Sciences, Dharwad, Karnataka, India.

Vipula Vyas, Forest Protection Division, Arid Forest Research Institute, Jodhpur, India.

Yandigeri M.S., Division of Germplasm Conservation and Utilisation, ICAR – National Bureau of Agricultural Insect Resources, Bengaluru, Karnataka, India.

About the Editors

Dr Akshay Kumar Chakravarthy, Head and Principal Scientist (Rtd), Division of Crop Protection, is an author of many books and over 320 scientific papers and 60 chapters in Entomology and Natural History. He is Fellow of the Entomological Society of India. His interests include insects, birds, bats, rodents and mammals. With nearly four decades of experience in teaching, research and extension, Dr Chakravarthy has been the investigator for over 35 research projects and has guided more than 28 students for post-graduation. Recently, he has published books titled *Innovative Pest Management Approaches for the 21st Century* (Foreword by Prof. Dr Remzi Atlihan, Turkey) *New Horizons in Insect Science: Towards Sustainable Pest Management* (Foreword by Prof. Hiroshi Honda, Japan), *Economic and Ecological Significance of Arthropods in Diversified Ecosystems: Sustaining Regulatory Mechanisms* (Foreword by Fani Hatijina, Greece), *Arthropod Diversity and Conservation in the Tropics and Sub-tropics* (Foreword by Prof. Madhav Gadgil, India), *The Black Spotted, Yellow Borer, Conogethes punctiferalis Guenee and Allied Species* (Foreword by Prof. Alma Solis, USA) and *Experimental Techniques in Host Plant Resistance* (Foreword by Prof. Michael C. Smith, USA) by Springer, *Genetic Methods and Tools for Managing Crop Pests* (Foreword by Dr Srinivasan Ramaswamy, Taiwan) and *Advances in Commercial Flowers*, AAP-Apple Academic Press.

Dr Pathour Rajendra Shashank, Senior Scientist, Division of Entomology, ICAR-Indian Agricultural Research Institute, is a researcher who is mainly working on insect taxonomy, biodiversity, molecular diagnosis and invasive insect pests. He has been involved in describing 25 new species, three new genera, 15 new records to India and also 25 new host records, 185 new range records, one monograph, seven checklists for Lepidoptera and Cicadellidae groups. His expertise in DNA barcoding of insect pests generated more than 750 DNA barcodes which aid in authentic identification of domestic and invasive insects. He is responsible for collection, curation and identification of order Lepidoptera in IARI-National Pusa Collection for agricultural insect pests. From 2013 to 2025, he has provided Insect Pest Diagnostic Services for more than 45 national institutes/ universities by identifying more than 2650 specimens. Dr Shashank has 13 years of teaching, research and extension experience in Entomology and he is a recipient of a DST INSPIRE fellowship, DST-Start Up Research Grant (Young Scientist), NAAS Associateship, ESI–Young Entomologist Award. He has received many prestigious awards from different societies, and has published about 95 research articles in peer-reviewed journals. He is also conceiver and managing editor of *Indian Entomologist*, an online magazine, and Editor-in-chief of *Hexapoda*, the journal published by the Entomological Society of India (ESI), New Delhi to promote insect science and technology in India. Shashank is associated with ESI actively as a member and in other capacities, serving the ESI in its various activities for many years.

1 Evolution of Invasiveness in Arthropods: Divergence is the Key

Pathour Rajendra Shashank[1]*, Akshay Kumar Chakravarthy[2] and Santhosh Naik[1]

[1]Division of Entomology, ICAR – Indian Agricultural Research Institute, New Delhi, India; [2]Division of Crop Protection, ICAR – Indian Institute of Horticultural Research, Bengaluru, Karnataka, India (Retired)

Abstract

The evolution of invasiveness in arthropod populations is poorly understood. Gene flow, founder effects, population size and genetic structure may contribute to genetic drift. The unique characteristics of the introduced habitats may also influence evolutionary changes in arthropods, including invasiveness. Much remains to be known about the diverse ways in which several factors shape the genetic structure of invasive populations. Several arthropods become invasives when moved by humans or other routes outside their native habitat range. Many such species adapt to exploit their new environments. Introduced arthropods can have diverse effects on closely related native species through competition, dispersal ability, escape from natural enemies, disease transmission or predation. Short generation times and rapid responses to landscape and climatic changes enable arthropods to adapt quickly to changing food resources and environmental conditions. Each arthropod assemblage may prefer a specific successional stage of a habitat or the environmental conditions. The genetic structure of invasive populations may not undergo wide changes, but there may be substructuring of the populations. This is an important and interesting aspect of arthropod biology and ecology for future research.

Keywords: Arthropods, invasiveness, evolution, genetics, gene flow, founder effects, adaptation

1.1 Introduction

Invasive alien species are spreading at unprecedented rates globally, primarily due to human activities such as trade, travel, land use changes, and climate change. Over 37,000 species have been introduced worldwide, with around 3,500 identified as harmful to biodiversity, ecosystems, and human well-being. The annual global cost of biological invasions has reached $423 billion, quadrupling since the 1970s, and projections suggest a 36% increase by 2050 (IPBES, 2023). Islands and terrestrial ecosystems are particularly vulnerable, with climate change exacerbating threats by expanding the range of invasive species and destabilizing native habitats. Invasive species contribute to feedback loops such as "invasional meltdown", intensified by habitat degradation and urbanization. These species, alongside other factors, are linked to nearly 60% of global extinctions, with 90% occurring on islands. Furthermore, the economic costs of invasions have increased fourfold each decade,

*Corresponding author: spathour@gmail.com

DOI: 10.1079/9781836991977.0001

and their impact on quality of life is negative in 85% of documented cases (Huang *et al.*, 2011; IPBES, 2023).

Arthropods comprising insects, mites, and crustaceans are among the most successful invasive organisms, raising serious concerns for global ecology, agriculture, and biodiversity. Their evolutionary adaptability enables them to colonize novel habitats, outcompete native species, and rapidly adjust to changing environmental conditions. Despite their significant global impact, the evolutionary mechanisms underlying arthropod invasiveness remain insufficiently explored. Arthropods are the most diverse and ancient animal group on Earth, with around 1,302,810 recorded species, including 45,770 fossils. Arthropods range from microscopic to several metres long and comprise 80% of all known animal species, with insects representing about three-fourths of the total (Chapman, 2009). Originating during the Cambrian period, arthropods are considered macrophyletic, although their precise evolutionary relationships remain debated (Resh and Carde, 2009). Arthropods dominate terrestrial ecosystems, thriving in diverse habitats. Their segmented bodies, often with tagmata (crustaceans have two, myriapods lack them), chitinous exoskeletons, and complex internal systems contribute to their success. Their respiratory systems vary, including gills, skin, tracheae, and book lungs, and reproduction is typically dioecious, though some reproduce parthenogenetically (Chakravarthy and Sridhara, 2016; Chakravarthy *et al.*, 2016).

The invasion process in arthropods typically progresses through transport, introduction, establishment, and spread. Arthropods are often inadvertently transported via global trade, shipping, or agriculture (Liebhold *et al.*, 2012). During the introduction phase, survival depends on overcoming new abiotic and biotic challenges, with success influenced by propagule pressure, the number and frequency of introduced individuals (Simberloff, 2009). Once established, they form self-sustaining populations, aided by rapid evolutionary adaptation and phenotypic plasticity that enable exploitation of new resources (Sakai *et al.*, 2001; Lee, 2002). The spread phase is driven by dispersal ability and further human facilitation, often accelerated by synergistic interactions among invaders, termed "invasion meltdown" (Simberloff and Von Holle, 1999; Hulme, 2009). Invasive arthropods displace native species, alter habitats, transmit diseases, and cause substantial ecological and economic damage (Kenis *et al.*, 2009). Continuous evolutionary changes often increase their fitness, persistence, and impact (Prentis *et al.*, 2008).

Effective management requires robust national policies aligned with global efforts such as the Kunming–Montreal Global Biodiversity Framework, which targets halving the rate of invasive introductions by 2030. Yet, over 80% of countries lack specific legislation and under-invest in control efforts. Enhanced regulations, especially for online trade, development of voluntary industry codes, and improved biosecurity protocols are critical. Global treaty coordination (CBD, WTO, IMO, CITES), inclusion of Indigenous knowledge, and investment in technology, awareness, and economic incentives such as subsidies or penalties are essential for an effective response. The ongoing biodiversity crisis marked by habitat loss, extinctions, and biological invasions is a direct outcome of human activity (IPBES, 2016, 2019).

This chapter explores the hypothesis that genetic, phenotypic, and ecological divergence drives the evolution of invasiveness in arthropods. Key evolutionary forces including founder effects, genetic drift, gene flow, local adaptation, and population substructuring shape divergence from ancestral populations. The interaction between evolutionary history and ecological opportunity leads to diverse invasion outcomes, depending on invader traits and ecosystem vulnerabilities. Arthropods, with short generation times, high fecundity, and varied reproductive modes, are ideal models for studying rapid evolutionary shifts. Traits such as phenotypic plasticity, pre-adaptation, and selection after introduction enable them to achieve ecological dominance. Understanding these mechanisms aids in forecasting invasions and developing effective control strategies. This chapter synthesizes existing knowledge, highlights case studies of divergent invasion patterns, and identifies future research priorities in evolutionary biology, invasion ecology, and population genetics.

1.2 The Nature of Arthropod Invasiveness

Biological invasions occur when species establish themselves beyond their native geographic ranges, often through natural dispersal or human-mediated transport, leading to negative impacts on ecosystems, biodiversity, and human welfare. Unlike simple colonization, invasions typically disrupt ecological balance. Adaptation, the genetic change in populations due to natural selection, enables invasive species to thrive in novel environments. The evolutionary dynamics of invasions remain underexplored despite their increasing frequency and intensity (Lee, 2002). Evolutionary processes such as natural selection, genetic drift, and gene flow can rapidly alter the genetic characteristics of invasive populations, influencing their success and interaction with native species (Filchak *et al.*, 2000; Reznick and Ghalambor, 2001). Genetic attributes like additive genetic variance, epistasis, hybridization, and genetic trade-offs contribute significantly to invasion potential (Lee, 2002). For instance, initial selection may favour dispersal ability, followed by adaptation to local conditions. Rapid evolutionary changes have been documented in response to environmental gradients, and in some cases, invasions have led to reproductive isolation and even speciation (Lee, 1999). Given the mounting ecological threats posed by invasions, integrating evolutionary genetics into biosecurity and management is essential (Lee, 2002).

1.2.1 Defining invasiveness

Invasive species are non-native arthropods that, once introduced into environments beyond their natural range, establish self-sustaining populations, spread without further human intervention, and cause significant ecological, economic, or health-related impacts. Examples include *Linepithema humile* (Argentine ant), *Tuta absoluta* (tomato leaf miner), and *Aedes albopictus* (Asian tiger mosquito). Invasiveness is a complex trait influenced by ecological compatibility, reproductive potential, dispersal ability, and evolutionary plasticity. Such species often possess traits such as high reproductive rates, broad environmental or host tolerance, and absence of natural enemies, enabling them to disrupt native biodiversity, agricultural productivity, or ecosystem services (Blackburn *et al.*, 2011; Liebhold *et al.*, 2016).

1.2.2 Key traits promoting invasiveness in arthropods

Invasive arthropods often share specific biological, ecological, and behavioural characteristics that enable them to successfully establish, proliferate, and spread in new environments. These invasive species pose a major threat to natural and agricultural ecosystems worldwide, where they act as competitors, predators, parasites, or vectors of pathogens for native species (Kenis *et al.*, 2009). Such key traits facilitate their ability to colonize, adapt, and compete effectively with closely related local species. The following biological traits contribute to the success of invasive species once introduced beyond their native ranges.

1.2.2.1 High reproductive capacity

High reproductive capacity is a key factor driving the rapid evolution and successful establishment of invasive insects. Producing a large number of offspring increases survival chances and genetic variation, which in turn fuels adaptive evolution under new environmental pressures. For example, the invasive spotted wing drosophila (*Drosophila suzukii*) exhibits exceptionally high fecundity, laying hundreds of eggs over its lifetime. This reproductive advantage has enabled it to rapidly adapt to diverse climates and outcompete native fruit flies across North America and Europe (Asplen *et al.*, 2015). Similarly, the fall armyworm (*Spodoptera frugiperda*), known for its high reproductive rate, has quickly spread across Africa and Asia, evolving pesticide resistance and adapting to various host plants in a short time (Nagoshi *et al.*, 2019). Invasive arthropods often combine high reproductive capacity with short generation times and elevated fecundity, allowing them to establish and dominate new areas rapidly. For instance, *Tuta absoluta* completes up to 12 generations annually, with each female laying 250–300 eggs, posing a severe threat to tomato crops (Desneux *et al.*, 2010). The silverleaf whitefly *Bemisia tabaci* lays over 300 eggs per female and spreads plant viruses efficiently (De Barro *et al.*, 2011). *Drosophila suzukii* develops in 8–10 days, swiftly infesting fresh fruits (Hauser, 2011). Social insects like the Argentine ant (*L. humile*) form massive multi-queen colonies that outcompete native species (Holway *et al.*, 2002). These traits collectively make such species challenging to manage and emphasize the critical need for early detection and rapid-response biosecurity strategies.

1.2.2.2 Environmental tolerance

Environmental tolerance plays a crucial role in the evolution and success of invasive insects, enabling them to survive, reproduce, and spread across diverse and often unpredictable conditions. A broad tolerance to temperature, humidity, and host availability allows these species to establish in new regions and adapt to variable climates, thereby accelerating evolutionary responses to novel environmental pressures. For example, the Asian tiger mosquito (*Aedes albopictus*) has shown remarkable adaptability to both tropical and temperate environments, aided by its ability to enter diapause and withstand cold temperatures, which has facilitated its spread throughout Europe and North America (Medley, 2010). Similarly, *Drosophila suzukii*, originally native to Southeast Asia, has successfully colonized a wide range of fruit-growing regions worldwide due to its tolerance of varied climatic conditions and its ability to shift between wild and cultivated hosts (Asplen *et al.*, 2015). Invasive arthropods frequently exhibit such broad environmental tolerance, allowing them to thrive across a wide range of ecological settings from tropical ecosystems to temperate zones. For instance, *Solenopsis invicta* (red imported fire ant) demonstrates high thermal tolerance and soil adaptability, enabling its spread across the Americas, Asia, and Australia (Tschinkel, 2006). *Halyomorpha halys* (brown marmorated stink bug) has invaded temperate regions by entering diapause and seeking shelter during cold winters (Cira *et al.*, 2016), while *Solenopsis invicta* has also adapted to cooler montane environments, allowing it to colonize high-altitude areas (Lytle *et al.*, 2020). This environmental flexibility not only gives invasive arthropods a competitive edge over native species but also complicates management strategies, underscoring the importance of predictive, climate-based risk assessments in biosecurity planning.

1.2.2.3 Generalist feeding behaviors

Generalist feeding behaviour plays a significant role in the evolution and invasive success of insect species by enabling them to utilize a wide range of host plants or prey across diverse environments. This dietary flexibility enhances their ability to establish in new habitats, adapt to varying ecological conditions, and sustain populations even when preferred hosts are absent. For instance, the brown marmorated stink bug (*Halyomorpha halys*) is a highly polyphagous pest that feeds on over 100 plant species, including fruits, vegetables, and ornamentals. Its broad host range has facilitated rapid expansion throughout North America and Europe, leading to substantial agricultural losses (Leskey *et al.*, 2012). Similarly, the silverleaf whitefly (*Bemisia tabaci*) feeds on more than 600 plant species and transmits numerous plant viruses, making it one of the most damaging invasive agricultural pests globally (Oliveira *et al.*, 2001). Many invasive arthropods display such generalist feeding behaviours, allowing them to exploit a wide array of host resources across various ecosystems. This adaptability increases their invasive potential by enhancing survival, establishment, and spread. For example, *Spodoptera frugiperda* (fall armyworm) feeds on more than 186 plant species across 42 families, contributing to its rapid global expansion (Goergen *et al.*, 2016; van der Gaag and Straten, 2017; CABI (Centre for Agriculture and Bioscience International), 2019). Likewise, *Lycorma delicatula* (spotted lanternfly) feeds on over 70 host plants, causing severe economic damage in invaded regions (Barringer *et al.*, 2015). This broad dietary range enables these species to thrive in novel environments and evade control measures focused on specific host associations, further complicating pest management.

1.2.2.4 Dispersal mechanisms

Different dispersal mechanisms and structural adaptations are crucial to the evolution and success of invasive insects, enabling rapid colonization, range expansion, and adaptation to new environments. Invasive species disperse actively by strong flight or passively through wind, water, and human-mediated transport, with traits like well-developed wings and lightweight bodies enhancing mobility. For example, *Spodoptera frugiperda* (fall armyworm) uses strong flight and prevailing wind currents for seasonal migrations, promoting gene flow and rapid invasion across Africa and Asia, aided further by human activities (Westbrook *et al.*, 2016; Nagoshi *et al.*, 2019; Boaventura *et al.*, 2020). Similarly, *Drosophila suzukii* disperses widely due to robust flight (Asplen *et al.*, 2015), while *Solenopsis invicta* (red imported fire ant) uses nuptial flights for aerial queen dispersal (Tschinkel, 2006). Passive dispersal through trade facilitates spread of pests such as *Lycorma delicatula* (Barringer *et al.*, 2015). Liebhold *et al.* (2016) emphasize that invasion pathways often drive spread more than species traits. Moreover, global cultivation of host plants such as *Eucalyptus* has accelerated invasions (Hurley. *et al.*, 2016), and invasive insects are spreading faster than before due to evolving pathways (Garnas *et al.*, 2016; Roques *et al.*, 2016). While many invasions are accidental through commodities, biological control efforts can also contribute (Elliott *et al.*, 1996; Boettner *et al.*, 2000). Species such as *Lymantria dispar* utilize wind-assisted larval dispersal (Tobin *et al.*, 2012), and *Stegobium paniceum* spreads globally on trade goods (Rees, 2004). These diverse dispersal strategies enhance gene flow, population connectivity, and evolutionary adaptability, increasing invasive potential and complicating management.

1.2.2.5 Evasion of natural enemies

Evasion of natural enemies is crucial for the evolution and success of invasive insects, as it reduces pressures from predators, parasitoids, and pathogens in new environments, enabling rapid population growth and adaptation. This "enemy release" allows invasive species to reallocate energy from defence to reproduction and dispersal (Keane and Crawley, 2002). For example, the emerald ash borer (*Agrilus planipennis*) has thrived in North America due to the absence of its native natural enemies (Poland and McCullough, 2006), and the Argentine ant (*Linepithema humile*) dominates many regions by escaping native predators (Holway *et al.*, 2002). Such enemy release often leads to evolutionary shifts such as increased fecundity (Blossey and Nötzold, 1995). The brown marmorated stink bug (*Halyomorpha halys*) has rapidly spread in North America and Europe largely free from Asian parasitoids (Leskey *et al.*, 2012), while the Asian long horned beetle (*Anoplophora glabripennis*) causes extensive tree damage due to lack of natural enemies in introduced areas (Haack *et al.*, 2010). Similarly, *L. dispar* initially expanded due to missing parasitoids, though some biological control now exists (Boukouvala *et al.*, 2022). This evasion enhances invasion success but complicates management, highlighting the importance of understanding enemy interactions in controlling invasive insects.

1.2.2.6 Phenotypic plasticity and rapid evolution

Phenotypic plasticity and rapid evolution are key drivers of the success and adaptability of invasive insects, enabling them to quickly respond to novel environmental conditions and selective pressures. Phenotypic plasticity, the ability to alter physiology, morphology, or behaviour without genetic change, provides immediate advantages in new habitats, while rapid evolution involves genetic adaptation over short timescales that enhances survival and reproduction. For example, the spotted wing drosophila (*Drosophila suzukii*) displays high plasticity by adjusting reproductive timing and cold tolerance, supporting its spread across North America and Europe (Asplen *et al.*, 2015). Likewise, the fall armyworm (*Spodoptera frugiperda*) has rapidly evolved insecticide resistance and adapted to a variety of host plants, underscoring its evolutionary flexibility (Nagoshi *et al.*, 2019; Boaventura *et al.*, 2020). Invasive species often thrive due to such adaptive traits, supported by high genetic variation or epigenetic mechanisms (Richards *et al.*, 2006; Lande, 2016). For instance, *Myzus persicae* (green peach aphid) exhibits plasticity in development and reproduction, enabling survival under diverse climatic and host conditions (van Emden *et al.*, 1969; Margaritopoulos *et al.*, 2007), while *Ceratitis capitata* (Mediterranean fruit fly) evolves thermal and reproductive traits to expand into temperate regions (Ricalde et al., 2012). Similarly, *Leptinotarsa decemlineata* (Colorado potato beetle) has evolved resistance to numerous insecticides within decades (Alyokhin et al., 2008), and *Stator limbatus* demonstrates plastic responses that buffer selection when adapting to new hosts (Amarillo-Suárez and Fox, 2006). These combined mechanisms allow invasive insects to overcome ecological barriers, resist control measures, and establish successfully across varied environments (Cook and Crozier, 1995; Sakai et al., 2001; Lee, 2002; Beggs et al., 2011; Hufbauer et al., 2012; Ricalde et al., 2012; Eyer et al., 2019).

1.2.2.7 Parthenogenesis or asexual reproduction

Parthenogenesis, or asexual reproduction, significantly contributes to the evolution and success of invasive insects by enabling rapid population growth and colonization without the need for mates. This strategy allows a single individual to establish a new population, which is particularly advantageous in low-density or recently invaded areas where mate availability is limited. Asexual reproduction also preserves successful genotypes that are well-adapted to local conditions, enhancing establishment and spread. The little fire ant, *Wasmannia auropunctata* (Roger, 1863) reproduces both sexually and asexually, with invasive populations often relying on clonal reproduction to dominate new tropical and subtropical habitats (Foucaud et al., 2009). Similarly, *Aphis nerii* (oleander aphid), rapidly produces large populations that efficiently colonize host plants, increasing its invasive potential (Simon et al., 1996). *Linepithema humile* (Argentine ant) forms massive supercolonies characterized by unicoloniality and reduced genetic diversity, supported in part by queen parthenogenesis (Tsutsui and Suarez, 2003). *Aphis gossypii* (cotton aphid) reproduces mainly via thelytokous parthenogenesis under favourable conditions, leading to explosive outbreaks on crops such as cotton, cucurbits, and citrus (Wang et al., 2016). Moreover, parthenogenesis can help invasive species overcome genetic bottlenecks, maintain viable populations despite low genetic diversity, and adapt quickly through mutation accumulation. These traits collectively enhance colonization success and complicate pest management, making parthenogenesis a powerful evolutionary tool for many invasive arthropods.

1.2.2.8 Dormant or resistant life stages

Dormant or resistant life stages play a crucial role in the evolution and success of invasive insect species by enhancing their ability to survive unfavourable conditions, disperse over long distances, and establish in new environments. These stages such as diapausing eggs, pupae, or encysted larvae enable insects to withstand environmental stressors such as cold, drought, and chemical treatments, increasing survival during transport and colonization (Denlinger, 2002). For example, the gypsy moth (*L. dispar*) exhibits egg diapause, allowing populations to endure harsh winters and synchronize hatching with favourable conditions (Gray, 2004). Similarly, the red imported fire ant (*Solenopsis invicta*) forms resistant queen and worker stages that survive flooding and facilitate long-distance dispersal (Tschinkel, 2006). These resistant stages also preserve genetic variation over time, providing evolutionary flexibility to adapt to new hosts, climates, and habitats. Consequently, dormancy and resistance traits are often selected in invasive species, significantly contributing to their ecological plasticity and invasiveness (Grevstad and Coop, 2015). Bale et al. (2002) demonstrated that rising temperatures accelerate insect growth and survival, further aiding invasive spread, though ecological factors can influence these effects. Dormant stages like diapause, cysts, or egg cases serve as biological buffers, enabling arthropods to survive winter, desiccation, or transport, thereby enhancing their dispersal potential. Many invasive insects possess such resistant life stages; for instance, *Hyphantria cunea* overwinters as pupae in silken cocoons, supporting establishment in cold regions (Yang et al., 2006). *Lymantria dispar* lays eggs undergoing obligate diapause, surviving freezing temperatures and spreading via infested outdoor materials (Keena, 2016). Additionally, *Aedes albopictus* produces drought-resistant eggs viable for months, facilitating passive dispersal through used tyres and containers (Medlock et al., 2005). These resilient stages not only support colonization but also complicate control efforts.

1.2.2.9 Exploitative competition

Exploitative competition plays a pivotal role in the evolution and success of invasive insect species by enabling them to efficiently utilize available resources, often to the detriment of native species. In this form of competition, invasive insects consume shared resources more rapidly or thoroughly, thereby reducing availability for native competitors and gaining a competitive edge (Tilman, 1982). For instance, the invasive Argentine ant (*Linepithema humile*) outcompetes native ant species by swiftly exploiting food sources and nesting sites, resulting in the displacement of native ants (Holway, 1999). This efficient resource utilization not only fosters population growth in invasive species but also drives evolutionary adaptations such as enhanced foraging efficiency, accelerated development, and increased reproductive output (Tsutsui *et al.*, 2000). Over time, these traits enable invasive insects to exploit novel environments more effectively, facilitating their establishment and spread (Keane and Crawley, 2002). Moreover, exploitative competition can exert selective pressures on native species, potentially reshaping community dynamics and disrupting ecosystem functioning (Callaway and Ridenour, 2004). For example, the rapid ecological replacement of a native bumblebee by an invasive species illustrates the profound impact exploitative competition can have on local biodiversity (Morales *et al.*, 2013). These traits may also pre-adapt invasive species to swiftly occupy new ecological niches, though their evolutionary origins are often shaped by selection pressures encountered in novel environments. Thus, the capacity to outcompete native species through resource exploitation represents a fundamental evolutionary mechanism driving the invasiveness of many insect species.

1.3 Evolutionary Mechanisms Shaping Invasiveness

Invasive arthropods evolve through a dynamic interplay of evolutionary mechanisms, founder effects, genetic drift, gene flow, hybridization, and natural selection that drive adaptive divergence and ecological flexibility. Traits such as high fecundity, environmental tolerance, phenotypic plasticity, and efficient dispersal equip them to colonize and dominate new habitats. Human-induced pressures such as habitat fragmentation, pesticide use, and climate change accelerate these evolutionary processes. Understanding these mechanisms is essential for predicting invasion risks and developing evolution-informed management and biosecurity strategies.

1.3.1 Founder effects and genetic drift

Invasive populations often originate from a small number of founders, leading to reduced genetic diversity and strong founder effects. While this bottleneck can limit adaptability, it may also result in genetic drift that accelerates divergence, especially if combined with rapid population expansion. Founder effects and genetic drift are key evolutionary mechanisms that significantly influence the invasiveness of insect species by altering their genetic makeup during the early stages of colonization. The founder effect occurs when a small number of individuals establish a new population in a novel environment, carrying with them only a fraction of the genetic diversity of the source population. This limited gene pool can lead to rapid fixation of alleles, including those that may confer advantages in the new habitat. Genetic drift, the random fluctuation of allele frequencies in small populations, further amplifies this effect by reducing genetic variation and potentially allowing deleterious or neutral traits to persist. Together, these mechanisms can inadvertently favour traits such as high reproductive potential, broad diet, or stress tolerance, enhancing the insect's capacity to invade and establish. A notable example is the fire ant (*Solenopsis invicta*), which was introduced to the United States from South America. The initial founding population exhibited reduced genetic diversity, yet underwent rapid expansion and adaptation in its introduced range, partly due to founder effects and subsequent genetic drift (Shoemaker *et al.*, 2006). Similarly, invasive populations of the gypsy moth (*L. dispar*) in North America have shown significant genetic differences from their native populations, likely shaped by these evolutionary processes (Keena *et al.*, 2008); it was however found that serial founder effects slow range expansion in an invasive social insect, the Asian honey bee (*Apis cerana*) in Australia (Hagan *et al.*, 2024). These examples underscore how founder effects and genetic drift, though often associated with reduced fitness, can paradoxically facilitate invasiveness by streamlining genetic traits beneficial under novel environmental pressures.

1.3.2 Gene flow and hybridization

Gene flow and hybridization play crucial roles in the evolution and adaptability of invasive insect species. Gene flow, the movement of genes between populations, increases genetic diversity, counteracts founder effects, and enhances adaptive potential in novel environments (Ellstrand and Schierenbeck, 2000; Facon *et al.*, 2008). In invasive insects, repeated introductions from genetically distinct populations often result in admixture, producing novel gene combinations that favour invasion success (Kolbe *et al.*, 2004). Hybridization, the interbreeding between genetically divergent populations or species, can further facilitate evolutionary innovation. For example, hybridization between native and introduced populations can generate hybrid vigour or heterosis, improving survival and reproductive success (Hufbauer *et al.*, 2012). The red imported fire ant (*Solenopsis invicta*) in the USA exemplifies this, showing admixture between multiple introduced lineages that enhanced its invasive performance (Shoemaker *et al.*, 2006). Similarly, hybridization among *Drosophila subobscura* populations has driven rapid clinal adaptations in invasive ranges (Ayala *et al.*, 1989).

However, gene flow can also introduce maladaptive alleles, potentially hinder local adaptation if not balanced by natural selection (Lenormand, 2002). Yet, it often boosts evolutionary potential, especially when hybrid genotypes express novel phenotypes or increased environmental tolerance. Introduced arthropods may hybridize with native or previously introduced species, creating unique genotypes. Gene flow between divergent lineages can lead to transgressive traits, promoting ecological novelty and invasiveness. For instance, hybridization between invasive *Helicoverpa armigera* and native *Helicoverpa zea* in the Americas has produced novel genotypes with enhanced pest status and resistance to *Bacillus thuringiensis* (Bt) crops (Tay *et al.*, 2013; Anderson *et al.*, 2018). Additionally, gene flow among *Solenopsis invicta* populations in the USA has facilitated the spread of the polygyne (multi-queen) social form, which is linked to greater colony survival and invasiveness (Shoemaker *et al.*, 2006). Together, gene flow and hybridization contribute significantly to the evolutionary flexibility of invasive insects, highlighting the importance of genetic monitoring in pest management strategies.

1.3.3 Natural selection and adaptive divergence

Natural selection plays a central role in the evolution of invasive insects by favouring traits that enhance survival and reproduction in novel environments. Adaptive divergence occurs when invasive populations evolve traits distinct from their source populations in response to local selective pressures such as climate, predators, or host plant availability (Lee, 2002). For example, invasive populations of *Drosophila subobscura* in North and South America rapidly evolved latitudinal clines in wing size, mirroring those of their European ancestors and demonstrating parallel adaptive evolution (Huey *et al.*, 2000). Similarly, the Formosan termite (*Coptotermes formosanus*) exhibits thermal tolerance adaptations in different introduced regions, improving its survival across diverse climates (Evans *et al.*, 2013). These rapid evolutionary changes often occur despite reduced genetic variation caused by founder effects, underscoring the influence of strong directional selection and phenotypic plasticity (Prentis *et al.*, 2008). Thus, adaptive divergence driven by natural selection enables invasive insects to establish, spread, and persist across varied habitats. Selective pressures in new environments such as the absence of predators, novel host availability, and abiotic stressors can drive adaptive divergence. Rapid evolution has been documented in several invasive insects, including shifts in behaviour, thermal tolerance, and developmental rates occurring within just a few generations. Natural selection and adaptive divergence are critical evolutionary mechanisms that shape invasiveness by enabling populations to survive and thrive under novel ecological conditions. Upon invading new habitats, insects face unfamiliar challenges such as new predators, competitors, climates, and host plants. Natural selection favours individuals with traits that improve fitness under these conditions, leading to divergence from ancestral populations.

For instance, *Drosophila subobscura*, introduced from Europe to the Americas, showed rapid changes in wing size and chromosomal inversion frequencies linked to local climates, clear examples of adaptive divergence driven by natural selection (Huey *et al.*, 2000). Likewise, the invasive Argentine ant (*Linepithema humile*) has undergone genetic and behavioural changes, such as reduced intraspecific aggression and increased cooperation within supercolonies, facilitating its spread across diverse habitats (Tsutsui *et al.*, 2000). These examples illustrate how selection pressures in novel environments drive divergence from source populations, promoting ecological fitting and niche expansion key components of invasiveness. Understanding these evolutionary processes is essential for predicting and managing biological invasions.

1.3.4 Ecological and behavioural divergence

Ecological and behavioural divergence plays a pivotal role in the success of invasive arthropods. Changes in habitat preference, resource use, and reproductive or foraging behaviours enable these species to exploit new ecological niches and escape native constraints. Reduced biotic pressures such as predation or competition in invaded environments often lead to ecological release, allowing invasive species to expand their range and adapt rapidly. Behavioural plasticity further supports this divergence, enabling flexible responses to novel conditions. Together, these shifts enhance colonization efficiency, resilience, and the capacity to outcompete native species, making ecological and behavioural divergence a key driver of arthropod invasiveness.

1.3.5 Habitat preference and niche shifts

Invasive insects often exhibit shifts in habitat preference or ecological niche as they colonize new environments, a process central to their evolutionary success. These niche shifts may include expansion into novel climatic zones, host plants, or ecological roles not occupied in their native range (Broennimann *et al.*, 2007; Guisan *et al.*, 2014). Such changes can result from phenotypic plasticity, which allows individuals to exploit new conditions without genetic changes, or through rapid local adaptation driven by selection pressures in the invaded range (Lee, 2002; Sexton *et al.*, 2009). For example, the invasive ladybird *Harmonia axyridis* demonstrates broader climatic tolerance and habitat use in its introduced range compared with its native distribution, likely due to a combination of plastic and genetic factors (Bahlai *et al.*, 2015). Similarly, the Argentine ant (*Linepithema humile*) has shifted from aggressive interspecific behaviour in its native range to forming large cooperative supercolonies in introduced habitats, reflecting both ecological and evolutionary transformations (Holway *et al.*, 2002). These shifts challenge the assumption of niche conservatism, the idea that species retain ancestral ecological traits over time, and illustrate that ecological niches can evolve rapidly during invasion (Petitpierre *et al.*, 2012). Understanding these shifts is critical for predicting invasive spread and designing effective management strategies.

Many invasive arthropods also show shifts in habitat preference by adapting to human-modified or disturbed landscapes. Divergence in host plant use, diurnal rhythms, or microhabitat selection is common and may result from both plasticity and evolutionary change. Habitat preference and niche shifts play a key role in the ecological and behavioural divergence of invasive insects. Many invasive arthropods display remarkable adaptability by altering their habitat preferences, particularly favouring anthropogenic environments that offer new ecological opportunities with fewer competitors or predators. These shifts often lead to changes in behaviours such as host plant utilization, daily activity patterns (diurnal or nocturnal), and microhabitat selection. For instance, the brown marmorated stink bug (*Halyomorpha halys*), native to East Asia, has expanded its ecological niche in North America and Europe by exploiting a wide range of cultivated and wild host plants, many absent from its native range (Leskey *et al.*, 2012). Likewise, the Argentine ant (*Linepithema humile*) has demonstrated behavioural flexibility in nesting and foraging, enabling it to dominate urban landscapes and displace native ant species (Holway *et al.*, 2002). These behavioural and ecological shifts not only facilitate invasion success but also promote divergence from native populations, potentially leading to local adaptation or even incipient speciation. Understanding these dynamics is essential for managing invasive species and predicting their impacts on native biodiversity.

1.3.6 Competition and ecological release

Competition and ecological release are central to understanding the rapid evolution of invasive insect species in novel environments. When these insects are introduced to new regions, they often escape their natural predators, parasites, and competitors, leading to what ecologists call "ecological release". This absence of biotic resistance allows invasive populations to proliferate and adapt, often outcompeting native species and evolving traits that enhance survival and reproduction. For example, the emerald ash borer (*Agrilus planipennis*), native to Asia, has devastated North American ash trees since its accidental introduction, benefiting from a lack of natural enemies and causing widespread ecological damage (Herms and McCullough, 2014). Similarly, the spotted lanternfly (*Lycorma delicatula*), an invasive planthopper in the USA, has rapidly expanded due to a broad host range and minimal predation, resulting in significant agricultural losses (Barringer *et al.*, 2015). Another notable case is the African fig fly (*Zaprionus indianus*), which has invaded the Americas and now competes with native *Drosophila* species; studies show it exhibits rapid thermal adaptation and outcompetes natives under changing climatic conditions (Bergland *et al.*, 2014). These examples highlight how reduced competition and ecological freedom in new habitats enable invasive insects to evolve quickly and disrupt local ecosystems, underscoring the importance of vigilant biosecurity and ecological monitoring.

Invasive species often experience reduced competition in their new environments, which allows them to diverge behaviourally from their native counterparts. In the absence of predators or competitors, natural selection may favour individuals that are bolder, more fecund, or more dispersive. Competition and ecological release are thus key drivers of the ecological and behavioural divergence of invasive insect species in novel environments. Upon establishing in new habitats, invasive insects frequently encounter diminished interspecific competition and a lack of natural enemies, conditions conducive to ecological release. This release enables invasive populations to exploit a wider range of resources and adopt novel behaviours not typically observed in their native ranges. For instance, the Argentine ant (*Linepithema humile*), native to South America, forms expansive supercolonies in invaded regions such as California due to the absence of competing ant species and natural predators. This behavioural shift, which includes increased aggression toward native ants and cooperative foraging strategies, enhances both its invasiveness and ecological impact (Holway *et al.*, 2002). Similarly, the invasive fruit fly (*Drosophila subobscura*) in North America has exhibited shifts in reproductive timing and dispersal behaviour compared with its native European populations, likely driven by altered selective pressures and the absence of native competitors (Huey *et al.*, 2000). These examples demonstrate how ecological release in novel environments can favour traits such as increased fecundity, boldness, and dispersal ability, leading to rapid behavioural evolution in invasive species. Such divergence underscores the complex interplay between ecological context and evolutionary responses in biological invasions.

1.3.7 Landscape genetics and spatial substructuring

Competition and ecological release are central mechanisms driving the evolution of invasive insect species in new environments. When introduced into non-native habitats, invasive insects often escape their natural predators, parasites, and competitors, experiencing what is termed ecological release. This freedom allows them to allocate more energy to growth, reproduction, and dispersal, often resulting in rapid population expansion and evolutionary changes that enhance their invasiveness. For example, the emerald ash borer (*Agrilus planipennis*), native to Asia, has caused widespread destruction of ash trees across North America due to the absence of its natural parasitoids, leading to unchecked population growth and significant host tree mortality (Herms and McCullough, 2014). Similarly, the spotted lanternfly (*Lycorma delicatula*) has spread aggressively in the USA, damaging grapevines and hardwood trees while facing minimal predation (Barringer *et al.*, 2015). Another case is the African fig fly (*Zaprionus indianus*), which competes with native *Drosophila* species in North and South America. Research by Bergland *et al.* (2021) demonstrated that even temporary interspecific competition with this invader can leave a lasting legacy effect, altering the evolutionary trajectory of native species. These examples illustrate how ecological release from natural enemies and competition with native species drive evolutionary change, shaping both the success and ecological impact of invasive insects.

Even within introduced ranges, invasive populations often exhibit spatial substructuring due to fragmented landscapes, microclimatic variation, or host specialization. These localized populations may evolve independently, resulting in divergent lineages with varying degrees of invasiveness. Understanding these patterns through landscape genetics is critical for predicting invasive spread. Invasive insects frequently show spatial genetic structuring influenced by landscape fragmentation, microclimate, and host preferences, factors that affect gene flow and promote local adaptation. Landscape genetics, a discipline integrating population genetics, landscape ecology, and spatial statistics provides valuable insights into how these elements shape invasive population dynamics. For example, the emerald ash borer (*Agrilus planipennis*) has caused extensive ash tree mortality across North America. Landscape genetic studies reveal significant population substructuring driven by fragmented forests, urban development, and limited natural dispersal, further complicated by human-mediated long-distance transport (Bray *et al.*, 2011). These factors have led to the emergence of isolated populations with varying genetic diversity and invasion dynamics. Similarly, the red imported fire ant (*Solenopsis invicta*) in the USA exhibits distinct genetic clusters shaped by environmental barriers such as rivers and elevation, as well as anthropogenic influences such as land use and climate zones (Ascunce *et al.*, 2011). This spatial structuring affects colony organization, aggressiveness, and ecological impact. Another example is the brown marmorated stink bug (*Halyomorpha halys*), which shows genetic differentiation among populations in North America along urban-rural gradients and microclimatic conditions, influencing its distribution and host utilization patterns (Valentin *et al.*, 2017). These cases demonstrate how landscape genetics enhances our understanding of invasive insect spread and supports the development of targeted management strategies.

1.3.8 Human-mediated selection and rapid evolution

Anthropogenic factors such as pesticide use, urbanization, and climate change exert strong directional selection on arthropods, significantly shaping their evolutionary trajectories. Resistance to insecticides has evolved independently in multiple invasive lineages, illustrating rapid, parallel divergence under human-imposed selection pressures. Human-mediated selection plays a pivotal role in driving rapid adaptive changes in invasive insects subjected to intense anthropogenic influences. For instance, the invasive mosquito *Aedes aegypti*, a primary vector of dengue and Zika viruses, has independently acquired mutations in the voltage-gated sodium channel gene (kdr mutations) across different regions, enabling survival in heavily sprayed urban environments (Smith *et al.*, 2016). Similarly, the Colorado potato beetle (*Leptinotarsa decemlineata*) exemplifies extraordinary evolutionary capacity, having developed resistance to over 50 different pesticides under intense agricultural chemical regimes (Alyokhin *et al.*, 2008).

Urbanization further influences evolutionary trajectories in invasive insects. For example, the ant *Tapinoma sessile* exhibits notable changes in colony structure and behaviour when adapting to urban habitats compared with rural ones, highlighting behavioural and ecological shifts driven by anthropogenic habitat modification (Menke *et al.*, 2010). Climate change also acts as a strong selective force, as demonstrated by the invasive moth *Operophtera brumata*, which has expanded its range and advanced seasonal emergence in response to warming temperatures, thereby enhancing its survival and reproductive success (Jepsen *et al.*, 2008). Collectively, these cases underscore that human activities not only facilitate the spread of invasive species but also catalyse rapid evolutionary responses through consistent and strong directional selection. Global trade, agriculture, and urbanization create novel environmental pressures and dispersal pathways that accelerate the evolution of invasive insects. For example, the fall armyworm (*Spodoptera frugiperda*), originally native to the Americas, has rapidly expanded across Africa and Asia, evolving resistance to multiple insecticides within a few years due to intense chemical control measures (Boaventura *et al.*, 2020). Likewise, the diamondback moth (*Plutella xylostella*) has developed widespread insecticide resistance globally, directly linked to human-mediated selection in agricultural landscapes (Furlong *et al.*, 2013). Another notable case is the invasive Asian tiger mosquito (*Aedes albopictus*), which has quickly adapted to temperate climates, a process facilitated by urban heat islands and human transport networks (Kotsakiozi *et al.*, 2017). These examples illustrate that human actions accelerate both the spread and rapid evolutionary change of invasive insects, complicating management efforts and highlighting the urgent need for integrated pest management strategies that account for ongoing evolutionary dynamics.

1.4 Case Studies

1.4.1 Arthropod invasions

Arthropod invasions are widespread and pose significant threats to ecosystems, native species, human health and food security (Venette and Hutchison, 2021). Direct impacts include crop destruction, such as maize consumption by the fall armyworm (*Spodoptera frugiperda*) in Africa, and biodiversity loss, exemplified by *Philornis downsi* infestations on Darwin's finches in the Galápagos Islands (Koop *et al.*, 2020). Indirect effects involve disease transmission, as seen with the Asian tiger mosquito (*Aedes albopictus*), a global vector of dengue viruses (Rezende *et al.*, 2020). Dr C.S. Elton (1958) was among the first to document insect invasions in tropical regions, and later reviews have expanded on the ecological consequences of arthropod invasions (Venette and Hutchison, 2021). Arthropods display remarkable diversity in feeding strategies and adaptations, which reflect complex genomic transformations across their evolutionary history. However, linking these phenotypic traits to genomic changes remains challenging. Major evolutionary transitions in arthropods such as body segmentation, the independent use of terrestrial and aquatic habitats, the evolution of flight, and insect metamorphosis have been shaped by genomic innovations (Thomas *et al.*, 2020). These changes likely involve new gene formation via duplication or *de novo* evolution (Santos *et al.*, 2017), gene loss (Porter and Crandall, 2003), and shifts in gene families associated with phenotypic traits. Thomas *et al.* (2020) identified three key genomic processes driving arthropod evolution: gene duplications and losses, protein domain rearrangements, and amino acid substitutions. Understanding these genomic mechanisms is essential for unravelling the evolution of invasiveness in arthropods.

1.4.2 Ants invasions

Cremer *et al.* (2008) investigated the evolution of invasiveness in the garden ant *Lasius neglectus*, a species exhibiting traits characteristic of the "invasive ant syndrome", a phenomenon common among the most ecologically disruptive ant invaders. This syndrome involves intranidal mating (queens mate within the nest), cooperative dispersal with workers, and the formation of expansive, interconnected supercolonies that can outcompete native species and disrupt local ecosystems. Though ecologically dominant, such colonies are sometimes viewed as evolutionary dead-ends due to their low genetic variability (Queller and Strassmann, 1998; Queller, 2000). *L. neglectus* exemplifies this syndrome through its non-dispersive mating, extensive nest networks with mutual tolerance, and lack of aggression among nests contrasting with smaller, territorial colonies. Its invasive success is attributed to a combination of factors, including parasite release post-introduction, pre-adapted traits such as small body size, high queen numbers per colony, efficient nestmate recognition, and low variability in cuticular hydrocarbon profiles. These traits, present even in its native range, coupled with limited dispersal and a highly viscous population structure, facilitated the emergence of supercoloniality. Under favourable conditions, such colonies can span several kilometres (Giraud *et al.*, 2002), displacing native ant species and negatively impacting broader biodiversity (Passera, 1994).

1.4.3 Haematophagous arthropods

Haematophagous arthropods such as Diptera, fleas, ticks, bugs, and mosquitoes are introduced into new areas through human movement, trade, land and water transport, wind, and materials such as tyres, with birds, cattle, dogs, horses, rats, reptiles, and pigs serving as hosts. Their economic and ecological impacts are well documented since the Palaeolithic era (Athni *et al.*, 2021), but historical and social dimensions remain underexplored. The intensification of trade and land-use changes during the 5th and 6th centuries significantly influenced their invasions. These arthropods, including fleas, mites, and houseflies, act as vectors for arboviruses causing diseases such as yellow fever, dengue, COVID-19, Zika, and chikungunya in humans and pets (Cuthbert *et al.*, 2021a, b), resulting in major health and economic burdens. Rising international trade, climate and land-use change further exacerbate vector-borne diseases (Lin *et al.*, 2012). Industrialized nations experience higher invasive species rates (Pyšek *et al.*, 2011), with vector-borne diseases contributing to substantial economic losses (Ahmed *et al.*, 2022); Aedes mosquitoes are among the costliest, with aquatic and semi-aquatic invaders potentially costing $311 billion globally (Roiz *et al.*, 2024). Urbanization, population pressure, and heat island effects facilitate invasions, with niche shifts and inversion helping these species adapt to human-modified habitats. Effective mitigation requires strong quarantine, surveillance, political commitment, and socio-ethnological strategies. Increased connectivity expands host and geographic ranges, as seen in *Aedes albopictus* (Medley *et al.*, 2014), while ticks requiring blood at all life stages over 3–4 years spread via vertebrate host movement and livestock trade. Human-created habitats, such as water containers, support mosquito breeding, and changes in life-history traits further enhance invasive potential (Cuthbert *et al.*, 2023).

1.4.4 Aquatic arthropods

Freshwater ecosystems play vital and dynamic roles in nutrient and water cycling, Earth's cooling processes, and the conservation of biodiversity among marine and terrestrial floral and faunal elements, thereby supporting livelihoods (Dudgeon *et al.*, 2006; Simberloff *et al.*, 2013). However, these ecosystems are highly vulnerable to invasive species and biological invasions, which can significantly disrupt their ecological functions. Case studies have highlighted the adverse impacts of invasions and climate change on water bodies (Bright, 1998; Goldsmith, 2012). For example, the zebra mussel (*Dreissena polymorpha*) can transform river ecosystems by shifting food webs from pelagic-based to littoral-based systems through intense filtration of algae, leaving minimal food for pelagic zooplankton (Strayer, 2010). Invasive species are also known to restructure freshwater food webs and alter hydrological regimes, as seen in the introduction of *Lates niloticus* in East African lakes, which led to the extinction of numerous native fish species (Goldsmith, 2012). Aquatic invasive plants further disrupt communities; for instance, water hyacinth (*Eichhornia crassipes*) makes wetlands boggy with limited light penetration, while water milfoil (*Myriophyllum spicatum*) forms dense surface mats that reduce light availability and native plant diversity (Madsen *et al.*, 1991). Many aquatic invasive plants produce wind-dispersed seeds to enhance their spread and remain aloft, and asexual reproduction such as apomictic parthenogenesis is a common trait among successful aquatic invertebrates and plants (Dodson and Fry, 2001). Species with wide geographic ranges and broad environmental tolerance, especially among fish and aquatic invertebrates, tend to become more dominant invaders than those with restricted ranges (Kolar and Lodge, 2002). Overall, key features of successful aquatic invaders include asexual reproduction, rapid growth, high reproductive frequency and rate, and a broad tolerance to environmental disturbances.

1.4.5 Crop pests

Two fall armyworm (FAW; *Spodoptera frugiperda*) corn and rice strains were identified based on host plant association (Pashley *et al.*, 1985). Genetic studies revealed slight differences between the strains, including variations in mating behaviour, timing, and reproductive isolation barriers (Saldamando-Benjumea *et al.*, 2014). Single-marker analyses of populations from South Africa, India, Myanmar, and Southern China showed genetic similarity to West African populations (Nagoshi *et al.*, 2020). Genomic studies from Ethiopia, Kenya, South Africa, and Louisiana support an African origin, while multigenetic analyses suggest multiple introductions from both African and Asian sources. Whole-genome analyses are still needed to clarify invasion pathways. Differences in host preference, circadian rhythms, and mating compatibility have been observed, although pheromone blend attraction appears insufficient to cause assortative mating. Tay *et al.* (2023) reviewed the ecology, evolution, and management of the fall armyworm. Genomic data show higher-than-expected diversity in native and invasive populations. Strain-specific expansions of chemosensory and detoxification genes may aid adaptation to host plants and insecticides. Frequent gene translocations in invasive populations (Zhang *et al.*, 2020) and varying selection pressures across regions, including differing natural enemies, suggest complex evolutionary dynamics. Hybridization is common in both native and non-native ranges. The role of genetic structure in thermal tolerance adaptation remains underexplored. However, Logan *et al.* (2020) demonstrated low heritability in thermal performance traits in *Harmonia axyridis*, indicating limited evolutionary capacity under climate change. Similarly, *Drosophila suzukii* showed rapid local adaptation, with higher oviposition and offspring performance on native fruit substrates, driven by genetic rather than plastic effects (Olazcuaga *et al.*, 2022), suggesting that variable selection can still allow swift local adaptation in natural populations.

1.5 Implications for Management and Biosecurity

Understanding the evolutionary divergence of invasive arthropods has direct applications in management, as discussed below.

1.5.1 Identifying invasion pathways and source populations

Evolutionary analyses, particularly those based on molecular phylogenetics and population genetics, can help trace the origin and introduction routes of invasive insects. By comparing genetic profiles of populations across native and invaded ranges, researchers can identify source populations and multiple introduction events. For example, the global spread of the brown marmorated stink bug (*Halyomorpha halys*) has been tracked using mitochondrial haplotypes, revealing multiple introductions from distinct native populations in East Asia into Europe and North America (Valentin *et al.*, 2017). Such knowledge enables targeted surveillance at ports and high-risk entry points.

1.5.2 Predicting future range expansions

Identifying invasion pathways and source populations of invasive insects is vital for effective management and biosecurity, enabling early detection, targeted prevention, and control to limit spread. Understanding how invasive species enter new regions via international trade, transport networks, or accidental commodity introductions guides surveillance and quarantine priorities. Genetic analyses pinpoint source populations, revealing origins, introduction history, and adaptive traits that inform risk assessments and management strategies. For example, the red imported fire ant (*Solenopsis invicta*) in the USA was traced to multiple South American introductions using mitochondrial DNA and microsatellites, uncovering complex invasion routes influencing its genetic diversity and success (Ascunce *et al.*, 2011). Similarly, studies on the emerald ash borer (*Agrilus planipennis*) revealed multiple introductions from eastern Asia to North America, underscoring the need to regulate wood packaging and plant trade to prevent further spread (Bray *et al.*, 2011). These molecular tools, combined with pathway analysis, enhance biosecurity programmes by enabling more precise monitoring, prevention, and rapid response, thereby reducing ecological and economic impacts. Moreover, evolutionary adaptations such as thermal tolerance and phenotypic plasticity affect colonization capacity. Modelling potential distributions under climate change, informed by evolutionary data predicts range shifts, for instance cold-tolerant genotypes of *Drosophila suzukii* expanding northward, and highlights the need for proactive monitoring in vulnerable areas (Fraimout *et al.*, 2017).

1.5.3 Developing targeted genetic or ecological controls

Targeted genetic and ecological control strategies are essential for managing invasive insects sustainably and effectively. Genetic methods such as sterile insect technique (SIT), gene drives, and RNA interference (RNAi) suppress populations by disrupting reproduction or survival. For instance, SIT has reduced *Ceratitis capitata* populations through the release of sterilized males (Dyck *et al.*, 2021), while CRISPR-based gene drives show promise in controlling *Aedes aegypti* by spreading genes that impair vector competence (Burt, 2014). Ecological controls, such as introducing co-evolved natural enemies, also play a key role. The release of *Tamarixia radiata* to manage *Diaphorina citri* has successfully suppressed populations and reduced pesticide use (Grafton-Cardwell *et al.*, 2013). Integrating these approaches allows for precise, species-specific interventions. Understanding genetic divergence and adaptive traits is critical for tools such as RNAi and gene drives, which depend on knowledge of population-level genetic diversity to avoid resistance, for example in *Anopheles* mosquito control (Carballar-Lejarazú and James, 2018). Ecological strategies also benefit from insights into host-parasitoid coevolution. These advances support more effective biosecurity and long-term invasive species management.

1.5.4 Monitoring emergence of novel invasive phenotypes

Monitoring the emergence of novel invasive phenotypes in insect populations is essential for proactive management and biosecurity, as these traits often reflect rapid adaptation to new environments, resistance to control measures, or increased invasiveness. Such phenotypic changes may involve shifts in behavior, physiology, life history, or morphology that help invasive species overcome ecological or anthropogenic barriers. For instance, *Drosophila suzukii* evolved a serrated ovipositor that enables it to lay eggs in ripening fruit, unlike other *Drosophila*, greatly amplifying its impact on agriculture (Atallah *et al.*, 2014). Similarly, *Aedes albopictus* has demonstrated plasticity in diapause and cold tolerance, allowing it to expand into temperate regions beyond its tropical origin (Medley *et al.*, 2014). Rapid evolution in invasive populations can also result in traits such as pesticide resistance or host range expansion. For example, pyrethroid resistance in *Aedes aegypti* has evolved independently in multiple regions, with genomic surveillance guiding local insecticide policies (Smith *et al.*, 2016). Tracking these changes through field monitoring, molecular tools, and ecological modelling enables early detection, informed risk assessment, and adaptive management. Incorporating evolutionary insights into biosecurity planning enhances our capacity to anticipate invasions and mitigate their impacts effectively.

1.6 Conclusion

The evolution of invasiveness in arthropods is a complex, ongoing process shaped by genetic divergence, ecological dynamics, and behavioural flexibility. Key evolutionary mechanisms such as founder effects, genetic drift, gene flow, hybridization, and natural selection interact with environmental filtering and anthropogenic pressures to produce novel trait combinations that enhance invasion success. Traits such as high fecundity, broad environmental tolerance, generalist feeding behaviour, dispersal efficiency, phenotypic plasticity, parthenogenesis, and dormant life stages collectively empower invasive species to overcome ecological barriers and rapidly establish in new habitats. Ecological release, spatial substructuring, and behavioural shifts often driven by reduced biotic resistance and human-altered landscapes further reinforce their adaptability and competitive edge.

As biological invasions intensify under the influence of globalization and climate change, understanding the evolutionary and ecological mechanisms behind invasiveness becomes crucial. Invasive arthropods, from crop pests to aquatic invertebrates, demonstrate a wide array of invasion strategies and genomic adaptations. Insights from landscape genetics, molecular surveillance, and ecological modelling reveal how habitat fragmentation, pesticide use, and urbanization accelerate evolutionary change. A divergence-centric framework not only deepens our understanding of invasion dynamics but also enhances our capacity to predict and manage them. Addressing the global invasion crisis demands the integration of scientific research with robust biosecurity policies, international cooperation, and evolution-informed management strategies to build resilient ecosystems and protect biodiversity and human well-being.

Acknowledgements

We thank Dr V.V. Ramamurthy, Former professor, Division of Entomology, ICAR-IARI, New Delhi for critical reading and suggestions for improvement of the manuscript. Dr Shashank acknowledges support from the project titled “Development of pest and disease management strategies for major crops (CRSCIARISIL2014033265)”.

References

Ahmed, D.A., Hudgins, E.J., Cuthbert, R.N., Kourantidou, M., Diagne, C. *et al.* (2022) Managing biological invasions: The cost of inaction. *Biological Invasions* 24(7), 1927–1946. DOI: 10.1007/s10530-022-02755-0.

Alyokhin, A., Baker, M., Mota-Sanchez, D., Dively, G. and Grafius, E. (2008) Colorado potato beetle resistance to insecticides. *American Journal of Potato Research* 85(6), 395–413. DOI: 10.1007/s12230-008-9052-0.

Amarillo-Suárez, A.R. and Fox, C.W. (2006) Population differences in host use by a seed-beetle: Local adaptation, phenotypic plasticity and maternal effects. *Oecologia* 150, 247–258.

Anderson, C.J., Oakeshott, J.G., Tay, W.T., Gordon, K.H., Zwick, A. *et al.* (2018) Hybridization and gene flow in the mega-pest lineage of moth, Helicoverpa. *Proceedings of the National Academy of Sciences USA* 115(19), 5034–5039. DOI: 10.1073/pnas.1718831115.

Ascunce, M.S., Yang, C.C., Oakey, J., Calcaterra, L., Wu, W.J. *et al.* (2011) Global invasion history of the fire ant *Solenopsis invicta*. *Science* 331(6020), 1066–1068. DOI: 10.1126/science.1198734.

Asplen, M.K., Anfora, G., Biondi, A., Choi, D.S., Chu, D. *et al.* (2015) Invasion biology of spotted wing drosophila (*Drosophila suzukii*): A global perspective and future priorities. *Journal of Pest Science* 88(3), 469–494. DOI: 10.1007/s10340-015-0681-z.

Atallah, J., Teixeira, L., Salazar, R., Zaragoza, G. and Kopp, A. (2014) The making of a pest: The evolution of a fruit-penetrating ovipositor in *Drosophila suzukii*. *Proceedings of the Royal Society B: Biological Sciences* 281(1781), 20132840. DOI: 10.1098/rspb.2013.2840.

Athni, T.S., Shocket, M.S., Couper, L.I., Nova, N., Caldwell, I.R. *et al.* (2021) The influence of vector-borne disease on human history: Socio-ecological mechanisms. *Ecology Letters* 24(4), 829–846. DOI: 10.1111/ele.13675.

Ayala, F.J., Serra, L. and Prevosti, A. (1989) A grand experiment in evolution: The *Drosophila subobscura* colonization of the americas. *Genome* 31(1), 246–255. DOI: 10.1139/g89-042.

Bahlai, C.A., Colunga-Garcia, M., Gage, S.H. and Landis, D.A. (2015) The role of exotic ladybeetles in the decline of native ladybeetle populations: Evidence from long-term monitoring. *Biological Invasions* 17, 1005–1024. DOI: 10.1007/s10530-014-0772-4.

Bale, J.S., Masters, G.J., Hodkinson, I.D., Awmack, C., Bezemer, T.M. *et al.* (2002) Herbivory in global climate change research: Direct effects of rising temperature on insect herbivores. *Global Change Biology* 8(1), 1–16. DOI: 10.1046/j.1365-2486.2002.00451.x.

Barringer, L.E., Donovall, L.R., Spichiger, S.E., Lynch, D. and Henry, D. (2015) The first New World record of *Lycorma delicatula* (Insecta: Hemiptera: Fulgoridae). *Entomological News* 125(1), 20–23. DOI: 10.3157/021.125.0105.

Beggs, J.R., Brockerhoff, E.G., Corley, J.C., Kenis, M., Masciocchi, M. *et al.* (2011) Ecological effects and management of invasive alien vespidae. *BioControl* 56, 505–526. DOI: 10.1007/s10526-011-9389-z.

Bergland, A.O., Behrman, E.L., O'Brien, K.R. and Schmidt, P.S. (2021) Competitive history shapes rapid evolution in a seasonal climate. *Proceedings of the National Academy of Sciences USA* 118(13), e2015772118. DOI: 10.1073/pnas.2015772118.

Bergland, A.O., Behrman, E.L., O'Brien, K.R., Schmidt, P.S. and Petrov, D.A. (2014) Genomic evidence of rapid and stable adaptive oscillations over seasonal time scales in drosophila. *PLoS Genetics* 10(11), e1004775.

Blackburn, T.M., Pyšek, P., Bacher, S., Carlton, J.T., Duncan, R.P. *et al.* (2011) A proposed unified framework for biological invasions. *Trends in Ecology and Evolution* 26(7), 333–339. DOI: 10.1016/j.tree.2011.03.023.

Blossey, B. and Nötzold, R. (1995) Evolution of increased competitive ability in invasive nonindigenous plants: A hypothesis. *Journal of Ecology* 83(5), 887–889. DOI: 10.2307/2261425.

Boaventura, D., Bolzan, A., Padovez, F.E., Okuma, D.M., Omoto, C. *et al.* (2020) Detection of a ryanodine receptor target-site mutation in diamide insecticide resistant fall armyworm, *Spodoptera frugiperda*. *Pest Management Science* 76(1), 47–54. DOI: 10.1002/ps.5505.

Boettner, G.H., Elkinton, J.S. and Boettner, C.J. (2000) Effects of a biological control introduction on three nontarget native species of saturniid moths. *Conservation Biology* 14(6), 1798–1806. DOI: 10.1111/j.1523-1739.2000.99193.x.

Boukouvala, M.C., Kavallieratos, N.G., Skourti, A., Pons, X., Alonso, C.L. *et al.* (2022) *Lymantria dispar* (L.) (Lepidoptera: Erebidae): Current status of biology, ecology, and management in Europe with notes from North America. *Insects* 13(9), 854. DOI: 10.3390/insects13090854.

Bray, A.M., Bauer, L.S., Poland, T.M., Haack, R.A., Cognato, A.I. *et al.* (2011) Genetic analysis of emerald ash borer (*Agrilus planipennis* Fairmaire) populations in Asia and North America. *Biological Invasions* 13, 2869–2887. DOI: 10.1007/s10530-011-9970-5.

Bright, C. (1998) *Life out of Bounds: Bioinvasion in a Borderless World*. WW Norton & Company, New York, p. 287.

Broennimann, O., Treier, U.A., Müller-Schärer, H., Thuiller, W., Peterson, A.T. *et al.* (2007) Evidence of climatic niche shift during biological invasion. *Ecology Letters* 10(8), 701–709. DOI: 10.1111/j.1461-0248.2007.01060.x.

Burt, A. (2014) Heritable strategies for controlling insect vectors of disease. *Philosophical Transactions of the Royal Society B: Biological Sciences* 369(1645), 20130432. DOI: 10.1098/rstb.2013.0432.

CABI (Centre for Agriculture and Bioscience International) (2019) Datasheet on *Spodoptera frugiperda* (fall armyworm). In: *CABI Compendium*. CABI Digital Library. DOI: 10.1079/cabicompendium.29810.

Callaway, R.M. and Ridenour, W.M. (2004) Novel weapons: Invasive success and the evolution of increased competitive ability. *Frontiers in Ecology and the Environment* 2(8), 436–443. DOI: 10.1890/1540-9295(2004)002[0436:NWISAT]2.0.CO;2.

Carballar-Lejarazú, R. and James, A.A. (2018) Population modification of *Anopheline* species to control malaria transmission. *Pathogens and Global Health* 111(8), 424–435. DOI: 10.1080/20477724.2018.1427192.

Chakravarthy, A.K. and Sridhara, S. (2016) *Economic and ecological significance of arthropods in diversified ecosystems*. Springer Nature, Singapore, pp. 1–422.

Chakravarthy, A.K., Kammar, V. and Shashank, P.R. Chakravarthy, A.K. and Sridhara, S. (eds). (2016) *Arthropods: Evolution and Ecology*. Springer, pp. 1–17.

Chapman, A.D. (2009) *Numbers of Living Species in Australia and the World (PDF)*, 1st edn. Australian Biological Resources Study. Canberra, pp. 1–80.

Cira, T.M., Venette, R.C., Aigner, J., Kuhar, T., Mullins, D.E. *et al.* (2016) Cold tolerance of *Halyomorpha halys* across geographic and temporal scales. *Environmental Entomology* 45(2), 484–491. DOI: 10.1093/ee/nvv220.

Cook, J.M. and Crozier, R.H. (1995) Sex determination and population biology in the hymenoptera. *Trends in Ecology & Evolution* 10(7), 281–286. DOI: 10.1016/0169-5347(95)90011-X.

Cremer, S., Ugelvig, L.V., Drijfhout, F.P., Schlick-Steiner, B.C., Steiner, F.M. *et al.* (2008) The evolution of invasiveness in garden ants. *PLoS ONE* 3(12), e3838. DOI: 10.1371/journal.pone.0003838.

Cuthbert, R.N., Bartlett, A.C., Turbelin, A., Haubrock, P.J., Diagne, C. *et al.* (2021a) Economic costs of biological invasions in the United Kingdom. *NeoBiota* 67, 299–328. DOI: 10.3897/neobiota.67.59743.

Cuthbert, R.N., Pattison, Z., Taylor, N.G., Verbrugge, L., Diagne, C. *et al.* (2021b) Global economic costs of aquatic invasive alien species. *Science of The Total Environment* 775, 145238. DOI: 10.1016/j.scitotenv.2021.145238.

Cuthbert, R.N., Darriet, F., Chabrerie, O., Lenoir, J., Courchamp, F. *et al.* (2023) Invasive hematophagous arthropods and associated diseases in a changing world. *Parasites & Vectors* 16(1), 291.

De Barro, P.J., Liu, S.S., Boykin, L.M. and Dinsdale, A.B. (2011) *Bemisia tabaci*: A statement of species status. *Annual Review of Entomology* 56(1), 1–19. DOI: 10.1146/annurev-ento-112408-085504.

Denlinger, D.L. (2002) Regulation of diapause. *Annual Review of Entomology* 47(1), 93–122. DOI: 10.1146/annurev.ento.47.091201.145137.

Desneux, N., Wajnberg, E., Wyckhuys, K.A., Burgio, G., Arpaia, S. *et al.* (2010) Biological invasion of European tomato crops by *Tuta absoluta*: Ecology, geographic expansion and prospects for biological control. *Journal of Pest Science* 83, 197–215. DOI: 10.1007/s10340-010-0321-6.

Dodson, S.I. and Fry, D.G. (2001) Cladocera and other branchiopoda. In: *Ecology and Classification of North American Freshwater Invertebrates*, 2nd edn. Academic Press, Boston, pp. 849–913.

Dudgeon, D., Arthington, A.H., Gessner, M.O., Kawabata, Z.I., Knowler, D.J. *et al.* (2006) Freshwater biodiversity: Importance, threats, status and conservation challenges. *Biological Reviews* 81(2), 163–182. DOI: 10.1017/S1464793105006950.

Dyck, V.A., Hendrichs, J. and Robinson, A.S. (2021) *Sterile Insect Technique: Principles and Practice in Area-Wide Integrated Pest Management*. Taylor & Francis, p. 1216.

Elliott, N., Kieckhefer, R. and Kauffman, W. (1996) Effects of an invading coccinellid on native coccinellids in an agricultural landscape. *Oecologia* 105, 537–544. DOI: 10.1007/BF00330017.

Ellstrand, N.C. and Schierenbeck, K.A. (2000) Hybridization as a stimulus for the evolution of invasiveness in plants. *Proceedings of the National Academy of Sciences USA* 97(13), 7043–7050. DOI: 10.1073/pnas.97.13.7043.

Elton, C.S. (1958) The ecology of invasions by animals and plants. *Progress in Physical Geography* 31(6), 559–666. DOI: 10.1177/03091333070870.

Evans, T.A., Forschler, B.T. and Grace, J.K. (2013) Biology of invasive termites: A worldwide review. *Annual Review of Entomology* 58(1), 455–474. DOI: 10.1146/annurev-ento-120811-153554.

Eyer, P.A., Blumenfeld, A.J. and Vargo, E.L. (2019) Sexually antagonistic selection promotes genetic divergence between males and females in an ant. *Proceedings of the National Academy of Sciences USA* 116(48), 24157–24163. DOI: 10.1073/pnas.1906568116.

Facon, B., Pointier, J.P., Jarne, P., Sarda, V. and David, P. (2008) High genetic variance in life-history strategies within invasive populations by way of multiple introductions. *Current Biology* 18(5), 363–367. DOI: 10.1016/j.cub.2008.01.063.

Filchak, K.E., Roethele, J.B. and Feder, J.L. (2000) Natural selection and sympatric divergence in the apple maggot *Rhagoletis pomonella*. *Nature* 407(6805), 739–742. DOI: 10.1038/35037578.

Foucaud, J., Orivel, J., Fournier, D., Delabie, J.H., Loiseau, A. *et al.* (2009) Reproductive system, social organization, human disturbance and ecological dominance in native populations of the little fire ant, *Wasmannia auropunctata*. *Molecular Ecology* 18(24), 5059–5073. DOI: 10.1111/j.1365-294X.2009.04440.x.

Fraimout, A., Debat, V., Fellous, S., Hufbauer, R.A., Foucaud, J. *et al.* (2017) Deciphering the routes of invasion of *Drosophila suzukii* by means of ABC random forest. *Molecular Biology and Evolution* 34(4), 980–996. DOI: 10.1093/molbev/msx050.

Furlong, M.J., Wright, D.J. and Dosdall, L.M. (2013) Diamondback moth ecology and management: Problems, progress, and prospects. *Annual Review of Entomology* 58, 517–541. DOI: 10.1146/annurev-ento-120811-153605.

Garnas, J., Auger-Rozenberg, M.A., Roques, A., Bertelsmeier, C. and Wingfield, M.J. (2016) Complex patterns of global spread in invasive insects: Eco-evolutionary and management consequences. *Biological Invasions* 18, 935–952. DOI: 10.1007/s10530-016-1082-9.

Giraud, T., Pedersen, J.S. and Keller, L. (2002) Evolution of supercolonies: The Argentine ants of southern Europe. *Proceedings of the National Academy of Sciences USA* 99(9), 6075–6079. DOI: 10.1073/pnas.092694199.

Goergen, G., Kumar, P.L., Sankung, S.B., Togola, A. and Tamo, M. (2016) First report of outbreaks of the fall armyworm *Spodoptera frugiperda* (J.E. Smith) (Lepidoptera, Noctuidae), a new alien invasive pest in West and Central Africa. *PLoS ONE* 11(10), e0165632. DOI: 10.1371/journal.pone.0165632.

Goldsmith, F.B. (2012) *Monitoring for Conservation and Ecology*, Vol. 3. Springer Science & Business Media, p. 275.

Grafton-Cardwell, E.E., Stelinski, L.L. and Stansly, P.A. (2013) Biology and management of Asian citrus psyllid, vector of the huanglongbing pathogens. *Annual Review of Entomology* 58, 413–432. DOI: 10.1146/annurev-ento-120811-153542.

Gray, D.R. (2004) The gypsy moth life stage model: Landscape-wide estimates of gypsy moth establishment using a multi-generational phenology model. *Ecological Modelling* 176(1–2), 155–171. DOI: 10.1016/j.ecolmodel.2003.11.010.

Grevstad, F.S. and Coop, L.B. (2015) The consequences of photoperiodism for organisms in new climates. *Ecological Applications* 25(6), 1506–1517. DOI: 10.1890/14-2071.1.

Guisan, A., Petitpierre, B., Broennimann, O., Daehler, C. and Kueffer, C. (2014) Unifying niche shift studies: Insights from biological invasions. *Trends in Ecology and Evolution* 29(5), 260–269. DOI: 10.1016/j.tree.2014.02.009.

Haack, R.A., Hérard, F., Sun, J. and Turgeon, J.J. (2010) Managing invasive populations of Asian longhorned beetle and citrus longhorned beetle: A worldwide perspective. *Annual Review of Entomology* 55, 521–546. DOI: 10.1146/annurev-ento-112408-085427.

Hagan, T., Ding, G., Buchmann, G., Oldroyd, B.P. and Gloag, R. (2024) Serial founder effects slow range expansion in an invasive social insect. *Nature Communication* 15, 3608. DOI: 10.1038/s41467-024-47894-1.

Hauser, M. (2011) A historic account of the invasion of *Drosophila suzukii* (Matsumura) (Diptera: Drosophilidae) in the continental United States, with remarks on their identification. *Pest Management Science* 67(11), 1352–1357. DOI: 10.1002/ps.2265.

Herms, D.A. and McCullough, D.G. (2014) Emerald ash borer invasion of North America: History, biology, ecology, impacts, and management. *Annual Review of Entomology* 59, 13–30. DOI: 10.1146/annurev-ento-011613-162051.

Holway, D.A. (1999) Competitive mechanisms underlying the displacement of native ants by the invasive Argentine ant. *Ecology* 80(1), 238–251. DOI: 10.1890/0012-9658(1999)080[0238:CMUTDO]2.0.CO;2.

Holway, D.A., Lach, L., Suarez, A.V., Tsutsui, N.D. and Case, T.J. (2002) The causes and consequences of ant invasions. *Annual Review of Ecology and Systematics* 33(1), 181–233. DOI: 10.1146/annurev.ecolsys.33.010802.150444.

Huang, D., Haack, R.A. and Zhang, R. (2011) Does global warming increase establishment rates of invasive alien species? A centurial time series analysis. *PLoS ONE* 6, e24733.

Huey, R.B., Gilchrist, G.W., Carlson, M.L., Berrigan, D. and Serra, L. (2000) Rapid evolution of a geographic cline in size in an introduced fly. *Science* 287(5451), 308–309. DOI: 10.1126/science.287.5451.308.

Hufbauer, R.A., Facon, B., Ravigné, V., Turgeon, J., Foucaud, J. *et al.* (2012) Anthropogenically induced adaptation to invade (AIAI): Contemporary adaptation to human-altered habitats within the native range can promote invasions. *Evolutionary Applications* 5(1), 89–101. DOI: 10.1111/j.1752-4571.2011.00211.x.

Hulme, P.E. (2009) Trade, transport and trouble: Managing invasive species pathways in an era of globalization. *Journal of Applied Ecology* 46(1), 10–18. DOI: 10.1111/j.1365-2664.2008.01600.x.

Hurley., B.P., Garnas, J.R., Wingfield, M.J., Branco, M. and Richardson, D.M. (2016) Increasing numbers and intercontinental spread of invasive insects on eucalypts. *Biological Invasions* 18, 921–933. DOI: 10.1007/s10530-016-1081-x.

IPBES (2016) Summary for policymakers of the assessment report of the intergovernmental science-policy platform on biodiversity and ecosystem services on pollinators, pollination and food production. In: *Secretariat of the Intergovernmental Science-Policy Platform on Biodiversity and Ecosystem Services*. Bonn, Germany, pp. 1–36.

IPBES (2019) The global assessment report on biodiversity and ecosystem services of the intergovernmental science-policy platform on biodiversity and ecosystem services. In: *IPBES Secretariat*. Bonn, Germany, pp. 1–4.

IPBES (2023) Summary for policymakers of the thematic assessment report on invasive alien species and their control of the intergovernmental science-policy platform on biodiversity and ecosystem services. In: *IPBES Secretariat*. Bonn, Germany, pp. 1–56.

Jepsen, J.U., Hagen, S.B., Ims, R.A. and Yoccoz, N.G. (2008) Climate change and outbreaks of the geometrids *Operophtera brumata* and *Epirrita autumnata* in subarctic birch forest: Evidence of a recent outbreak range expansion. *Journal of Animal Ecology* 77(2), 257–264. DOI: 10.1111/j.1365-2656.2007.01339.x.

Keane, R.M. and Crawley, M.J. (2002) Exotic plant invasions and the enemy release hypothesis. *Trends in Ecology and Evolution* 17(4), 164–170. DOI: 10.1016/S0169-5347(02)02499-0.

Keena, M.A., Côté, M.J., Grinberg, P.S. and Wallner, W.E. (2008) World distribution of female flight and genetic variation in *Lymantria dispar* (Lepidoptera: Lymantriidae). *Environmental Entomology* 37(3), 636–649. DOI: 10.1603/0046-225X(2008)37[636:WDOFFA]2.0.CO;2.

Keena, M.A. (2016) Inheritance and world variation in thermal requirements for egg hatch in *Lymantria dispar* (Lepidoptera: Erebidae). *Environmental Entomology* 45(1), 1–10. DOI: 10.1093/ee/nvv163.

Kenis, M., Auger-Rozenberg, M.A., Roques, A., Timms, L., Péré, C. *et al.* (2009) Ecological effects of invasive alien insects. *Biological Invasions* 11(1), 21–45. DOI: 10.1007/s10530-008-9318-y.

Kolar, C.S. and Lodge, D.M. (2002) Ecological predictions and risk assessment for alien fishes in North America. *Science* 298(5596), 1233–1236. DOI: 10.1126/science.1075753.

Kolbe, J.J., Glor, R.E., Rodríguez Schettino, L., Lara, A.C., Larson, A. *et al.* (2004) Genetic variation increases during biological invasion by a Cuban lizard. *Nature* 431(7005), 177–181. DOI: 10.1038/nature02807.

Koop, J.A.H., Causton, C.E., Bulgarella, M., Cooper, E. and Heimpel, G.E. (2020) Population structure of a nest parasite of Darwin's finches within its native and invasive ranges. *Conservation Genetics* 22(1), 11–22. DOI: 10.007/x10592-020-01315-0.

Kotsakiozi, P., Richardson, J.B., Pichler, V., Favia, G., Martins, A.J. *et al.* (2017) Population genomics of the Asian tiger mosquito, *Aedes albopictus*: Insights into the recent worldwide invasion. *Ecology and Evolution* 7(23), 10143–10157. DOI: 10.1002/ece3.3514.

Lande, R. (2016) Evolution of phenotypic plasticity in colonizing species. In: *Invasion Genetics: The Baker and Stebbins Legacy*. pp. 165–174. DOI: 10.1002/9781119072799.ch9.

Lee, C.E. (1999) Rapid and repeated invasions of fresh water by the salt water copepod *Eurytemora affinis*. *Evolution* 53, 1423–1434. DOI: 10.1111/j.1558-5646.1999.tb05407.x.

Lee, C.E. (2002) Evolutionary genetics of invasive species. *Trends in Ecology and Evolution* 17(8), 386–391. DOI: 10.1016/S0169-5347(02)02554-5.

Lenormand, T. (2002) Gene flow and the limits to natural selection. *Trends in Ecology & Evolution* 17(4), 183–189. DOI: 10.1016/S0169-5347(02)02497-7.

Leskey, T.C., Hamilton, G.C., Nielsen, A.L., Polk, D.F., Rodriguez-Saona, C. *et al.* (2012) Pest status of the brown marmorated stink bug, *Halyomorpha halys* in the USA. *Outlooks on Pest Management* 23(5), 218–226. DOI: 10.1564/23oct07.

Liebhold, A.M., Brockerhoff, E.G., Garrett, L.J., Parke, J.L. and Britton, K.O. (2012) Live plant imports: The major pathway for forest insect and pathogen invasions of the US. *Frontiers in Ecology and the Environment* 10, 135–143. DOI: 10.1890/110198.

Liebhold, A.M., Yamanaka, T., Roques, A., Augustin, S., Chown, S.L. *et al.* (2016) Global compositional variation among native and non-native regional insect assemblages emphasizes the importance of pathways. *Biological Invasions* 18, 893–905. DOI: 10.1007/s10530-016-1079-4.

Lin, H., Yang, L., Liu, Q., Wang, T., Hossain, S.R. *et al.* (2012) Time series analysis of Japanese encephalitis and weather in Linyi City, China. *International Journal of Public Health* 57(2), 289–296.

Logan, M.L., Minnaar, I.A., Keegan, K.M. and Clusella-Trullas, S. (2020) The evolutionary potential of an insect invader under climate change. *Evolution* 74(1), 132–144. DOI: 10.1111/evo.13862.

Lytle, A.J., Costa, J.T. and Warren II, R.J. (2020) Invasion and high-elevation acclimation of *Solenopsis invicta* in the southern Blue Ridge Escarpment. *PLoS ONE* 15(5), e0232264. DOI: 10.1371/journal.pone.0232264.

Madsen, J.D., Bloomfield, J.A., Eichler, L.W., Boylen, C.W. and Sutherland, J.W. (1991) The decline of native vegetation under dense Eurasian watermilfoil canopies. *Journal of Aquatic Plant Management* 29, 94–99.

Margaritopoulos, J.T., Malarky, G., Tsitsipis, J.A. and Blackman, R.L. (2007) Microsatellite DNA and behavioural studies provide evidence of host-mediated speciation in *Myzus persicae* (Hemiptera: Aphididae). *Biological Journal of the Linnean Society* 91(4), 687–702. DOI: 10.1111/j.1095-8312.2007.00828.x.

Medley, K.A. (2010) Niche shifts during the global invasion of the Asian tiger mosquito, *Aedes albopictus* Skuse (Culicidae), revealed by reciprocal distribution models. *Global Ecology and Biogeography* 19(1), 122–133. DOI: 10.1111/j.1466-8238.2009.00497.x.

Medley, K.A., Jenkins, D.G. and Hoffman, E.A. (2014) Human-aided and natural dispersal drive gene flow across the range of an invasive mosquito. *Molecular Ecology* 24(2), 284–295. DOI: 10.1111/mec.12925.

Medlock, J.M., Snow, K.R. and Leach, S. (2005) Potential transmission of West Nile virus in the British Isles: An ecological review of candidate mosquito bridge vectors. *Medical and Veterinary Entomology* 19(1), 2–21. DOI: 10.1111/j.0269-283X.2005.00547.x.

Menke, S.B., Booth, W., Dunn, R.R., Schal, C., Vargo, E.L. *et al.* (2010) Is it easy to be urban? Convergent success in urban habitats among lineages of a widespread native ant. *PLoS ONE* 5(2), e9194. DOI: 10.1371/journal.pone.0009194.

Morales, C.L., Arbetman, M.P., Cameron, S.A. and Aizen, M.A. (2013) Rapid ecological replacement of a native bumble bee by invasive species. *Frontiers in Ecology and the Environment* 11(10), 529–534. DOI: 10.1890/120321.

Nagoshi, R.N., Goergen, G., Tounou, K.A., Agboka, K., Koffi, D. *et al.* (2019) Genetic comparisons of fall armyworm populations from 11 countries spanning sub-Saharan Africa provide insights into strain composition and migratory behaviors. *Scientific Reports* 9(1), 8311. DOI: 10.1038/s41598-019-44744-9.

Nagoshi, R.N., Htain, N.N., Boughton, D., Zhang, L., Xiao, Y. *et al.* (2020) Southeastern Asia fall armyworms are closely related to populations in Africa and India, consistent with common origin and recent migration. *Scientific Reports* 10(1), 1421. DOI: 10.1038/s41598-020-58249-3.

Olazcuaga, L., Foucaud, J., Deschamps, C., Loiseau, A., Claret, J.L. *et al.* (2022) Rapid and transient evolution of local adaptation to seasonal host fruits in an invasive pest fly. *Evolution Letters* 6(6), 490–505. DOI: 10.1002/evl3.304.

Oliveira, M.R.V., Henneberry, T.J. and Anderson, P. (2001) History, current status, and collaborative research projects for *Bemisia tabaci*. *Crop Protection* 20(9), 709–723. DOI: 10.1016/S0261-2194(01)00108-9.

Pashley, D.P., Johnson, S.J. and Sparks, A.N. (1985) Genetic population structure of migratory moths: The fall armyworm (Lepidoptera: Noctuidae). *Annals of the Entomological Society of America* 78(6), 756–762. DOI: 10.1093/aesa/78.6.756.

Passera, L. D.F, W. (ed.). (1994) *Characteristics of Tramp Species*. Westview Press, Boulder, Colorado, pp. 23–43.

Petitpierre, B., Kueffer, C., Broennimann, O., Randin, C., Daehler, C. *et al.* (2012) Climatic niche shifts are rare among terrestrial plant invaders. *Science* 335(6074), 1344–1348. DOI: 10.1126/science.12159.

Poland, T.M. and McCullough, D.G. (2006) Emerald ash borer: Invasion of the urban forest and the threat to North America's ash resource. *Journal of Forestry* 104(3), 118–124. DOI: 10.1093/jof/104.3.118.

Porter, M.L. and Crandall, K.A. (2003) Lost along the way: The significance of evolution in reverse. *Trends in Ecology and Evolution* 18, 541–547. DOI: 10.1016/S0169-5347(03)00244-1.

Prentis, P.J., Wilson, J.R.U., Dormontt, E.E., Richardson, D.M. and Lowe, A.J. (2008) Adaptive evolution in invasive species. *Trends in Plant Science* 13(6), 288–294. DOI: 10.1016/j.tplants.2008.03.004.

Pyšek, P., Jarošik, V. and Pergl, J. (2011) Alien plants introduced by different pathways differ in invasion success: Unintentional introductions as a threat to natural areas. *PLoS ONE* 6, e24890. DOI: 10.1371/journal.pone.0024890.

Queller, D.C. and Strassmann, J.E. (1998) Kin selection and social insects. *BioScience* 48, 165–175. DOI: 10.2307/1313262.

Queller, D.C. (2000) Pax argentinica. *Nature* 405, 519–520. DOI: 10.1038/35014705.

Rees, D. (2004) *Insects of Stored Products*. CSIRO Publishing, Collingwood, Australia, pp. 1–192.

Resh, V.H. and Carde, R.T. (2009) *Encyclopedia of Insects*. Academic Press, Cambridge, MA, USA, pp. 1–1132.

Rezende, H.R., Romano, C.M., Claro, I.M., Caleiro, G.S., Sabino, E.C. *et al.* (2020) First report of *Aedes albopictus* infected by dengue and zika virus in a rural outbreak in Brazil. *PLoS ONE* 15(3), e0229847. DOI: 10.1371/journal.pone.0229847.

Reznick, D.N. and Ghalambor, C.K. (2001) The population ecology of contemporary adaptations: What empirical studies reveal about the conditions that promote adaptive evolution. *Microevolution Rate, Pattern, Process* 183–198. DOI: 10.1007/978-94-010-0585-2_12.

Ricalde, M.P., Nava, D.E., Loeck, A.E. and Donatti, M.G. (2012) Temperature-dependent development and survival of Brazilian populations of the Mediterranean fruit fly, *Ceratitis capitata*, from tropical, subtropical and temperate regions. *Journal of Insect Science* 12(1), 33. DOI: 10.1673/031.012.3301.

Richards, C.L., Bossdorf, O., Muth, N.Z., Gurevitch, J. and Pigliucci, M. (2006) Jack of all trades, master of some? On the role of phenotypic plasticity in plant invasions. *Ecology Letters* 9(8), 981–993. DOI: 1111/j.1461-0248.2006.00950.x.

Roiz, D., Pontifes, P.A., Jourdain, F., Diagne, C., Leroy, B. *et al.* (2024) The rising global economic costs of invasive Aedes mosquitoes and Aedes-borne diseases. *Science of the Total Environment* 933, 173054. DOI: 10.1016/j.scitotenv.2024.173054.

Roques, A., Auger-Rozenberg, M.A., Blackburn, T.M., Garnas, J., Pyšek, P. *et al.* (2016) Temporal and interspecific variation in rates of spread for insect species invading Europe during the last 200 years. *Biological Invasions* 18, 907–920. DOI: 10.1007/s10530-016-1080-y.

Sakai, A.K., Allendorf, F.W., Holt, J.S., Lodge, D.M. and Molofsky, J. (2001) The population biology of invasive species. *Annual Review of Ecology and Systematics* 32, 305–332. DOI: 10.1146/annurev.ecolsys.32.081501.114037.

Saldamando-Benjumea, C.I., Estrada-Piedrahíta, K., Velásquez-Vélez, M.I. and Bailey, R.I. (2014) Assortative mating and lack of temporality between corn and rice strains of *Spodoptera frugiperda* (Lepidoptera, Noctuidae) from Central Colombia. *Journal of Insect Behavior* 27, 555–566. DOI: 10.1007/s10905-014-9451-7.

Santos, M.C., Le Bouquin, A., Crumiere, A.J.J. and Khila, A. (2017) Taxon-restricted genes at the origin of a novel trait allowing access to a new environment. *Science* 358(6361), 386–390. DOI: 10.1126/science.aan274.

Sexton, J.P., McIntyre, P.J., Angert, A.L. and Rice, K.J. (2009) Evolution and ecology of species range limits. *Annual Review of Ecology, Evolution and Systematics* 40(1), 415–436. DOI: 10.1146/annurev.ecolsys.110308.120317.

Simberloff, D. (2009) The role of propagule pressure in biological invasions. *Annual Review of Ecology, Evolution, and Systematics* 40, 81–102. DOI: 10.1146/annurev.ecolsys.110308.120304.

Shoemaker, D.D., Ahrens, M.E. and Ross, K.G. (2006) Molecular phylogeny of fire ants of the *Solenopsis saevissima* species-group based on mtDNA sequences. *Molecular Phylogenetics and Evolution* 38(1), 200–215. DOI: 10.1016/j.ympev.2005.07.014.

Simberloff, D. and Von Holle, B. (1999) Positive interactions of nonindigenous species: Invasional melt-down. *Biological Invasions* 1, 21–32. DOI: 10.1023/A:1010086329619.

Simberloff, D., Martin, J.L., Genovesi, P., Maris, V., Wardle, D.A. *et al.* (2013) Impacts of biological invasions: What's what and the way forward. *Trends in Ecology and Evolution* 28(1), 58–66. DOI: 10.1016/j.tree.2012.07.013.

Simon, J.C., Martinez-Torres, D., Latorre, A., Moya, A. and Hebert, P.D. (1996) Molecular characterization of cyclic and obligate parthenogens in the aphid *Rhopalosiphum padi* (L.). *Proceedings of the Royal Society of London. Series B: Biological Sciences* 263(1369), 481–486. DOI: 10.1098/rspb.1996.0072.

Smith, L.B., Kasai, S. and Scott, J.G. (2016) Pyrethroid resistance in *Aedes aegypti* and *Aedes albopictus*: Important mosquito vectors of human diseases. *Pesticide Biochemistry and Physiology* 133, 1–12. DOI: 10.1016/j.pestbp.2016.03.005.

Strayer, D.L. (2010) Alien species in fresh waters: Eecological effects, interactions with other stressors, and prospects for the future. *Freshwater Biology* 55(1), 152–174. DOI: 10.1111/j.1365-2427.2009.02380.x.

Tay, W.T., Meagher, R.L., Czepak, C. and Groot, A.T. (2023) *Spodoptera frugiperda*: Ecology, evolution, and management options of an invasive species. *Annual Review of Entomology* 68(1), 299–317. DOI: 10.1146/annurev-ento-120220-102548.

Tay, W.T., Soria, M.F., Walsh, T., Thomazoni, D., Silvie, P. *et al.* (2013) A brave new world for an old-world pest: *Helicoverpa armigera* (Lepidoptera: Noctuidae) in Brazil. *PLoS ONE* 8(11), e80134. DOI: 10.1371/journal.pone.0080134.

Thomas, G.W., Dohmen, E., Hughes, D.S., Murali, S.C., Poelchau, M. *et al.* (2020) Gene content evolution in the arthropods. *Genome Biology* 1–14. DOI: 10.1186/s13059-019-1925-7.

Tilman, D. (1982) *Resource Competition and Community Structure (no.17)*. Princeton University Press, p. 296.

Tobin, P.C., Kean, J.M., Suckling, D.M., McCullough, D.G., Herms, D.A. *et al.* (2012) Determinants of successful arthropod eradication programs. *Biological Invasions* 16(2), 401–414. DOI: 10.1007/s10530-013-0529-5.

Tschinkel, W. (2006) *The Fire Ants*. Harvard University Press, pp. 1–723.

Tsutsui, N.D., Suarez, A.V., Holway, D.A. and Case, T.J. (2000) Reduced genetic variation and the success of an invasive species. *Proceedings of the National Academy of Sciences USA* 97(11), 5948–5953. DOI: 10.1073/pnas.100110397.

Tsutsui, N.D. and Suarez, A.V. (2003) The colony structure and population biology of invasive ants. *Conservation Biology* 17(1), 48–58. DOI: 10.1046/j.1523-1739.2003.02018.x.

Valentin, R.E., Nielsen, A.L., Wiman, N.G., Lee, D.H. and Fonseca, D.M. (2017) Global invasion network of the brown marmorated stink bug, *Halyomorpha halys*. *Scientific Reports* 7, 9866. DOI: 10.1038/s41598-017-10315-z.

van der Gaag, D.J. and Straten, M. (2017) Assessment of the potential impact of *American Spodoptera* species for the European Union. In: *Netherlands Food and Consumer Product Safety Authority (NVWA)*. Utrecht, Catharijnesingel, the Netherlands, p. 59.

van Emden, H.F., Eastop, V.F., Hughes, R.D. and Way, M.J. (1969) The ecology of myzus persicae. *Annual Review of Entomology* 14, 197–270. DOI: 10.1146/annurev.en.14.010169.001213.

Venette, R.C. and Hutchison, W.D. (2021) Invasive insect species: Global challenges, strategies & opportunities. *Frontiers in Insect Science* 1, 650520. DOI: 10.3389/finsc.2021.650520.

Wang, Y., Zhang, H., Li, H. and Liu, T. (2016) Biology and population dynamics of the cotton aphid, *Aphis gossypii* Glover (Hemiptera: Aphididae), under different temperature and humidity conditions. *Journal of Insect Science* 16(1), 1–8. DOI: 10.1093/jisesa/iew027.

Westbrook, J.K., Nagoshi, R.N., Meagher, R.L., Fleischer, S.J. and Jairam, S. (2016) Modeling seasonal migration of fall armyworm moths. *International Journal of Biometeorology* 60, 255–267. DOI: 10.1007/s00484-015-1022-x.

Yang, Z.Q., Wang, X.Y., Gould, J.R. and Zhang, Y.N. (2006) Seasonal cycles and diapause of *Hyphantria cunea* (Lepidoptera: Arctiidae) in northern China. *Biological Control* 39(3), 345–353. DOI: 10.1016/j.biocontrol.2006.05.010.

Zhang, L., Liu, B., Zheng, W., Liu, C., Zhang, D. *et al.* (2020) Genetic structure and insecticide resistance characteristics of fall army worm populations invading China. *Molecular Ecology Resources* 20(6), 1682–1696. DOI: 10.1111/1755-0998.13219.

2 Biological Invasions in South Asia: An Overview

Akshay Kumar Chakravarthy[1]*, Mogili Ramaiah[2] and Kumara A.D.N.T.[3]
[1]Division of Crop Protection, ICAR – Indian Institute of Horticultural Research, Bengaluru, Karnataka, India (Retired); [2]Division of Entomology, ICAR – Indian Agricultural Research Institute, New Delhi, India; [3]Department of Biosystems Technology, Faculty of Technology, South Eastern University of Sri Lanka, University Park, Oluvil, Sri Lanka

Abstract

Biological invasions represent a significant and growing challenge to biodiversity conservation, ecosystem integrity, agricultural productivity, and human well-being across South Asia. This region, encompassing a diverse range of ecosystems from high-altitude Himalayan landscapes to tropical forests, wetlands, and coastal zones is increasingly vulnerable to the introduction and spread of invasive alien species (IAS). This chapter provides current knowledge on IAS in South Asian countries, highlighting patterns of introduction, spread, and ecological impacts. Despite a noticeable increase in research over the past two decades, particularly on invasive plants, significant gaps remain, especially in tropical regions and countries such as Afghanistan and the Maldives, which lack comprehensive data. This chapter also outlines major invasive faunal species affecting terrestrial and aquatic ecosystems, as well as recently emerging invasive insect pests that threaten food security. Compared with global leaders in IAS management, South Asian countries often lack the institutional capacity, data infrastructure, and policy frameworks necessary for effective prevention and control. Strengthening regional collaboration, enhancing early detection and rapid response systems, and investing in targeted research are critical for mitigating the threats posed by biological invasions in this ecologically and economically vital region.

Keywords: Invasive Alien Species (IAS), South Asia, biological invasions, biodiversity loss, ecological impact, food security, sustainable management

2.1 Introduction

South Asia comprises eight countries: Afghanistan, Bangladesh, Bhutan, India, Maldives, Nepal, Pakistan, and Sri Lanka. The region is bordered to the north by the Himalayas and to the south by the Indian Ocean. It encompasses several biodiversity hotspots, including the Himalayas, the Western Ghats, and the Indo-Burma region. However, increasing human population, urbanization, and climate change have exacerbated problems associated with invasive species (Mungi *et al.*, 2018). With the current pace of global trade, travel, and tourism, the threat from invasive species is expected to intensify (Early *et al.*, 2016). The Intergovernmental Science-Policy Platform on Biodiversity and Ecosystem Services (IPBES,

*Corresponding author: chakravarthyakshay@gmail.com

DOI: 10.1079/9781836991977.0002

2019) has also highlighted that biodiversity in South Asia is under significant threat, emphasizing the urgent need for effective invasive species management strategies in the region (Bhatta *et al.*, 2023).

South Asia is one of the most biologically diverse regions in the world, home to an incredible variety of plant and animal species, and accounting for a significant proportion of the world's biological diversity (Briggs, 1996). However, the region's landscape is undergoing rapid transformation, not only environmentally but also in its political and economic systems. For example, ten of the more than 330 invasive species in India have collectively cost the country approximately $127.3 billion (₹8.3 trillion) over the past 60 years. The economic impact of biological invasions in South Asia is particularly severe due to high population densities, socio-economic challenges, and weaker economies. Diagne *et al.* (2021) have documented the financial burden of invasive species in South Asia, which mirrors trends seen in other tropical regions, such as Southeast Asia, where land-use and environmental changes are also accelerating. According to Achard *et al.* (2002), at the current rate of deforestation and degradation, only 25% of the forest cover in Southeast Asia is expected to remain. South Asia spans approximately 6.2 million km^2, accounting for nearly 12% of Asia's landmass and 3.5% of the global land surface. This vast region features highly diverse climatic conditions, soils, vegetation types, and land-use patterns. It includes a range of agro-climatic zones, mountain ranges, deserts, rivers, grasslands, lakes, oceans, and forests. The region comprises islands, peninsular areas, arid zones, and mountainous countries, encompassing tropical, subtropical, and temperate landscapes.

There remains a significant knowledge gap regarding invasive species, and current cost estimates associated with their impacts are often imprecise. Analyses by researchers and other stakeholders suggest that India is experiencing greater economic losses from semi-aquatic and aquatic invasive species compared with terrestrial ones (Ghosh, 2022). The effects of biological invasions in tropical regions may rival, or even exceed, the economic and ecological impacts of climate change. While the biodiversity crisis in the tropics is largely driven by climate change, invasive species also play a major role, alongside deforestation, pollution, and habitat loss. Evergreen tropical forest tracts are increasingly vulnerable to invasions due to expanding agricultural activities and other anthropogenic pressures. Forest fragmentation and degradation further exacerbate the issue, contributing to the displacement of wildlife. Species such as tigers, elephants, and leopards are often forced out of their natural habitats, leading to increased human–wildlife conflicts, a trend frequently reported in newspapers, television, and other media outlets. Illegal wildlife trade and ongoing habitat destruction continue to threaten native species, a situation mirrored in Southeast Asia (Clements *et al.*, 2008). According to Sodhi *et al.* (2010), a substantial proportion of biodiversity in South and Southeast Asia is at risk of being lost in the near future.

The impact of biological invasions on biodiversity and other sectors in South Asia remains poorly documented. Much of the available information is published in local or regional newspapers and journals, which are often not indexed in global databases. One major barrier to reporting and documenting invasive species in the region, particularly in India is the linguistic diversity. Valuable data published in regional languages can be difficult to access through online searches. The Invasive Species Specialist Group (ISSG) of the IUCN has developed the Global Invasive Species Database, which provides some information on the issue. In recent years, both governmental and non-governmental organizations have begun publishing reports on the impacts of biological invasions in South Asia. However, it is only over the past two to three decades that invasive species databases and related documentation have become more prevalent in academic literature, government records, and global reports. According to Kolar and Lodge (2001), all non-native species that have spread beyond their point of introduction and successfully established themselves in new environments are considered invasive. The number of invasive and naturalized species on the South Asian mainland has increased significantly, and this trend correlates with the size of the country, an important factor influencing the extent of biological invasions (Bhatta *et al.*, 2023).

2.2 Global Scenario

Globally, over 1.75 million species have been identified as invasive. Among these, approximately 270,000 are plants, 950,000 are insects, 9,000 are birds, 19,000 are fish, and 4,000 are mammals (Rana *et al.*, 2017). Biological invasions have imposed enormous economic and ecological costs and, in many cases, have proven extremely difficult or even impossible to manage due to a range of complex factors. Various control methods such as chemical, physical, and biological have been deployed to contain invasive species, yet there has been no universally successful approach to invasive species management. Hanley and Roberts (2019) discuss economic valuation methods for managing invasives, highlighting their detrimental effects on landscapes and human well-being. Mooney and Cleland (2001) also emphasized the ecological, environmental, and social impacts of invasive species.

Beyond invasives, global ecosystems are simultaneously threatened by climate change, pollution, biodiversity loss, and habitat degradation. According to Khatri *et al.* (2023), global warming is likely to exacerbate biological invasions, as invasive species often thrive better than native species under elevated temperature conditions. The rapid spread of alien species presents a major challenge for governments worldwide, as many invasives reproduce at exponential rates and adapt swiftly to new environments. Their proliferation disrupts ecosystem functions by altering nutrient cycling, gas exchange, hydrology, and the composition and structure of native vegetation. In fact, biological invasions are responsible for over 20% of global species extinctions and are a major contributor to biodiversity loss (Rana, 2024).

Despite widespread recognition of IAS as a major threat to biodiversity and ecosystem services, documented economic costs related to their prevention, control, and associated damage particularly in the context of biodiversity and conservation remain surprisingly scarce. Equally concerning is the underrepresentation of IAS as a leading threat to global agriculture and human health. While studies such as those by Nuñez *et al.* (2020), Eschen *et al.* (2021), and Vilà *et al.* (2011) underscore the severity of these impacts, alien species are still seldom prioritized in these critical sectors. However, several recent reports indicate that some of the most substantial economic costs of biological invasions stem from agricultural losses and were the main component of reported costs for Africa (Diagne *et al.*, 2021b), the Americas (Crystal-Ornelas *et al.*, 2021; Heringer *et al.*, 2021), and Asia (Kirichenko *et al.*, 2021; Liu *et al.*, 2021; Watari *et al.*, 2021). In addition, significant human health-related costs have been attributed to mosquitoes of the genus *Aedes*, which were major contributors to reported economic burdens in Brazil (Adelino *et al.*, 2021), Central America (Heringer *et al.*, 2021), and Singapore (Haubrock *et al.*, 2021b). These findings emphasize the urgent need for broader recognition of IAS impacts and more targeted, sector-specific responses at national and global levels.

Over the past five decades, invasive alien species (IAS) have caused substantial economic damage globally, with estimated costs reaching a minimum of US$1.288 trillion (in 2017 US dollars), and these figures continue to rise steadily (Diagne *et al.*, 2021a). Accurately understanding and quantifying the economic impacts of IAS is crucial, especially as the rate of new introductions and associated damages shows no sign of slowing (Seebens *et al.*, 2017; Essl *et al.*, 2020). Moreover, both the current and projected numbers and types of IAS vary significantly across different ecosystems (van Kleunen *et al.*, 2015; Essl *et al.*, 2020), leading to considerable spatial and temporal variation in their ecological and economic impacts (Angulo *et al.*, 2021b; Diagne *et al.*, 2021a). Preventing the introduction of alien species remains a crucial strategy for mitigating the impacts of biological invasions, especially in tropical regions (Rana *et al.*, 2017).

2.3 Invasions in South Asia

A country-wise list of major biological invasions reported across different South Asian nations is presented below.

2.3.1 Afghanistan

Afghanistan, covering an area of 652,225 km², is a landlocked country bordered by Iran, Pakistan, Tajikistan, Turkmenistan, Uzbekistan, and China. Biogeographically divided into 17 regions, the country harbours a diverse assemblage of plant and animal taxa, owing to its strategic position at the intersection of the Indo-Malayan and Palaearctic realms. This biological richness stems from significant continental processes that began approximately 50 million years ago, following the collision of the Eurasian and Indian tectonic plates, an event that led to the formation of major mountain ranges such as the Hindu Kush, Himalayas, and Karakoram. Afghanistan's terrain varies from western deserts to alpine forests in the northeast.

Despite its ecological significance, Afghanistan has limited documented information on biodiversity, invasive species, and naturalized flora due to over three decades of armed conflict, which severely hampered scientific research and environmental monitoring. Currently, only 26 plant species have been recorded as invasive and naturalized within the country. This is the lowest reported among South Asian nations. Of the approximately 5,000 native plant species in Afghanistan, about 1.9% are considered naturalized (Bhatta *et al.*, 2023). Common invasive plant species include *Lantana camara* L., *Parthenium hysterophorus* (Mart.) Solms, *Chromolaena odorata* (King and Rob.), *Mimosa pudica* L., and *Argemone mexicana*. Control measures primarily involve chemical interventions by using herbicides; mechanical methods remain labour-intensive and are only effective in small-scale applications.

In addition to invasive plants, the country faces growing threats from the illegal wildlife trade, which increases the risk of biological invasions. Between 1990 and 2000, Afghanistan experienced a 75% rise in wildlife trafficking, with animal products being traded for consumption, the pet industry, fashion, and cosmetics (Bennett, 2006). The expansion of such trade is strongly linked to a heightened risk of introducing invasive species into native ecosystems (Wingard, 2010). Effective management of invasive species in Afghanistan necessitates the development of comprehensive biodiversity assessments, the establishment of national databases, strengthened border and trade controls, and the implementation of integrated pest management strategies. Public awareness and community engagement are also essential to support early detection and rapid response efforts.

2.3.2 Bangladesh

Bangladesh covers an area of 144,000 km^2 and is characterized by a deltaic plain and a hilly region. It is bordered with India along most of its borders and Burma in the southeast and more than 50 transboundary rivers feed into the country, forming the world's second largest riverine drainage basin and the largest delta of the world (Hofer and Messerli, 2006). The country enjoys a tropical climate with high rainfall, humidity, and temperature. The deltaic plain is often subjected to flooding and it is formed from the Ganges and the Brahmaputra rivers and their tributaries. The *National Biodiversity Strategy and Action Plan (NBSAP) 2016–2021* of Bangladesh identifies several invasive alien plant (IAP) species present in the country. These include *Pontederia crassipes* (water hyacinth), *Chromolaena odorata* (Siam weed), *Mikania cordata* (Assam lota), *Croton bonplandianum* (jungle tulsi), *Lantana camara* (lantana), *Ageratum conyzoides* (Billy goat weed), *Cajanus scarabaeoides* (wild kulthi), *Commelina obliqua* (jotakansira), *Convolvulus arvensis* (bindweed), *Evolvulus nummularius* (roundleaf bindweed), *Hyptis suaveolens* (pignut), *Ipomoea carnea* (pink morning glory), *Ludwigia adscendens* (water primrose), and *Mimosa pudica* (common sensitive plant) (DFE, 2016).

A review of IAS in northern Bangladesh by Akter and Zuberi (2009) reported seven invasive alien plants, five of which align with the NBSAP list, while also including *Mikania micrantha* and *Parthenium hysterophorus*. A more comprehensive assessment by Mukul *et al.* (2020) identified 69 invasive alien species in Bangladesh, comprising 46 plant species, 16 fish species, and five insect species. According to the Global Invasive Species Database (ISSG, 2015), notable invasive insects in Bangladesh include the silverleaf whitefly (*Bemisia tabaci*), the Asian citrus psyllid (*Diaphorina citri*, a vector of huanglongbing disease), and two invasive ant species: *Paratrechina longicornis* and *Tapinoma melanocephalum*. Further data from CABI's Invasive Species Compendium (ISC) list 651 records of invasive species relevant to Bangladesh (CABI ISC, 2022). The Global Register of Introduced and Invasive Species (GRIIS) database (accessed October 2020) reports 107 naturalized plant species in the country (Shrestha *et al.*, 2022). Additional invasive species reported by Hossain and Pasha (2001) include *Acacia auriculiformis* (northern black wattle), *Eucalyptus camaldulensis* (red gum), *Leucaena leucocephala* (leucaena), *Acanthospermum hispidum* (bristly starbur), *Senna occidentalis* (senna), *Cestrum diurnum* (day jessamine), and *Alternanthera ficoidea* (Joseph's coat). Within the Sundarbans mangrove forest, a World Heritage Site, Biswas *et al.* (2007) documented 23 alien species, including four identified as invasive.

The fall armyworm (FAW), *Spodoptera frugiperda*, first detected in Bangladesh in 2018, has caused substantial damage to maize and other crops, posing a serious threat to national food security (Lamsal *et al.*, 2020). Similarly, the rugose spiralling whitefly (*Aleurodicus rugioperculatus*), first identified in 2019, has spread rapidly and now affects a wide range of host plants, including coconut, banana, and guava. In total there are 61 species of invasives introduced into Bangladesh, while 2.8% are naturalized species, there being non-significant differences between the invasive and naturalized plant species (Bhatta *et al.*, 2023). The country has porous boundaries with its neighbouring countries and the phytosanitary and quarantine infrastructural facilities need to be stepped up. The country is undergoing drastic modifications in the political system, policies, land use and land cover attributes (Mukul *et al.*, 2011).

2.3.3 Bhutan

Bhutan, located in the Himalayan mountain range with an area of 38,394 km^2 and situated between China and India, harbours 101 invasive plant species, second only to India. The trend in the number of naturalized plant species in Bhutan parallels that of India, with invasive species representing approximately 1.9% of the flora (Bhatta *et al.*, 2023). A total of 30 invasive species were recorded in Bhutan as of 2014 (NBC, 2014). However, more recent assessments indicate a significant increase, with 101 invasive species currently reported (Dorjee *et al.*, 2020). Several major invasive alien plant species have been recorded in Bhutan, many of which are also widespread in India, including *Ageratum conyzoides*, *Eupatorium adenophorum*, *Eupatorium odoratum*, *Galinsoga parviflora*, *Lantana camara*, *Opuntia* spp., *Leucaena leucocephala*, *Mikania micrantha*, and *Parthenium hysterophorus* (Parker, 1992; NBC2011). Recent invasive pest incursions in Bhutan include the fall armyworm (*Spodoptera frugiperda*), a highly destructive species posing a significant threat to maize and other staple crops. Another major pest is the diamondback moth (*Plutella xylostella*), which is known for causing severe damage to cole crops. In addition to these insect pests, invasive plant species such as *Ageratina adenophora* and *Lantana camara* are increasingly problematic in forested areas and pastures, where they suppress native vegetation and disrupt ecosystem balance. Nythri and Rinchen (2025) provided a comprehensive review of the status and geographic distribution of invasive alien plant species in Bhutan.

2.3.4 India

India, as one of the fastest-growing economies, exports a wide range of commodities to 190 countries worldwide. The liberalization of trade in agricultural and other products, along with the increased movement of plant materials such as vegetables, fruits, ornamentals, planting materials, and seeds between nations, has significantly heightened the risk of introduction of insect pests, nematodes, plant pathogens, and invasive weeds into the country. Several important invasive species have become established in India across different taxonomic groups. Among plants, *Lantana camara*, *Parthenium hysterophorus* (communist weed), and *Prosopis juliflora* are particularly widespread. The giant African land snail (*Achatina fulica*) is a notable invasive mollusc. Significant insect pests include chilli black thrips, coffee berry borer, fall armyworm (*Spodoptera frugiperda*), tomato pinworm, and spiralling whitefly. Invasive fish species such as *Clarias gariepinus* (African catfish) and *Oreochromis mossambicus* (Mozambique tilapia) have been introduced into

Indian waters. Additionally, the red seaweed *Kappaphycus alvarezii* has spread along coastal regions, raising ecological concerns. Sandilyan (2015, 2022) catalogued invasive alien species in India, documenting a total of 171 species across various ecological groups. His compilation includes 54 invasive plant species and 56 aquatic species comprising plants, fishes, marine organisms, and microbes. In agroecosystems, he identified 47 invasive species, including insects, viruses, nematodes, fungi, bacteria, and fishes. Additionally, 14 invasive species were recorded from other faunal groups such as insects, cnidarians, molluscs, fishes, amphibians, reptiles, birds, and mammals. The National Biodiversity Authority has also undertaken assessments to evaluate the impacts of invasive species on the islands of India, with a particular focus on the Andaman group of islands.

Despite the existence of a robust regulatory framework, India continues to experience frequent incursions of alien invasive insect pests. The precise pathways of entry remain difficult to ascertain, largely due to India's porous borders with neighbouring countries (Shah *et al.*, 2020). Over the past decade, at least 15 alien insect pests have been formally reported within Indian territory. Notable recent invaders include the nesting whitefly (*Paraleyrodes pseudonaranjae*), apple leaf blotch miner (*Leucoptera malifoliella*), cassava mealybug (*Phenacoccus manihoti*), fall armyworm (*Spodoptera frugiperda*), rugose spiralling whitefly (*Aleurodicus rugioperculatus*), tomato pinworm (*Phthorimaea* (*Tuta*) *absoluta*), chilli black thrips (*Thrips parvispinus*) papaya mealybug (*Paracoccus marginatus*), eucalyptus gall wasp (*Leptocybe invasa*), and the Erythrina gall wasp (*Quadrastichus erythrinae*), all of which have caused significant economic losses (Shashank *et al.*, 2015; Sushil *et al.*, 2015; Sushil, 2016, 2018; Chakravarthy *et al.*, 2017; Sundararaj & Selvaraj, 2017; Sharanabasappa *et al.*, 2018; Shylesha *et al.*, 2018; Chandrika *et al.*, 2019; Josephrajkumar *et al.*, 2019; Joshi *et al.*, 2020; Nagaraju, 2021; Sampathkumar *et al.*, 2021, 2023; Sushil *et al.*, 2021; Basavaraj *et al.*, 2022a, b; Sushil *et al.*, 2022). More recently, Kumaranag *et al.* (2025) reported the first occurrence of the invasive small hive beetle from India. Currently, 173 invasive alien species have been documented in India, spanning 117 genera and 44 families, which together constitute approximately 1% of the nation's flora (Reddy, 2008).

India is prone to invasive threats due to lack of synergies between Institutions and Ministries. Hence, clear-cut roles and strong collaboration between ministries/departments as well as regional cooperation are required. Effective prevention, early detection, and rapid response plans for invasive pests must be developed. A coordinated network involving lead institutions such as ICAR and SAUs should facilitate information exchange, monitoring, and prevention. Strengthening the Directorate of Plant Protection, Quarantine, and Storage (DPPQS) manpower, infrastructure, and quarantine enforcement is critical. Tools such as detection kits, DNA barcoding, and remote microscopy should be explored for quick pest identification. An Agristat database and pictorial guides on potential invasives should be shared with customs and relevant agencies. Proactive biological control strategies and data on invasion pathways must inform policy decisions. Sharing and expedited import permits for biocontrol agents are essential to enable timely releases. Regulation of online and illegal sales of plants and arthropods should be enforced, with nurseries and seed companies providing annual certifications against introducing invasives. E-sensors at airports and ports, along with public awareness signage, can help prevent pest entry.

2.3.5 Maldives

The Maldives is an archipelagic nation in the Indian Ocean, located approximately 475 km south of India. It comprises 26 distinct reef systems, including atolls, islands, and coral line reefs, spanning an exclusive economic zone of 916,000 km^2 and territorial waters of 63,000 km^2. These waters encompass about 3.1% of the world's coral reefs. The archipelago hosts 248 coral species, with affinities to the Chagos Archipelago and Seychelles, and mangroves cover approximately 6% of the area, although research on these ecosystems is limited. Coastal lagoons have experienced eutrophication, partly due to tourism, resulting in proliferation of seagrass meadows. Of the roughly 1,200 recorded fish species, 40 are threatened. The Maldives is a critical habitat for sea turtles, with five of seven regional species present, including the critically endangered Hawksbill sea turtle (*Eretmochelys imbricata*). Nearly 200 seabird species, mostly migratory, also inhabit the islands. Prominent invasive fauna include the blood-sucker lizard (*Calotes versicolor*), Pinta giant tortoise (*Chelonoidis abingdonii*), Chiriqui harlequin, and golden toad (*Incilius periglenes*), while invasive flora include *Euphorbia heterophylla*, *Cuscuta* spp., *Ageratum*, *Lantana camara*, *Leucaena leucocephala*, and *Casuarina equisetifolia*. A recent, significant invasive pest issue in the Maldives involves the Coconut Hispid Beetle (*Brontispa longissima*). This beetle infests coconut trees and other palms, particularly young palms under 5 years old. The Maldives has been struggling with these infestations for some time. The beetle was introduced in the late 1990s from ornamental palms imported from Malaysia and Indonesia. A severely affected resort in Maldives has been reported to have incurred a direct economic loss of over US$200,000 within a period of 3 years. Stevens and Froman (2021) provide an overview of the region's ecology and the impacts of tourism. Lenzner *et al.* (2020) highlighted trade and transport as primary drivers for invasive species introduction to island ecosystems, with land use and land cover changes promoting their establishment and spread, especially in oceanic islands, atolls, and continental islands. Urgent implementation of stringent phytosanitary, quarantine, and regulatory measures is needed to manage biological invasions in the Maldives.

2.3.6 Myanmar

Myanmar (formerly Burma), covering approximately 678,577 km^2, is predominantly characterized by central lowlands surrounded by high mountain ranges. The country shares porous and largely unregulated borders that facilitate open commerce and trade, but also enable the unmonitored introduction and spread of invasive species. Many invasives in Myanmar have been introduced intentionally. Although geographically part of Southeast Asia, Myanmar is closely connected to South Asia through trade and transportation networks. Public awareness regarding the ecological and economic impacts of invasive species remains low (Than, 2011), highlighting the need for enhanced outreach on detection and management. Myanmar's quarantine and phytosanitary systems require comprehensive review and alignment with international standards, including standardization of phytosanitary measures (Haseeb *et al.*, 2011). Common invasive species include *Acacia auriculiformis*, *Acacia cunninghamii*, *Leucaena leucocephala*, *Prosopis juliflora*, *Lantana* spp., the giant African land snail (*Achatina fulica*), two fish species, one beetle species, and a marine borer (*Teredo* sp.). Singh *et al.* (2023) reported the introduction of 98 invasive species along altitudinal gradients within the Indo-Burma biodiversity hotspot, with up to 59% originating from the Americas. Many invasive plants are concentrated in the tropical regions, dominated by species such as *Ageratina adenophora*, *Ageratina conyzoides*, *Bidens pilosa*, *Chromolaena odorata*, *Lantana camara*, and *Mikania micrantha*. Despite ongoing plant invasions, studies on biological invasions in Myanmar remain sparse or lacking. A long-term, comprehensive management and prevention plan is urgently needed to address the rapid spread of invasive species across the Indo-Burma and Manipur regions. The fall armyworm (*Spodoptera frugiperda*), a serious agricultural pest, arrived in Myanmar in 2018 and has since spread widely, posing a significant threat to maize and other crops (Yee *et al.*, 2019). The South American tomato leaf miner, detected in Myanmar in 2019, is a key pest of tomato production, causing considerable damage to tomato crops (Yule *et al.*, 2021). The Global Register of Introduced and Invasive Species (GRIIS) checklist for Myanmar (2020) provides a comprehensive account of introduced and invasive species reported from the country (Crystal-Ornelas *et al.*, 2021).

2.3.7 Nepal

Nepal, a landlocked nation nestled in the foothills of the Himalayas, covers an area of 147,181 km^2. It shares its northern border with Tibet (China) and is surrounded on the other three sides by India. The country is divided into three major regions: the Himalayan region, the mid-hill region, and the Tarai region, with extensive water systems draining south into India. Nepal exhibits remarkable diversity in geography, temperature, and climate, making it a suitable habitat for a wide range of organisms from various biogeographic regions. However, biological invasions have emerged as a significant challenge, affecting biodiversity conservation, agricultural productivity, and ecosystem services (PQPMC, 2019). Invasive pests contribute to reduced agricultural yields and income, thereby posing serious threats to the country's economy, food security, and native biodiversity. Despite the severity of the issue, invasive species remain a relatively neglected topic in Nepal (Shrestha, 2016).

Invasive plant species documented in Nepal include *Argemone mexicana, Parthenium hysterophorus, Mimosa pudica, Mikania micrantha, Alternanthera philoxeroides*, and *Erigeron karvinskianus* (Shrestha *et al.*, 2017). Invasive fauna comprise the lizard *Calotes versicolor*, mosquito fish (*Gambusia affinis*), golden apple snail (*Pomacea canaliculata*), giant African snail (*Lissachatina fulica*), potato tuber moth (*Phthorimaea operculella*), and leaf miners (*Liriomyza* spp.). A global survey on pest and pathogen invasions across 124 countries identified Nepal as one of the top ten nations most at risk from IAS in the agriculture sector, relative to its gross domestic product (Paini *et al.*, 2016). To date, a total of 69 alien faunal species have been reported in Nepal, including 20 insect species (Budha, 2014). Among the fauna, the most important invasive insect pests are potato tuber moth (*Phthorimaea operculella*), tomato leafminers (*Liriomyza* spp.), and tomato pin worm (*Phthorimaea absoluta*) (Bajracharya *et al.*, 2016) (Table 2.1). A significant recent invasive pest in Nepal is the fall armyworm (*Spodoptera frugiperda*). It was first detected in Nepal in May 2019 and has since spread widely, affecting maize production (Bajracharya *et al.*, 2019). Recently, Shrestha and Shrestha (2021) provided detailed information on the patterns and processes of alien plant species invasions in Nepal.

Table 2.1. List of important invasive pests introduced in Nepal (Author's own table)

Common name	Scientific name	Year of introduction
African giant land snail	*Achatina fulica*	1930–1940
San Jose scale	*Quadraspidiotus perniciosus*	1962
Potato tuber moth	*Phthorimaea operculella*	1966
Chinese citrus fly	*Bactrocera minax*	2014
Tomato pinworm	*Phthorimaea absoluta*	2016
Fall armyworm	*Spodoptera frugiperda*	2019

2.3.8 Pakistan

Pakistan, situated between the Middle East and Asia, encompasses an area of 796,090 km² and is geographically divided into five regions: the Thar Desert and lower Indus Valley, the Balochistan Plateau, the Northeast Frontier region, the Northern region, and the Indus Valley. The invasive flora documented in the region includes *Prosopis juliflora, Parthenium hysterophorus, Lantana camara, Eucalyptus* spp., and *Leucaena leucocophala*. Invasive fauna of concern includes Nile tilapia (*Oreochromis niloticus*), various carp species, the apple snail (*Pomacea maculata*), and catfish (*Bagarius bagarius*). A significant recent invasive pest in Pakistan is the fall armyworm (FAW), *Spodoptera frugiperda*. It was first reported in Sindh, Pakistan, in 2019, and has since become a major concern for farmers, posing economic and agricultural challenges (Bhatti *et al.*, 2021). To mitigate biological invasions, stringent enforcement of national policies and regulations aligned with international phytosanitary standards is imperative (Khan and Hasan, 2019). Furthermore, the current monitoring and management frameworks require enhancement to effectively address invasive species threats, consistent with challenges faced by other South Asian nations. Farooq and Zarif (2024) provided detailed information on invasive species in Pakistan, discussing their impacts, associated challenges, and management strategies.

2.3.9 Sri Lanka

Sri Lanka is an island nation with a land area of approximately 65,610 km², complemented by territorial waters and an Exclusive Economic Zone (EEZ) encompassing around 517,000 km². The country's high biological diversity is largely attributed to its varied topography and climatic conditions. Sri Lanka's natural ecosystems and habitats include forests, grasslands, freshwater and marine wetlands, rivers, streams, mangroves, and coral reefs, all of which support a rich array of flora and fauna. The invasive flora includes *Mayaca* sp., *Indwigia* sp., *Panicum trichocladum*, *Setaria barbata*, and *Cissus* sp. The invasive plants include herbs, shrubs, creepers, and trees. Almost all plant species reported as invasives in Sri Lanka are vascular plants. The invasive fauna includes clown knife fish (*Chitala ornata*), walking catfish (*Clarias batrachus*), Mozambique tilapia (*Oreochromis mossambicus*), carp (*Cyprinus* sp.), slider turtle (*Trachemys pectoralis*), house mouse (*Mus musculus*), ship rat (*Rattus rattus*), feral dog (*Canis* sp.), garden slug (*Deroceras reticulatum*), and others.

Several researchers have documented the presence and proliferation of invasive alien faunal species (IAS) in Sri Lanka, particularly over the past two decades. Bambaradeniya (2002) identified 20 invasive alien fauna species spreading across natural and semi-natural ecosystems in various bioclimatic zones of the country (Table 2.2). Notably, ten of these species are listed among the top 100 of the world's worst invasive alien species (IUCN-ISSG, 2001). An updated list of 12 of the most troublesome invasive alien faunal species in Sri Lanka appears in the *National Atlas of Sri Lanka* (Wijesekara and Bambaradeniya, 2007). This list is a subset of those compiled earlier by Bambaradeniya (2002) and Marambe *et al.* (2003)

Several exotic invertebrates are confined to agricultural habitats and act as pests. For instance, multiple species of European garden slugs (*Deroceras reticulatum*, *D. leave*, *D. karuanae*) have established in agricultural areas of the central highlands above 1000 m elevation (Naggs, 2002; Mordan *et al.*, 2003; Raheem and Naggs, 2006; Kumuburegama & Ranawana,

Table 2.2. The list of invasive alien fauna recorded in Sri Lanka (Bambaradeniya, 2002)

Species	Category	Common name	Scientific name
Vertebrates	Freshwater fish	Rainbow trout[1,2]	*Oncorhynchus mykiss*
		Clown knife fish[1,2]	*Chitala ornate*
		Plecostomus cat fish[2] (tank cleaner)	*Pterygoplichthys multiradiatus*
		Walking cat fish[1]	*Clarias batrachus*
		Guppy[2]	*Poecilia reticulata*
		Western mosquitofish	*Gambusia affinis*
		Mozambique tilapia[1]	*Oreochromis mossambicus*
		Carp	*Cyprinus carpio*
		Snakeskin gourami[2]	*Trichogaster pectoralis*
	Turtle	Red-eared slider turtle[1]	*Trachemys scripta*
	Mammals	House mouse	*Mus musculus*
		Ship rat[1,2]	*Rattus rattus*
		Feral cat[1]	*Felis catus*
		Feral dog[1]	*Canis familiaris*
		Feral buffalo[1,2]	*Bubalus bubalis*
Invertebrates	Molluscs	Apple snail[1,2]	*Pomacea diffusa*
		Giant African snail[1,2]	*Lissachatina fulica*
		Field slug[2]	*Laevicaulis alte*
		Garden slug[1]	*Deroceras reticulatum*
		Garden slug	*Deroceras caruanae*

[1]Species identified in the National Atlas of Sri Lanka (Wijesekara and Bambaradeniya, 2007); [2]species listed by Marambe *et al.* (2003).

2007). The potato cyst nematode (*Globodera rostochiensis*), native to South America, became a serious pest in Sri Lanka's Nuwara Eliya District in the late 1980s (Ekanayake, 1999). The beet/potato leaf miner (*Liriomyza huidobrensis*), first reported in 1997, and *L. sativae* are now significant pests in vegetable cultivation (Ekanayake, 1999; Marambe *et al.*, 2003). Other recently introduced pests include the paddy root-knot nematode (*Meloidogyne graminicola*) and panicle mite (*Steneotarsonemus spinki*), both of which affect rice crops (Marambe *et al.*, 2003). The coconut mite (*Aceria guerreronis*), originally reported in Mexico in 1965, was found in Sri Lanka in 1997, while *Acaphylla theae* has been identified as an exotic pest of tea (Marambe *et al.*, 2003).

Further problematic species include the lesion nematode (*Pratylenchus vubrus*), fruit flies (*Bactrocera dorsalis* and *B. zonata*) (Ekanayake, 1999), the spiralling whitefly (*Aleurodicus dispersus*) (Marambe *et al.*, 2003) and papaya mealybug (*Paracoccus marginatus*). Silva and Kurukulasuriya (2010) and Marambe *et al.* (2011) provided a comprehensive review on invasive alien fauna in Sri Lanka, covering their national list, introduction, spread, impacts, and management. The most recently documented invasive pest in Sri Lanka is the Indian red scorpion (*Hottentotta tamulus*). This scorpion was first recorded in Sri Lanka in 2019 and was not previously known to inhabit the island. It was confirmed by a researcher to be the highly venomous Indian red scorpion Kovařík *et al.*, 2019a, b). Another invasive pest documented in Sri Lanka, and a significant agricultural threat, is the fall armyworm (*Spodoptera frugiperda*), which was first detected in all major maize growing areas in Sri Lanka in October 2018 and the new pest was confirmed as the fall armyworm by the International Plant Protection Convention (IPPC) in January 2019.

2.4 Conclusion and Future Perspectives

South Asia, a region of exceptional biodiversity, is increasingly threatened by invasive alien species, yet research and management efforts remain limited and largely descriptive. While inventories of invasive species are growing, there is a critical lack of data on their ecological, economic, and health impacts, with key processes such as nutrient cycling, productivity, and pollination rarely studied. Management strategies remain largely fragmented, with limited documentation or evaluation of the effectiveness of existing policies. Biological control is recognized globally as a sustainable management tool, but receives insufficient attention in South Asia, despite its potential to mitigate the spread of high-impact invaders. To protect the region's rich ecosystems and livelihoods, it is essential to shift focus toward impact-based research, prioritize species and areas most at risk, and strengthen regional collaboration and community awareness. Leveraging global experiences and investing in targeted biological control research will be vital to developing effective, long-term solutions to the invasive species crisis in South Asia.

Acknowledgements

Dr Mogili Ramaiah acknowledges support from Dr Mukesh K. Dhillon, Head; Dr Shashank P.R, In charge, National Pusa Collection and other administrative staff of ICAR – Indian Agriculture Research Institute, New Delhi, India.

References

Achard, F., Eva, H.D., Stibig, H.J., Mayaux, P., Gallego, J. *et al.* (2002) Determination of deforestation rates of the world's humid tropical forests. *Science* 297, 999–1002.

Acharya, R., Akintola, A.A., Malekera, M.J., Kamulegeya, P., Nyakunga, K.B. *et al.* (2021) Genetic relationship of fall armyworm (*Spodoptera frugiperda*) populations that invaded Africa and Asia. *Insects* 12(5), 439.

Adelino, J.R.P., Haubrock, P.J., Cuthbert, R.N., Diagne, C., Courchamp, F. *et al.* (2021) The economic costs of biological invasions in Brazil: A first assessment. *NeoBiota* 67, 349–374. DOI: 10.3897/neobiota.67.59185.

Akter, A. and Zuberi, M.I. (2009) Invasive alien species in Northern Bangladesh: Identification, inventory and impacts. *International Journal of Biodiversity and Conservation* 1(5), 129–134.

Angulo, E., Diagne, C., Ballesteros-Mejia, L., Adamjy, T., Ahmed, D.A. *et al.* (2021) Non-english languages enrich scientific knowledge: The example of economic costs of biological invasions. *Science of The Total Environment* 775, e144441. DOI: 10.1016/j.scitotenv.2020.144441.

Bajracharya, A.S.R., Mainali, R.P., Bhat, B., Bista, S., Shashank, P.R. *et al.* (2016) The first record of South American tomato leaf miner, *Tuta absoluta* (Meyrick 1917) (Lepidoptera: Gelechiidae) in Nepal. *Journal of Entomology and Zoology Studies* 4(4), 1359–1363.

Bajracharya, A.S.R., Bhat, B., Sharma, P., Shashank, P.R., Meshram, N.M. *et al.* (2019) First record of fall army worm *Spodoptera frugiperda* (JE Smith) from Nepal. *Indian Journal of Entomology* 81(4), 635–639.

Bambaradeniya, C.N.B. (2002) The status and implications of invasive alien species in Sri Lanka. *Zoos' Print Journal* 17(11), 930–935.

Basavaraj, K.A.G., Sreenivas, P.R., Badari Prasad, R.R. and Rachana, R.R. (2022a) First report of invasive thrips, *Thrips parvispinus* (Karny) (Thysanoptera: Thripidae) infesting chilli, *Capsicum annuum* L. in Kalaburagi, Karnataka, India. *Journal of Experimental Zoology India* 25(1), 191–194.

Basavaraj, T., Kumar, R. and Hiremath, A.S. (2022b) Invasive insect pests and their impact on Indian agriculture. *Journal of Agricultural Research* 60(4), 412–421.

Bennett, E.L. (2006) Consuming the world's wildlife. *Chinese National Geography* X, 126–135.

Bhatta, S., Shrestha, B.B. and Pyšek, P. (2023) Invasive alien plants in South Asia: Impacts and management. *NeoBiota* 88, 135–167. DOI: 10.3897/neobiota.88.104118.

Bhatti, Z., Ahmed, A.M., Khatri, I., Rattar, Q., Rajput, S. *et al.* (2021) First report of morphometric identification of *Spodoptera frugiperda* (JE Smith) (Lepidoptera: Noctuidae) an invasive pest of maize in Southern Sindh, Pakistan. *Asian Journal of Agriculture and Biology* 1–8. DOI: 10.35495/ajab.2020.03.169.

Biswas, S.R., Choudhury, J.K., Nishat, A. and Rahman, M.M. (2007) Do invasive plants threaten the Sundarbans mangrove forest of Bangladesh? *Forest Ecology and Management* 245(1–3), 1–9.

Briggs, J.C. (1996) Tropical diversity and conservation. *Conservation Biology* 10, 713–718.

Budha, P.B. (2014) Invasive alien fauna of Nepal: Current situation and future perspectives. *Proceedings*. Available at: https://www.researchgate.net/publication/319860259

CABI (2022) *Invasive species compendium*. Available at: https://www.cabi.org/isc

Chakravarthy, A.K., Kumar, K.P., Sridhar, V., Prasannakumar, N.R., Nitin, K.S. *et al.* (2017) Incidence, hosts and potential areas for invasion by Rugose Spiralling Whitefly, *Aleurodicus rugiopersculatus* Martin (Hemiptera: Aleyrodidae) in India. *Pest Management in Horticultural Ecosystems* 23, 41–49.

Chandrika, M., Josephrajkumar, A., Merin, B., Arya, K., Prathibha, P.S. *et al.* (2019) First record of the non-native neotropical nesting whitefly, *Paraleyrodes minei* laccarino on coconut palms in India and its coexistence with Bondar's nesting whitefly *Paraleyrodes bondari* Peracchi. *Current Science* 117, 515–519.

Clements, R., Ng, P.K.L., Lu, X., Ambu, S., Schilthuizen, M. *et al.* (2008) Using biogeographical patterns of endemic land snails to improve conservation planning for limestone karsts. *Biological Conservation* 141(11), 2751–2764.

Crystal-Ornelas, R., Hudgins, E.J., Cuthbert, R.N., Haubrock, P.J., Fantle-Lepczyk, J. *et al.* (2021) Economic costs of biological invasions within North America. In: *The Economic Costs of Biological Invasions around the World*, Vol. 67. NeoBiota, pp. 485–510. DOI: 10.3897/neobiota.67.58038.

Department of Forests and Environment (DFE) (2016) *National Biodiversity Strategy and Action Plan of Bangladesh 2016–2021*. Ministry of Environment and Forests, Government of Bangladesh, Dhaka.

Diagne, C., Leroy, B., Vaissière, A.C., Gozlan, R.E., Roiz, D. *et al.* (2021) High and rising economic costs of biological invasions worldwide. *Nature* 592(7855), 571–576.

Dorjee, K., Johnson, S.B., Buckmaster, A.J. and Downey, P.O. (2020) Weeds in the land of gross national happiness: Knowing what to manage by creating a baseline alien plant inventory for Bhutan. *Biological Invasions* 22(10), 2899–2914. DOI: 10.1007/s10530-020-02306-5.

Early, R., Bradley, B.A., Dukes, J.S., Lawler, J.J., Olden, J.D. *et al.* (2016) Global threats from invasive alien species in the twenty-first century and national response capacities. *Nature Communications* 7, 12485.

Ekanayake, H.M.R.K. (1999) Alien invasive fauna and agricultural importance and their control. In: *Proceedings of the First National Workshop on Alien Invasive Species in Sri Lanka*, Ministry of Forestry and Environment, Sri Lanka, pp. 57–59.

Eschen, R., Britton, K., Brockerhoff, E.G., Burgess, T.I., Dalley, V. *et al.* (2021) Towards estimating the economic cost of invasive alien species to African crop and livestock production. *CABI Agriculture and Bioscience* 2(1), 1–12. DOI: 10.1186/s43170-021-00038-7.

Essl, F., Lenzner, B., Bacher, S., Bailey, S., Capinha, C. *et al.* (2020) Drivers of future alien species impacts: An expert-based assessment. *Global Change Biology* 26, 4880–4893. DOI: 10.1111/gcb.15199.

Farooq, M. and Zarif, S. (2024) Invasive species in Pakistan: Impacts, challenges and management strategies. *Pakistan Journal of Forestry* 74(1), 18–23.

Ghosh, S. (2022) The cost of invasive species bears heavy on Indian economy, finds study. *Mongabay Series: Climate Connections, Economy*,

Global Register of Introduced and Invasive Species (GRIIS) (2020) *Bangladesh checklist*. Available at: https://cloud.gbif.org/griis/resource?r=griis-bangladesh

Hanley, N. and Roberts, M. (2019) The economic benefits of invasive species management. *People and Nature* 1, 124–137. DOI: 10.1002/pan3.31.

Haseeb, M., Kairo, M. and Flowers, R.W. (2011) New approaches and possibilities for invasive pest identification using web-based tools. *American Entomologist* 57(4), 223–226.

Haubrock, P.J., Cuthbert, R.N., Diagne, C., Courchamp, F. and Liu, C. (2021) The economic costs of invasive species in Asia. *NeoBiota* 67, 113–138. DOI: 10.3897/neobiota.67.58147.

Heringer, G., Haubrock, P.J., Cuthbert, R.N., Diagne, C., Courchamp, F. *et al.* (2021) The economic costs of biological invasions in Central America and the Caribbean. *NeoBiota* 67, 511–538. DOI: 10.3897/neobiota.67.59184.

Hofer, T. and Messerli, B. (2006) *Floods in Bangladesh: History, Dynamics and Rethinking the Role of the Himalayas*. United Nations University Press, Tokyo, Japan.

Hossain, M.K. and Pasha, M.K. (2001) Alien invasive plants in Bangladesh and their impacts on the ecosystem. In: *Assessment and Management of Alien Species That Threaten Ecosystems, Habitats and Species*. Secretariat of the Convention on Biological Diversity, Montreal, Canada. DOI: 10.18590/euscorpius.2019.vol2019.iss283.1.

Invasive Species Specialist Group (ISSG) (2015) *Global invasive species database*. Available at: https://www.iucngisd.org/gisd/

IPBES (2019) *Global assessment report on biodiversity and ecosystem services of the Intergovernmental Science-Policy Platform on Biodiversity and Ecosystem Services*. IPBES Secretariat. DOI: 10.5281/zenodo.3831673.

IUCN-ISSG (2001) *100 of the World's Worst Invasive Alien Species*. Invasive Species Specialist Group, IUCN, Auckland.

Josephrajkumar, A., Mohan, C., Babu, M., Krishna, A., Krishnakumar, V. *et al.* (2019) First record of the invasive Bondar's Nesting Whitefly, *Paraleyrodes bondari* Peracchi on coconut from India. *Phytoparasitica* 47, 333–339. DOI: 10.1007/s12600-019-00741-2.

Joshi, S., Pai, S.G., Deepthy, K.B., Ballal, C.R. and Watson, G.W. (2020) The cassava mealybug, *Phenococcus manihoti* Matile-Ferrero (Hemiptera: Coccomorpha: Pseudococcidae) arrives in India. *ZooTaxa* 4772, 191–194.

Khan, S. and Hasan, M.-U. (2019) Climate classification of Pakistan. *International Journal of Economic and Environmental Geology* 10(2), 60–71.

Khatri, K., Negi, B., Bargali, K. and Bargali, S.S. (2023) Effects of different concentrations of leaf residues of *Ageratina adenophora* on seed germination and growth behavior of two native tree species of Kumaun Himalaya, India. *Waste and Biomass Valorization* 15(2), 923–943. DOI: 10.1007/s12649-023-02213-5.

Kirichenko, N., Kenis, M., Baranchikov, Y. and Grégoire, J.C. (2021) Economic impact of invasive alien species on agriculture in Russia. *Biological Invasions* 23(3), 725–740. DOI: 10.1007/s10530-020-02400-8.

Kolar, C.S. and Lodge, D.M. (2001) Progress in invasion biology: Predicting invaders. *Trends in Ecology and Evolution* 16(4), 199–204.

Kovařík, F., Lowe, G., Ranawana, K.B. and Jayarathne, V.A.D. (2019a) First record of the Indian red scorpion *Hottentotta tamulus* (Scorpiones: Buthidae) from Sri Lanka. *Euscorpius* 279, 1–7.

Kovařík, F., Ranawana, K.B., Jayarathne, V.A., Hoferek, D. and Šťáhlavský, F. (2019b) Scorpions of Sri Lanka (Arachnida: Scorpiones). Part III. *Heterometrus yaleensis* sp. n. (Scorpionidae). *Heterometrus Yaleensis Sp* (283), 1–13.

Kumaranag, K.M., Garain, P.K., Chandran, N., Shashank, P.R., Kukkamudi, M.R. *et al.* (2025) First report of invasive small hive beetle, *Aethina tumida* Murray (Coleoptera: Nitidulidae) from India. *Journal of Apicultural Research* 1–11. DOI: 10.1080/00218839.2025.2478328.

Kumburegama, S. and Ranawana, K.B. (2007) Economic importance of pest snails and slugs in vegetable growing areas in Sri Lanka. In: *Proceedings of the Annual Research Sessions*, University of Peradeniya, Sri Lanka, p. 127.

Lamsal, S., Sibi, S. and Yadav, S. (2020) Fall armyworm in South Asia: Threats and Management. *Asian Journal of Advances in Agricultural Research* 13(3), 21–34. DOI: 10.9734/AJAAR/2020/v13i330106.

Lenzner, B., Latombe, G., Capinha, C., Bellard, C., Courchamp, F. *et al.* (2020) What will the future bring for biological invasions on islands? An expert-based assessment. *Frontiers in Ecology and Evolution* 8, 280.

Liu, X., Blackburn, T.M., Song, T., Li, X., Huang, C. *et al.* (2021) Risks of biological invasions on the belt and road. *Current Biology* 29(3), 499–505. DOI: 10.1016/j.cub.2018.12.036.

Marambe, B., Amarasinghe, L. and Gamage, G. (2003) Sri Lanka country report. In: Pallewatta, N., Reaser, J.K. and Gutierrez, A.T. (eds) *Invasive Alien Species in South-Southeast Asia: National Reports and Directory of Resources*. The Global Invasive Species Programme, Cape Town, pp. 91–100.

Marambe, B., Silva, P., Ranwala, S., Gunawardena, J., Weerakoon, D. *et al.* (2011) Invasive alien fauna in Sri Lanka: National list, impacts and regulatory framework. In: Veitch, C.R., Clout, M.N. and Towns, D.R. (eds) *Island Invasives: Eradication and Management*. IUCN, pp. 445–450.

Mooney, H.A. and Cleland, E.E. (2001) The evolutionary impact of invasive species. *Proceedings of the National Academy of Sciences USA* 98(10), 5446–5451.

Mordan, P., Naggs, F., Ranawana, K., Kumburegama, S. and Grimm, B. (2003) A guide to the pest and exotic gastropods of Sri Lanka. Department of Zoology, The Natural History Museum.

Mukul, S.A., Arfin-Khan, M.A.S. and Uddin, M.B. (2011) Invasive alien species of Bangladesh. In: Pullaiah, T. and Ielmini, M.R. (eds) *Invasive Alien Species: Observations and Issues from Around the World, Volume 2: Issues and Invasions in Asia and the Pacific Region*. Wiley, Chichester.

Mukul, S.A., Uddin, M.B., Rashid, A.Z.M.M. and Fox, J. (2020) Invasive alien species in Bangladesh: Status, impacts and management. *Environmental Challenges* 1, 100002.

Mungi, N.A., Coops, N.C., Ramesh, K. and Rawat, G.S. (2018) How global climate change and regional disturbance can expand the invasion risk? Case study of Lantana camara invasion in the Himalaya. *Biological Invasions* 20(7), 1849–1863. DOI: 10.1007/s10530-018-1666-7.

Nagaraju, D.K. (2021) Rugose Spiralling Whitefly (RSW), *Aleurodicus rugioperculatus* Martin (Hemiptera: Aleyrodidae): An urban nuisance. *Insect Environment* 24, 156–157.

Naggs, F. (2002) *Molluscan pests in Sri Lanka: Voracious exotics having a major and rapidly increasing impact on agriculture*. Department of Zoology, The Natural History Museum, London.

National Biodiversity Centre (2014) *National biodiversity strategies and action plan of Bhutan*. Thimphu: Royal Government of Bhutan. Available at: https://www.cbd.int/doc/world/bt/bt-nbsap-v4-en.pdf

National Biodiversity Centre (NBC) (2014) *Invasive Alien Species of Bhutan*. Ministry of Agriculture and Forests, Royal Government of Bhutan, Thimphu.

National Biodiversity Centre (NBC) (2011) *Bhutan Biodiversity Action Plan. Ministry of Agriculture and Forests*. Royal Government of Bhutan, Thimphu.

Nuñez, M.A., Pauchard, A., Ricciardi, A., Blackburn, T.M., Carlton, J.T. *et al.* (2020) Invasive species and the anthropocene: An independent review of the science. *Biological Invasions* 22(3), 1533–1549. DOI: 10.1007/s10530-020-02246-0.

Paini, D.R., Sheppard, A.W., Cook, D.C., De Barro, P.J., Worner, S.P, *et al.* (2016) Global threat to agriculture from invasive species. *Proceedings of the National Academy of Sciences USA* 113, 7575–7579.

Parker, C. (1992) *Weeds of Bhutan*. National Plant Protection Centre, Simtokha, Royal Government of Bhutan.

Plant Quarantine and Pesticide Management Centre (PQPMC) (2019) *Status of Invasive Alien Species in Nepal*. Ministry of Agriculture and Livestock Development, Kathmandu.

Raheem, D. and Naggs, F. (2006) *An Illustrated Guide to the Land Snails of Sri Lanka Natural Forest and Cultivated Habitats*. Department of Zoology, The Natural History Museum, London.

Rana, R.S. (2024) Biological invasions and biodiversity loss: Global patterns, impacts and management challenges. *Environmental Science and Policy* 152, 103–112.

Rana, S.S., Kumar, S., Sharma, N. and Angiras, N.N. (2017) Herbicidal management of *Parthenium hysterophorus* L. in grassland ecosystem of Himachal Pradesh. *Contemporary Research in India* 7, 30–35.

Reddy, C.S. (2008) Catalogue of invasive alien flora of India. *Life Science Journal* 5, 84–89.

Sampathkumar, M., Mohan, M., Shylesha, A.N., Joshi, S., Venkatesan, T. *et al.* (2021) Occurrence of cassava mealybug, *Phenacoccus manihoti* Matile Ferrero (Pseudococcidae: Hemiptera), a new invasive pest on cassava in India and prospects for its classical biological control opportunities in India. *Current Science* 120, 432–435.

Sampathkumar, M., Shylesha, A.N., Gani, M., Khurshid, R., Venkatesan, T. *et al.* (August 2023) Occurrence of invasive insect pest, apple leaf blotch miner (ALBM), *Leucoptera malifoliella* (Costa) (Lyonetiidae: Lepidoptera) in the Jammu and Kashmir UT of India (Lyonetiidae), 1–7.

Sandilyan, S. (2015) Highways of India: A heaven to invasive alien plants. *Science India* 33–36.

Sandilyan, S. (2022) Alien fish species in Indian inland wetlands: Current status and future challenges. *Wetland Ecology and Management* 30, 423–437.

Shah, K.K., Tiwari, I., Tripathi, S., Subedi, S. and Shrestha, J. (2020) Invasive alien plant species: A threat to biodiversity and agriculture in Nepal. *Agriways* 8(2), 62–73. DOI: 10.38112/agw.2020.v08i01.008.

Sharanabasappa, K., Kalleshwarswamy, C.M., Ashokan, R., Mahadevaswamy, H.M. and Maruthi, H.B. (2018) First report of fall armyworm, *Spodoptera frugiperda* (J.E. Smith) (Lepidoptera: Noctuidae): An alien invasive pest on maize in India. *Pest Management in Horticultural Ecosystems* 24, 23–29.

Shashank, P.R., Chandrashekhar, K., Meshram, N.M. and Sreedevi, K. (2015) Occurrence of *Tuta absoluta* (Lepidoptera: Gelechiidae), an invasive pest from India. *Indian Journal of Entomology* 74, 323–329.

Shrestha, B.B. (2016) Invasive alien plant species in Nepal: Status and management. *Botanica Orientalis* 10, 1–9.

Shrestha, B.B., Ranjit, J.D. and Siwakoti, M. (2017) Status of invasive alien plant species in Nepal. In: *National Workshop on Conservation and Utilization of Agricultural Plant Genetic Resources in Nepal*. Nepal.

Shrestha, B.B., Bhatta, S., Shrestha, U.B. and Pysek, P. (2022) Global register of introduced and invasive species (GRIIS): Bangladesh. *Invasive Alien Species Network* IUCN.

Shrestha, B.B. and Shrestha, K.K. (2021) Invasions of alien plant species in Nepal: Patterns and process. In: Pullaiah, T. and Ielmini, M.R. (eds) *Global Biodiversity*, Vol. 4. Wiley. DOI: 10.1002/9781119607045.ch20.

Shylesha, A.N., Jalali, S.K., Gupta, A., Varshney, R.*et al.* (2018) Studies on new invasive pest *Spodoptera frugiperda* (J.E. Smith) (Lepidoptera: Noctuidae) and its natural enemies. *Journal of Biological Control* 32, 1–7.

Silva, P. and Kurukulasuriya, M. (2010) Invasive alien fauna in Sri Lanka – Introduction, spread, impacts and management. In: Marambe, B., Silva, P., Wijesundara, S. and Atapattu, N. (eds) *Invasive Alien Species in Sri Lanka – Strengthening Capacity to Control Their Introduction and Spread*. Biodiversity Secretariat of the Ministry of Environment, Sri Lanka, pp. 39–61.

Singh, M.S., Tripathi, O.P., Yadav, H.S., Singha, L.B. and Basumatary, D. (2023) *Assessment of non-woody alien and invasive plant species along the altitudinal gradients of Indo-Burma biodiversity hotspot in Manipur, India*. Environmental Sciences Europe. DOI: 10.1007/s42535-023-00649-7.

Singh, S., Sharma, J.H., Udikeri, A. and Ansari, H. (2020) Invasive insects in India. *IntechOpen*. DOI: 10.5772/intechopen.91986.

Sodhi, N.S., Posa, M.R.C., Lee, T.M., Bickford, D., Koh, L.P. *et al.* (2010) The state and conservation of Southeast Asian biodiversity. *Biodiversity and Conservation* 19, 317–328. DOI: 10.1007/s10531-009-9607-5.

Stevens, G.M.W. and Froman, N. (2021) The Maldives archipelago. In: Sheppard, C. (ed.) *World Seas: An Environmental Evaluation. Volume II: The Indian Ocean to the Pacific*. School of Life Sciences, University of Warwick, United Kingdom. Available at: https://www.researchgate.net/publication/330046267

Sundararaj, R. and Selvaraj, K. (2017) Invasion of rugose spiralling whitefly, *Aleurodicus rugioperculatus* Martin (Hemiptera: Aleyrodidae): A potential threat to coconut in India. *Phytoparasitica* 45. DOI: 10.1007/s12600-017-0567-0.

Sushil, S.N., Verghese, A., Viraktamath, C.A., Ballal, C.R. and Nagaraju, D.K. (2015) *South American Tomato Pinworm, Tuta absoluta (Meyrick) (Lepidoptera: Gelechiidae)*. Directorate of Plant Protection Quarantine and Storage, Ministry of Agriculture and Farmers Welfare, Government of India.

Sushil, S.N. (2016) Emerging invasive insect pests in India and their management. *Indian Journal of Entomology* 78(3), 187–196.

Sushil, S.N. (2018) Invasive insect pests: Threats to Indian agriculture and biosecurity. *Current Science* 115(9), 1706–1714.

Sushil, S.N., Nagaraju, D.K. and Srivastava, R.P. (2021) Safeguarding Indian agriculture through plant quarantine regulations: Emerging issues and way forward. *Journal of Ecofriendly Agriculture* 16(2), 97–105.

Sushil, S.N., Sampathkumar, M., Mohan, M., Nagaraju, D.K. and Singh, J.P. (2022) Recent events of invasive and migratory pests and lessons learnt. *Indian Journal of Entomology* 84.

Than, W. (2011) Invasive alien species in Myanmar: Current status and management challenges. *Biodiversity Journal of Asia* 5, 45–52.

Than, W.W. (2011) *Information Gaps in Management of Invasive Alien Species in Myanmar*. World Forest Institute.

Van Kleunen, M., Dawson, W., Essl, F., Pergl, J., Winter, M. *et al.* (2015) Global exchange and accumulation of non-native plants. *Nature* 525, 100–103. DOI: 10.1038/nature14910.

Vilà, M., Espinar, J.L., Hejda, M., Hulme, P.E., Jarošík, V. *et al.* (2011) Ecological impacts of invasive alien plants: A meta-analysis of their effects on species, communities and ecosystems. *Ecology Letters* 14(7), 702–708. DOI: 10.1111/j.1461-0248.2011.01628.x.

Watari, Y., Komine, H. and Kiritani, K. (2021) Economic impacts of invasive alien species in Japan: A review. *Ecological Research* 36(1), 1–12. DOI: 10.1111/1440-1703.12123.

Wijesekara, G.A.W. and Bambaradeniya, C.N.B. (2007) Invasive alien species. In: *The National Atlas of Sri Lanka*, 2nd edn. Survey Department, Colombo, Sri Lanka, pp. 95–97.

Wingard, J. (2010) *Wild fauna trade in Afghanistan*. Technical Report. DOI: 10.13140/RG.2.2.17310.82248.

Yee, K.N., Aye, M.M., Htain, N.N., Oo, A.K., Kyi, P.P. *et al.* (2019) First detection report of the fall armyworm *Spodoptera frugiperda* (Lepidoptera: Noctuidae) on maize in Myanmar. Pest report, International Plant Protection Convention. Available at: https://www.ippc.int/pestreport/2019/01/11

Yule, S., Htain, N.N., Oo, A.K., Sotelo-Cardona, P. and Srinivasan, R. (2021) Occurrence of the South American tomato leaf miner, *Tuta absoluta* (Meyrick), in southern Shan, Myanmar. *Insects* 12(11), 962.

3 Invasives in Tropical Forests

Tambat B.[1]*, Sringeswara A.N.[2] and Chaithra G.N.[3]
[1]Department of Forestry and Environmental Science, University of Agricultural Sciences, GKVK, Bengaluru, India; [2]Mahatma Gandhi Botanical Garden, University of Agricultural Sciences, GKVK, Bengaluru, India; [3]Department of Biotechnology, College of Agriculture, University of Agricultural Sciences (B), Hassan, India

Abstract

Invasive species in tropical forests lead to alterations in community of vegetation, composition and modified systems of native regeneration. This potentially impacts local and regional biodiversity. Endemic species are threatened by the invasives in tropical forests. Impacts of multiple invasions are greater than individual invasion effects on native plants' function and structure, soil nutrients and herbivory. Plant invasions force higher herbivory on already declining native plants thereby exhausting the sustenance of native plants. It can cause a future decline of herbivores, which are an important food resource for charismatic carnivores in ecosystems. Invasives are a threat to global biodiversity and forests are not exceptions. A few species of insect invasives can have profound impacts on forest trees. In this chapter, aspects of biological invasions in forests and their impact on native vegetation has been reviewed.

Keywords: Invasion, tropical forests, vegetation, Western Ghats

3.1 Introduction

Forests in the tropics are known for being the most species-rich ecosystems on Earth and are of paramount importance for human welfare. They are characterized by high biological diversity and are the source of varied forms of food, medicinal plants and myriad other useful products. Forests are a store house of diverse and unique indigenous cultures. Tropical forests play a crucial role in regulating global climate conditions, maintaining water cycles and safeguarding the earth against floods, droughts and erosion.

Highly diverse tropical forests occur in all three continents and are concentrated in lowland areas with high and evenly distributed rainfall but with the greatest diversity usually occurring in northwest South American forests. Tree and liana species richness is the greatest in the upper Amazon and non-tree species richness is the greatest in the northern Andean foothills and southern Central America, suggesting conservation priority for these areas. Since different taxa show different distributional patterns, herbs, epiphytes, trees and large vertebrates must be considered for tropical conservational planning (Gentry, 1992).

Biological invasions are considered as a key factor in all major ecosystems for bringing environmental changes. Four main factors are the abiotic constituents of the new habitat, the population of newly born invasives, biological

*Corresponding author: btambat@gmail.com

 Biological Invasions in the Tropics: Monitoring, Mechanism and Management (eds Akshay Kumar Chakravarthy and Pathour Rajendra Shashank)
DOI: 10.1079/9781836991977.0003

features of the invasive species and the introduced habitat. A majority of these changes are brought about by humans (Pyšek and Richardson, 2006). In a given biological invasion, the above factors contribute to varying extents and it is important to determine their relative importance (Catford *et al.*, 2009). According to Fine (2002) and Denslow and DeWalt (2008) tropical humid forests are robust enough to resist plant invasions. Typically, high native diversity in rainforests means there are very few empty habitat-niches, giving biotic resistance to invading alien plant species.

Recently Chong *et al.* (2021) critically analysed the biological invasions in the tropics and compared them with temperate regions. They indicated relatively weak evidence suggests that tropical and temperate regions differ in invasives' invisibility and in the traits that determine invader success and impacts. It implied that propagule pressure and the traits that promote adaptation to disturbances are generally favourable to invasions in both tropical and temperate regions. Chong *et al.* (2021) emphasize the urgent need for greater investment and regional cooperation in the study, prevention, and management of biological invasions in the tropics.

3.2 What is Invasion?

The term invasive alien species refers to those organisms that have moved from their native habitat to a new location, where they cause significant harm to the environment, economic system and/or human health (Richardson *et al.*, 2000; Adhikari, 2005; Beck *et al.*, 2008). In other words, invasive alien species are those species whose ranges have been altered either unintentionally (such as accidental introductions) or intentionally (such as purposeful introductions) by humans. This results in self-replacing populations over several life cycles and thus spread over large areas in the new habitat (Pyšek and Richardson, 2010; Richardson and Rejmánek, 2011; Abrahams *et al.*, 2019). A successful biological invasion is one in which the alien species is capable of overcoming abiotic and biotic barriers and establishing a self-sustaining population (called naturalization) for a considerable distance away from the original place of introduction along the introduction-naturalization-invasion continuum (Richardson *et al.*, 2000; Richardson and Pyšek, 2012; Gioria and Osborne, 2014). They occur in practically all kinds of terrestrial ecosystems (Vitousek, 1994) and comprise 20% or more of the continental flora and 50% or more of the island flora (Seabloom *et al.*, 2006; Sardans *et al.*, 2017). Among the different kinds of terrestrial ecosystems, montane ecosystems are at a lower risk of plant invasions due to various ecological filters faced by the species as they move along high altitude (Pauchard *et al.*, 2009; Alexander *et al.*, 2011; McDougall *et al.*, 2011; Pathak *et al.*, 2019).

3.3 Ecological Impact

Invasion is a global problem that threatens natural ecosystems. Biological invasions are an important threat to global biodiversity and have significant effects in a diverse ecosystem (Seebens *et al.*, 2017; Stinca *et al.*, 2020). Invasives are the second-largest threat (after habitat destruction) to biodiversity (Drake *et al.*, 2003; Wekhanya, 2016) and one of the major drivers of global change (Sardans *et al.*, 2017). After habitat deterioration and fragmentation, invasives are the factors for biodiversity loss (Simberloff *et al.*, 2013; Dar *et al.*, 2019) as they pose detrimental impacts on the native local biodiversity, ecosystems, economy and human health (Hejda and Pyšek, 2009; Ramírez-Albores *et al.*, 2019). Invasive species impact habitats by affecting fire regimes and geomorphology (Fei *et al.*, 2014; Gaertner *et al.*, 2014) and have substantial impacts on ecosystem structure and functioning, such as reduction in native species diversity (Pyšek *et al.*, 2012).

Numerous mechanisms have been detected by which plants alter the physical, chemical, and biological features of soils. The invasive shrub *Lantana camara* L. alters the flora and soils in tropical dry deciduous forests of central India (Lone *et al.*, 2022). Plant invasion represents one of the most significant drivers of global change that threaten natural habitats (Bazzichetto *et al.*, 2018), impact on greenhouse gas fluxes (Tamura and Tharayil, 2014) and alternative ecosystem functioning and stability

(Strayer, 2012). Invasive plants can construct a favourable niche through host-microorganism interaction and rhizosphere effect that directly mediates the soil physicochemical features, that may lead to suppression or stimulation on the plant measures, if any, on invasion (Stefanowicz *et al.*, 2019; Zhao *et al.*, 2019).

3.4 Invasion and Fragmented Landscapes

Forest deterioration alters the availability of light, modifying the nature of inter-specific competition (Carson and Schnitzer, 2008; Denslow and DeWalt, 2008) and potentially exposes these forest patches to invasion of exotic plants. Massive deforestation is predominantly driven by expansion of agriculture, and these areas have a high degree of disturbance and abiotic constituents that differ from tropical forests (Malhi *et al.*, 2014; Curtis *et al.*, 2018). Even with their significant role, tropical forests are restricted to a small area globally. Though most of the tropical forests with closed canopy are comparatively weed-free, plants have strong impact on tropical forests where canopy structure is open or fragmented (Fig. 3.1).

An abiotic situation in tropical agroecosystems (e.g. high temperature, low humidity, imbalanced soil nutrients, etc.) facilitates the spread of ruderal alien plants which are transported and introduced by humans by networks of roads (With, 2002; Von der Lippe and Kowarik, 2007; Padmanaba and Sheil, 2014; Fee *et al.*, 2017), resulting in alien weed-covered banks with disturbed roadsides (Fee *et al.*, 2017). The altered agricultural tracts supply an ongoing source of propagules and facilitate invasives in forest remnants embedded within these landscapes. Plant invasions in tropical forests correlate with variables related to propagule pressure (Edward *et al.*, 2009; Dawson *et al.*, 2014). If propagule pressure is high enough, seed swamping of new environments can lead to successful alien establishment, regardless of the conditions of the invaded system (Lockwood *et al.*, 2005; Colautti *et al.*, 2006).

Commercial selective logging and the effects of fragmentation modify the structure of rainforest due to timber removal and disturbance-related tree mortality (Laurance *et al.*, 2002; Malhi *et al.*, 2014; Gibbs and Salmon, 2015). The decrease in canopy cover increases light in the forest understorey, which influences the abiotic situations of remnant forest patches, compared with continuous, undisturbed forest (Laurance *et al.*, 2002). Gradients from forest edge to interior have been recorded for temperature, humidity, wind intensity and soil properties, with the most extreme differentiation seen at forest edges, known as edge effects (Laurance

Fig. 3.1. Invasion of *Senna spectabilis* (DC) H S Irwin & Barneby in naturally open and fragmented tropical forest in Bandipur Tiger Reserve, Karnataka, India (Author's own photo).

et al., 2002). In disturbed tropical forests, more exotic plants have been recorded at forest edges, driven by a combination of increased propagule pressure and soil disturbance associated with forest edges (Peters, 2001; Harper *et al.*, 2005; Edward *et al.*, 2009; Dawson *et al.*, 2014; Döbert *et al.*, 2017).

The negative impacts of fragmentation on indigenous communities are higher in small, separated fragments and accumulate over time (Gonzalez, 2000; Ferraz *et al.*, 2003; Hill and Curran, 2003; Haddad *et al.*, 2015), with the understorey tree community strongly affected by trees producing few viable seeds (i.e. extinction debt; Stride *et al.*, 2018). So, forest remnants are especially susceptible to invasions due to decreased biotic resistance arising from the loss of native forest species. Evidence for biotic resistance to alien plant invasion in tropical rainforests is limited (Fine, 2002; Levine *et al.*, 2004; Denslow and DeWalt, 2008), with no consensus currently (Teo *et al.*, 2003; Brown *et al.*, 2006). Thus, it is yet not clear whether biotic interactions between native and exotic species (i.e. biotic resistance from the native community and suppression of native growth by invading aliens) play a significant role in mediating invasion in tropical forests.

South-East Asia has the highest rates of deforestation (Achard *et al.*, 2002; Sodhi *et al.*, 2004) and logging (primarily trees from the family Dipterocarpaceae; Hansen *et al.*, 2008; Lewis *et al.*, 2015; Brearley *et al.*, 2016) in the world (Bryan *et al.*, 2013; Gaveau *et al.*, 2017), particularly due to the spread of oil palm (*Elaeis guineensis* Jacq.) cultivation in Malaysia and Indonesia. This spread leads to a highly fragmented forest tract with uncultivated, heavily logged forest patches (Reynolds *et al.*, 2011). These remnant forests therefore contain a far lower diversity of plants and animals, compared with established growth continuous forests, if little core forest area remains within the remnants (Lucey *et al.*, 2017). Especially, forest remnants are significant for conserving biodiversity and ecosystem services (Yeong *et al.*, 2016; Lucey *et al.*, 2017; Scriven *et al.*, 2017; Fleiss *et al.*, 2020). Alien species found in tropical forests are not well studied. Waddell *et al.* (2020) showed that fragmentation and local disturbance enhance the invasion of alien species in rainforest within anthropogenic landscapes. So, Waddell *et al.* (2020) suggests for better management of alien populations within oil palm plantations to reduce exotic invasions in forest patches.

3.5 Is Invasion Pattern and Impact of Exotic and Non-exotic Species the Same?

Elton (1958) hypothesized that exotic species invade areas of low species diversity more easily than those with high species diversity. A number of researchers have attempted to test this invasion theory (Case, 1990; Law and Morton, 1996; Rejmanek, 1996; Tilman, 1997). Several other explanations based on a limiting resource model, co-evolution and food-web relation have been proposed for the apparent inverse relationship between biological diversity and the probability of invasion. Tambat *et al.* (2001) conducted a study in sacred groves (both government and community managed forest) of Kodagu (12°N, 75°E), central Western Ghats, South India. The study indicated that the proportion of the exotic (*Chromolena odorata* (L.) R.M.King and H.Rob. and *Lanatana camara* L., Fig. 3.2A and B) and the non-exotic (*Senegalia rugata* (Lam.) Britton & Rose and fern sp., Fig. 3.2C and D) invaders decreased significantly with increase in the species diversity of the quadrats (Fig. 3.3) conforming to the prediction of the invasion theory. Further, regeneration of the seedlings of the native vegetation decreased with increase in the proportion of the exotics (Fig. 3.4A and B). On the other hand, regeneration of the native vegetation was unaffected by the proportion of the non-exotic invaders (Fig. 3.4C and D). This pattern was also reflected in the species diversity of the seedling population. Thus, it is evident that though the phenomenon of biological invasion by the two categories of the species can follow the general principles postulated by Tilman and others, the actual effect on the ecosystem processes may depend upon the nature of the invaders. Hence, they caution that the effects of invaders on ecosystem processes cannot be generalized. They emphasize the implications of these results on the management of invader species in tropical forest ecosystems.

Fig. 3.2. Exotic and non-exotic invasives in tropical forests, India (Author's own photos)

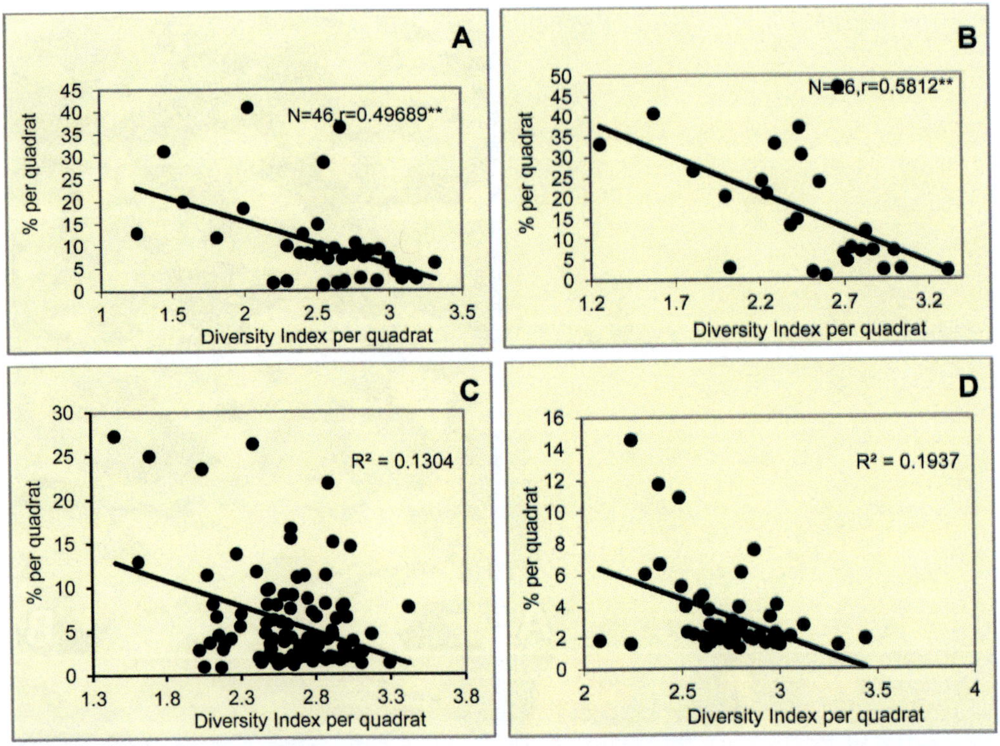

Fig. 3.3. Species diversity per quadrat versus proportion of exotic and non-exotic invasion. (A) *Chromolaena odorata* (B) *Lantana camara* (C) *Senegalia rugata* (D) Fern sp. (source: Tambat *et al*., 2001)

3.6 Insect Invasives

Insects are the most specious and diverse set of biological entities on Earth. Liebhold *et al.* (2023) categorized all insect invasives in forests into four groups, viz. borers, defoliators, sap-feeding insects and seed eaters. Most species of non-native forest insects have little effect on forest trees. But a few species may have greater impact on trees. For instance, the infestation of the teak defoliator *Hyblaea puera* in tropical, deciduous forest-patches. However, it is often difficult to quantify the damage of insect pests in forests. A few species of forest pest such as the soft and hard scales (*Coccidae*), aphids, whiteflies and other sap-feeding insects can facilitate infection of tree pathogens.

Mechanisms by which insect pests were inadvertently introduced in tropical forests are varied and referred to as "invasion pathways" (Liebhold *et al.*, 2023). Forest insects invasion can occur through one of four routes, via global movement of (a) wood, (b) plants and plant parts, (c) hitchhiking and (d) intentional introductions. A few forest insects were introduced naturally, for example via wind or water (Liebhold *et al.*, 2023). These days advanced quarantine and prevention measures are at quarantine stations globally. However, in some countries, regulation of plant imports is still not updated.

A significant number of invasive insects have been introduced intentionally, mostly as biological agents. Kenis *et al.* (2017) reported that 6158 insects in the form of parasitoids were introduced into new areas against 588 forest insects. Of these, 172 pests have been suppressed with some success. The data on accumulation of invasive insects indicated that the rate of non-native forest insects has not slowed down during the last century (Brockerhoff and Liebhold, 2017). For example, invasions of Asian bark beetle species have continued to increase. Imports of wood commodities from Asia into North America is comparatively recent

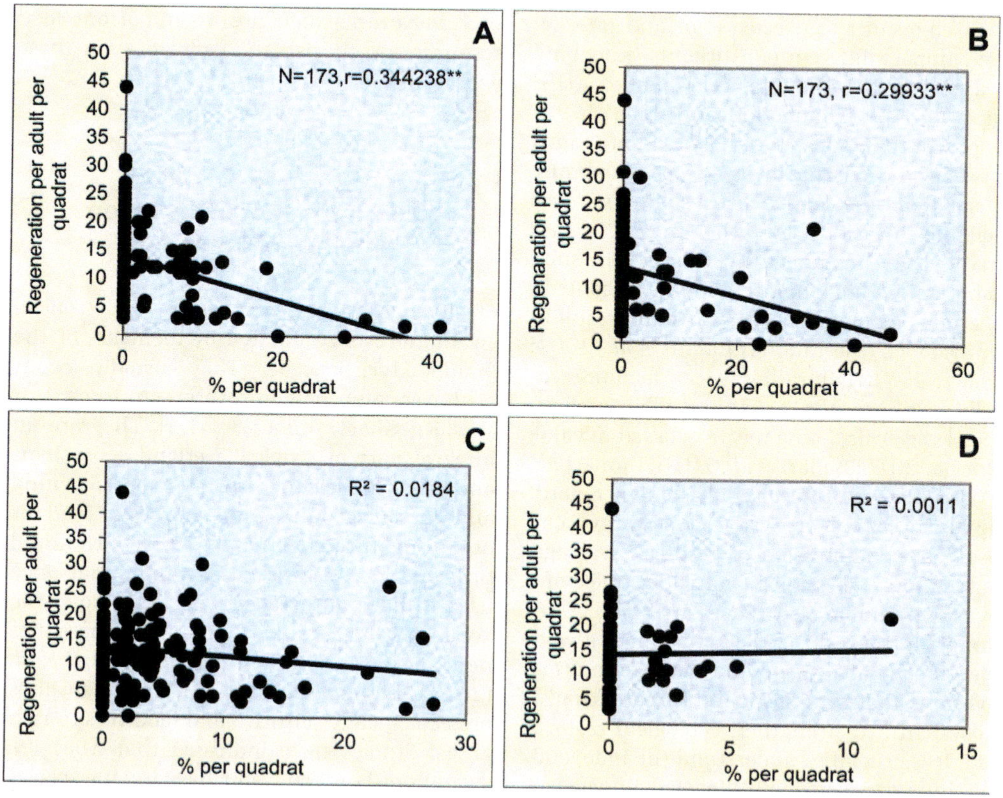

Fig. 3.4. Regeneration of tree species per adult per quadrat in relation to the proportion of exotic and non-exotic invasion (A) *Chromolaena odorata* (B) *Lantana camara* (C) *Senegalia rugata* (D) Fern sp. (source: Tambat *et al*., 2001)

and species pools have not decreased (Liebhold *et al.*, 2017). Live plants and wood are the two principal invasion pathways for forest invasive insects. Fumigation or treatment of plants/wood with pesticides may decrease the probability that pests will be introduced to new forest patches. Biosecurity measures have successfully reduced risks of pest introductions in legally imported forest trees/saplings (Liebhold and Griffin, 2016).

Rates of spread vary considerably among non-native forest insect pests. Understanding the mechanisms by which a pest impacts forest trees can facilitate their management. Liebhold *et al.* (2023) cites examples to this effect. Chemical insecticides are now rarely used to restrict/suppress populations of insect invasive in forests. This is because of environmental and economic concerns. Climate change may also affect the impacts of insect invasives on forests.

3.7 Impact of Multiple Invasions

Invasion by multiple species in the same area may lead to biotic simplification and homogenization, further causing biodiversity to plummet and necessitating a holistic approach for ecological restoration. Cases of single species invasions and their impacts on native ecosystems are typically and globally recognized (Vilà *et al.*, 2011; Pyšek *et al.*, 2012), and management strategies are mostly attempted for controlling single invasions (Gooden *et al.*, 2009; Love *et al.*, 2009; Tekiela and Barney, 2017). However, a cohort of invasive species often co-occur in an ecosystem and interactively impact different constituents (Kuebbing *et al.*, 2013, 2014; Mahla and Mlambo, 2019; Xu *et al.*, 2020). Simberloff and Holle (1999) proposed that such synergistic interactions between co-occurring invasive species magnify impacts on the native ecosystem and can facilitate invasions by other novel species (the invasion meltdown hypothesis). In such complex invasion impact dynamics, controlling only one invasive species will not result in the intended restoration (Chapuis *et al.*, 2004; Prior *et al.*, 2018); rather, it can facilitate co-occurring invasions to manifest their impacts, progressively depleting native ecosystem value (Prasad, 2010; Ballari *et al.*, 2016; Dana *et al.*, 2019). The resultant discouragement regarding unsuccessful ecosystem restoration can result in an impression of resource wastage, and reinforce a pessimistic narrative of leaving invasions unmanaged (Pearson *et al.*, 2016; Mungi and Qureshi, 2018). It is imperative to scientifically investigate and manage multiple invasions that synergistically threaten native ecosystems, to strengthen ecological restoration. It is crucial to understand the independent, interactive and multifaceted impacts of co-occurring invasions in an ecosystem and subsequently prioritize the management sites and strategies.

Rastogi *et al.* (2023) indicated that compositional change in the native plant assemblage reduced richness and abundance of native plants, increased potassium content in the soil, and increased herbivory on native plants in Kanha Tiger Reserve, India. These effects were more pronounced with co-occurring invasive plants, and larger than other key ecological determinants, indicating a homogenized alternate state of an ecosystem. Additive impacts of multiple invasions were more impactful than their individual impacts on native vegetation structure, composition, soil nutrients and herbivory. Rastogi *et al.* (2023) infer potential negative feedback, where plant invasions force higher herbivory on already declining native plants thereby exhausting the sustenance of native plants. It can cause a future decline of herbivores, which are an important food resource for charismatic carnivores in these ecosystems.

3.8 Impact of Invasion on Freshwater Swamps

Freshwater *Myristica* swamps are tropical swamp forests dominated by members of the family Myristicaceae. They usually occur at lower altitudes in evergreen forests of Western Ghats, India (Fig. 3.5). They are an integral part of tropical wetland ecosystems and are the serving habitats for spawning and nursery grounds, and sources of food for organisms (Brown *et al.*, 1979; Wharton and Brinson, 1979). *Myristica* swamps that once used to have an extensive network along the Western Ghats, due to manmade activities now appear as fragment patches. Gadgil *et al.* (2011) highlighted that *Myristica* swamps are ecologically sensitive and biodiversity-rich areas. Studies have indicated that *Myristica* swamps of Western Ghats have unique species composition and often behave as islands even though they occur amidst evergreen forests. In swamps, disturbances alter the microclimate, which facilitates the invasion of adjacent non-swampy species into the swampy habitat. This invasion results in higher richness and diversity within the swamps. Consequently, the similarity between swampy and non-swampy areas increases. However, this process reduces the densities of native swampy tree species. Studies indicate that disturbance reduces density and reproductive fitness of species, perhaps due to changes in microclimate, illegal felling, over-harvesting and altered pollination and dispersal mechanism (Gazoul *et al.*, 1998; Kadaval and Parthasarathy, 2001; Tambat *et al.*, 2004, 2005; Krishna *et al.*, 2009). The freshwater swamps are known to shelter a relatively high number of rare, endemic and economically important tree species. The disturbance and invasion of non-swampy species

Fig. 3.5. Freshwater *Myristica* swamps, an endangered tropical swamp forest characterized by waterlogged conditions and plants with breathing roots (Author's own photo).

is likely to threaten the endemic species in the swamps to extinction risk.

3.9 Impact of Invasion on Soil Microbial Diversity

Research is now directed towards the role of soil biota, such as arbuscular mycorrhizal fungi, ammonia-oxidizing bacteria and ammonia-oxidizing archaea on plant invasion, because of their vital capacity in regulating plant communities and soil functioning that can in turn provide positive feedback against plant invasion (Zhang *et al.*, 2019). However, studies have shown plant invasion influences soil microbial community assembly and co-occurrence network complexity. Previous studies showed that plant invasion greatly changes soil pH (Ouyang *et al.*, 2022), nutrient availability and organic matter quality (Xu *et al.*, 2015). Exploring the ecological strategy of co-occurring species helps to reveal the mechanism of microbial community assembly (Jiao *et al.*, 2020). For example, rhizobia can act as a medium for "hitchhiking" dispersal along fungal mycelial networks to the peanut rhizosphere and triggers peanut–rhizobium nodulation, demonstrating the role of mutualism in microbial community assembly (Zhang *et al.*, 2020). Zhang *et al.* (2020) also demonstrated that plant invasion strengthened the association among soil microbial species due to the increase of negative links for their adaptability to invaded habitats. Plant invasion can strongly affect both biotic and abiotic factors of aboveground and belowground communities, and alter co-occurrence networks and assembly processes of microbial communities.

Mōsō bamboo (*Phyllostachys edulis* (Carrière) J.Houz.), a local invasive species widely distributed in East and Southeast Asia (Xu *et al.*, 2015), has invaded subtropical forests. Caixia *et al.* (2023) showed that this bamboo invasion increases soil bacterial and fungal diversity and altered their community composition. This invasion also widened the habitat niche breadths and increased the relative contributions of deterministic processes in determining bacterial and fungal community assembly. Soil pH and available nutrients (DOC, NH_4+-N and NO_3–-N) were key drivers mediating the bacterial and fungal community assembly. More investigations are required to understand the mechanisms underlying assembly processes in bacterial and fungal communities in soil under invasion situations.

3.10 Summary

Invasion of forest remnants is likely to rise in future due to continuing deforestation, leading to changes in plant community composition and altered patterns of native regeneration, thereby potentially impacting local and regional biodiversity. Disturbance to freshwater swamps leading to invasion by non-swampy species is likely to threaten the native endemic species and thereby influence ecosystem composition. Impacts of multiple invasions are greater than individual invasion which impacts on the native vegetation structure, composition, soil nutrients and herbivory. Plant invasions force higher herbivory on already declining native plants thereby exhausting the sustenance of native plants. It can cause a future decline of herbivores, which are an important food resource for charismatic carnivores in these ecosystems. Invasion also widens habitat niche breadths and increases the relative contributions of deterministic processes in determining bacterial and fungal community assembly. In a natural resource management context, the way in which invasive species move through a landscape should be considered in the design of multifunctional landscapes, to ensure that these species do not undermine the conservation of forest remnants.

References

Abrahams, B., Sitas, N. and Esler, K.J. (2019) Exploring the dynamics of research collaborations by mapping social networks in invasion science. *Journal of Environmental Management* 229, 27–37. DOI: 10.1016/j.jenvman.2018.06.051.

Achard, F., Eva, H.D. and Stibig, H.J. (2002) Determination of deforestation rates of the world's humid tropical forests. *Science* 297, 999–1002. DOI: 10.1126/science.1070656.

Alexander, J.M., Kueffer, C., Daehler C.C., Edwards, P.J., Pauchard, A. *et al.* (2011) Assembly of nonnative floras along elevational gradients explained by directional ecological filtering. In: *Proceedings of National Academy of Sciences USA*, pp. 656–661 (Vol. 108). DOI: 10.1073/pnas.1013136108.

Ballari, S.A., Anderson, C.B. and Valenzuela, A.E. (2016) Understanding trends in biological invasions by introduced mammals in southern South America: A review of research and management. *Mammal Review* 46(3), 229–240. DOI: 10.1111/mam.12065.

Bazzichetto, M., Malavasi, M., Barták, V., Acosta, A.T.R., Moudrý, V. *et al.* (2018) Modelling plant invasion on Mediterranean coastal landscapes: An integrative approach using remotely sensed data. *Landscape and Urban Planning* 171, 98–106. DOI: 10.1016/j.landurbplan.2017.11.006.

Beck, K.G., Zimmerman, K., Schardt, J.D., Stone, J., Lukens, R.R. *et al.* (2008) Invasive species defined in a policy context: Recommendations from the federal invasive species advisory committee. *Invasive Plant Science and Management* 1, 414–421. DOI: 10.1614/IPSM-08-089.1.

Brearley, F.Q., Banin, L.F. and Saner, P. (2016) The ecology of the Asian dipterocarps. *Plant Ecological Divers* 9, 429–436. DOI: 10.1080/17550874.2017.1285363.

Brockerhoff, E.G. and Liebhold, A.M. (2017) Ecology of forest insect invasions. *Biological Invasions* 19, 3141–6159. DOI: 1007/s10530-017-1514-1.

Brown, K.A., Scatena, F.N. and Gurevitch, J. (2006) Effects of an invasive tree on community structure and diversity in a tropical forest in Puerto Rico. *Forest Ecology and Management* 226, 145–152. DOI: 10.1016/j.foreco.2006.01.031.

Brown, S.L., Brinson, M.M. and Lugo, A.E. (1979) Structure and function of riparian wetlands. In: Johnson, R.R. and McCormick, J.F. (eds) *Strategies for Protection and Management of Floodplains, Wetlands and Other Riparian Ecosystems*. WO-12, Washington, D.C., USA, p. 1731.

Bryan, J.E., Shearman, P.L. and Asner, G.P. (2013) Extreme differences in forest degradation in Borneo: Comparing practices in Sarawak, Sabah, and Brunei. *PLoS ONE* 8, e69679. DOI: 10.1371/journal.pone.0069679.

Caixia, L., Chunying, Z., Liang, W., Jing, Z., Qizan, W. *et al.* (2023) Mōsō bamboo invasion changes the assembly process and interactive relationship of soil microbial communities in a subtropical broadleaf forest. *Forest Ecology and Management* 536, 120901. DOI: 10.1016/j.foreco.2023.120901.

Carson, W.P. and Schnitzer, S.A. (2008) *Tropical Forest Community Ecology*. Wiley-Blackwell, Hoboken.

Case, T.K. (1990) Invasion resistance arises in strongly interacting species-rich model competition communities. In: *Proceedings of National Academy of Sciences USA*, pp. 9610–9614 (Vol. 87). DOI: 10.1073/pnas.87.24.9610.

Catford, J.A., Jansson, R. and Nilsson, C. (2009) Reducing redundancy in invasion ecology by integrating hypotheses into a single theoretical framework. *Diversity and Distribution* 15, 22–40. DOI: 10.1111/j.1472-4642.2008.00521.x.

Chapuis, J.L., Frenot, Y. and Lebouvier, M. (2004) Recovery of native plant communities after eradication of rabbits from the subantarctic Kerguelen Islands, and influence of climate change. *Biological Conservation* 117(2), 167–179. DOI: 10.1016/S0006-3207(03)00290-8.

Chong, K.Y., Corlett, R.T., Nuñez, M.A., Chiu, J.H., Courchamp, F. *et al.* (2021) Are terrestrial biological invasions different in the tropics? *Annual Review of Ecology, Evolution, and Systematics* 52, 219–314. DOI: 10.1146/annurev-ecolsys-012021-095454.

Colautti, R.I., Grigorovich, I.A. and MacIsaac, H.J. (2006) Propagule pressure: A null model for biological invasions. *Biological Invasions* 8, 1023–1037. DOI: 10.1007/s10530-005-3735-y.

Curtis, P.G., Slay, C.M., Harris, N.L., Tyukavina, A. and Hansen, M.C. (2018) Classifying drivers of global forest loss. *Science* 361, 1108–1111. DOI: 10.1126/science.aau3445.

Dana, E.D., García-de-Lomas, J., Verloove, F. and Vilà, M. (2019) Common deficiencies of actions for managing invasive alien species: A decision-support checklist. *NeoBiota* 48, 97–112.

Dar, J.A., Subashree, K., Sundarapandian, S., Saikia, P., Amit, K. *et al.* (2019) Invasive species and their impact on tropical forests of central India: A review. In: Garkoti, S.C., Bloem, S.K.V., Fulé, P.Z. and Semwal, R.L. (eds) *Tropical Ecosystems: Structure, Functions and Challenges in the Face of Global Change*. Springer Nature Singapore, pp. 69–109.

Dawson, W., Burslem, D.F.R.P. and Hulme, P.E. (2014) Consistent effects of disturbance and forest edges on the invasion of a continental rain forest by alien plants. *Biotropica* 47, 27–37. DOI: 10.1111/btp.12183.

Denslow, J.S. and DeWalt, S.J. (2008) Exotic plant invasions in tropical forests: Patterns and hypotheses. In: Carson, W.P. and Schnitzer, S.A. (eds) *Tropical Forest Community Ecology*. Wiley-Blackwell, Hoboken, pp. 409–426.

Döbert, T.F., Webber, B.L., Sugau, J.B., Dickinson, K.J.M. and Didham, R.K. (2017) Logging, exotic plant invasions, and native plant reassembly in a lowland tropical rain forest. *Biotropica* 50, 254–265.

Drake, S.J., Weltzin, J.F. and Parr, P.D. (2003) Assessment of non-native invasive plant species on the United States Department of Energy Oak Ridge National Environmental Research Park. *Castanea* 68, 15–30.

Edward, E., Munishi, P.K.T. and Hulme, P.E. (2009) Relative roles of disturbance and propagule pressure on the invasion of humid tropical forest by *Cordia alliodora* (Boraginaceae) in Tanzania. *Biotropica* 41, 171–178.

Elton, C.S. (1958) *The Ecology of Invasions of Animals and Plants*. Springer New York, p. 181.

Fee, C.G., Tui, L.C., Bin, C.S. and Hoy, C.K. (2017) *Pictorial Guide to Common Weeds of Plantations and their Control*, 2nd ed. Agricultural Crop Trust, Bonn.

Fei, S., Phillips, J. and Shouse, M. (2014) Biogeomorphic impacts of invasive species. *Annual Review of Ecology Evolution and Systematics* 45, 69–87.

Ferraz, G., Russell, G.J., Stouffer, P.C., Bierregaard, R.O., Pimm, S.L. *et al.* (2003) Rates of species loss from amazonian forest fragments. In: *Proceedings of National Academy of Sciences USA*, pp. 14069–14073 (Vol. 100). DOI: 10.1073/pnas.2336195100.

Fine, P.V.A. (2002) The invasibility of tropical forests by exotic plants. *Journal of Tropical Ecology* 18(5), 687–705.

Fleiss, S., Waddell, E., Ola, B.B., Banin, L.F., Benedick, S. *et al.* (2020) Conservation set-asides improve carbon storage and support associated plant diversity in certified sustainable oil palm plantations. *Biological Conservation* 248, 108631.

Gadgil, M., Daniels, R.J., Ganeshaiah, K.N., Prasad, S.N., Murthy, M.S.R. *et al.* (2011) Mapping ecologically sensitive, significant and salient areas of Western Ghats: Proposed protocols and methodology. *Current Science* 100(2), 175–182.

Gaertner, M., Reinette, B., Mariska, T.B., Cang, H., Jane, M. *et al.* (2014) Invasive plants as drivers of regime shifts: Identifying high-priority invaders that alter feedback relationships. *Diversity and Distributions* 20, 733–744.

Gaveau, D.L.A., Salim, M. and Arjasakusuma, S. (2017) Deforestation and industrial plantations development in Borneo. *Scientific Reports* 6, 32017.

Gazoul, J., Listen, K.A. and Boyle, T.J.B. (1998) Disturbance induced density-dependent seed set in *Shorea siamensis* (Dipterocarpaceae) a tropical forest tree. *Journal of Ecology* 86, 462–473.

Gentry, A.H. (1992) Tropical forest biodiversity: Distribution pattern and their conservation significance. *OIKOS* 63, 19–28.

Gibbs, H.K. and Salmon, J.M. (2015) Mapping the world's degraded lands. *Applied Geography* 57, 12–21.

Gioria, M. and Osborne, B.A. (2014) Resource competition in plant invasions: Emerging patterns and research needs. *Frontiers in Plant Science* 5, 501. DOI: 10.3389/fpls.2014.00501.

Gonzalez, A. (2000) Community relaxation in fragmented landscapes: The relation between species richness, area and age. *Ecological Letters* 3, 441–448.

Gooden, B., French, K. and Turner, P.J. (2009) Invasion and management of a woody plant, *Lantana camara* L., alters vegetation diversity within wet sclerophyll forest in south-eastern Australia. *Forest Ecology and Management* 257, 960–967. DOI: 10.1016/j.foreco.2008.10.040.

Haddad, N.M., Brudvig, L.A., Clobert, J., Davies, K.F., Gonzalez, A. *et al.* (2015) Habitat fragmentation and its lasting impact on earth's ecosystems. *Science Advances* 1(2), e1500052.

Hansen, M.C., Stehman, S.V., Potapov, P.V., Loveland, T.R., Townshend, J.R. *et al.* (2008) Humid tropical forest clearing from 2000 to 2005 quantified by using multitemporal and multiresolution remotely sensed data. *Proceedings of National Academy of Sciences USA* 105(27), 9439–9444.

Harper, K.A., Macdonald, S.E., Burton, P.J., Chen, J., Brosofske, K.D. *et al.* (2005) Edge influence on forest structure and composition in fragmented landscapes. *Conservation Biology* 19, 768–782.

Hejda, M.P. and Pyšek, J.V. (2009) Impact of invasive plants on the species richness, diversity and composition of invaded communities. *Journal of Ecology* 97, 393–403.

Hill, J.L. and Curran, P.J. (2003) Area, shape and isolation of tropical forest fragments: Effects on tree species diversity and implications for conservation. *Journal of Biogeography* 30, 1391–1403.

Jiao, S., Yang, Y., Xu, Y., Zhang, J. and Lu, Y. (2020) Balance between community assembly processes mediates species coexistence in agricultural soil microbiomes across eastern China. *The ISME Journal* 14, 202–216. DOI: 10.1038/s41396-019-0522-9.

Kadaval, K. and Parthasarathy, N. (2001) Population analysis of *Alphonsea sclerocarpa* Thw. (Annonaceae) in the Kalyaran Hills of Eastern Ghats India. *International Journal of Ecology and Environmental Science* 27, 51–54.

Kenis, M., Hurley, B.P., Hajek, A.E. and Cock, M.J. (2017) Classical biological control of insect pests of trees: Facts and figures. *Biological Invasions* 19, 3401–3417.

Krishna, U., Kanta, B.S., Dibyendu, A., Ratul, B. and John, L.N. (2009) Regeneration ecology and population status of a critically endangered and endemic tree species (*Ilex khasiana* Purk.) in north-eastern India. *Journal of Forestry Research* 20(3), 223–228.

Kuebbing, S.E., Nuñez, M.A. and Simberloff, D. (2013) Current mismatch between research and conservation efforts: The need to study co-occurring invasive plant species. *Biological Conservation* 160, 121–129.

Kuebbing, S.E., Classen, A.T. and Simberloff, D. (2014) Two co-occurring invasive woody shrubs alter soil properties and promote subdominant invasive species. *Journal of Applied Ecology* 51, 124–133.

Laurance, W.F., Vasconcelos, H.L., Stouffer, P.C. and Laurance, S.G. (2002) Ecosystem decay of Amazonian forest fragments: A 22-year investigation. *Conservation Biology* 16, 605–618.

Law, R. and Morton, R.D. (1996) Permanence and the assembly of ecological communities. *Ecology* 77, 762–775.

Levine, J.M., Adler, P.B. and Yelenik, S.G. (2004) A meta-analysis of biotic resistance to exotic plant invasions. *Ecological Letters* 7, 975–989.

Lewis, S.L., Edwards, D.P. and Galbraith, D. (2015) Increasing human dominance of tropical forests. *Science* 349, 827–832.

Liebhold, A.M. and Griffin, R.L. (2016) The legacy of Charles Marlatt and efforts to limit plant pest invasions. *American Entomologist* 62(4), 218–227.

Liebhold, A.M., Brockerhoff, E.G. and Kimberley, M. (2017) Depletion of heterogeneous source species pools predicts future invasion rates. *Journal of Applied Ecology* 54, 1968–1977.
Liebhold, A.M., Brockerhoff, E.G. and McCullough, D.G. (2023) Forest insect invasions and their management. In: Allison, J.D., Paine, T.D., Slippers, B. and Wingfield, M.J. (eds) *Forest Entomology and Pathology*. Springer, Switzerland, pp. 789–810.
Lockwood, J., Cassey, P. and Blackburn, T.M. (2005) The role of propagule pressure in explaining species invasions. *Trends in Ecology and Evolution* 20, 223–228.
Lone, P.A., Dar, J.A., Kothandaraman, S. and Khan, M.L. (2022) Invasive shrub *Lantana camara* L. alters the flora and soils in tropical dry deciduous forests of Central India. *Botropica* 54, 1412–1427. DOI: 10.1111/btp.13112.
Love, A., Suresh, B. and Babu, C.R. (2009) Management of Lantana, an invasive alien weed, in forest ecosystems of India. *Current Science* 97, 1421–1429.
Lucey, J.M., Palmer, G., Yeong, K.L., Edwards, D.P., Senior, M.J. *et al*. (2017) Reframing the evidence base for policy-relevance to increase impact: A case study on forest fragmentation in the oil palm sector. *Journal of Applied Ecology* 54, 731–736.
Mahla, N. and Mlambo, D. (2019) Influence of two co-occurring invasive plant species on resident woody species and surface soil properties in Chipinge Safari Area, Zimbabwe. *Tropical Ecology* 60, 129–139. DOI: 10.1007/s42965-019-00016-1.
Malhi, Y., Gardner, T.A., Goldsmith, G.R., Silman, M.R. and Zelazowski1, P. (2014) Tropical forests in the anthropocene. *Annual Review of Environment and Resources* 39, 125–159.
McDougall, K.L., Khuroo, A.A., Loope, L.L., Parks, C.G. and Pauchard, A. (2011) Plant invasions in mountains: Global lessons for better management. *Mountain Research and Development* 31, 380–387. DOI: 10.1659/MRD-JOURNAL-D-11-00082.1.
Mungi, N.A. and Qureshi, Q. (2018) On the history, politics and science of invasion ecology. *DIALOGUE: Science, Scientists and Society* 1, 1–16. DOI: 10.29195/dsss.01.01.0009.
Ouyang, M., Tian, D., Pan, J., Chen, G., Su, H. *et al.* (2022) Moso bamboo (*Phyllostachys edulis*) invasion increases forest soil pH in subtropical China. *Catena* 215, 106339. DOI: 10.1016/j.catena.2022.106339.
Padmanaba, M. and Sheil, D. (2014) Spread of the invasive alien species *Piper aduncum* via logging roads in Borneo. *Tropical Conservation Science* 7, 35–44.
Pathak, R., Negi, V.S., Rawal, R.S. and Bhatt, I.D. (2019) Alien plant invasion in the Indian Himalayan Region: State of knowledge and research priorities. *Biodiversity and Conservation* 28, 3073–3102.
Pauchard, A., Kueffer, C., Dietz, H., Daehler, C.C., Alexander, J. *et al.* (2009) Ain't no mountain high enough: Plant invasions reaching new elevations. *Frontiers in Ecology and the Environment* 7(9), 479–486.
Pearson, D.E., Ortega, V.K., Runyon, J.B. and Butler, J.L. (2016) Secondary invasion: The bane of weed management. *Biological Conservation* 197, 8–17. DOI: 10.1016/j.biocon.2016.02.029.
Peters, H. (2001) *Clidemia hirta* invasion at the Pasoh Forest Reserve: An unexpected plant invasion in an undisturbed tropical forest. *Biotropica* 33, 60–68.
Prasad, A.E. (2010) Effects of an exotic plant invasion on native understory plants in a tropical dry forest. *Conservation Biology* 24(3), 747–757. DOI: 10.1111/j.1523-1739.2009.01420.x.
Prior, K.M., Adams, D.C., Klepzig, K.D. and Hulcr, J. (2018) When does invasive species removal lead to ecological recovery? Implications for management success. *Biological Invasions* 20, 267–283. DOI: 10.1007/s10530-017-1542-x.
Pyšek, P. and Richardson, D.M. (2006) The biogeography of naturalization in alien plants. *Journal of Biogeography* 33, 2040–2050.
Pyšek, P. and Richardson, D.M. (2010) Invasive species, environmental change and management, and health. *Annual Review of Environment and Resources* 35, 25–55.
Pyšek, P., Jarošík, V., Hulme, P.E., Pergl, J., Hejda, M. *et al.* (2012) A global assessment of invasive plant impacts on resident species, communities and ecosystems: The interaction of impact measures, invading species' traits and environment. *Global Change Biology* 18(5), 1725–1737.
Ramírez-Albores, J.E., Badano, E.I., Flores, J., Flores, J.L. and Yañez, L. (2019) Scientific literature on invasive alien species in a megadiverse country: Advances and challenges in Mexico. *NeoBiota* 48, 113–127. DOI: 10.3897/neobiota.48.36201.
Rastogi, R., Qureshi, Q., Shrivastava, A. and Jhala, Y.V. (2023) Multiple invasions exert combined magnified effects on native plants, soil nutrients and alters the plant-herbivore interaction in dry tropical forest. *Forest Ecology and Management* 531, 120781.

Rejmanek, M. (1996) Species richness and resistance to invasions. In: Orians, G.H., Dirzo, R. and Cushman, J.H. (eds) *Biodiversity and Ecosystem Processes in Tropical Forests*. Springer, Berlin, pp. 153–172.

Reynolds, G., Payne, J., Sinun, W., Mosigil, G., Walsh, R.P.D. *et al.* (2011) Changes in forest land use and management in Sabah, Malaysian Borneo, 1990–2010, with a focus on the Danum valley region changes in forest land use and management in Sabah, Malaysian Borneo, 1990–2010, with a focus on the Danum Valley region. *Philosophical Transactions of Royal Society, Biological Sciences* 366, 3168–3176. DOI: 10.1098/rstb.2011.0154.

Richardson, D.M. and Rejmánek, M. (2011) Trees and shrubs as invasive alien species – A global review. *Diversity and Distributions* 17(5), 788–809.

Richardson, D.M. and Pyšek, P. (2012) Naturalization of introduced plants: Ecological drivers of biogeographical patterns. *New Phytologist* 196, 383–396.

Richardson, D.M., Allsopp, N., D'Antonio, C.M., Milton, S.J. and Rejmánek, M. (2000) Plant invasions: The role of mutualisms. *Biological Reviews of the Cambridge Philosophical Society* 75(1), 65–93.

Sardans, J., Bartrons, M., Margalef, O., Gargallo-Garriga, A., Janssens, I.A. *et al.* (2017) Plant invasion is associated with higher plant–soil nutrient concentrations in nutrient-poor environments. *Global Change Biology* 23(3), 1282–1291.

Scriven, S.A., Carlson, K.M., Hodgson, J.A., McClean, C.J., Heilmayr, R. *et al.* (2017) *The impact of RSPO membership on avoiding biodiversity losses in oil palm landscapes*. Science-for-policy brief. SEnSOR programme. Available at: http://www.sensorproject.net/wp-content/uploads/2017/09/SEnSOR-HCV-Project-Two-Page-Report-FINAL.pdf

Seabloom, E.W., Williams, J.W., Slayback, D., Stoms, D.M., Viers, J.H. *et al.* (2006) Human impacts, plant invasion, and imperiled plant species in California. *Ecological Applications* 16, 1338–1350.

Seebens, H., Blackburn, T.M., Dyer, E.E., Genovesi, P., Hulme, P.E. *et al.* (2017) No saturation in the accumulation of alien species worldwide. *Nature Communications* 8, 14435.

Simberloff, D. and Holle, B.V. (1999) Positive interactions of nonindigenous species: Invasional meltdown. *Biological Invasions* 1, 21–32.

Simberloff, D., Martin, J.L., Genovesi, P., Maris, V., Wardle, D.A. *et al.* (2013) Impacts of biological invasions: What's what and the way forward. *Trends in Ecology and Evolution* 28, 58–66.

Sodhi, N.S., Koh, L.P., Brook, B.W. and Ng, P.K.L. (2004) Southeast Asian biodiversity: An impending disaster. *Trends in Ecology and Evolution* 19, 654–660.

Stefanowicz, A.M., Stanek, M., Majewska, M.L., Nobis, M. and Zubek, S. (2019) Invasive plant species identity affects soil microbial communities in a mesocosm experiment. *Applied Soil Ecology* 136, 168–177.

Stinca, A., Carmelo, M.M., Leonardo, R., Valentina, L.A.L., Wolfgang, L. *et al.* (2020) Italian vascular flora: New findings, updates and exploration of floristic similarities between regions. *Diversity* 13(11), 600. DOI: 10.3390/d13110600.

Strayer, D.L. (2012) Eight questions about invasions and ecological system functioning. *Ecology Letters* 15, 1199–1210. DOI: 10.1111/j.1461-0248.2012-01817.x.

Stride, G., Thomas, C.D., Benedick, S., Hodgson, J.A., Jelling, A. *et al.* (2018) Contrasting patterns of local richness of seedlings, saplings, and trees may have implications for regeneration in rainforest remnants. *Biotropica* 50, 1–9.

Tambat, B., Ravikanth, G., Nageshwara Rao, M., Chaithra, G.N., Ramesh, B.T. *et al.* (2004) *Report on Conservation of the Myristica Swamps – The highly threatened and unique ecosystem in the Western Ghats, India*. 2003 BP Conservation Programme, Silver award winning project, UK.

Tambat, B., Rajanikanth, G., Ravikanth, G., Ganeshaiah, K.N., Shaanker, R.U. *et al.* (2005) Seedling mortality in two vulnerable tree species in the sacred groves of Western Ghats, South India. *Current Science* 88, 350–352.

Tambat, B.S., Channamallikarjuna, V., Rajanikanth, G., Uma Shaanker, R. and Ganeshaiah, K.N. (2001) Biological diversity and invasion: Testing the invasion theory. In: Ganeshaiah, KN., Uma Shaanker, R. and Bawa, K.S. (eds) *Tropical Ecosystem: Structure, Diversity and Human Welfare. Oxford and IBH, Publ. Co.* Pvt Ltd, New Delhi, pp. 97–100.

Tamura, M. and Tharayil, N. (2014) Plant litter chemistry and microbial priming regulate the accrual, composition and stability of soil carbon in invaded ecosystems. *New Phytologist* 203, 110–124.

Tekiela, D.R. and Barney, J.N. (2017) Invasion shadows: The accumulation and loss of ecological impacts from an invasive plant. *Invasive Plant Science Management* 10(1), 1–8.

Teo, D.H.L., Tan, H.T.W., Corlett, R.T., Wong, C.M. and Lum, S.K.Y. (2003) Continental rain forest fragments in Singapore resist invasion by exotic plants. *Journal of Biogeography* 30, 305–310.

Tilman, D. (1997) Community invisibility, recruitment limitation, and grassland biodiversity. *Ecology* 78, 81–92.

Vilà, M., Espinar, J.L., Hejda, M., Hulme, P.E., Jarôsík, V. *et al*. (2011) Ecological impacts of invasive alien plants: A meta-analysis of their effects on species, communities and ecosystems. *Ecological Letters* 14, 702–708.

Vitousek, P.M. (1994) Beyond global warming: Ecology and global change. *Ecology* 75, 1861–1876.

Von der Lippe, M. and Kowarik, I. (2007) Long-distance dispersal of plants by vehicles as a driver of plant invasions. *Conservation Biology* 21, 986e996.

Waddell, E.H., Banin, L.F., Fleiss, S., Hill, J.K., Hughes, M. *et al*. (2020) Land-use change and propagule pressure promote plant invasions in tropical rainforest remnants. *Landscape Ecology* 35, 1891–1906.

Wekhanya, M.N.N. (2016) *The effect of invasive species Lantana camara on soil chemistry at Ol-Donyo Sabuk National Park, Kenya*. (Doctoral dissertation, Kenyatta University).

Wharton, C.H. and Brinson, M.M. (1979) Characteristics of southeastern river system. In: Johnson, R.R. and Cormick, J.F. (eds) *Strategies for Protection of Management of Flood Plains Wetlands and Other Riparian Ecosystems. U.S. Forest Survey General Technical Report W.O.12*. pp. 32–40.

With, K.A. (2002) The landscape ecology of invasive species. *Conservation Biology* 16, 1192–1203.

Xu, L., Lou, Q., Cheng, C., Lu, M. and Sun, J. (2015) Gut-associated bacteria of *Dendroctonus valens* and their involvement in verbenone production. *Microbial Ecology* 70, 1012–1023.

Xu, Q.Y., Wang, D., Quan, G.M., Xiang, H.M. and Zhang, J.E. (2020) Invasive *Chromolaena odorata* species specifically affects growth of its co-occurring weeds. *Annals of the New York Academy of Sciences USA* 1470(1), 57–66.

Yeong, K.L., Reynolds, G. and Hill, J.K. (2016) Enrichment planting to improve habitat quality and conservation value of tropical rainforest fragments. *Biodiversity and Conservation* 25, 957–973.

Zhang, P., Li, B., Wu, J. and Hu, S. (2019) Invasive plants differentially affect soil biota through litter and rhizosphere pathways: A meta-analysis. *Ecological Letters* 22(1), 200–210.

Zhang, W., Li, X.G., Sun, K., Tang, M.J., Xu, F.J. *et al*. (2020) Mycelial network-mediated rhizobial dispersal enhances legume nodulation. *ISME Journal* 14, 1015–1029.

Zhao, M., Lu, X., Zhao, H., Yang, Y., Hale, L. *et al*. (2019) *Ageratina adenophora* invasions are associated with microbially mediated differences in biogeochemical cycles. *Science of the Total Environment* 677, 47–56. DOI: 10.1016/j.scitotenv.2019.04.330.

4 Invasive Species in Tea Ecosystems

Somnath Roy[1], Chandra Ghosh[2*], Abhay Kumar Pandey[3*], Sarbani Roy[1] and Sourajit Bayen[1]

[1]Tocklai Tea Research Institute, Tea Research Association, Jorhat, Assam, India; [2]Department of Tea Science, University of North Bengal, Darjeeling, West Bengal, India; [3]Department of Mycology and Microbiology, Tea Research Association, North Bengal Regional R&D Centre, Nagrakata, Jalpaiguri, West Bengal, India

Abstract

Tea plantations, sustain biodiversity when managed in an agro-ecological context with good agricultural practices. This, however, renders the tea gardens suitable for invasive species too. Tea is the preferred drink worldwide. Tea has bolstered dramatic growth in international trade and transportation. The foreign trade and transportation enhanced the possibility of new exotic alien species in the tropical tea ecosystem. A majority of the taxonomic groups are represented by invasives distributed in the tea plantations and are introduced purposely as well as accidentally for cultivation. Exotic plant species invasions represent serious risks to ecological stability and biodiversity. These plant species have the potential to alter vital components of the carbon, nitrogen, water and other cycling processes of the soil. Besides these, the invasive species also act like a weed and compete with the tea plants for space, light, moisture and nutrients and ultimately cause low productivity and economic loss. In tea plants, fungal pathogens are the major disease-causing elements along with pests causing huge economic losses. Invasives to different tea growing regions cause havoc to tea agroecosystems affecting yield and quality. This chapter lists various invasive alien species of tea plantations and their impact on tea ecosystems.

Keywords: Tea ecosystem, invasive weed, invasive pests, invasive phytopathogens

4.1 Introduction

The term "Invasive Alien Species" refers to species that become established in natural or semi-natural habitats, act as a catalyst for change, and pose a threat to the biological diversity of the nation where they are introduced (IUCN, 2000). Non-native or exotic species appearing outside their natural habitat ranges and their spreading capacity are categorized as alien species. Numerous alien species provide significant assistance to agroforestry systems. However, when transferred purposefully or accidentally outside of natural landscapes into new places where they have the capacity to establish, invade and outcompete native species, select aliens become invasive (Yan *et al.*, 2001; Singh, 2005). Invasive species are found extensively throughout the world's ecosystems and encompass all types of living things. The most prevalent forms of invasive aliens in terrestrial ecosystems include plants, animals and insects. While several exotic species have increased agricultural output and are now an essential component of the global economy, others have become major ecological issues (Sujay *et al.*, 2010).

*Corresponding authors: amichandra07@gmail.com and abhaykumarpandey.ku@gmail.com

DOI: 10.1079/9781836991977.0004

Differences in the requirements and ways of intake of native and non-native plants result in modification in soil profile, composition, nutrients and moisture availability. The plant species have the potential to alter vital components of the carbon, nitrogen, water, and other cycling processes of the soil (Ehrenfeld, 2003; Singh, 2005). Invasives can compete with tea plants for space, light, moisture and nutrients ultimately causing low productivity and economic losses.

Historically, tea farming was established in British colonial India and later marketed as an export-oriented industry. When the first shipment of Indian tea was delivered to London for auction in 1839, tea export from India officially started. Since that time, India has exported tea to the majority of countries in the tea-consuming regions, including Europe, America, West Asia, Africa and Australia. The foreign trade and transportation enhanced the possibility of new exotic alien species. A majority of the taxonomic groups are represented by invasives distributed across tea plantations and introduced purposely as well as accidentally for cultivation. Like other economic crops, tea is also affected by weed invasion. In a tea field in Sri Lanka, the yield was reduced by 5–10% in seed tea and 5–9% in clonal tea when weeding was not done for 4–6 months (Wettasinghe, 1971). Invasion of weeds in tea fields also affects different important field operations. Some weeds act as shelters for harmful insect pests and help spread deadly diseases. A thick cover of weeds provides suitable habitat for fungal growth by increasing relative humidity. The unchecked growth of weeds can reduce crop yield by up to 50–70%. From April to September, the timespan is congenial for weed overgrowth because of high rainfall and temperature in North East India. Eighteen weeks without weeding in a young tea plantation led to plant death (Deka and Barua, 2015). In young tea fields, weeds eliminate up to 270 kg nitrogen per ha, incurring a huge loss in plant yields. In addition, exotic species such as *Parthenium hysterophorus* L. causes harm to human health (Ghosh, 2006). Diseases caused by *Parthenium* include asthma and atopic dermatitis. The detrimental health effects of *Parthenium* weed are attributed to the sesquiterpene lactones and in particular parthenin in the plant is toxic to farm animals and responsible for allergic diseases in humans. Contact with the plant causes dermatitis and respiratory malfunction in humans, and dermatitis in cattle and domestic animals. It also is responsible for bitter milk disease in livestock when their fodder is contaminated with *Parthenium* leaves.

Numerous environmental and other agroecological issues, such as massive infestations, are associated with tea farming. Tea gardens attacked by fungal diseases may face yield and quality losses. Each part of the tea plant, i.e. leaf, stem and root are infected by these fungal pathogens (Pandey *et al.*, 2021). Exotic fungal and bacterial pathogens substantially disturb native plant species susceptible to them, potentially resulting in alterations in community composition. Invasive organisms that infest tea include sucking pests, chewing pests, mites, fungi, algae, weeds and others (Kabir and Ghosh, 2018).

4.2 Features of Invasive Species

By changing environmental features and ecological processes, invasive species have the capacity to transform ecosystems (Denóbile *et al.*, 2023). Exotic plants that successfully invade have typically developed high reproductive capacities. Many seeds and seedlings can be produced quickly by exotic plants. For example, *Ambrosia artemisiifolia* L. (common ragweed) can produce up to 8000 seeds (Zhao, 2003). Invasives such as *Mikania micrantha* Kunth (bitter vine) can by means of vegetative propagation grow from underground rhizomes even from the detritus parts of the plant (Rejmanek and Richardson, 1996). *Mikania micrantha* is a serious threat to plantation crops, especially tea and rubber. It is a climbing perennial weed (Asteraceae) native to central and South America, and has been counted in the 100 worst invasives in the world (Holm *et al.*, 1979). As a fast-growing vine it climbs over other vegetation, blocking sunlight, smothering forests and preventing forest tree regeneration. It forms dense tangled infestations in pastures, plantations and disturbed forests. It is known to suppress the underlying vegetation. *Mikania micrantha* covers the tree crown, affecting the photosynthesis of the trees and habitats of birds. The alien plants that are successful invaders usually have specific traits for propagation. A majority of the seeds of invasives are small, light and equipped with wings or pappus that allow them to fly long distances in the wind. Many invasive foreign plants of the genus *Amaranthus* are spread by seeds in guano (bird droppings). Numerous studies have demonstrated that when foreign animals are introduced to a new environment, their morphology, behaviour and genetic characteristics change quickly, even making leap changes, to develop the features essential for their survival and development. When entering arid regions with high temperatures, for instance, certain C3 plants will convert into C4 plants by the photosynthetic pathway, which will be advantageous to their survival, development and successful invasion (Wang *et al.*, 2011).

The ability of a plant to invade is closely associated with ability to develop quickly from the vegetative to the flowering stage, its capacity to compete with other species using unique strategies such as the release of allelochemicals, and its ability to germinate in diverse environments (Baker, 1974). Explosive growth, a quick generation period, tolerance for mutilation, and a broad range of adaptive potential are further characteristics that help a plant to succeed as an invader (Vijay, 2015).

4.3 Pathways of Invasion

Human-mediated activities that move organisms between different regions around the world, such as trade and tourism, are one of the key routes for species introduction into these ecosystems (Denóbile *et al.*, 2023). Increased globalization makes it easier for alien species to be introduced in more places around the planet. Investigations revealed that invasives are often introduced through international trade as immigrants or impurities in commodities and packaging materials. Major sources for introducing exotic biological entities are the online trade in pets and ornamental plants. Due to uncontrolled/unnoticed online trade, invasive alien species frequently escape or are released into the wild. Surprisingly, most nations, especially the developing ones, have little ability to stop such introductions of exotic species (Early *et al.*, 2016). Invasive plants primarily spread along roadways, in agricultural fields and throughout natural landscapes, with the aid of wind, birds and wildfires, especially in tourist hotspots. The invasive species are more likely to thrive in areas with a combination of sunlight and moisture (Tiwari *et al.*, 2022). Several years of research has revealed that humans – their actions and the infrastructure they build – are major vectors for the spread of invasive plant species (Szumańska *et al.*, 2021). Biological invasions are linked with foreign trade. In addition to unintentional human-mediated channels, plants can also be introduced and spread accidentally. In this instance, plants or their propagules are unintentionally conveyed as stowaways or pollutants (Montagnani *et al.*, 2022). According to Keane and Crawley's enemy release hypothesis, plant species face a decrease in regulation by other natural enemies after being introduced to an exotic environment, resulting in a rapid rise in spread and abundance (Keane and Crawley, 2002). On the other hand, Feng *et al.*'s hypothesis of evolution of nitrogen allocation states that the key to a successful invasion of an exotic species is that plants reduce cell wall nitrogen allocations to defences while increasing allocations to photosynthesis. This will accelerate the growth of single-plant biomass and result in an invasive's success (Feng *et al.*, 2009). Several investigations revealed that allelopathy offers a way of widespread distribution and mass proliferation for successful invasion (Wang *et al.*, 2011).

4.4 Impacts of Invasive Alien Species

With global scale, the effect of habitat loss and degradation may not be as severe as the potential harm done via invasives to native species and ecosystems (IUCN, 2000). *Ageratum conyzoides* (Billygoat-weed) has invaded agricultural fields. It interferes with crops and causes yield reductions of major staple crops. A major reason of biodiversity extinction at the ecological level is the spread of invasive species, which harm the structure, functionality and services offered by the ecosystems they occupy (Bang *et al.*, 2022). The genetic makeup of native populations, the behaviour of native animals, alteration of phylogenetic differences across communities, modification of trophic networks and the breakdown of biogeographic realms are all effects of invasive alien species. *Pontederia crassipes* (formerly *Eichhornia crassipes*) the common water hyacinth, is an aquatic, invasive flowering plant. This species is native to Brazil, but is now widely established in tropical and subtropical regions around the world. It invades and takes over entire water bodies in the tea gardens within a short period, depriving the local inhabitants of the aquatic ecosystem of resources and nutrients. This invasive species forms dense colonies that block sunlight and crowd out native species. It drains oxygen from the water, leading to death of fishes. Additionally, they affect the richness and abundance of native species, raising the risk of their extinction (Pyšek *et al.*, 2020). Numerous invasive alien species affect ecosystem functioning and service delivery by modifying nutrient and pollutant cycling, hydrology, habitat system and disturbance regimes. Invasive alien species may cause a significant effect on human health and the economy (Szumańska *et al.*, 2021; Montagnani *et al.*, 2022).

Many invasive species compete to extirpate their new neighbours rather than becoming the dominant species in their native ecosystems. One popular argument for the extraordinary success of invasive plants is that they have managed to get away from the predators that keep them in check, allowing them to compete to the fullest extent possible (Callaway and Aschehoug, 2000). Differences in the requirements and ways of both acquisition and consumption of native and non-native plants result in changes in soil structure, profile, composition, nutrient content, moisture availability and so on. These plant species have the potential to alter vital components of the carbon, nitrogen, water and other cycling processes of the soil (Ehrenfeld, 2003; Singh, 2005; Charles and Dukes, 2007). Invasive plants that fix nitrogen also leach chemicals that hinder nitrogen fixation by other species. They release compounds that modify the availability of nutrients such as nitrogen and phosphorus, and change topsoil erosion or fire frequency. Invasive plant species have been proven to disrupt hydrological cycles by influencing evapotranspiration rates and timing of evaporation, water runoff and water table levels. Impacts are highest when invaders differ from natives in features including the rate of transpiration, foliage area index, photosynthetic tissue biomass, rooting depth and phenology (Charles and Dukes, 2007). The most dangerous result of not limiting these pathways is that the farming sector suffers as a result of invasive alien plant invasion and expansion with severe financial losses (Montagnani *et al.*, 2022; Rai, 2022). Certain exotic plants like *Crotolaria pallida* have high potential to invade in a new land (Fonseca *et al.*, 2006). Successful invasion of exotic species such as *Rhus typhina* might have a negative impact on soil physicochemical and microbiological properties (Qu *et al.*, 2021). Invasive weed species in tea ecosystems are summarized in Table 4.1 and Fig. 4.1 (A–I).

4.5 Invasive Pests in Tea Ecosystem

The modification in agro-climatic conditions in north-east India and physiognomy of tea estates has transformed the pest infestation and status. Insects such black inch worm, *Hyposidra talaca* (Lepidoptera: Geometridae), hairy caterpillar (*Arctornis submarginata*) (Lepidoptera: Lymantriidae), tea looper (*H. infixaria*) (Lepidoptera: Geometridae), green weevil (*Astycus* sp.) (Coleoptera: Curculionoidea) (Fig. 4.2A), cocoa tussock moth (*Orgya postica*) (Lepidoptera: Erebidae) (Fig. 4.2B), the moth *Arna bipunctapex* (Erebidae) and white scale insects for example mealybugs (Hemiptera) have invaded the tea ecosystem.

Geometrid looper populations comprising *H. talaca*, *H. infixaria*, *Ascotis* sp., *Cleora* sp. and *Ectropis* sp. (Fig. 4.3A–J) and a few other species have become a threat to the tea plantations in north-east India, affecting crop productivity. The primary reason for increase in the incidence of these pests during the 1950s was its migration to tea from forests (Das, 1957). Feeding of pests leads to heavy defoliation of tea bushes across the year; *H. talaca* being more dominant than *Buzura suppressaria* (tea pest), as reported by Roy *et al.* (2017). This sylvan moth has invaded tea plantation in agro-forest conditions, and must have been present in the stretch of the nearby region from where it migrated to the tea estates. It is likely that with passage of time and heating up of the higher latitudes under global warming that *H. talaca* has migrated northward to reach

Table 4.1. Invasive weed species reported in tea ecosystem (Author's own table).

Sl. No.	Plant Name	Family	Native	Impact
1.	*Ageratum conyzoides* L.	Asteraceae	Tropical America	It is a common agricultural weed (Thinley and Sonam, 2022).
2.	*Argemone mexicana* L.	Papaveraceae	Tropical Central & South America	Allelochemicals produced by this plant may affect germination, growth and chlorophyll content of other plant (Namkeleja *et al.*, 2014).
3.	*Blumea lacera* (Burm.f.) DC.	Asteraceae	Tropical America	Different parts of the plant have allelopathic effect (Oudhia *et al.*, 1998).
4.	*Calotropis gigantea* (L.) W.T.Aiton	Apocynaceae	Tropical Africa	The plant has a high tendency to persist near unmanaged crop fields, and as a result, it may cause crop damage through allelopathy (Hassan *et al.*, 2015).
5.	*Chromolaena odorata* (L.) R.M.King & H.Rob.	Asteraceae	Tropical & Subtropical America	The plant changes soil microbial characteristics as well as the unstable soil C and N fractions, which may affect C and N cycle (Wei *et al.*, 2017).
6.	*Celosia argentea* L.	Amaranthaceae	Tropical Africa	In India and China, the plant is considered as an unmanageable weed (Grant, 1954).
7.	*Crotalaria pallida* Aiton	Fabaceae	Tropical America	The exotic plant has high potential to invade in a new land (Fonseca *et al.*, 2006).
8.	*Datura innoxia* Mill.	Solanaceae	Tropical America	It has been regarded as the most noxious invasive plant species by many workers because it competes fiercely with native crops, forms dense monospecific stands, and causes native plant seedling growth to be delayed (Mardare *et al.*, 2022).

Continued

Table 4.1. Continued

Sl. No.	Plant Name	Family	Native	Impact
9.	*Datura metel* L.	Solanaceae	Tropical America	It has been declared as a deadly poisonous weed in Australia, and considered as one of the main weeds present in Ghana and India (Holm *et al.*, 1979).
10.	*Grangea maderaspatana* (L.) Poir.	Asteraceae	Tropical South America	Leaf juice of the invasive weed can be used as antispasmodic (Jana, 2016)
11.	*Ipomoea pes-tigridis* L.	Convolvulaceae	Tropical East Africa	These weeds have such a strong vegetative development that once their weight becomes excessive, they damage the crops and result in significant crop loss (Shrikhandia *et al.*, 2017).
12.	*Ipomoea quamoclit* L.	Convolvulaceae	Tropical America	It is an invasive alien weed of West Bengal (Sekar *et al.*, 2012).
13.	*Lantana camara* L.	Verbenaceae	Tropical America	Bushes of this plant may modify plant community composition through influencing the growth of native tree seedlings (Joshi *et al.*, 2015).
14.	*Ludwigia octovalvis* (Jacq.) P.H.Raven	Onagraceae	Tropical Africa	The plant is sometimes considered as an invasive species (Natural Resources Conservation Service, 2025).
15.	*Melilotus albus* Medik.	Fabaceae	Europe	The species has the potential to increase invasive species and lower the number of local species by altering soil nutrient condition (Conn *et al.*, 2018).
16.	*Mimosa pudica* L.	Fabaceae	Brazil	The plant is known as a very common invasive species in maize, sugarcane, rubber, tea, soybean farm (Parsons and Cuthbertson, 2001).
17.	*Opuntia stricta* (Haw.) Haw.	Cactaceae	Tropical America	The flower of this invasive plant can be used to treat asthma (Jana, 2016).

Continued

Table 4.1. Continued

Sl. No.	Plant Name	Family	Native	Impact
18.	*Parthenium hysterophorus* L.	Asteraceae	Tropical & Subtropical America	It generates an alellopathic chemical that hinders the growth of other plants, resulting in a terrible failure of crop production (Thinley and Sonam, 2022).
19.	*Pontederia crassipes* Mart.	Pontederiaceae	Tropical South America	The plant is one of the most common invasive plants in India which causes huge ecological problems (Datta *et al.*, 2021).
20.	*Scoparia dulcis* L.	Scrophulariaceae	Tropical America	The common tea garden weed propagates effectively and rapidly through seeds which makes them invasive in nature (Jain and Singh, 1989).
21.	*Sida acuta Burm*.f.	Malvaceae	Tropical America	It might be effective in preventing streambank erosion (Kumar *et al.*, 1996).
22.	*Solanum torvum* Sw.	Solanaceae	West Indies	This plant invades a variety of ecosystems (Cuda *et al.*, 2002).
23.	*Sonchus oleraceus* L.	Asteraceae	Mediterranean	The weed is considered as an invasive weed (CABI, 2020).
24.	*Synedrella nodiflora* (L.) Gaertn.	Asteraceae	West Indies	It is an invasive species which can cause negative impact on seed germination of other plants by allelopathy (Ghayal *et al.*, 2013).
25.	*Tridax procumbens* L.	Asteraceae	Tropical Central America	The plant has become a troublesome invasive species around the world and it is considered as an economically important weed (Reed, 1977).
26.	*Triumfetta rhomboidei* Jacq.	Malvaceae	Tropical America	The competitive invasive plant can prevent the establishment of native plants (Motooka, 2003).

Continued

Table 4.1. Continued

Sl. No.	Plant Name	Family	Native	Impact
27.	*Urena lobata* L.	Malvaceae	Tropical Africa	The plant invades perennial crop fields and act as an ecological threat (Jana, 2016).
28.	*Xanthium strumarium* L.	Asteraceae	Tropical America	It is a very common invasive species present in different crop fields (Venodha, 2016).
29.	*Erigeron canadensis* L.	Asteraceae	Tropical South America	The plant is considered as an alien invasive plant (Wiese *et al*., 1995).
30.	*Leucaena leucocephala* (Lam.) de Wit	Fabaceae	Mexico	As soon as it takes hold, it dispenses with local vegetation and may help provide the right conditions for the introduction of more destructive invaders (Shrikhandia *et al*., 2017).
31.	*Amaranthus spinosus* L.	Amaranthaceae	Mexico to Tropical America	The invasive weed is helpful in phytoremediation (Chinmayee *et al*., 2012).
32.	*Amaranthus viridis* L.	Amaranthaceae	Tropical South America	Shoot of the invasive plant is used as a vegetable by tea garden workers (Ghosh, 2006).
33.	*Alternanthera philoxeroides* (Mart.) Griseb.	Amaranthaceae	Trinidad to North Argentina	It is one of the worst plant pests in the world as it has potentiality to invade both in terrestrial and aquatic habitats (ISSG, 2016).
34.	*Mikania micrantha* Kunth	Asteraceae	Tropical and Subtropical America	The plant is one of the worst 100 invasive species of the world and it is considered as a serious threat to ecosystems (Vijay, 2015).

Terai-Dooars of the Darjeeling Himalaya (Das, 1957) and north-east India. Switch over of hosts from sylvan vegetation (*Schima wallichii*, *Lagerestroemia speciosa*, *L. indica*, *Shorea robusta*, *Derris robusta*, *Acacia catechu*, *Mikania* sp. and some others) to tea plantations may be due to phenomena induced by changing host preference under the pathways of entraining.

The invasive tea spiny whitefly, *Aleurocanthus camelliae* Kanmiya and Kasai, has recently been spreading through tea plantations in Japan. By 2008, *A. camelliae* had spread to Kyoto, Shiga, Nara and Mie prefectures in the Kinki district and subsequent outbreaks followed in four other prefectures, Shimane, Fukuoka, Saitama and Gifu, in 2009–2010

Fig. 4.1. Invasive weed species in tea ecosystems. A. *Lantana camara*. B. *Chromolaena odorata*. C. *Calotropis gigantea*. D. *Mikania micrantha*. E. *Ageratum conyzoides*. F. *Xanthium strumarium*. G. *Mimosa pudica*. H. *Parthenium hysterophorus*. I. *Datura metel*. (Author's own image).

Fig. 4.2. Two common invasive pests in tea. A. Green weevil (*Astycus* sp.) and B. caterpillars of cocoa tussock moth (*Orgya postica*). (Author's own image).

(Kasai *et al.*, 2010). Other outbreaks have since occurred in Oita (Oita Agriculture, Forestry and Fisheries Research Center, 2010), Okayama (Okayama Prefecture, 2010), Hyogo (Hyogo Plant Protection Office, 2010), Aichi (Aichi Prefecture, 2010), and Shizuoka (Shizuoka

Fig. 4.3. Invasive looper species in tea plantations, viz. A. Looper of *H. talaca*. B. Female moth of *H. talaca*. C. Looper of *H. infixaria*. D. Female moth of *H. infixaria*. E. Looper of *Ascotis* sp. F. Moth of *Ascotis* sp. G. Looper of *Cleora* sp. H. Moth of *Cleora* sp. I. Looper of *Ectropis* sp. J. Moth of *Ectropis* sp. (Author's own image).

Plant Protection Office, 2010). Yamashita and Yoshiyasu (2010) suggested that this *A. camelliae* range expansion occurred via nymph transfer on tea saplings, rather than via adult migratory flight.

4.6 Invasive Phytopathogens Causing Diseases in Tea Plants

In tea growing regions, many diseases are associated with growth and performance of cultivated tea. In particular, diseases associated with foliage directly affect the harvest, while root and stem diseases impact the persistence of tea plants. Worldwide, grey blight, brown blight, blister blight, root rots and stem cankers are considered as the most serious diseases of tea plants (Pandey *et al.*, 2021). Bacterial blight caused by a pathogenic *Acidovorax avenae* sub sp. *avenae* was native to Japan in 1983 (Takikawa *et al.*, 1988), was accidentally introduced to China (Geng *et al.*, 2017) and in Taiwan, it was being first detected in 2020 by Sung *et al.* (2020). Currently, bacterial blight is one of the major threats to the tea industry of China and Taiwan causing 30–40% yield losses in production. Bacterial blight was recently introduced in India and is rapidly spreading in tea gardens of north-east and eastern India, causing potential yield losses (Pandey *et al.*, 2024).

At present, in Sri Lanka blister blight is a major disease hindering tea production (Sinniah *et al.*, 2012). The pathogen *Exobasidium vexans* was introduced from India in 1855 (Tubbs, 1946). The pathogen *Fusarium solani* causes twig dieback in Taiwan (Muraleedharan and Chen, 1997) and branch canker in Sri Lanka (Sinniah *et al.*, 2017). In India, this pathogen was introduced after 2015 (Kumhar and Babu, 2019) and at present it is causing huge losses in tea production in both organic and conventional tea gardens (Pandey *et al.*, 2023a). More recently, Pandey *et al.*, 2023b reported three new *Pestalotiopsis*-like species causing grey blight in India. These species include *Pseudopestalotiopsis thailandica*, *Pseudopestalotiopsis camelliae sinensis* and *Neopestalotiopsis natalensis*, which were native to Taiwan and China and have now been introduced in India and are causing significant yield losses in tea production.

4.7 Conclusion

Through the course of this chapter, we have brought to light the several ways in which invasive alien species might invade, both intentionally and accidentally, as well as the effects that these invasions have on the tea ecosystem and on human health. It is possible to draw the conclusion from the review that an agroecosystem, such as tea plantations, can facilitate the growth of an invasion hotspot if it is not maintained through the implementation of good agricultural practices. The review provides a list of several invasive alien species that are found in tea plantations. Based on this information, it is evident that many of the existing weeds in the tea environment are members of invasive species. The invasion of weeds is primarily detrimental to a cash crop like tea since it results in decreased productivity and economic losses. Nevertheless, there is still a paucity of information regarding the actual invasion rate of each species and the specific impacts that each species has on the tea ecosystem. For this reason, there is an urgent need for substantial research on this matter.

Acknowledgement

Authors are thankful to Professor Ananda Mukhopadhyay for his contribution in the improvement of this book chapter.

References

Aichi Prefecture (2010) Special report on forecast of pest occurrence. No. 2. Aichi Prefecture 2 (in Japanese).

Baker, H.G. (1974) The evolution of weeds. *Annual Review of Ecology, Evolution, and Systematics* 5, 1–24.

Bang, A., Cuthbert, R.N., Haubrock, P.J., Fernandez, R.D., Moodley, D. *et al.* (2022) Massive economic costs of biological invasions despite widespread knowledge gaps: A dual setback for India. *Biological Invasions* 24, 2017–2039. DOI: 10.1007/s10530-022-02780-z.

CABI (2020) *Sonchus oleraceus*, (common sow thistle). In: *Invasive Species Compendium*. CAB International, Wallingford, UK. Available at: www.cabi.org/isc

Callaway, R.M. and Aschehoug, E.T. (2000) Invasive plants versus their new and old neighbors: A mechanism for exotic invasion. *Science* 290, 521–523. DOI: 10.1126/science.290.5491.521.

Charles, H. and Dukes, J.S. (2007) Impacts of invasive species on ecosystem services. *Biological Invasions* 193, 217–237. DOI: 10.1007/978-3-540-36920-2_13.

Chinmayee, M.D., Mahesh, B., Pradesh, S., Mini, I. and Swapna, T.S. (2012) The assessment of phytoremediation potential of invasive weed *Amaranthus spinosus* L. *Applied Biochemistry and Biotechnology* 167(6), 1550–1559. DOI: 10.1007/s12010-012-9657-0.

Conn, J.S., Werdin-Pfisterer, N.R., Beattie, K.L. and Densmore, R.V. (2018) Ecology of invasive *Melilotus albus* on Alaskan Glacial River Floodplains. *Arctic, Antarctic and Alpine Research* 43(3), 343–354. DOI: 10.1657/1938-4246-43.3.343.

Cuda, J.P., Parker, P.E., Coon, B.R., Vasquez, F.E. and Harrison, J.M. (2002) Evaluation of exotic *Solanum* spp. (Solanales: Solanaceae) in Florida as host plants for the leaf beetles *Leptinotarsa defecta* L. Texan (Coleoptera: Chrysomelidae). *Florida Entomologist* 85(4), 599–610. DOI: 10.1653/0015-4040(2002)085[0599:EOESSS]2.0.CO;2.

Das, G.M. (1957) Pests in relation to environment. *Two and a Bud* 4, 14–16.

Datta, A., Maharaj, S., Prabhu, G.N., Bhowmik, D., Marino, A. *et al.* (2021) Monitoring the spread of water hyacinth (*Pontederia crassipes*): Challenges and future developments. *Frontiers in Ecology and Evolution* 9, 1–8. DOI: 10.3389/fevo.2021.631338.

Deka, J. and Barua, I.C. (2015) Weeds of tea fields and their control. In: *National Seminar on Plant Protection in Tea: Recent Advances, Conference paper*, pp. 26–27. DOI: 10.13140/RG.2.1.5150.9845.

Denóbile, C., Chiba de Castro, W.A. and Silva Matos, D.M.D. (2023) Public health implications of invasive plants: A scientometric study. *Plants* 12(3), 661. DOI: 10.3390/plants12030661.

Early, R., Bradley, B.A., Dukes, J.S., Lawler, J.J., Olden, J.D. *et al.* (2016) Global threats from invasive alien species in the twenty-first century and national response capacities. *Nature Communications* 7, 12485. DOI: 10.1038/ncomms12485.

Ehrenfeld, J.G. (2003) Effects of exotic plant invasions on soil nutrient cycling processes. *Ecosystems* 6(6), 503–523. DOI: 10.1007/s10021-002-0151-3.

Feng, Y.L., Lei, Y.B., Wang, R.F., Callaway, R.M., Valiente-Banuet, A. *et al.* (2009) Evolutionary tradeoffs for nitrogen allocation to photosynthesis versus cell walls in an invasive plant. *Proceedings of the National Academy of Sciences USA* 106(6), 1853–1856. DOI: 10.1073/pnas.0808434106.

Fonseca, R., Guimarães, P., Morbiolo, S., Scachetti-Pereira, R. and Peterson, A. (2006) Predicting invasive potential of smooth crotalaria (*Crotalaria pallida*) in Brazilian national parks based on African records. *Weed Science* 54(3), 458–463. DOI: 10.1614/WS-05-059R4.1.

Geng, G.M., Zhao, Y.Q., Zhu, C.C., Wang, M.X., Tian, Y.L. *et al.* (2017) First report of bacterial leaf blight of tea (*Camellia sinensis*) caused by *Acidovorax avenae* in China. *Plant Diseases* 101(10), 1819. DOI: 10.1094/PDIS-02-17-0240-PDN.

Ghayal, N., Dhumal, K., Deshpande, N., Ruikar, A. and Phalgune, U. (2013) Phytotoxic effects of leaf leachates of an invasive weed *Synedrella nodiflora* and characterization of its allelochemical. *International Journal of Pharmaceutical Science Review and Research* 19(1), 79–86.

Ghosh, C. (2006) Biology of tea garden weeds in Darjeeling district of West Bengal, India. PhD thesis, University of North Bengal, West Bengal, India.

Grant, W.F. (1954) A cytological study of *Celosia argentea*, *C. argentea* var. *cristata*, and their hybrids. *Botanical Gazette. The University of Chicago Press* 115(4), 323–336.

Hassan, L.M., Galal, T.M., Farahat, E.A. and El-Midany, M.M. (2015) The biology of of *Calotropis procera* (Aiton) WT. *Trees: Structure and Function* 29(2), 311–320. DOI: 10.1007/s00468-015-1158-7.

Holm, L.G., Pancho, J.V., Herberger, J.P. and Plucknett, D.L. (1979) *A Geographical Atlas of World Weeds*. Wiley and Sons, New York, USA.

Hyogo Plant Protection Office (2010) Special report on forecast of pest occurrence. No. 1. Hyogo Prefecture, p. 3.

ISSG (2016) Global Invasive Species Database (GISD). Invasive Species Specialist Group of the IUCN Species Survival Commission.

IUCN (2000) *IUCN Guidelines for the Preservation of Biodiversity Loss caused by Alien Invasive Species*. Gland, Switzerland.

Jain, R. and Singh, M. (1989) Factors affecting goatweed (*Scoparia dulcis*) seed germination. *Weed Science* 37(6), 766–770. Available at: http://www.jstor.org/stable/4044996

Jana, B. (2016) Common invasive medicinal plant species in east midnapore district, west bengal, india. *Annals of Clinical and Laboratory Research* 4(1), 1–13. DOI: 10.21767/2386-5180.100060.

Joshi, A.A., Mudappa, D. and Shankar Raman, T.R. (2015) Invasive alien species in relation to edges and forest structure in tropical rainforest fragments of the Western Ghats. *Journal of Tropical Ecology* 56(2), 233–244.

Kabir, S.E. and Ghosh, C. (2018) Tea cultivation in the Indian Himalayan region. In: Bera, S. and Singh, B. (eds) *Plant Diversity in the Himalaya Hotspot Region*, Vol. II. Mahendra Pal Singh, Dehra Dun, pp. 743–767.

Kasai, A., Yamashita, K. and Yoshiyasu, Y. (2010) Tea-infesting population of the citrus spiny whitefly, *Aleurocanthus spiniferus* (Homoptera: Aleyrodidae), does not accept citrus leaves. *Japanese Journal of Applied Entomology and Zoology* 54(3), 140–143. DOI: 10.1303/jjaez.2010.140.

Keane, R.M. and Crawley, M.J. (2002) Exotic plant invasions and the enemy release hypothesis. *Trends in Ecology and Evolution* 17(4), 164–169. DOI: 10.1016/S0169-5347(02)02499-0.

Kumar, R., Ambasht, R.S., Srivastava, A.K. and Srivastava, N.K. (1996) Role of some riparian wetland plants in reducing erosion of organic carbon and selected cations. *Ecological Engineering* 6, 227–239.

Kumhar, K.C. and Babu, A. (2019) Biocontrol potency of *Trichoderma* isolates against tea (*Camellia* sp.) pathogens and their susceptibility towards fungicides. *International Journal of Chemical Studies* 7, 4192–4195.

Mardare, G., Lazar, L. and Malutan, T. (2022) Potential application of invasive plant species *Datura innoxia* for the scopolamine extracts of the plant organs and analysis using UV–VIS spectrophotometry. *Forests* 13(10), 1555. DOI: 10.3390/f13101555.

Montagnani, C., Gentili, R., Brundu, G., Caronni, S. and Citterio, S. (2022) Accidental introduction and spread of top invasive alien plants in the European Union through human-mediated agricultural pathways: What should we expect? *Agronomy* 12(2), 423. DOI: 10.3390/agronomy12020423.

Motooka, P. (2003) Weeds of Hawaii's pastures and natural areas; an identification and management guide. In: *Weeds of Hawaii's Pastures and Natural Areas; an Identification and Management Guide*. College of Tropical Agriculture and Human Resources, University of Hawaii, Manoa, USA.

Muraleedharan, N. and Chen, Z.M. (1997) Pests and diseases of tea and their management. *Journal of Plantation Crops* 25, 15–43.

Namkeleja, H.S., Tarimo, M.T.C. and Ndakidemi, P.A. (2014) Allelopathic effects of *Argemone mexicana* to growth of native plant species. *American Journal of Plant Sciences* 5(9), 1336–1344. DOI: 10.4236/ajps.2014.59147.

Natural Resources Conservation Service (2025) *PLANTS Database*. United States Department of Agriculture. Available at: https://plants.usda.gov (accessed 28 May 2025).

Oita Agriculture, Forestry and Fisheries Research Center (2010) Forestry and fisheries research center. No. 4. Ohita Prefecture, p. 2.

Okayama Prefecture (2010) Special report on forecast of Pest Occurrence. No. 1. Okayama Prefecture, p. 3.

Oudhia, P., Kolhe, S.S. and Tripathi, R.S. (1998) Allelopathic effect of *Blumea lacera* L. on rice and common Kharif weeds. *Oryza* 35, 175–177.

Pandey, A.K., Sinniah, G., Babu, A. and Tanti, A. (2021) How the global tea industry copes up with fungal diseases- Challenges and opportunities. *Plant Disease* 105(7), 1868–1879. DOI: 10.1094/PDIS-09-20-1945-FE.

Pandey, A.K., Samota, M.K., Tanti, A.J. and Babu, A. (2023a) *Trichoderma reesei* induces defense-related biochemical markers associated with resistance to *Fusarium* dieback in tea crop. *Biological Control* 180(43), 105200. DOI: 10.1016/j.biocontrol.2023.105200.

Pandey, A.K., Hubbali, M.V., Dutta, P. and Babu, A. (2023b) Characterization and identification of fungicide insensitive Pestalotiopsis like species pathogenic to tea crop in India. *World Journal of Microbiology and Biotechnology* 39, 34. DOI: 10.1007/s11274-022-03474-3.

Pandey, A.K., Varghese, S. and Babu, A. (2024) First report of bacterial leaf blight on tea: An emerging threat to the Indian tea industry. *European Journal of Plant Pathology* 169, 131–136. DOI: 10.1007/s10658-024-02812-5.

Parsons, W.T. and Cuthbertson, E. (2001) *Noxious Weeds of Australia*, Second Edition. CSIRO Publishing, Collingwood, p. 698.

Pyšek, P., Hulme, P.E., Simberloff, D., Bacher, S., Blackburn, T.M. *et al.* (2020) Scientists' warning on invasive alien species. *Biological Reviews of the Cambridge Philosophical Society* 95(6), 1511–1534. DOI: 10.1111/brv.12627.

Qu, T., Du, X., Peng, Y., Guo, W., Zhao, C. *et al.* (2021) Invasive species allelopathy decreases plant growth and soil microbial activity. *PLoS ONE* 16(2), e0246685. DOI: 10.1371/journal.pone.0246685.

Rai, P.K. (2022) Environmental degradation by invasive alien plants in the Anthropocene: Challenges and prospects for sustainable restoration. *Anthropocene Science* 1(1), 5–28. DOI: 10.1007/s44177-021-00004-y.

Reed, C.F. (1977) *Economically Important Foreign Weeds*. Agricultural Research Service, United States Department of Agriculture, Washington, D.C, p. 746.

Rejmanek, M. and Richardson, D.M. (1996) What attributes make some plant species more invasive? *Ecology* 77(6), 1655–1661.

Roy, S., Das, S., Handique, G., Mukhopadhyay, A. and Muraleedharan, N. (2017) Ecology and management of the black inch worm, *Hyposidra talaca* Walker (Geometridae: Lepidoptera) infesting *Camellia sinensis* (Theaceae): A review. *Journal of Integrative Agriculture* 16, 60345–60347. DOI: 10.1016/S2095-3119(16)61573-3.

Sekar, K.C., Manikandan, R. and Srivastava, S.K. (2012) Invasive alien plants of Uttarakhand Himalaya. *Proceedings of the National Academy of Sciences, India Section B: Biological Sciences* 82(3), 375–383.

Shizuoka Plant Protection Office (2010) Special report on forecast of pest occurrence. No. 1. Shizuoka Prefecture, p. 2.

Shrikhandia, P., Shrikhandia, S.P. and Bhatia, S. (2017) Studies on the impact of invasive alien species of family convolvulaceae, fabaceae and amaranthaceae in Rajouri district of Jammu and Kashmir. *International Journal of Recent Scientific Research* 8(11), 22056–22062. DOI: 10.24327/ijrsr.2017.0811.1191.

Singh, J.S. (2005) Invasive alien species and biodiversity in India. *Current Science* 88(4), 539.

Sinniah, G.D., Alagiyawadu, U. and Wasantha, K.L. (2012) An assessment key for tea blister blight: Development and validation. In: *Proceedings of 4th Symposium on Plantation Crop Research*, Colombo, Sri Lanka, pp. 135–144.

Sinniah, G.D., Munasinghe, C.E., Mahadevan, N., Jayasinghe, S.K. and Kulatunga, D.C.M. (2017) Recent incidence of collar canker and dieback of tea (*Camellia sinensis*) caused by *Fusarium solani* species complex in Sri Lanka. *Australasian Plant Disease Notes* 12(1), 41. DOI: 10.1007/s13314-017-0262-5.

Sujay, Y.H., Sattagi, H.N. and Patil, R.K. (2010) Invasive alien insects and their impact on agroecosystem. *Karnataka Journal of Agricultural Science* 23(1), 26–34.

Sung, I., Lai, Y. and Hung, C. (2020) First report of *Acidovorax avenae* subsp. *avenae* causing bacterial leaf blight of tea in Taiwan. *Journal of Plant Pathology* 102(8), 981. DOI: 10.1007/s42161-020-00547-w.

Szumańska, I., Lubińska-Mielińska, S., Kamiński, D., Rutkowski, L., Nienartowicz, A. *et al.* (2021) Invasive plant species distribution is structured by soil and habitat type in the city landscape. *Plants* 10(4), 773. DOI: 10.3390/plants10040773.

Takikawa, Y., Ando, Y., Hamaya, E., Tsuyumu, S. and Goto, M. (1988) Identification of the pathogens responsible for bacteriosis of tea plant occurred in 1983. *Annual Phytopathological Society of Japan* 54, 224–228.

Thinley, U. and Sonam, T. (2022) *A study and survey on key invasive plant species in Southwestern Bhutan*. Tarayana Foundation, Thimpu, Bhutan.

Tiwari, S., Mishra, S.N., Kumar, D., Kumar, B., Vaidya, S.N. *et al.* (2022) Modelling the potential risk zone of *Lantana camara* invasion and response to climate change in Eastern India. *Ecological Processes* 11, 10. DOI: 10.1186/s13717-021-00354-w.

Tubbs, F.R. (1946) Blister blight. *Tea Quarterly* 18, 90.

Venodha, P.M. (2016) Landscape degradation and restoration – A planning approach. *International Journal of Environmental Science and Development* 7(3), 229–233. DOI: 10.18178/IJESD.

Vijay, B. (2015) Invasive plant *Mikania micrantha* a major threat to biological diversity – A review. International Journal of Advance Research in Science and Engineering 4(5), 48–52.

Wang, C., Zhu, M., Chen, X. and Bo, Q. (2011) Review on allelopathy of exotic invasive plants. *Procedia Engineering* 18, 240–246. DOI: 10.1016/j.proeng.2011.11.038.

Wei, H., Xu, J., Quan, G., Zhang, J. and Qin, Z. (2017) Invasion effects of *Chromolaena odorata* on soil carbon and nitrogen fractions in a tropical Savanna. *Ecosphere* 8(5). DOI: 10.1002/ecs2.1831.

Wettasinghe, D.T. (1971) Report of the research officer for the low country station for 1970. *Annual Report Tea Research Institute* 126–142.

Wiese, A.F., Salisbury, C.D. and Bean, B.W. (1995) Downy brome (*Bromus tectorum*), jointed goat grass (*Aegilops cylindrica*) and horseweed (*Conyza canadensis*) control in fallow. *Weed Technology* 9(2), 249–254.

Yamashita, K. and Yoshiyasu, Y. (2010) Control effects of the spraying of petroleum oil against the over-wintering larvae of the citrus spiny whitefly, *Aleurocanthus spiniferus* (Quaintance), in tea fields. *Annual Report of Kansai Plant Protection* 52, 157–159.

Yan, X., Zhenyu, L., Gregg, W.P. and Dianmo, L. (2001) Invasive species in China – An overview. *Biodiversity Conservation* 10(8), 1317–1341. DOI: 10.1023/A:1016695609745.

Zhao, M. (2003) Integrated control and counter measure of external invasion biology. *Agro-Environment and Development* 5, 38–39.

5 Invasive Arthropods of the Indian Himalayan Region

Rahul Kumar[1*†], Stuti[2†] and Manish Bharadwaj[3]
[1]Department of Zoology, Sheodeni Sao College (Magadh University), Kaler, India; [2]Department of Zoology, Banaras Hindu University, Varanasi, India; [3]Department of Zoology, Satyendra Narayan Sinha College (Magadh University), Warisaliganj, India

Abstract
The Himalaya is one of the world's richest biodiversity hotspots and a treasure of unfathomable bioresources. A significant area of Himalaya falls under the Indian territory and is known as Indian Himalayan Region. It is home to approximately 30.16% of the total Indian fauna, of which arthropods account for more than 87%. Being one of the most vulnerable ecosystems, Himalaya faces threat of biodiversity loss from climate change and anthropogenic disturbances. The problem of invasive alien species in Himalayas is also a manifestation of the ongoing climate change and anthropogenic perturbations. Invasive alien species are non-native species whose introduction and spread threatens the local biodiversity of a region. Invasive alien species cause a significant economic loss worldwide. Globally, India holds the second position among highest economic cost-bearing countries due to invasive alien species. Invasive arthropods are responsible for a major proportion of these losses. The present chapter provides an overview of the invasive arthropod species of India and a detailed account of agriculturally important invasive arthropod species reported from different parts of the Indian Himalayan Region. Descriptions of a few such arthropods have also been provided which are likely to become invasive in the near future or pose a threat of spreading over a larger area in the Indian Himalayan Region. Taxonomic position, morphology, life cycle, distribution, hosts, damage symptoms and natural enemies of each of these species are discussed. Very limited information is available regarding the research on the problem of invasive arthropod species in the Himalayan region. More studies need to be conducted to assess the status and impacts of the invasive arthropod species in the Himalaya.

Keywords: Anthropogenic perturbations, biodiversity, climate change, Himalaya, invasive alien species

5.1 Introduction

> Wherever the European had trod, death seems to pursue the aboriginal. We may look to the wide extent of the Americas, Polynesia, the Cape of Good Hope, and Australia, and we find the same result.
>
> – Charles Darwin, The Voyage of the Beagle (1839)

This is not only true for the aboriginals, but also for the native species of different animals and plants which are facing threat from invasion of alien species. Crosby (2015), in his book, *Ecological Imperialism*, has provided some of the most fascinating accounts of the introduction of different plants and animals by Europeans from their native habitats to foreign lands and their effects. Humans have been intentionally

*Corresponding author: rahuldayanand33@gmail.com

†Both Rahul Kumar and Stuti have contributed equally

 Biological Invasions in the Tropics: Monitoring, Mechanism and Management (eds Akshay Kumar Chakravarthy and Pathour Rajendra Shashank)
DOI: 10.1079/9781836991977.0005

or unintentionally spreading different species around the earth for centuries. Anthropogenic perturbations provide opportunities to the invasive alien species to expand their habitat range (Myers, 1993). Human population density is an important factor responsible for increase in species richness of an invasive alien species in an area (Spear *et al.*, 2013). About one-sixth of the global land surface is vulnerable to invasion by alien species (Lamsal *et al.*, 2018). It is a global problem. According to the Convention of Biological Diversity (2002), "an invasive alien species refers to an alien species whose introduction and/or spread threatens the biological diversity of the region/habitat" where "an alien species is a species, subspecies or lower taxon introduced outside its natural past or present distribution, which includes any part, gametes, seeds, eggs or propagules of such species that might survive and subsequently reproduce". Invasion of alien species has many deleterious effects on the ecosystem. It is a threat to local biodiversity. It alters the dynamics of community, and changes its structural and functional attributes (Yurkonis *et al.*, 2005). It affects ecosystem services (Tilman, 1999). Invasive species are resilient, have short life cycles, have broad host range, benefit from mutualism with other species, exhibit high dispersal power and fecundity, are voracious feeders and have abilities to withstand a wide range of environmental conditions (Fig. 5.1) (Singh *et al.*, 2020). Most invasive alien species have a broad range of physiological niche (Higgins and Richardson, 2014).

According to the Invasive Species of India Report published by National Biodiversity Authority, Ministry of Environment, Forests and Climate Change, Government of India, there are 173 invasive alien species reported from India. There are 27 species of arthropods which are invasive. This is the highest number among all invasive animal species (Fig. 5.2). Out of these 27 species, 23 have been reported from agricultural ecosystems, two are marine and two are reported from major islands. Among these 27, 24 are insects with only one arachnid and two crustaceans (Table 5.1) (Sandilyan, 2020; Singh *et al.*, 2020). Bharti *et al.* (2016) has discovered more invasive arthropod species in specific localities of India but more studies are required to assess their effects on the economy and environment. Recently *Thrip parvispinus* has been in the limelight for spreading fast from one part of India to another (Timmanna *et al.*, 2022a, b). More studies are being conducted on this species. Invasive alien species have special economic significance. These are responsible

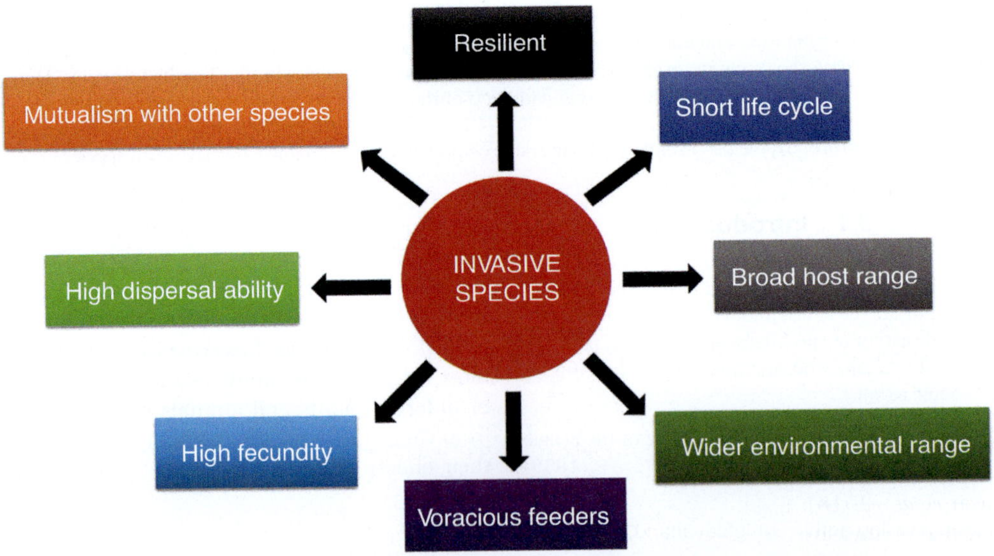

Fig. 5.1. Characteristic features of an invasive species. (Author's own image)

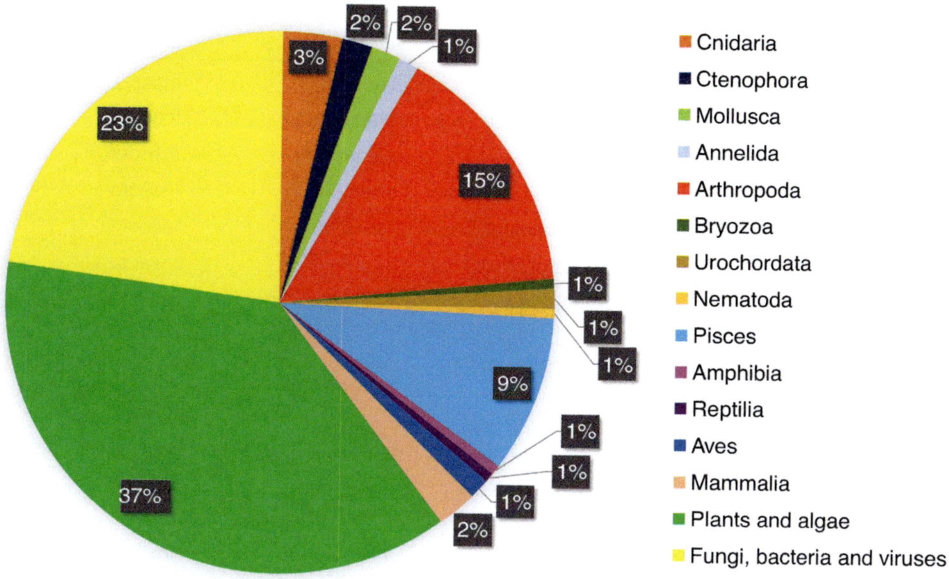

Fig. 5.2. Percentage of different groups of invasive alien species from India. (Author's own image)

for incurring huge economic losses globally (Encarnação *et al.*, 2021). In an estimate, it was found that biological invasion cost US$ 1.4 trillion in the 1990s which is approximately 5% of the global economy (Pimentel *et al.*, 2001). Biological invasions have been termed biological pollution by Reshi and Khuroo (2012). Among invasion cost-bearing countries, India stands at the second position, with the USA being first in the list. Of the total economic cost incurred by India by the invasive alien species, invasive animals are responsible for 35% (Fig. 5.2). Surprisingly, all these animal costs were caused by insects only (Bang *et al.*, 2022; Renault *et al.*, 2022). Therefore, it is important to study the biology of invasive arthropods.

Invasive alien species pose a grave threat to significant areas of global biodiversity hotspots (Early *et al.*, 2016). The Himalayan region is not spared this global threat. Global climate change is the leading factor behind the invasion of alien species from the regions of lower altitude to the areas at higher altitude such as the Himalaya (Simberloff, 2000; Alexander *et al.*, 2016; Lamsal *et al.*, 2018). More studies have been conducted on invasive plant alien species of the Himalayan region than any animal taxa (Sekar, 2012; Lamsal *et al.*, 2018; Pathak *et al.*, 2019). As mentioned above, among all animals and plants combined, arthropods cause maximum damage to the economy of India, and most of these arthropods have been reported from agricultural ecosystems. Some of these arthropods have also invaded areas of the Himalayan region in India (Sharma and Gavkare, 2017; Gupta *et al.*, 2019; Singh *et al.*, 2020). The need for assessment of the status and impact of invasive arthropods in the Himalaya, especially in view of expected climate change has been highlighted by some workers (Tewari *et al.*, 2017). Therefore, a comprehensive account of the recorded invasive species or the species with potential of becoming invasive from the Indian Himalayan Region is being provided in subsequent sections of this chapter.

5.2 The Himalaya

The Himalaya spans over 2400 km over India, Pakistan, Nepal, Bhutan and China and it has the highest peaks and supporting largest glaciers on the planet with a vast reservoir of Oriental, Palearctic and Ethiopian biodiversity. More than ten mountains have elevations of 8120

Table 5.1. Invasive arthropods of India. (Author's own table)

S.No.	Scientific name	Common name	Order	Class
1.	*Icerya purchasi*	Cottony cushion	Hemiptera	Insecta
2.	*Eriosoma lanigerum*	Woolly aphid	Hemiptera	Insecta
3.	*Orthezia insignis*	Lantana bug	Hemiptera	Insecta
4.	*Aleurodicus dispersus*	Spiralling whitefly	Hemiptera	Insecta
5.	*Paracoccus marginatus*	Papaya mealybug	Hemiptera	Insecta
6.	*Heteropsylla cubana*	Subabul psyllid	Hemiptera	Insecta
7.	*Phenacoccus solenopsis*	Solenopsis mealybug	Hemiptera	Insecta
8.	*Bemisia argentifolii* (*Bemisia tabaci* biotype B)	Silver leaf whitefly	Hemiptera	Insecta
9.	*Pineus pini*	Pine woolly aphid	Hemiptera	Insecta
10.	*Quadraspidiotus perniciosus*	San Jose scale	Hemiptera	Insecta
11.	*Aleurothrixus floccosus*	Woolly whitefly	Hemiptera	Insecta
12.	*Aleurodicus rugioperculatus*	Rugose spiralling whitefly	Hemiptera	Insecta
13.	*Aleurotrachelus atratus*	Neotropical whitefly	Hemiptera	Insecta
14.	*Pseudococcus jackbeardsleyi*	Banana mealybug	Hemiptera	Insecta
15.	*Anoplolepis gracilipes*	Yellow crazy ant	Hymenoptera	Insecta
16.	*Quadrastichus erythrinae*	Erythrina gall wasp	Hymenoptera	Insecta
17.	*Leptocybe invasa*	Blue gum chalcid	Hymenoptera	Insecta
18.	*Liriomyza trifolii*	Serpentine leaf miner	Diptera	Insecta
19.	*Citripestis eutraphera*	Mango borer	Lepidoptera	Insecta
20.	*Phthorimaea operculella*	Potato tuber moth	Lepidoptera	Insecta
21.	*Spodoptera frugiperda*	Fall armyworm	Lepidoptera	Insecta
22.	*Tuta absoluta*	South American tomato leaf miner	Lepidoptera	Insecta
23.	*Plutella xylostella*	Diamondback moth	Lepidoptera	Insecta
24.	*Hypothenemus hampei*	Coffee berry borer	Coleoptera	Insecta
25.	*Aceria guerreronis*	Coconut eriophyid mite	Acari	Arachnida
26.	*Jassa marmorata*	-	Amphipoda	Crustacea
27.	*Penaeus vannamei*	-	Decapoda	Crustacea

metres or higher, with Mount Everest (8848 m), K2 (8611 m), Kangchenjunga (8586 m), Dhaulagiri (8176 m) and Nanga Parbat (8126 m) among them. Furthermore, the Himalaya is also termed the "water tower region of the world" as it has the largest constellation of glaciers outside the polar caps; these glaciers feed some of the largest rivers of the continent including the Indus, Sutlej, Ganges and Brahmaputra, which originate from the region that is also globally known as the "water tank of the world". The Himalaya is known as "Abode of the Snow" in

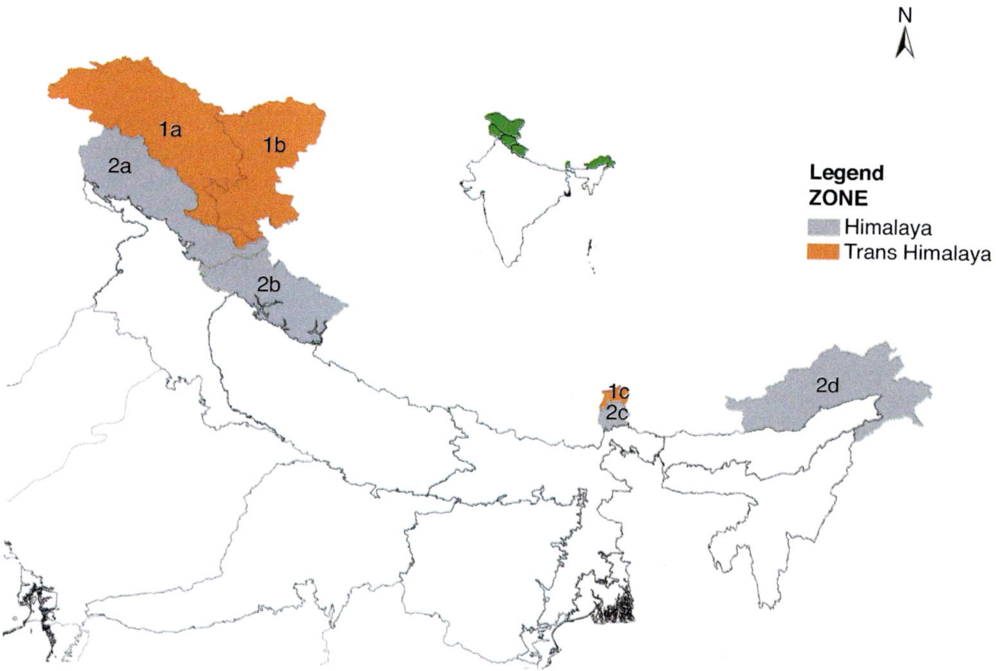

Fig. 5.3. Biogeographic zones (*N*=2) and provinces (*N*=7) in the Indian Himalayan Region. (Author's own image)

India and "Sagarmatha" in Nepal, which means "Goddess of the Universe" or "Forehead of the Sky" in the local language.

The major Himalayan Mountain system in India stretches through six states from Jammu and Kashmir in the west to Arunachal Pradesh in the east (Rodgers and Panwar, 1988) and is known as Indian Himalayan Region (IHR) (Fig. 5.3). It passes through five states of India and two union territories, viz. Jammu and Kashmir, Ladakh, Himachal Pradesh, Uttarakhand, Sikkim, a small part of West Bengal (Darjeeling and Kalimpong) and Arunachal Pradesh. Considering the distinctiveness of similar ecosystem characteristics, biome representation and biodiversity composition within each zone, Rodgers *et al.* (2002) broadly divided India into 10 biogeographic zones: Himalaya, Trans-Himalaya, Gangetic Plains, Desert, Deccan Peninsula, Semi-Arid, Northeast, Western Ghats, Coasts and Islands. These biogeographic zones are classified into 26 biotic provinces, each reflecting a distinct ecosystem separated by geographic and ecological barriers, species composition, climatic conditions and thus changes in the environment. The Indian Himalaya has been classified into two biogeographic zones and seven provinces depending on differences in physiographic, climatic and eco-biological characteristics. These two zones, known as Himalaya and Trans-Himalaya, are organized into seven biotic provinces as 1a, 1b, 1c, 2a, 2b, 2c and 2d, possessing a total area of 395,485 km^2 (Fig. 5.3).

5.2.1 Himalayan ecosystem

Himalaya, with a geographic expanse of over 595,000 km^2 (area of 537,000 km^2 in Indian Himalayan Region), plays a critical role in the long-term development of India and the subcontinent. This vast mountainous environment is one of the world's richest biodiversity hotspots and the source of Asia's ten major rivers, supplying food, water, electricity and carbon stock to almost two billion people. The magnificent

Himalaya is globally renowned for its biophysical variety and sociocultural legacy. Right from the foothills or 'Shivaliks' the region extends up to the Tibetan Plateau, the topographic variability of this ecosystem provides habitats for a range of faunal as well as floral life forms. Himalayan ranges are the meeting point for Oriental, Palaearctic and Ethiopian realms, which can be further identified to cover African and Euro-Mediterranean, Sino-Tibetan, Malayo-Burman and Austro-Polynesian, and elements from 77° in the west to 84° in the eastern region. The proportion of unique and rare taxa is quite characteristic in the whole Indian Himalayan Region. Vegetation of Himalaya is categorized into tropical, subtropical, temperate and alpine, which is shaped by enormous topographic variation (from 500m to over 8000m) and thus elevation and precipitation. The region is home to more than 20,000 species of plants and more than 30,000 species of animal, this richness in biodiversity in the region is often attributed to the climate and biogeography. The vegetation varies from a low to 500m in Shivalik to 4200m in Great Himalayan belt. The forest mainly consists of sal, pine, oak and rhododendron species as climax forest species from 300 to 2,500m. This region's faunal composition is distinct and very rich and diverse in the Indian subcontinent. Many species are specific to the region, including the Himalayan Bearded Vultures, Himalayan Black Bear, Himalayan Ibex, Himalayan Lynx, Himalayan Musk Deer, Himalayan Tahr, Kashmir Stag, Red Panda, Snow Leopard, Takin and Yak. A total of 30,377 animal species (including 372 species of protozoa) are known from the Indian Himalayan region, accounting for approximately 30.16% of the total Indian fauna (100,762 species). With 26,392 species, arthropods constitute approximately 87.9% of all Indian Himalayan fauna (excluding protozoa) (Chandra *et al.*, 2018) (Fig. 5.4).

5.2.2 Drivers and challenges to biodiversity loss in Indian Himalayan Region

The Himalaya is nature's most generous and flexible renewable resource, providing a broad spectrum of socioeconomic, cultural and environmental ecosystem services. Additionally, the Himalaya is a complex and diverse ecosystem with respect to biophysical features, sustaining the lives of millions of people. It governs the climate for most of Asia and is revered across the world for its spiritual, sacred and philosophical

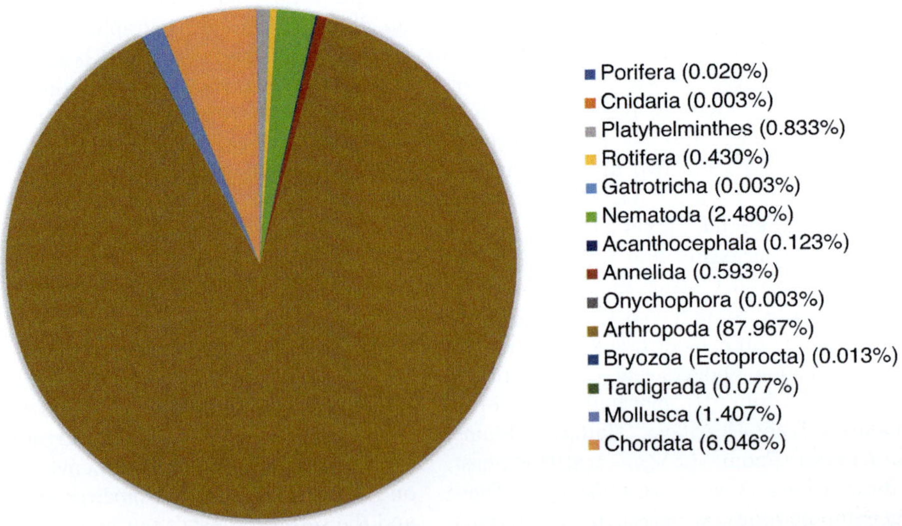

Fig. 5.4. Percentage of occurrence of animal phyla in the Indian Himalayan region. (Author's own image)

elements. Despite the numerous advantages and values provided by this ecosystem, it is facing enormous threats in most of its region of influence. As an accepted theory, biological diversity declines with increasing elevation within mountain areas, while endemism typically rises owing to geographic topographic separation (Körner, 2004; Salick *et al.*, 2004). The adverse consequences of climatic events will inevitably increase with a negative cumulative impact on groundwater recharge, river hydrology, natural disasters and biodiversity and human livelihoods over the next few decades (Nijssen *et al.*, 2001; Parmesan, 2006; Bates *et al.*, 2008; Ma *et al.*, 2009). The spatial distribution of wetland areas and freshwater habitats, species phenology and tree-line shifts are all governed by the availability of water, which also affects physiological, metabolic and reproductive activities of organisms (He *et al.*, 2006; Parmesan, 2006; Bates *et al.*, 2008). The Himalaya, as one of susceptible and threatened ecosystems, is threatened by biodiversity loss due to both climate change and anthropogenic disturbances (Field and Barros, 2014). Another significant factor responsible for loss of biodiversity is change in land use patterns owing to huge anthropogenic factors. The potential biological cumulative consequences associated with rapid climate change also result in unfathomable and unintentional loss of biodiversity. As mentioned above, the invasive species problem is a manifestation of climate change and anthropogenic disruptions, and it is a severe danger to local biodiversity (Myers, 1993; Simberloff, 2000; Alexander *et al.*, 2016; Lamsal *et al.*, 2018).

5.3 Invasive Arthropods of the Indian Himalayan Region

Based on the distribution records of the invasive arthropod species of India (Table 5.1), seven species have been reported from different states falling under the Indian Himalayan Region. For West Bengal, only Darjeeling and Kalimpong fall under the Indian Himalayan Region. Hence, only these two parts of West Bengal have been considered (not the entire state). All these agriculturally important invasive arthropod species have been described in the subsequent section. Apart from these species, *Thrips parvispinus* has been reported to be invasive on a large scale in different parts of India in recent years and it may turn invasive in Himalaya too in the near future. Also, some species of the genus *Aceria* and *Tetranychus* are known to be invasive in India and outside India respectively. Various other species of these genera have been reported as serious pests from the Himalaya. Therefore, these arthropods are also agriculturally significant and have also been described subsequently. Pine forests are common in the Indian Himalayan region. *Pineus pini* is a well-documented invasive species of pine trees from south India. This species poses a threat to the pine forests of the Himalaya in future and has been described here as well. Bharti *et al.* (2016) has reported 19 ant species from the Indian Himalayan Region which are known to be invasive (Table 5.2).

5.3.1 Description of agriculturally important invasive arthropod species of the Indian Himalayan Region

5.3.1.1 Phthorimaea operculella (*Zeller*)

It was introduced with imported potato from Italy to India in 1937 and has spread throughout India (Singh *et al.*, 2020). It is distributed in almost all parts of the world.

Common name: Potato tuber moth.

Scientific name: *Phthorimaea operculella* (Zeller).

Taxonomy: Phylum Arthropoda, Class Insecta, Order Lepidoptera, Family Gelechiidae, Genus *Phthorimaea*, Species *P. operculella* (Zeller).

Hosts: Brinjal, tomato, black nightshade, silverleaf nightshade, chilli pepper, tobacco, cape gooseberry, field ground cherry, prickly nightshade, Jimson weed. Although PTM can be found in all of these crops and weeds, it specifically completes its life cycle including egg-laying, larval development and pupation on these four host plants, i.e. potato, tomato, brinjal, and tobacco, according to a study by Das and Raman in 1994.

Description: The adult moths are small, grey to off-white in colour, measuring around 0.94 cm in length, with a wingspan of approximately 1.27 cm. Males have 2–3 dark spots on

Table 5.2. Invasive arthropods reported from the Indian Himalayan region. (Author's own table)

Well established agriculturally important invasive arthropod species	Arthropods with potential to become invasive in Himalayas	Invasive ant species (after Bharti *et al.*, 2016)
1. *Phthorimaea operculella*	1. *Aceria* sp.	1. *Cardiocondyla nuda*
2. *Plutella xylostella*	2. *Tetranychus* sp.	2. *Cardiocondyla wroughtonii*
3. *Eriosoma lanigerum*	3. *Pineus pini*	3. *Cerapachys biroi*
4. *Quadraspidiotus perniciosus*	4. *Thrips parvispinus*	4. *Hypoponera coninis*
5. *Bemisia argentifolii* (=*Bemisia tabaci* biotype B)		5. *Hypoponera ragusai*
6. *Tuta absoluta*		6. *Monomorium destructor*
7. *Spodoptera frugiperda*		7. *Monomorium loricola*
		8. *Monomorium pharaonis*
		9. *Ochetellus glaber*
		10. *Paratrechina longicornis*
		11. *Solenopsis geminata*
		12. *Tapinoma melanocephalum*
		13. *Technomyrmex albipes*
		14. *Tetramorium bicarinatum*
		15. *Tetramorium caespitum*
		16. *Tetramorium caldarium*
		17. *Tetramorium lanuginosum*
		18. *Tetramorium simillimum*
		19. *Tetramorium tonganum*

their forewings, while females display an "X" marking. Both pairs of wings have fringed edges, and the moths rest with their wings held close to their bodies, giving them a slender appearance. The adult moths have a lifespan of 1–2 weeks and are most active during dawn and dusk when they feed on nectar. They can travel up to 0.25 km between crops to infest plants or tubers, often through the transportation of infested tubers. The PTM eggs are spherical, ranging in size from 0.5 × 0.35 mm, and their colour can vary from white or yellowish to light brown. Females prefer to lay their eggs on foliage, soil, plant debris or exposed tubers, with foliage being their preferred substrate. Larvae of PTM are typically light brown with a distinct brown head. Mature larvae can reach a length of approximately 0.94 cm and may exhibit a pink or greenish colouration. Sexual dimorphism becomes apparent in the 3rd larval stage, with incipient sexual structures becoming visible. By the 4th larval stage, males can be distinguished from females by the presence of elongated yellowish testes in the 5th and 6th abdominal segments. The pupae of PTM are smooth, brown and around 0.84 cm in length. They are often covered in fine soil and debris. Male and female pupae have clear distinctions, with males characterized by a longer distance between the incision located between the 8th and 9th abdominal segments and the tip of the abdomen.

Life cycle: The life cycle of the potato tuber moth (PTM) consists of four stages: adult, egg, larva and pupa. Adult PTM are predominantly inactive during the day and engage in egg-laying at night. Oviposition typically begins 1–3 days after emergence and lasts for 4–9 days. The number of eggs laid and their longevity are directly influenced by the female's nutrition (Fenemore, 1977, 1978; Gubbaiah and Thontadarya, 1977). Eggs are laid singly or in batches near buds, cracks, fissures or peeled potato skins, with a range of 38–290 eggs per batch. If potato foliage is available, adults do not lay eggs in the soil. The incubation period for the eggs varies from 5 to 34 days, depending on temperature. The PTM larvae progress through four stages, with the time between stages influenced by temperature. The larval period ranges from 15 to 17 days, depending on temperature conditions. Normal larval activity, including feeding and moulting, occurs within a temperature range of 11.1–39.4°C. The pupal stage lasts for 6–9 days. Adult PTM can emerge from the soil at depths of up to 10.16 cm. After emerging, the moths mate, and females seek suitable hosts to lay their eggs on. The moths exhibit activity levels at temperatures between 14.4 and 15.5°C, and while they can crawl at 11.1°C, they are unable to fly. In North India, up to 13 generations per year have been observed, with fecundity ranging from 80 to 118 eggs and a life cycle duration of 17–24 days in summer and 25–40 days in winter. Studies have noted that a majority of male and female moths emerge during the light phase of the day (Chauhan and Verma, 1985). The PTM life cycle can vary depending on geographic location and seasonal factors.

Damage: The PTM causes substantial damage to both plant foliage and tubers due to its feeding behaviour. Larvae of PTM create mines in the leaves, petioles and terminal shoots, resulting in wilting symptoms. Once the tubers have formed, the larvae enter the tubers and feed on the internal tissue, leading to the formation of tunnels, thus making it unsuitable for use as seed or for human consumption. Furthermore, larvae can bore into tubers stored in storage facilities too. The extensive tunnelling and feeding within the tubers not only make them unfit for use but also create entry points for microbial infections, leading to rotting. The degree of damage inflicted on stored tubers can vary significantly, ranging from 20–85%. This highlights the severe impact of PTM infestation on potato crops and storage systems.

Distribution: Maharashtra, Madhya Pradesh, Himachal Pradesh, Uttarakhand, Jharkhand, Bihar, West Bengal, Tamil Nadu, Karnataka, North-eastern hill states and plateau region.

Natural enemies: Various parasitoids and predators have been investigated as potential biological control agents against pests. One such parasitoid is *Copidosoma koehleri* Blanchard, which targets the eggs and larvae of pests. Another parasitoid, *Chelonus blackburnii* (Cameron), is an exotic species that has shown efficacy in controlling pest populations. Additionally, *Chrysoperla zastrowi sillemi* (Esben-Petersen), *Orius albidipennis* (Rueter), and *Labidura riparia* (Pallas) have been identified as beneficial organisms for pest management (Ballal *et al.*, 1989; Abbas *et al.*, 1993). These studies have underscored the potential utility of these parasitoids and predators in integrated pest management strategies, providing alternatives to conventional chemical approaches.

5.3.1.2 Plutella xylostella (*Linn.*)

In India, diamondback moth was first recorded in 1914 (Gupta *et al.*, 2019) on cruciferous vegetables. It is now distributed all over India wherever crucifers are grown.

Common name: Diamondback moth.

Scientific name: *Plutella xylostella* (Linn.).

Taxonomy: Phylum Arthropoda, Class Insecta, Order Lepidoptera, Family Plutellidae, Genus *Plutella*, Species *P. xylostella* (Linn.).

Hosts: Cabbage, cauliflower, radish, knoll (kholrabi), turnip, beetroot, mustard.

Description: The adult diamondback moth (DBM) is a slender and greyish-brown moth with well-defined antennae. Measuring approximately 6 mm in length and a wingspan of about 12–15 mm, forewings are brown or grey, adorned with prominent white or cream or light-brown band running along its back, forming one or more diamond-shaped patterns when the wings are folded at rest. This characteristic banding gives the moth its common name.

Life cycle: Researchers in India have conducted studies on the biology of the diamondback moth, and their findings have shown variations in different parameters due to environmental conditions and location. The DBM undergoes a complex life cycle consisting of distinct stages (Harika *et al.*, 2019). The DBM life cycle begins with the deposition of eggs. The eggs are usually laid on the undersides of leaves and vary in colour and size. The incubation period of the eggs lasts between 2–4 days, with an average duration of around 3 ± 0.5 days. After hatching, the DBM larvae enter the larval stages. The first instar larvae are tiny and have a white colouration with a dark brown head. They undergo rapid growth and development, lasting approximately 2.5 days on average. The larvae progress to the second instar, characterized by their larger size and yellowish-green colouration. The second instar larvae generally complete their development in about 2 days. During the third instar stage, they exhibit a light yellow body colour and a light brown head. This stage typically lasts around 1–3 days with an average of 1.75 ± 0.25 days. Lastly, the fourth instar, where they become dark green with a light brown head. They are covered with short hairs and grow in size until they are ready for pupation. The duration of the fourth instar varies from 2 to 4 days, with an average period of approximately 2.75 ± 0.25 days. Following the larval stages, the DBM larvae enter the pre-pupal stage. During this phase, the larvae become less active, cease feeding and undergo certain physiological changes in preparation for pupation. The pre-pupal stage typically lasts for 1–2 days. After the pre-pupal stage, the DBM larvae transform into pupae. The pupae are enclosed within thin silken cocoons, which may be formed on lower or outer leaves. Initially, the pupae appear pale green and gradually develop a light yellow colouration with brownish markings. The pupal stage generally lasts between 3–5 days. Finally, the adult stage marks the completion of the DBM life cycle. Adult males and females have differing lifespans, with males typically living for about 12 days and females for around 16 days (Harika *et al.*, 2019). DBM adults exhibit activity during the evening and throughout the night. The majority of adults emerge within the first 8 hours of the light phase and engage in mating during dusk on the same day of emergence (Gautam *et al.*, 2018). Following mating, female moths promptly initiate egg-laying, which lasts for a duration of 4 days. The peak period of oviposition typically occurs between 7:00 and 8:00 pm, with the majority of eggs being laid before midnight (Gautam *et al.*, 2018). The DBM life cycle is influenced by various factors such as temperature, humidity and availability of food sources.

Damage: The larvae of DBM exhibit distinct feeding behaviours and cause significant damage to various plant parts. Upon hatching, first-instar larvae immediately burrow into leaf tissue, transitioning to surface feeders in the second instar. Their voracious feeding habits result in the consumption of leaf material, leaving behind a thin, translucent layer resembling windows or "shot holes" on the leaf surface. These damaged leaves often harbour *P. xylostella* larvae and pupae. Severe infestations can lead to complete defoliation, where only the leaf veins remain (Saxena *et al.*, 1989).

Distribution: Assam, Bihar, Delhi, Haryana, Himachal Pradesh, Karnataka, Madhya Pradesh, Maharashtra, Punjab, Tamil Nadu, Uttar Pradesh, West Bengal.

Natural enemies: The DBM is targeted by a variety of parasites and predators throughout its life stages. *Brachymeria excarinata* Gahan and *Tetrastichus sokolowskii* Kurdj have been observed parasitizing DBM larvae, where *B. excarinata* Gahan exhibited a high level of parasitization. A study conducted in Bengaluru found that *A. plutellae* and *T. sokolowskii* cause mortality and parasitization of second instar DBM larvae. During the cold season, yellow wagtails (*Motacilla flava* Linnaeus) act as predators of DBM larvae. Additionally, ant species including *Tapinoma melanocephalum*, *Pheidole* sp. and *Camponotus sericeus* have been observed carrying away DBM larvae in field environments (Cherian and Basheer, 1939; Jayarathnam, 1977).

5.3.1.3 Eriosoma lanigerum (*Hausmann*)

It was introduced from China in the year 1889 (Singh *et al.*, 2020).

Scientific name: *Eriosoma lanigerum* (Hausmann).

Common name: Woolly apple aphid.

Taxonomy: Phylum Arthropoda; Class Insecta; Order Hemiptera; Suborder Sternorrhyncha; Superfamily Aphidoidea; Family Aphididae; Genus *Eriosoma*; Species *Eriosoma lanigerum*.

Hosts: *Cestrum nocturnum*, *Cotoneaster* sp., *Malus* sp., *Pyrus buccata*, *Pyrus communis*, *Pyrus malus*, *Pyrus polycarpa*, *Pyrus sylvestris*, *Pyrus* sp.

Description: The life cycle of woolly apple aphid shows endless complexity and includes several morphs. While in Himalaya it has an autoecious life cycle and has morphs of both virginoparae and sexuparae.

Apterous virginoparae/viviparous female: Small bodied, approximately 2 mm in length; dusky, reddish-brown in colour, covered with a white waxy secretion giving it a woolly appearance. The head possesses 6–10 wax plates on the upper side, with a central area surrounded by 6–10 cells. The antennae are pale yellow, consisting of 6 segments, measuring around 0.21–0.24 × the length of the body. The flagellum of the antennae has fine, spinulose striae, and the flagellar hairs are sparsely distributed. The rostrum extends up to the mid-coxae. Thoracic tergum, bearing 4 wax plates on each of the segment. Abdomen dorsally pale with 4 wax plates on each of the 1st–7th tergites. The legs are pale brown, and fine hairs can be observed on the femora and tibiae (Ghosh, 2008).

Alate virginoparae/viviparous female: Body smaller than the apterous females, elongated, measuring approximately 1.7 mm in length. Head is flat and smooth, with tiny dorsal cephalic hairs. The antennae have 6 segments and are about 0.40 times the length of the body. The third antennal segment is longer than the combined length of segments IV, V and VI; flagellum has fine spinulose imbrications. Rostrum extending up to the mid-coxae, and the length of the ultimate rostral segment is 1.3–1.4 times the length of the second hind tarsal segment, bearing around 10 accessory hairs. The dorsal side of the abdomen has paired marginal wax plates from 1st–8th tergites (Ghosh, 2008). This form is important for dispersal (Rahman and Khan, 1941; Fotedar and Kapur, 1943; Lal and Singh, 1947). It is also assumed that the production of the winged form is induced by overcrowding. Alate viviparous females are produced during different times of the year, and their peak population is also variable. However, in Kashmir alate virginoparae are produced during May–July and again during August and November; the earlier population gives birth to apterous nymphs whereas the later one produces sexuparae which lay apterous sexual nymphs (Fotedar and Kapur, 1943) which lack functional mouthparts.

Sexuparae: Theses are small bodied, winged, parthenogenetic and viviparous females. Antennae are six segmented with secondary rhinaria on the last four segments. Compound eyes are dark black and composed of multiple facets. They have paired wings, with the forewings featuring a distinct pterostigma. The branching of the media vein is forked and not visible near the base. The hind tarsi are two segmented, with the proximal segment being triangular and adorned with two long hairs, while the distal segment carries a pair of claws. Through a series of four moults, each individual progresses towards the final moult, leading to the emergence of adult alate sexuparae. Distinguishing the first and second instars of the alate sexuparae from the apterous virginoparae is challenging since the wing pads are not visible. However, in the third and fourth nymphal instars, the wing pads become highly conspicuous, enabling differentiation between these two morphs based on this major character. Later sexuparae produce sexuals, i.e. the male and ovipara. The rate of fecundity is lower in sexuparae as compared with virginoparae (Gautam, 1981). After giving birth to the sexuals, the abdomen of the sexupara shrinks (Gautam and Verma, 1983).

Sexuals: These are apterous males and females (oviparae). Mouth parts in both the morphs are degenerated. Antennae are five segmented throughout the nymphal instars as well as in the adults. Last two antennal segment with primary rhinaria. Compound eyes are three faceted. Fore, mid and hind tarsi single segmented with three pairs of long capitate hair, a pair of long hair and paired claws present. Males, being smaller in size and having a dark brown body, differ in body colouration from the reddish-brown females. In both sexual morphs, the antennae and legs appear transparent with yellowish tinge. The sexual individuals, including both nymphs and adults, possess degenerated mouthparts; however, they are able to survive and undergo various physiological changes during their development, such as moulting and maturing (Gautam and Verma, 1983). According to Fluiter (1948), it was observed that the sexuparae of woolly apple aphid have the ability to produce not only males and oviparae, but also intermediates with fully developed mouthparts similar to those of viviparae, or partially developed mouthparts.

Life cycle: *Eriosoma lanigerum*, follows an anholocyclic reproductive pattern, where it does not rely on host-alternation and can reproduce continuously on apple trees in several parts of the world. During the winter season, these aphids overwinter as nymphs or occasionally as adults by residing in plant roots or within bark crevices. While *E. lanigerum* is primarily anholocyclic, there are instances in various regions where a limited sexual phase occurs. This phase involves the production of sexual forms (sexuals) and eggs on apple trees. However, it is important to note that the eggs produced by the sexuals are non-viable and do not hatch, as documented by Blackman and Eastop (1994, 2000). According to Thakur (1970), *E. lanigerum* revealed that various aspects of its life cycle, such as nymphal development, adult longevity, fecundity and rate of larviposition, are influenced by temperature and the vigour of the host plant. In India, it was observed that winter nymphs

(December–January) take significantly longer for postembryonic development compared with early summer nymphs (March–April) when temperatures are higher and plant vigour is restored. Conversely, adult woolly apple aphids in India, observed to have shorter longevity and higher rates of larviposition during the summer months (March–April), while they exhibit longer longevity during autumn (October–November) (Thakur and Dogra, 1980). Seasonal variations in larviposition were also reported by Rahman and Khan (1941). In Kashmir, Fotedar and Kapur (1943) provided a brief description of the life cycle of *E. lanigerum*. They reported that apterae produce winged alate forms between May and July, which disappear after July. During August to November, a large number of winged alate forms, known as sexuparous alatae, return to the host plant and give rise to arostrate gamic forms. The female aphids, which are ovoid and light brown in colour, lay eggs (measuring 0.32 mm in length) near the base of the apple tree. Later Gautam and Verma (1983) reported the complete life history of sexuparae and sexual morphs of *E. lanigerum* from Shimla Hills.

Sexuparae: The average duration of the first, second, third and fourth nymphal instars is 3.95, 2.7, 2.75 and 4.25 days, respectively. The total nymphal period ranges between 12–15 days, with an average of 13.65 days. Adult individuals have a lifespan of 4–7 days, with an average of 5.4 days. The number of progeny produced by each adult ranges from 4 to 8 sexuals, with an average of 5.3 nymphs. The total lifespan of the sexupara stage ranges from 17 to 22 days, with an average duration of 19.05 days. After laying the sexuals, the abdomen of the sexupara undergoes shrinkage. The fecundity rate of the sexupara is observed to be lower compared with that of the virginopara, as documented by Gautam in 1981 and attributed to be due to morph differentiation (Gautam and Verma, 1983).

Sexuals (male): The average durations of the first, second, third and fourth nymphal instars are 1.45, 1.65, 2.25 and 1.65 days, respectively. The total nymphal period ranges between 6–9 days, with an average of 7 days. Their overall lifespan ranges between 8–12 days, with an average of 10.3 days. Males exhibit high activity levels compared with oviparae (Gautam and Verma, 1983).

Female/oviparae: The average durations of the first, second, third and fourth nymphal instars are 2.45, 1.2, 1.55 and 2.2 days, respectively. The total nymphal period is between 7–8 days, with an average of 7.4 days. The period between the last moult and egg laying ranges from 3 to 11 days, with an average of 5.5 days. Oviparae live longer than the male with a total lifespan of 10–16 days, with an average of 12.9 days (Gautam and Verma, 1983). Previous reports by Fotedar and Kapur (1943) in the Kashmir valley also indicated a male survival period of approximately 12 days and an ovipara lifespan of about 20 days. Oviparae lay a single egg, which occupies the entire abdomen in adult oviparae. They die during the process of egg laying. The freshly laid egg is reddish brown but later turns dark brown (Gautam and Verma, 1983).

Damage symptoms: *E. lanigerum* is a pest that affects various parts of apple trees, including the aboveground and underground woody tissues, causing damage to both the root and shoot systems. The aphids feed on sap, leading to the formation of galls, deformations and other abnormalities in the infested plant parts. These gall formations negatively impact the overall health and productivity of the apple tree, resulting in reduced yield quantity and quality (Thakur and Dogra, 1980). The presence of the colonies on the trunk, branches or twigs can cause deformations, blisters, splitting and bark swellings that resemble cancerous growths (Blackman and Eastop, 1994, 2000). Moreover, severe infestations can create a favourable environment for the colonization of basidiomycete fungi leading to secondary infection, particularly the development of root canker, further contributes to the deterioration of the plant's health and vitality (Molinari, 1986; Alfred Daniel *et al*, 2020).

Distribution: Himachal Pradesh, Meghalaya, Kashmir, Tamil Nadu, Uttarakhand.

Natural enemies: *Aphelinus mali* (Haldemann) *Aphelinus mali* (Haldemann), a host-specific parasitoid of *E. lanigerum*, was introduced in India in the 1930s. Its introduction resulted in an impressive 98% control of the woolly aphid in the Kullu Valley, Himachal Pradesh (Thakur and Dogra, 1980). *Coccinella septempunctata* L., a significant predator of woolly aphids, displays a feeding potential of up to 60% for males and up to 96% for females when it comes to consuming woolly aphid colonies (Singh, 2003). Other predators that play an important role in controlling woolly aphid populations include *Forficula auricularia* L., *Chrysoperla carnea* (Stephens), *Ballia ancharis* Muls., *Coleophora sanzeti* Muls., *Chilomenes bijugus infernalis* Muls., *Syrphus confrator* Wiedmann, *Coniocompsa indica* Withycombe, *Ancylonteryx punctate* Hay and *Exochomus uropygialis* (Muls.) (Rahman and Khan, 1941; Maurya, 2011). According to Asante (1997), *Verticillium lecanii* is the sole fungal pathogen documented to infect *E. lanigerum*.

5.3.1.4 *Quadraspidiotus perniciosus* (*Comstock*)

It was introduced from China in the year 1911 (Singh *et al.*, 2020).

Scientific name: *Quadraspidiotus perniciosus* (Comstock).

Common name: San Jose scale.

Taxonomy: Phylum: Arthropoda; Class: Insecta; Order: Hemiptera; Family: Diaspididae; Genus: *Quadraspidiotus*; Species: *Q. perniciosus*.

Hosts: San Jose scale is a major pest of rosaceous fruit trees, especially almond, apple, apricot, cherry, chestnut, grapes, peach, pear, plum, strawberry, raspberry walnut, and ornamental trees and shrubs.

Description: Crawlers, adult females and adult males are the developmental stages of *Q. perniciosus* that will help us to identify and monitor this insect pest. The adult females live under a small circular brown, smoky black or dark grey cover up to 1/12th of an inch (1–2 mm) in diameter, with a small protrusion in the centre. The scale covering is slightly convex, with dark concentric rings formed by the ducts present on the pygidium. Females inside the cover are reddish yellow in colour, neotenic, legless and wingless, their antennae are completely reduced or transformed into small, thick prominences. The head, abdomen and thorax are fused, and the whole body is dorso-ventrally flattened. Mounted specimens appear pear shaped to almost circular, ranging 0.8–1.2 mm in length. It lacks perivulvar pores that act as an important feature to distinguish it from its similar species (Golan, 2020). However, it is well described (Brookes and Hudson, 1968; Golan, 2020) and a key to the identification of *Q. perniciosus* and similar species of scale insects based on adult female morphology is provided by Golan (2020).

Males are free-living. The body colouration is yellowish-tan, very tiny about 1 mm in length, delicate, having well developed eyes, wings, 3 pairs of legs, with long, plumose antennae, but vestigial mouth parts and therefore survive only for a day. A prominent dark band across the back (dorsum) makes them distinguishable from other species of the genus (Golan, 2020). They undergo four developmental stages including two nymphal stages, prepupa and pupa stage; after emergence, they mate with the female which remains stationary under the scale, and die. A few elliptical-shaped scales (larger male immatures) may be interspersed among the otherwise round scales.

Crawlers are the live young produced by the female scales from her protective covering. Crawlers are small bodied, yellowish in colour, mobile with three pairs of legs, antennae and simple eyes. They can move to other parts of the plant and can be spread by the wind, birds, insects and other mechanical means. This life stage causes plant to plant infestation. Once the crawlers reach a new site on the bark, stem, fruit or leaf, it settles and feeds on the plant sap and produces a wax covering, which is initially white in colour (called white cap nymph) and gradually changes into black (black cap nymph) (Golan, 2020).

Life cycle: The life cycle of *Q. perniciosus* includes crawlers or nymph, adult male and female. Female is viviparous and can give birth to 200–400 nymphs within a month. Soon after emergence, the 1st instar nymph moves to a new feeding site, gets settled and feeds for around 12–24 hours. It secretes a white wax covering, grows inside and gradually changes its colour from black to dark grey. Female moults into an adult after two nymphal stages (1st and 2nd instar nymph). Whereas, the male acquire maturity later than female and develops after four moulting stages (1st instar nymph, 2nd instar nymph, pre-pupa and pupa). The pre-pupa and pupa are the non-feeding stages. After 1st nymphal moult, sexual dimorphism becomes prominent. The 1st instar nymph in the unfavourable condition overwinters in the form of black cap. Sometimes the 2nd instars and the adult female undergo hibernation (Golan, 2020). According to reports from ICAR and KVK, it has been observed that the hibernating nymphs of this species become active in March. The males, on the other hand, emerge in April and mate with the females. The females then start producing offspring in mid-May. The nymphal period lasts for about 20 days. In terms of longevity, the female insects have a longer lifespan compared with the males. Females can live for approximately 50–53 days, while the males typically die within 24–32 hours.

Damage symptoms: The San Jose scale feeds on plant juices by inserting its long, hollow, needle-like mouthparts into the host's tissue. The damaging stages are the nymphs and adult females, and they target various parts of the plant, such as fruits, twigs, stems and branches. This feeding activity leads to the discolouration of infested tissue, which can turn into a red or purple halo. Infestation by this scale insect causes progressive damage, including the gradual decline of infested twigs and stems, and in severe cases, it can result in the death of the entire tree. Infested fruits may show puncture marks and purple discolouration, while leaves can turn yellow and bark can crack with gum exudation. The infestation by San Jose scale poses a significant threat to plant health and vitality, leading to damage of fruit, twig and stem and mortality of the tree (Alfred Daniel *et al.*, 2020; Golan, 2020; Hamid *et al.*, 2022).

Distribution: Kashmir, Himachal Pradesh and Tamil Nadu.

Natural enemies: Promising results have been observed in the control of the San Jose scale on apple trees through the release of two exotic parasitoids, *Encarsia perniciosi* (Tower) and *Aphytis proclia* (Walker)/*Aphytis diaspidis* (Howard). These parasitoids were released at a rate of 2000 per infested tree, and studies have shown significant suppression of the scale population (Rao *et al.*, 1971; Singh, 1989). In Kashmir, the introduction of *E. perniciosi* and *A. proclia* resulted in a substantial increase in parasitism rates (Rao *et al.*, 1971). Further investigations on the biology demonstrated a remarkable multiplication rate of the parasitoids over 10 times. Additionally, other beneficial insects like *Coccinella infernalis* Muslant and *Aphelinus mali* (Haldeman) can also be employed in the control of San Jose scale (Alfred Daniel *et al.*, 2020). According to Rawat and Pawar (1991), several local natural enemies including *Aphytis* sp., *Novaproclia* (Walker), *Encarsia perniciosi* (Towers), and *Teleterbratus perversus* Compere and Zinna from Himachal Pradesh have been reported as effective in controlling San Jose scale. Four predator species, including *Chilocorus bijugus* Mulsant, *Coccinella septempunctata* Linnaeus, *Pharoscymnus flexibilis* Mulsant and *Sticholotis marginalis* Kapur, were identified as effective biocontrol agents against the pest in Jammu and Kashmir (Kumari *et al.*, 2021).

5.3.1.5 Bemisia argentifolii *Bellows & Perring*

There are taxonomic classification and naming controversies surrounding the silverleaf whitefly, formerly known as *Bemisia tabaci* biotype B. It was initially identified as a new strain of *B. tabaci* with distinct characteristics, such as inducing a silverleaf disorder and having a unique esterase profile (Costa and Brown, 1991; Bedford *et al.*, 1993). However, disagreements arose regarding its classification as a separate species, *Bemisia argentifolii*, with researchers questioning the basis for comparison with other biotypes (Bedford *et al.*, 1994; Bellows *et al.*, 1994). Further studies using morphological and genetic analyses revealed that *B. tabaci* consists of multiple races and genetic groups, with at least 34 morphologically indistinguishable species (Dinsdale *et al.*, 2010; De Barro *et al.*, 2011; Boykin and De Barro, 2014). These findings led to the conclusion that the separation of *B. tabaci* and *B. argentifolii* as different species should be discontinued (De Barro *et al.*, 2005). The decision is supported by researchers worldwide and the International Whitefly Symposium Network (CABI). It was introduced in India in 1999 (Singh *et al.*, 2020).

Scientific name: *Bemisia argentifolii* Bellows & Perring, *Bemisia tabaci* (B biotype), *Bemisia tabaci* Middle East Asia Minor 1.

Common name: Silver leaf whitefly, sweet potato whitefly.

Taxonomy: Phylum: Arthropoda; Class: Insecta; Order: Hemiptera; Family: Aleyrodidae; Genus: *Bemisia*; Species *B. argentifolii* Bellows & Perring.

Hosts: Tomato, squash, poinsettia, cucumber, eggplants, okra, beans and cotton.

Description: Adult whiteflies after emergence, initially have a soft whitish-yellow body. However, within a few hours, their wings develop an iridescent white appearance due to a powdery wax coating. The female whitefly measures about 0.96–1 mm in length from head to abdomen, while the male is slightly smaller at 0.82 mm. The whitefly eggs are oval-shaped and pearly white when freshly laid, but they darken over time. The eggs are arranged in a curved pattern around the female whitefly as she rotates using her stylet. They taper towards one end and measure approximately 0.21 mm in length and 0.096 mm in width. The broader end of the egg has a short stalk inserted into the leaf by the female during oviposition, which provides moisture to the egg. The distal end of the egg turns dark brown just before the first nymphal instar emerges. The first nymphal instar is known as the crawler. It is whitish-green in colour with two yellow spots called mycetomes visible in the abdomen. The crawler is oval-shaped, measuring around 0.27 mm in length and 0.14 mm in width. It has three pairs of well-developed four-segmented legs, three-segmented antennae and small eyes. The crawler often crawls on the lower surface of a leaf in search of a feeding site and gets settled for the subsequent stages. The second, third and fourth nymphal instars are immobile and have atrophied legs and antennae, as well as small eyes. They secrete a waxy substance at their body margins, aiding in their attachment to the leaf surface. These nymphs are flattened, oval-shaped and have a greenish-yellow colour. Their length ranges from 0.365 mm (second instar) to 0.662 mm (fourth instar). The mycetomes in these nymphs appear yellow. The subsequent nymphal stage, often referred to as the "pupal stage", is known as the red-eyed nymphal stage. There is no moult between the fourth nymphal instar and this stage, but there are distinct morphological differences. The red-eyed nymphal stage is characterized by larger, prominent red eyes compared with earlier nymphal instars. Additionally, it has a less flattened and more convex shape, with the sides appearing boat-shaped or wedged (McAuslane and Smith, 2015). On a smooth leaf the puparium lacks enlarged dorsal setae, but if the leaf is hairy, 2–8 long dorsal setae are present (Gangwar and Gangwar, 2018).

Life cycle: Whiteflies undergo a characteristic life cycle consisting of six stages: egg, four nymphal instars and adult. Factors such as temperature, relative humidity and host plants significantly influence the life cycle of whiteflies. *Bemisia argentifolii* deposits eggs on both the upper and lower leaf surfaces, with the number of eggs influenced by temperature. The eggs incubate for about 6–7 days. After hatching, the first instar (crawler) moves a short distance to feed on phloem sap; they start feeding, undergo moulting and enter the second nymphal stage, typically occurring within 2–3 days after hatching from the egg. The second and third nymphal stages each have a duration of approximately 2–3 days. Following these stages, the fourth nymphal stage, also known as the red-eyed nymphal stage, commences and lasts for about 5–6 days. Fully developed adults emerge from the pupal case through a T-shaped slit. Female whiteflies have a potential lifespan of 10–24 days, while males typically have a shorter lifespan of around 9–17 days. A single female has the capability to lay over 300 eggs throughout her lifetime depending on host plant and temperature (McAuslane and Smith, 2015).

Distribution: Throughout India including Himalaya.

Damage: Silver leaf whitefly can cause either direct or indirect damage. Directly by feeding on the phloem sap and reducing the available food to the plants, resulting in reduction of vigour and yield of plants. Feeding of the nymphs results in disorder symptoms, which include irregular ripening of tomato fruits. Indirect damage includes the development of sooty mould on the plant parts, a black fungus that grows on honeydew excreted by the whitefly nymphs and adults. Adult whiteflies also act as a vector of viral diseases and transmit the tomato yellow leaf curl virus. Another damage is cosmetic, resulting from the presence of adults and larvae on the leaves of ornamental plants (Singh *et al.*, 2020).

Natural enemies: The natural enemies of the silver leaf whitefly include parasitoids such as *Encarsia formosa* and *Eretmocerus eremicus*. In addition to parasitoids, fungal pathogens such as *Lecanicillium lecanii*, *Beauveria bassiana* and *Paecilomyces fumosoroseus* have also been found to be effective in managing whitefly populations (Singh *et al.*, 2020). In addition to the parasitoids and fungal pathogens mentioned earlier, various species of predatory mites have also demonstrated their effectiveness in its biological control. Among these predatory mites are *Amblyseius limonicus*, *A. swirskii* and *Transeius montdorensis* (Alfred Daniel *et al.*, 2020).

5.3.1.6 Spodoptera frugiperda (*J. E. Smith*)

Spodoptera frugiperda was introduced in India via Africa from America. First case in India was reported in 2018 (Singh *et al.*, 2020). Within 1 year, it spread to the whole of India including Himalaya. It has invaded even the mid and high hills of Himalaya (Paschapur *et al.*, 2022).

Scientific name: *Spodoptera frugiperda* (J. E. Smith).

Common name: Fall armyworm.

Taxonomy: Phylum: Arthropoda; Class: Insecta; Order: Lepidoptera; Family: Noctuidae; Genus: *Spodoptera*; Species: *S. frugiperda* (J. E. Smith).

Hosts: Brassicaceae, cotton, Cucurbitaceae, lucerne, maize, millets, onions, peanut, rice, sorghum, soybean, sugarcane, sweet potatoes, tomatoes, wheat and various ornamental plants.

Description: The adult male of *S. frugiperda* appears brownish grey in colour with body length of approximately 1.6 cm and a wingspan of about 3.7 cm. The forewings display a mottled pattern, combining light brown, grey and straw colours. In the discal cell of the forewing, three-quarters of the area is straw-coloured, while one quarter is dark brown, featuring triangular white spots at the wing's tip and near its centre. Female moths of this species measure about 1.7 cm in body length with a wingspan of approximately 3.8 cm. The forewings of females exhibit less distinct markings, ranging from a uniform greyish brown to a fine mottling of grey and brown. On the other hand, the hindwings are straw-coloured with a dark-brown margin (Luginbill, 1928; Sparks, 1979). The eggs measure 0.4 mm in diameter and 0.3 mm in height. Initially, they are pale yellow or creamish when laid, but they change to a light brown colour just before hatching. Egg masses are laid either in single layer or in multiple layers, on the underside of the leaves, or on top of the leaves. In a few cases, particularly on very young crops, eggs may be laid on the stem. The egg mass is usually covered with a protective layer of grey to pink scales from the female abdomen giving it a furry or mouldy appearance (CABI, 2019).

Typically, there are six larval instars, occasionally five. The larvae exhibit a colour range from light green to dark brown, with longitudinal stripes. At the sixth instar, larvae can grow up to 4.5 cm long. They possess eight prolegs and an additional pair of prolegs on the last abdominal segment. Upon hatching, the larvae are green with black lines and spots. As they mature, they may remain green or change to a brownish colour with black dorsal and spiracular lines. In cases of high population density and food scarcity, the final instar larvae may appear almost black in their armyworm phase due to crowding. Larger larvae are identifiable by an inverted Y-shaped yellow marking on the head, black dorsal pinaculae with long primary setae (two on each side of each segment within the pale dorsal zone) and four black spots arranged in a square on the last abdominal segment (Crumb, 1956; Levy and Habeck, 1976). The pupae are smaller in size compared with mature larvae, measuring around 1.3–1.5 cm in males and 1.6–1.7 cm in females in Mexico (CABI, 2019). Upon formation, the newly developed pupae display a greenish colour, which gradually changes to brownish and dark brown, eventually turning black just before emergence (Marri *et al.*, 2023). Pupation usually takes place in the soil, but in some cases, it may also occur in reproductive parts like mature maize ears. If the soil is too compacted, larvae may create a cocoon by webbing together leaf debris and other materials on the soil surface (CABI, 2019).

Life cycle: The life cycle of fall armyworm typically consists of four stages: egg, larva, pupa and adult stage. Adult fall armyworms engage in copulation shortly after emergence, typically occurring within a few hours. The peak copulation time was observed between 8 pm and 11 pm, with the highest activity at 9 pm. The preoviposition period ranged from 2 to 3 days, while the oviposition period ranged from 2 to 4 days. Female fall armyworms lived longer than males even without food. The adult lifespan was relatively short, with males living for 2–6 days and females for 2–9 days. Soon after emergence, FAW moths actively searched for food and then proceeded to mate. Afterward, female moths laid eggs and died within a few days. Adult females laid up to 1184 eggs, with an average of 469 ± 22 eggs per female. The incubation period of eggs was 2–3 days, and the hatching rate was around 80–87%. The total duration of larval development across the six instar stages was 16–18 days. Pupal duration varied from 8 to 14 days, with an average of 10.35 ± 0.26 days, and the pupal emergence rate ranged from 60–94%, with a mean of 80.25 ± 1.28%. Under laboratory conditions, the life cycle of males lasted 33–44 days, while that of females lasted 36–49 days (Marri *et al.*, 2023).

Damage symptoms: The fall armyworm (FAW) is a polyphagous pest known to infest 353 plant species across 76 plant families. While it attacks various plants, it primarily targets grasses, with maize being its preferred host. It can also cause economic damage to other cereals and millets. FAW inflicts harm on maize plants from seedling emergence to ear development stages (Suby *et al.*, 2020). The crop damage mainly results from larvae feeding on leaf tissue, and they may also bore into the growing point (bud, whorl, etc.), causing economic damage to the crop.

Distribution: Throughout India including Himalaya (Singh *et al.*, 2023b).

Natural enemies: Various predators, parasitoids and entomopathogens, have been documented to target *S. frugiperda*. This pest is susceptible to attacks from over 150 parasitoid and parasite species belonging to 14 families within the orders Hymenoptera, Diptera and Nematoda (Kumari *et al.*, 2021). Navik *et al.* (2021) observed a total of five parasitoid species parasitizing the eggs and larvae of *S. frugiperda* in the maize habitat. They also reported a higher rate of parasitism of *T. chilonis* on *S. frugiperda* eggs compared with *T. remus* in the Indian agroecosystem. Additionally, recent reports have also mentioned the presence of the egg-larval parasitoid *Chelonus formosanus* Sonan on *S. frugiperda* in maize from various regions in India (Firake and Behere, 2020; Gupta *et al.*, 2020).

5.3.1.7 Tuta absoluta *Meyrick*

Tuta absoluta is native to South America. In India, it was first reported from Maharashtra in 2014 (Shashank *et al.*, 2015; Singh *et al.*, 2020). Since then, it is fast spreading across India and has even reached the Himalaya (Rasheed *et al.*, 2018).

Scientific name: *Tuta absoluta* Meyrick.

Common name: Tomato pin worm, tomato leaf mining worm, South American tomato leaf miner.

Taxonomy: Phylum: Arthropoda; Class: Insecta; Order: Lepidoptera; Family: Gelechiidae; Genus: *Tuta*; Species: *T. absoluta* Meyrick.

Hosts: Brinjal, chili, pepper, potato, sweet pepper, tomato and tobacco.

Description: Adults of this species are predominantly active during the nighttime and tend to seek shelter between leaves or other suitable hiding spots during the daytime. *T. absoluta* are small moths, measuring approximately 5–7 mm in body length and having a wingspan of 10–14 mm. Its body colour is predominantly greyish or silvery grey, with dark patches on the forewings. The wing apex is fringed with speckled brown scales. The long, filliform antennae are marked with alternating bands of black and brown scales, while the labial palpi are noticeable, curved upwards and have a long and pointed apical segment. The correct identification is based on the closer examination of male genitalia; the male genitalia consist of valvae that are digitate and setose at the apex, with a prominent convex inner margin medially, broad uncus, U-shaped gnathos and a broad vinculum. The aedeagus is stout and elongated. On the other hand, the female genitalia possess a prominent signum on the corpus bursae (Shashank *et al.*, 2015). The adult female lays 250–280 eggs in her lifetime in 10–12 batches (Singh *et al.*, 2023a). The freshly laid eggs are small, smooth, shiny and oval or cylindrical in shape, initially pale creamy white in colour, which later turn yellow and black just before hatching (Halder *et al.*, 2019; Singh *et al.*, 2023a). According to Singh *et al.* (2023a), the larvae of *T. absoluta* undergo a colour transformation from whitish with a black head right after hatching to a greenish to pink hue, with an average length of 7.38 mm ranging between 7.10–7.65 mm. Sometimes freshly hatched larvae exhibit a light yellow to green colour, which darkens to a dark green shade as they mature and the pest can be identified by the characteristic dark band located at the posterior of the larva's head capsule (Ballal *et al.*, 2016).

During the pre-pupal stage, larvae appear lighter compared with their feeding larval stage (first to fourth instars). A distinguishing pink colouration develops on the dorsal surface of the last instar larvae. As the larva enters pre-pupa stage, it leaves its food source and proceeds to create silk cocoons either on the leaflets, stems or in the soil, depending on its habitat. However, if pupation takes place inside mined leaves or fruit, the pre-pupae do not construct cocoons (Singh *et al.*, 2023a). The pupae are octet in shape, initially green in colour, gradually transforming into chestnut brown to dark brown as they approach near-adult emergence (Singh *et al.*, 2023a). Pupation may occur inside a silken cocoon, which can be found either in the soil or on the leaf surface, within mines, or among plant debris (Ballal *et al.*, 2016).

Life cycle: *T. absoluta* undergoes four developmental stages during its life cycle, namely, egg, larva, pupa and adult. The adult female lays approximately 250–280 eggs throughout her lifetime, in 10–12 batches. Under laboratory conditions with a temperature of 26 ± 2°C and relative humidity of 75 ± 5%, the incubation period of eggs is 4.52 ± 0.68 days, and the hatching percentage ranges from 94–97% with an average of 95.48% (Singh *et al.*, 2023a). Similar results were reported by Shashank *et al.* (2015) on different host plants. Specifically, the fastest hatching occurred on tomato (3.75 days), followed by potato (5.10 days) and eggplant (6.10 days) (Singh *et al.*, 2023a). The average duration of the larval period is 12.33 days, with a range of 10–15 days. The individual larval instars (first, second, third and fourth) have average periods of 2.38, 3.95, 3.43 and 2.57 days, respectively (Singh *et al.*, 2023a). The average pupal period is 10.95 days, ranging from 10 to 12 days (Singh *et al.*, 2023a). This duration is in line with the findings of Desneux *et al.* (2010), who studied the biology of *T. absoluta* on tomato and reported a typical pupal duration of 9–11 days. Additionally, the results of Singh *et al.* (2023a) align with those of Polat *et al.* (2016), who observed a pupal period ranging from 7.52 to 20.62 days. The average development period of an adult is 12 days, with a range of 10–14 days. On average, the total lifespan of the moth is about 35.43 days, with a range spanning from 30 to 40 days (Singh *et al.*, 2023a). The tomato pinworm, *T. absoluta*, poses a potential threat due to its high reproductive capacity, shorter life cycle of 10–12 generations, and the reported ability to overwinter as an egg, pupa or adult (Sridhar *et al.*, 2014).

Damage symptoms: Larvae of *T. absoluta* feed on various plant parts such as leaves, buds, stalks and fruits. The larvae primarily consume the mesophyll content and soft parts of leaves and stems. Their feeding activity results in visible blotches on the leaf surface, galleries on leaves, and pinholes on fruits from the stalk end, filled with frass. The mines created by their feeding are visible on both sides of the leaf.

Distribution: Delhi, Uttarakhand, Himachal Pradesh, Meghalaya, Madhya Pradesh, Punjab, Andhra Pradesh, Telangana, Gujarat, Karnataka, Maharashtra, Rajasthan, Tamil Nadu, Uttar Pradesh, Chhattisgarh, Haryana.

Natural enemies: *Nesidiocoris tenuis* (Reuter) is a potential predator, predating on eggs and early instars of *T. absoluta* under field conditions (Ballal *et al.*, 2016). Four hymenopteran parasitoids were observed to be linked with *T. absoluta*, including *Trichogramma achaeae* Nagaraja and Nagarkatti, *Neochrysocharis formosa* (Westwood), *Habrobracon* sp. and *Goniozus* sp. *N. formosa* was the most abundant among these parasitoids (Ballal *et al.*, 2016).

5.3.2 Description of agriculturally important arthropods with potential to become invasive in the Indian Himalayan Region

5.3.2.1 Aceria *sp.*

Aceria guerreronis is already known to be invasive in India (Singh *et al.*, 2020). In the Indian Himalayan Region, *Aceria lycopersici* is emerging as a serious pest of tomato, chilli and capsicum (Paschapur *et al.*, 2022).

Scientific name: *Aceria* Kaeifer, 1944.

Common name: Eriophyid mite.

Taxonomy: Phylum: Arthropoda; Class: Arachnida; Order: Trombidiformes; Family: Eriophyidae; Genus: *Aceria*; Species: *Aceria* sp.

Hosts: Amaranthus (*Amaranthus tricolor*), ragweed agweed [*Ambrosia artemisiifolia* L. (Asteraceae)], olive, jasmine, mango, mulberry, Malvaceous plant hosts, garlic.

Description: The genus *Aceria* is classified under the family Eriophyidae, which represents about 85% of the superfamily Eriophyoideae. Eriophyids are extremely small mites, measuring less than 1 mm in length, typically ranging from 0.1 to 0.2 mm. They possess two pairs of legs located immediately behind the head. They are worm-like with elongated body, i.e. wedge-, cigar- or carrot-shaped and possess two pairs of legs immediately behind the head. At the caudal end, there is a ventral 'sucker' that enables the mite to attach to the substrate during feeding, moulting and other activities. Due to their short, piercing mouthparts, eriophyids are limited to feeding on the epidermal cells of their host plants (Grasswitz, 2012). To distinguish females from males, one needs to examine their external genitalia, which is located ventrally at the anterior of the opisthosoma (the portion of the mite body posterior to the legs). Females have a genital coverflap, a triangular plate hinged at the anterior margin, covering most of the genital area. In contrast, males have their genital area exposed except for a transverse ridge (Lindquist, 1996). Their eggs are oval, translucent and measure between 20–60 μm in diameter (Manson and Oldfield, 1996). Immature mites closely resemble adults but are smaller and lack external genitalia (Lindquist, 1996).

Life cycle: The development pattern is similar among all eriophyid mites but they often have variable and complex life cycles. In the simplest case, eggs are laid on leaves or in buds that subsequently go through two stages after hatching: larva (first instar) and nymph (second instar) stage before moulting into adults; a quiescent or resting stage typically precedes the two moults (Manson and Oldfield, 1996). Between each instar, the mites have 24-hour resting stages (Manson and Oldfield, 1996). The larva differs in size (smaller) than the nymph but otherwise exhibits a similar appearance (Lindquist, 1996). It is also reported that the life cycle of eriophyid mites consists of two nymphal instars, followed by a resting "pseudopupa" stage (Hull, 2014). The adult sex ratio of eriophyid mites is often biased towards females, with some cases reporting up to 95% of the population being female. Males transfer sperm indirectly by surrounding newly eclosed adult females with spermatophores, which are deposited on the leaf surface. In certain species, unfertilized eggs can develop into males. Females start laying eggs, approximately 50 each, within 1–7 days after emerging and have a lifespan of around 1 month. Under optimal conditions, this allows for the possibility of 6–8 generations per year. Some species of eriophyid mites exhibit two types of females: protogynes, the primary summer form, and deutogynes, an over-wintering form triggered by seasonal cues such as leaf-hardening and cooling temperatures. Deutogynes overwinter in protected sites like bark crevices and under bud-scales. They may be inseminated before entering their overwintering state but typically wait until the following spring to lay eggs (Manson and Oldfield, 1996).

Damage symptoms: Nearly all eriophyid mites exhibit host-specificity, with approximately 95% of species feeding exclusively on one host plant genus (Skoracka *et al.*, 2010). Eriophyid mite infestations and feeding behaviour cause various forms of plant injury such as russeting, gall formation, bronzing, browning, silvering or curling of leaves, along with deformed or stunted buds (Keifer, 1982). Furthermore, it is recently reported from India on Amaranthus causing crinkling, deformities and malformation of tender leaves, leading to a significant reduction in leaf size and overall stunting of the plants, shoots become more fibrous and crop yield is reduced (Sreekumar *et al.*, 2019). The presence of eriophyid mite (*Aceria guerreronis* Keifer) on coconut trees causes severe damage to nuts. Affected nuts may display oozing of gummy exudation from their surfaces, and uneven growth can lead to distortion and stunting of the nuts, ultimately reducing crop yield. In more severe cases, the nuts may become malformed, showing cracks and having a hardened husk (TNAU, 2015). Apart from their direct effects on plant growth, certain eriophyid mites have been identified as vectors of plant viruses such as wheat streak mosaic virus (Grasswitz, 2012). The tender leaves, buds and stem may also exhibit galls (Westphal and Manson, 1996).

Distribution: Himachal Pradesh, Uttarakhand, Tripura, Karnataka, Tamil Nadu, Kerala.

Natural enemies: Fungal pathogens such as *Hirsutella thompsonii* and *Verticillium lecanii* have shown great potential in effectively managing *Aceria guerreronis* Keifer (TNAU, 2015). Eriophyid mites have natural enemies that include predatory mites from different families, such as Phytoseiidae and Stigmaeidae, as well as the larval stage of certain hoverflies (family Syrphidae) and predatory midges (family Cecidomyiidae). Predatory hemipterans from the family Anthocoridae and certain species of coccinellid beetles also play a role in controlling these mites (Perring and McMurtry, 1996).

5.3.2.2 Tetranychus *sp.*

One species of the genus *Tetranychus* (*Tetranychus evansi*) is known to be globally invasive outside India (Navajas *et al.*, 2012). Various known and some unidentified species of this genus has been found to be a serious pest in the Indian Himalayan region. Among known species, *Tetranychus urticae* is the most widely reported from the Indian Himalayan region (Monica *et al.*, 2014; Premalatha *et al.*, 2016; Paschapur *et al.*, 2022) and has potential to become invasive.

Scientific name: *Tetranychus urticae* Koch.

Common name: Red spider mite, two spotted spider mite.

Taxonomy: Phylum: Arthropoda; Class: Insecta; Order: Trombidiformes; Family: Tetranychidae; Genus: *Tetranychus;* Species: *T. urticae*.

Hosts: Vegetable crops, fruit crops, viz. apple, bean, citrus, cucumber, pear, pepper, raspberry, strawberry, tomato, and flowering and ornamental plants.

Description: The adult female of the two-spotted spider mite is characterized by an oval-shaped body, measuring about 0.5 mm in length. Body colouration can vary, ranging from brown or orange-red to green, greenish-yellow, or almost translucent, with the latter being the most common. Females have an elliptical body with 12 pairs of dorsal setae (hairs). Overwintering females tend to be orange or orange-red, and their body contents (large dark spots) can often be seen through the transparent body wall. However, newly moulted mites may lack these spots as they are accumulations of body wastes. The adult males are also elliptical in shape, with the caudal end tapering and smaller in comparison to the female. The aedeagus has its knob axis parallel or forming a small angle with the shaft axis. Both adults generally have four pairs of legs, each adorned with long hairs (Fasulo and Denmark, 2016).

The eggs of the two-spotted spider mite are nearly transparent and round, measuring around 0.1–0.15 mm in diameter. When the eggs hatch, the first stage is called the larva, initially colourless but acquiring a yellowish or pinkish tint after feeding. The larva has three pairs of legs. Following the larval stage, there are two nymphal stages known as the protonymph and deutonymph, each with four pairs of prolegs. The duration of both the larval and nymphal stages depends on temperature and typically lasts between 1–3 days. Towards the end of each stage, there is a resting phase without feeding: the nymphochrysalis or protochrysalis between the larva and protonymph, the deutochrysalis between the two nymph stages, and the teliochrysalis between the deutonymph and adult stages (Sparks Jr and Riley, 2008).

When faced with shorter day lengths and cooler temperatures, *T. urticae* enters a diapause stage, with diapausing females or eggs serving as the primary overwintering form. During diapause, these mites suspend feeding and egg-laying activities, seeking refuge in tree bark crevices, clods of dirt and leaf litter (Tehri, 2014). It is important to note that specific species of spider mites may exhibit variations in their life cycle and development patterns.

Damage symptoms: Spider mite infestations can cause significant damage to plants, leading to various visible symptoms. Both nymphs and adults possess modified mouthparts, known as chelicerae, which allow them to pierce the epidermis of leaves and extract cell sap. This feeding behaviour results in characteristic symptoms such as light yellow spots, speckled appearance and stippling on the leaves. The mites primarily feed on the underside of leaves but can cover the entire leaf surface when populations are high. Their feeding activity damages spongy mesophyll cells and sometimes even the lowest palisade layer, leading to the destruction of chloroplasts and reduced photosynthesis activity. As a result, affected leaves may exhibit chlorotic spots at each feeding site, and prolonged infestations can cause leaf browning, wilting, deformity, desiccation and abscission. These detrimental effects on photosynthesis and overall plant health can result in stunted growth and reduced yield (Sandeepa *et al.*, 2017). Spider mites are typically more active and destructive during periods of strong sunlight and high temperatures, particularly in the hot and dry months of April to June (Sharma and Pati, 2012).

Distribution: Throughout India including Himalaya.

Natural enemies: Bacterial organisms, including *Yersinia entomophaga*, *Xenorhabdus* spp., *Pseudomonas entomophila*, *Burkholderia* spp., *Chromobacterium* spp., *Streptomyces* spp., *Saccharopolyspora* spp. and *Bacillus* spp., were reported to have potential in producing diverse metabolites that exhibit insecticidal properties and can help in control of *T. urticae* (Ruiu, 2015; Assouguem *et al.*, 2022). According to Osman *et al.* (2019) the fungal extracts of *Aspergillus melleus*, *Emericella nidulans*, *Chaetomium globosum* and *Aspergillus terreus* have shown effective control against female two-spotted spider mites and their eggs. Numerous predators, such as predatory mite species, viz. *Amblyseius swirskii* Athias-Henriot, *Typhlodromus rhenanoides* Athias Henriot, *Phytoseiulus persimilis* Athias-Henriot, *Typhlodromus phialatus* Athias Henriot, *Neoseiulus cucumeris* (Oudemans), *Neoseiulus stolidus* (Chaudhri), *Euseius scutalis* Athias Henriot and *Euseius stipulates* Athias Henriot, predatory gall midge species *Feltiella acarisuga* (Vallot), thrips species *Scolothrips longicornis* Priesner and coccinellid beetle *Stethorus punctillum* Weise, have been instrumental in effectively managing *Tetranychus urticae* and other spider mites (Assouguem *et al.*, 2022).

5.3.2.3 Pineus pini (*Macquart*)

Pineus pini is a well-documented invasive species which affects pine trees (*Pinus* sp., *Pinus patula*). It was reported from Nilgiri, Tamil Nadu in 1970 (Gupta *et al.*, 2019; Singh *et al.*, 2020). The Indian Himalayan Region is home to a large subtropical coniferous forest ecoregion consisting of pine trees as dominant species. Being invasive in nature, *Pineus pini* poses a threat to the Himalayan subtropical pine forests. Therefore, *Pineus pini* needs special consideration.

Scientific name: *Pineus pini* (Macquart).

Common name: Woolly pine aphid.

Hosts: *Pineus* spp.; *Pineus patula*

Taxonomy: Phylum: Arthropoda; Class: Insecta; Order: Hemiptera; Family: Adelgidae; Genus: *Pineus;* Species: *Pineus pini*.

Description: Adult *Pineus* sp. have a reddish-brown body that is typically concealed within a tangle of white wax threads. The wingless adult females are ovoid to turbinate in shape, ranging from 0.9 to 2.3 mm in length. Dorsally they have segmental chitinous plates covered with wax glands, which are of taxonomic significance. Adult *Pineus* sp. have an ovipositor, which is absent in the immature stages, and short, 1–3 segmented antennae which are often fused together. The adults exhibit greater fusion of dorsal plates, particularly on the head and anterior parts of the thorax. Winged adult female *Pineus* sp. resemble other winged aphids, with a clearly divided body into head, thorax and abdomen. They have well-developed legs and antennae, and their body length ranges from 1.0 to 1.9 mm. Unlike winged aphids, they possess an ovipositor similar to the apterous females. The winged adults have five segmented antennae, with large sensoria present on segments III, IV and V. They also have compound eyes, hyaline forewings with veins reddish in colour provide taxonomic insights.

Eggs of *P. pini* (Macquart), are laid within a nest of white wax fibres secreted by the wax pores of the adult female. Each egg is elongate-ovoid, approximately 0.2 mm long, and initially yellow in colour. As the embryo develops, the eggs gradually change their colour from brownish to red. The first-instar crawlers of *Pineus* sp. are very small, measuring between 0.2–0.4 mm in length, wingless, reddish-brown in colour and have an ovoid shape. These first-instar nymphs possess three pairs of legs, three segmented small setae and an elongated apical seta on the last antennal segment. Their eyes are represented by three ocelli on each side of the head (Carter, 1971).

The first-instar crawlers are small, reddish-brown in colour, wingless, ovoid in shape, ranging 0.2–0.4 mm in length. They have small legs with three-segmented setae. The last segment of their antennae bears a long apical seta, while their eyes are represented by three ocelli on each side of the head (Carter, 1971; CABI, 2021).

Life cycle: The life cycle of *Pineus pini* in Western Europe is anholocyclic, meaning it does not involve sexual reproduction. Throughout the year, colonies of *P. pini* can be found on the smaller stems and shoots of pine trees. There are typically three or more generations per year. The life cycle begins with the overwintering of immature forms from eggs that hatch in early October. These immature forms remain in their second or third instar until the following spring, when they become mature sistens. In March, the mature sistens start laying their eggs, with the egg clutches reaching their maximum size in the second half of April. Egg clutches are laid over a period of several weeks, with one egg being laid at a time. This results in significant variations in the hatching times of the first and last eggs that are laid. From these eggs, the spring generation of progredientes emerges. By the end of May, the progredientes develop into two distinct morphs: apterous progredientes and winged sexuparae. The apterous progredientes continue to feed and reproduce on the pine trees, while the winged sexuparae fly away. The flight period of the sexuparae extends from late May until the end of June. It is believed that most of the sexuparae settle on pine trees, although some may settle on fir (*Picea*) trees after an active flight period. This behaviour of settling on fir trees is considered a vestige of the holocycle.

Thus, the life cycle of *P. pini* in Western Europe involves overwintering of immature forms, egg laying by mature sistens in spring, hatching of eggs into progredientes, development into apterous progredientes and winged sexuparae, and the flight and settlement of sexuparae. This life cycle repeats itself multiple times within a year, with colonies consistently present on pine trees throughout the year (Influentialpoints.com).

Damage: The pine woolly aphid primarily feeds on the shoots of *Pinus* spp., and its infestation can result in tip dieback (Gupta *et al.*, 2019). The presence of these aphids leads to premature shedding of needles and a decrease in the length of infested needles, which ultimately reduces productivity and can cause up to 50% loss of growth increment (Day *et al.*, 2003). The severity of damage to trees is particularly pronounced in warm and dry conditions (Mailu *et al.*, 1978).

Distribution: Nilgiri, Tamil Nadu.

Natural enemies: The effectiveness of biological control methods in managing this pest varies. While there have been successful instances of controlling the pest through biological control in certain areas, there have also been cases where it has not been effective (Gupta *et al.*, 2019). It has been controlled by natural enemies such as predatory species of Anthocorid bugs, viz. *Tetraphleps raoi* Ghauri and *T. abdulghanii* Ghauri (Alfred Daniel *et al.*, 2020) and coccinellid beetle *Cheilomenes aurora* (Singh *et al.*, 2020). However, the control of *Pineus* aphids has been achieved through the successful introduction of natural predators such as *L. tapiae* and *L. nigraluna* (in USA), *Exochomus* spp. (in Kenya), coccinellid beetles (in Zimbabwe) and *Exochomus quadripustulatus* and *Diomus pumilo* (in Australia) (Kumari *et al.*, 2021).

5.3.2.4 Thrips parvispinus (*Karny*)

Thrips parvispinus is native to Thailand. It was reported from India in 2015 from Bengaluru (Tyagi *et al.*, 2015). It is fast spreading and high levels of infestations were recorded from Andhra Pradesh, Telangana and Karnataka (Basavaraj *et al.*, 2022; Timmanna *et al.*, 2022a). It has spread to central and north India, and heavy infestation of chilli crop has been reported from Madhya Pradesh, Chhattisgarh, Gujarat, Uttar Pradesh and Haryana recently (Lodaya *et al.*, 2022; Timmanna *et al.*, 2022b; Sethy *et al.*, 2023). Due to its fast spread from south to north India, *Thrips parvispinus* poses a threat to the Indian Himalayan Region too.

Scientific name: *Thrips parvispinus* (Karny).

Common name: Tobacco thrips, Taiwanese thrips and South East Asian Thrips.

Taxonomy: Phylum: Arthropoda; Class: Insecta; Order: Thysanoptera; Family: Thripidae; Genus: *Thrips*; Species: *T. parvispinus* (Karny).

Hosts: Beans, brinjal (eggplant), capsicum, chilli, green beans, papaya, pepper, potato, shallot, strawberry, tobacco, watermelon, Cucurbits belonging to different families, ornamentals, viz. *Anthurium*, *Chrysanthemum*, *Dahlia*, *Dipladenia*, *Gardenia*, *Ficus*, etc.

Description: Female and male thrips of *T. parvispinus* exhibit sexual dimorphism in size and colour. Females are approximately 1 mm in length and have a brown head and prothorax, while their meso- and metathorax are yellowish brown, and their abdomen is black. Their forewings are dark, with a light-coloured base. The third antennal segment and the base of the fourth and fifth segments are also light-coloured, appearing either yellow or white. Males are smaller, measuring about 0.6 mm in length, and are uniformly yellow in colour (Hutasoit *et al.*, 2017).

The legs are yellow, and segments III, IV and V of the antennae are also yellow. The forewing is brown with a pale base. The head is wider than it is long, and the small ocellar setae pair III are positioned on the front edges of the ocellar triangle. Postocular setae pairs I and III are slightly longer than ocellar setae pair III, while pair II is very small. The antennae have 7 segments, with segments III and IV constricted at the apex and featuring a forked sense cone. The seventh segment is small. The pronotum has 2 long setae pairs at the back angles and 3 pairs of setae along the posterior margin. The metanotum has median reticulations and weak internal markings, with long median setae positioned behind the front margin and lacking campaniform sensilla. The forewing has complete setae rows on the first and second veins, and the clavus has 5 marginal setae. Abdominal tergite II has 3 lateral marginal setae, while tergite VIII lacks a posteromarginal comb and only has a few microtrichia on the sides. The pleura and tergites of the abdomen do not have discal setae. Abdominal tergite II with 3 lateral marginal setae, VIII without posteromarginal comb, a few microtrichia laterally present. Abdominal sternite VII with median setae pair arising in front of posterior margin, II and VII without discal setae, III–VI with about 6–12 discal setae arranged in an irregular row. (DPPQ&S, 2022).

Life cycle: Thrips undergo a life cycle that includes the egg stage, followed by two larval stages, two pupal stages and ultimately, the adult stage. Thrips exhibit parthenogenesis, meaning they can reproduce without mating. Both mated and unmated females have the ability to lay eggs. The gravid females of *T. parvispinus* oviposit the eggs with the help of a saw-like ovipositor into leaves (Ananthakrishnan, 1984) and the incubation period lasts for 4–5 days. Larvae after emergence primarily feed on leaves and flowers during their developmental stage. Over a period of 4–5 days, they go through two moults, gradually maturing before entering the pupal stage. Pupal stage consists of two sub-stages: pre-pupal and pupal stage, which last for about 2–3 days each in the soil or leaf litter. Once the pupal stage is completed, thrips emerge as adults. The entire life cycle, from egg to adulthood, is typically completed within 13–14 days. *T. parvispinus* reproduction is sexual, with females laying an average of 15 eggs. Mated females have a lifespan of about 9 days, while adult males live for approximately 6 days under controlled greenhouse conditions on chilli plants (Hutasoit *et al.*, 2017).

Damage symptoms: *T. parvispinus* is a significant pest that causes direct damage to plants by feeding and laying eggs. They possess well-developed left mandibles, which they use to rasp and suck the sap from various plant parts. Additionally, thrips act as important vectors of Tospoviruses, transmitting them to economically important crops from different plant families worldwide (Marullo and Mound, 2002; Riley *et al.*, 2011).

Both the larval and adult stages cause harm by rasping and sucking plant sap. Adults predominantly colonize flowers and the underside of leaves, while larvae prefer the under-surface of the leaves. This specific preference for colonizing and feeding on leaves, particularly areas adjacent to veins, results in deep punctures and scratches on the underside of the leaves. As a consequence of scraping chlorophyll from the undersides and sucking cell sap, the corresponding portions on the upper side of leaves appear yellowish and blotchy, with reddish-brown discolouration visible on the under-surface. The damage caused by thrips leads to distorted leaf lamina with necrotic areas and yellow streaks. In severe infestations, new flesh on the leaves may dry up or become blighted.

On floral parts, thrips' feeding habits create brownish streaks on the petals due to scraping. Pollination can also be affected when thrips feed on pollen. The flowers may start drying and withering, ultimately impacting fruit set. Thrips' presence on floral parts can hinder the normal development of flowers and reduce the overall productivity of the plant (DPPQ&S, 2022).

Overall, thrips infestations on leaves and floral parts can have detrimental effects on plant health and reproduction, causing discolouration, deformation and reduced yield in severe cases. In papaya plants, thrips damage can result in secondary infections caused by the saprophytic fungus *Cladosporium* (Lim, 1989).

Distribution: Andhra Pradesh, Telangana, Karnataka, Madhya Pradesh, Haryana, Chhattisgarh, Gujarat. Fast spreading from south to north India.

Natural enemies: Two species of ladybird beetles, *Menochilus sexmaculatus* Fabricius and *Coccinella transversalis* Fabricius, along with the entomophagous fungus, *Lecanicillium lecanii* R. Zare & W. Gams, have been identified as potential natural enemies for controlling *T. parvispinus*. Prabaningrum *et al.* (2008) found that using *M. sexmaculatus* and *L. lecanii* was highly effective in managing *T. parvispinus* and also led to increased crop yield. Additionally, microbial biopesticides such as *Pseudomonas fluorescence* NBAIR-PFDWD or *Bacillus albus* NBAIR-BATP have proven to be successful in suppressing *T. parvispinus* when applied on flowers and fruits, as reported by Sridhar *et al.* (2021).

5.4 Conclusion

Arthropods account for a major proportion of invasive alien species occurring both globally and in India. They incur a huge economic cost to the country. The invasive arthropods known to thrive under warm tropical climates are now making a shift towards temperate regions due to climate change. This is the reason why many invasive arthropod species are frequently being reported from the Himalaya. This altitudinal shift has drastic consequences on the hill agricultural system and the local biodiversity. Anthropogenic disturbances such as overexploitation of forest resources, commercialization of eco-sensitive areas and land use patterns are destabilizing the environmental equilibrium of the Himalaya. This is evident from the recent rise in landslides, heavy floods and other disasters. The rise in numbers of outbreaks of invasive arthropod pest species is not an exception. It would not be wrong if the issue of invasive alien species is considered a side effect of globalization. A global problem like this needs a global solution. It is not only a threat to biodiversity but also to global food security. Combating the issue of invasive alien species outbreaks needs international regulations and its proper implementation. Countries need to take proper measures to check entry of species that are known to be invasive or have the potential to become invasive. As already discussed, we have information about only a few arthropod species which are invasive in the Indian Himalayan region. Owing to a limited number of studies in the Himalayan region, it may be possible that many instances of the occurrence of invasive arthropod species are underreported until a massive outbreak happens. Therefore, an extensive, well-planned and focused research in the Himalayan region is urgently needed.

References

Abbas, M.S.T., Abuo-Zeid, N.A. and Megahed, M.M. (1993) On the natural enemies of the potato tuber moth *Phthorimaea operculella* in Egypt. *Egyptian Journal of Agricultural Research* 71, 943–950.

Alexander, J.M., Lembrechts, J.J., Cavieres, L.A., Daehler, C., Haider, S. *et al.* (2016) Plant invasions into mountains and alpine ecosystems: Current status and future challenges. *Alpine Botany* 126, 89–103.

Alfred Daniel, J., Ashok, K., Pavithran, S. and Ranjith, M. (2020) A review on invasive insect pests in India and their predators and parasitoids. *Journal of Experimental Zoology India* 23, 987–1006.

Ananthakrishnan, T.N. (1984) *Bioecology of Thrips*. Indira Publishing House, Michigan, USA.

Asante, S.K. (1997) Natural enemies of the woolly apple aphid, *Eriosoma lanigerum* (Hausmann) (Hemiptera: Aphididae): A review of the world literature. *Plant Protection Quarterly* 12, 166–172.

Assouguem, A., Kara, M., Ramzi, A., Annemer, S., Kowalczyk, A. *et al.* (2022) Evaluation of the effect of four bioactive compounds in combination with chemical product against two spider mites *Tetranychus urticae* and *Eutetranychus orientalis* (Acari: Tetranychidae). *Evidence-Based Complementary and Alternative Medicine* 2022, 2004623.

Ballal, C.R., Gupta, A., Mohan, M., Lalitha, Y. and Verghese, A. (2016) The new invasive pest *Tuta absoluta* (Meyrick) (Lepidoptera: Gelechiidae) in India and its natural enemies along with evaluation of Trichogrammatids for its biological control. *Current Science* 110, 2155–2159.

Ballal, R., Singh, S.P., Jalali, S.K. and Kumar, P. (1989) Parasitising efficiency of *Chelonus blackburni* Cam. and *Copidosoma koehleri* Blanch. singly and in combination against the potato tuber moth. *Journal of Biological Control* 3, 4–6.

Bang, A., Cuthbert, R.N., Haubrock, P.J., Fernandez, R.D., Moodley, D. *et al.* (2022) Massive economic costs of biological invasions despite widespread knowledge gaps: A dual setback for India. *Biological Invasions* 24, 2017–2039.

Basavaraj, K., Sreenivas, A.G., Prasad, P.B. and Rachana, R.R. (2022) First report of invasive thrips, *Thrips parvispinus* (Karny) (Thysanoptera: Thripidae) infesting chilli, *Capsicum annuum* L. in Kalaburagi, Karnataka, India. *Journal of Experimental Zoology India* 25, 191–194.

Bates, B.C., Kundzewicz, Z.W., Wu, S. and Palutikof, J.P. (2008) *Climate Change and Water*. Intergovernmental Panel on Climate Change, IPCC Secretariat, Geneva, Switzerland.

Bedford, I.D., Briddon, R.W., Markham, P.G. and Brown, J.K. (1993) A new species of *Bemisia* or biotype of *Bemisia tabaci* (Genn) as a future pest of European agriculture. *Plant Health and the European Single Market, BCPC Monograph* 54, 381–386.

Bedford, I.D., Briddon, R.W., Brown, J.K., Rosell, R.C. and Markham, P.G. (1994) Geminivirus transmission and biological characterisation of *Bemisia tabaci* (Gennadius) biotypes from different geographic regions. *Annals of Applied Biology* 125, 311–325.

Bellows Jr, T.S., Perring, T.M., Gill, R.J. and Headrick, D.H. (1994) Description of a species of *Bemisia* (Homoptera: Aleyrodidae). *Annals of the Entomological Society of America* 87, 195–206.

Bharti, H., Bharti, M. and Pfeiffer, M. (2016) Ants as bioindicators of ecosystem health in Shivalik mountains of Himalaya: Assessment of species diversity and invasive species. *Asian Myrmecology* 8, 65–79.

Blackman, R.L. and Eastop, V.F. (1994) *Aphids on the World's Trees: An Identification and Information Guide*. CABI, Wallingford, UK.

Blackman, R.L. and Eastop, V.F. (2000) *Aphids on the world's crops: An identification and information guide*, 2nd edn. John Wiley & Sons, New York, USA.

Boykin, L.M. and De Barro, P.J. (2014) A practical guide to identifying members of the *Bemisia tabaci* species complex: And other morphologically identical species. *Frontiers in Ecology and Evolution* 2, 45. DOI: 10.3389/fevo.2014.00045.

Brookes, H.M. and Hudson, N.M. (1968) The identification and distribution of *Quadraspidiotus* species (Homoptera: Diaspididae) on pome and stone fruits in Australia. *Australian Journal of Entomology* 7, 90–100.

CABI (2019) *Spodoptera frugiperda* (fall armyworm). Available at: https://doi.org/10.1079/cabicompendium.29810 (accessed 28 July 2023).

CABI (2021) *Pineus pini* (pine woolly aphid). Available at: https://doi.org/10.1079/cabicompendium.41326 (accessed 29 July 2023).

Carter, C.I. (1971) Conifer woolly aphids (Aelgidae) in Britain. *Forestry Commission Bulletin* 42.

Chandra, K., Gupta, D., Gopi, K.C., Tripathy, B. and Kumar, V. (2018) *Faunal Diversity of Indian Himalaya*. Zoological Survey of India, Kolkata, India.

Chauhan, U. and Verma, L.R. (1985) Adult eclosion and mating behaviour of potato tubermoth, *Phthorimaea operculella* Zeller. *Journal of the Indian Potato Association* 12, 148–157.

Cherian, M.G. and Basheer, M. (1939) *Tetrastichus sokolowskii* Kurdj (family eulophidæ)—A larval parasite of *Plutella maculipennis* in South India. *Proceedings of Indian Academy of Sciences–Section B* 9, 87–98. DOI: 10.1007/BF03050540.

Convention of Biological Diversity (2002) Convention on Biological Diversity. Available at: https://www.cbd.int/ (accessed 4 July 2023).

Costa, H.S. and Brown, J.K. (1991) Variation in biological characteristics and esterase patterns among populations of *Bemisia tabaci*, and the association of one population with silverleaf symptom induction. *Entomologia Experimentalis et Applicate* 61, 211–219. DOI: 10.1111/j.1570-7458.1991.tb01553.x.

Crosby, A. (2015) *Ecological Imperialism: The Biological Expansion of Europe, 900–1900*, 2nd edn. Canto Classics. Cambridge University Press, Cambridge, UK. DOI: 10.1017/CBO9781316424032.

Crumb, S.E. (1956) *The Larvae of the Phalaenidae (Tech. Bull. No. 1135)*. United States Department of Agriculture, Washington DC, USA.

Darwin, C. (1839) *Journal of researches into the natural history and geology of the countries visited during the voyage of H.M.S. Beagle round the world* (Vol. 3). Henry Colburn, London.

Das, G.P. and Raman, K.V. (1994) Alternate hosts of the potato tuber moth, *Phthorimaea operculella* (Zeller). *Crop Protection* 13, 83–86.

Day, R.K., Kairo, M.T., Abraham, Y.J., Kfir, R., Murphy, S.T. *et al*. (2003) Biological control of homopteran pests of conifers in Africa. In: *Biological Control in IPM Systems in Africa*. CABI, Wallingford, UK, pp. 101–112.

De Barro, P.J., Trueman, J.W.H. and Frohlich, D.R. (2005) *Bemisia argentifolii* is a race of *B. tabaci* (Hemiptera: Aleyrodidae): The molecular genetic differentiation of *B. tabaci* populations around the world. *Bulletin of Entomological Research* 95, 193–203.

De Barro, P.J., Liu, S.S., Boykin, L.M. and Dinsdale, A.B. (2011) *Bemisia tabaci*: A statement of species status. *Annual Review of Entomology* 56, 1–19.

Desneux, N., Wajnberg, E., Wyckhuys, K.A., Burgio, G., Arpaia, S. *et al*. (2010) Biological invasion of European tomato crops by *Tuta absoluta*: Ecology, geographic expansion and prospects for biological control. *Journal of Pest Science* 83, 197–215.

Dinsdale, A., Cook, L., Riginos, C., Buckley, Y.M. and Barro, P.D. (2010) Refined global analysis of *Bemisia tabaci* (Hemiptera: Sternorrhyncha: Aleyrodoidea: Aleyrodidae) mitochondrial cytochrome oxidase 1 to identify species level genetic boundaries. *Annals of the Entomological Society of America* 103, 196–208.

DPPQ&S (2022) Management strategies for invasive thrips (*Thrips parvispinus*) in Chilli (ad-hoc). Available at: https://ppqs.gov.in/sites/default/files/adhoc_management_strategies_on_thrips_parvispinus.pdf (accessed 29 July 2023).

Early, R., Bradley, B.A., Dukes, J.S., Lawler, J.J., Olden, J.D. *et al*. (2016) Global threats from invasive alien species in the twenty-first century and national response capacities. *Nature Communications* 7, 12485.

Encarnação, J., Teodósio, M.A. and Morais, P. (2021) Citizen science and biological invasions: A review. *Frontiers in Environmental Science* 8, 602980.

Fasulo, T.R. and Denmark, H.A. (2016) Two-spotted spider mite, *Tetranychus urticae* Koch (Arachnida: Acari: Tetranychidae). University of Florida, IFAS Extension. Available at: https://doi.org/10.32473/edis-in307-2000 (accessed 29 July 2023).

Fenemore, P.G. (1977) Oviposition of potato tuber moth, *Phthorimaea operculella* Zell. (Lepidoptera: Gelechiidae); fecundity in relation to mated state, age, and pupal weight. *New Zealand Journal of Zoology* 4, 187–191.

Fenemore, P.G. (1978) Oviposition of potato tuber moth, *Phthorimaea operculella* Zell. (Lepidoptera: Gelechildae); the physical nature of the oviposition substrate. *New Zealand Journal of Zoology* 5, 591–599.

Field, C.B. and Barros, V.R. (2014) *Climate Change 2014 – Impacts, Adaptation and Vulnerability: Regional Aspects*. Cambridge University Press, Cambridge, UK.

Firake, D.M. and Behere, G.T. (2020) Natural mortality of invasive fall armyworm, *Spodoptera frugiperda* (JE smith) (Lepidoptera: Noctuidae) in maize agroecosystems of northeast India. *Biological Control* 148, 104303.

Fluiter, H.I. (1948) Over het optreden van virginopare en sexupare gevleugelden bij *Eriosoma lanigerum* Hausmann op e appelboompjes in Kassen. *Entomologische Berichten* 12, 209–211.

Fotedar, M. R. and Kapur, A. P. (1943) First record of sexual forms and oviparous reproduction of woolly apple aphids, *Eriosoma lanigerum* Hausmann from Kashmir, India. *Current Science* 12, 84–85.

Gangwar, R. K. and Gangwar, C. (2018) Lifecycle, distribution, nature of damage and economic importance of whitefly, *Bemisia tabaci* (Gennadius). *Acta Scientific Agriculture* 2, 36–39.

Gautam, D.C. (1981) Biology and chromosomal behaviour of woolly apple aphids (*Eriosoma lanigerum* Hausmann). MPhil Thesis, Himachal Pradesh University, Shimla, India.

Gautam, D. C. and Verma, L. R. (1983). Life history of sexuparae and sexual morphs of woolly apple aphid (*Eriosoma lanigerum* Hausmann). *Proceedings: Animal Sciences* 92, 247–251.

Gautam, M. P., Singh, H., Kumar, S., Kumar, V., Singh, G. and Singh, S. N. (2018) Diamondback moth, *Plutella xylostella* (Linnaeus) (Insecta: Lepidoptera: Plutellidae) a major insect of cabbage in India: A review. *Journal of Entomology and Zoology Studies* 6, 1394–1399.

Ghosh, L.K. (2008) *Handbook on Hemipteran Pests in India*. Zoological Survey of India, Kolkata, India.

Golan, K. (2020) Contributions to the knowledge of the San José scale (Hemiptera, Coccomorpha, Diaspididae) in Poland. *Polish Journal of Entomology* 89, 7–19.

Grasswitz, T.R. (2012) Biology and control of eriophyid mites with a case study of Aceria sp. on New Mexico olive (Forestiera pubescens Nutt. var. pubescens). In: Haase, D. L., Pinto, J. R., Riley, L. E.

(tech. coords.) *National Proceedings: Forest and Conservation Nursery Associations. RMRS-P-68.* USDA Forest Service, Rocky Mountain Research Station, Fort Collins, CO, pp. 86–89.

Gubbaiah, G. and Thontadarya, T.S. (1977) Bionomics of the potato tuberworm, *Gnorimoschema operculella* Zeller (Lepidoptera: Gelechiidae) in Karnataka. *Mysore Journal of Agricultural Sciences* 11, 380–386.

Gupta, A., Lalitha, Y., Varshney, R., Shylesha, A.N. and Achterberg, C. (2020) *Chelonus formosanus* Sonan (Hymenoptera: Braconidae), an egg-larval parasitoid of the invasive pest *Spodoptera frugiperda* (JE smith) (Lepidoptera: Noctuidae) amenable to laboratory mass production in India. *Journal of Entomology and Zoology Studies* 8, 1521–1524.

Gupta, N., Verma, S.C., Sharma, P.L., Thakur, M., Sharma, P. *et al.* (2019) Status of invasive insect pests of India and their natural enemies. *Journal of Entomology and Zoology Studies* 7, 482–489.

Halder, J., Kushwaha, D., Rai, A.B. and Singh, B. (2019) Biology and biorational management of *Tuta absoluta* (Meyrick) (Lepidoptera: Gelechiidae): A global challenge to tomato production. *Proceedings of the Zoological Society* 72, 107–110.

Hamid, I., Wani, A.R., Sofi, M.A., Pathania, S.S., Hamid, F. *et al.* (2022) Severity of San Jose scale (*Quadraspidiotus perniciosus* comstock) on apple tree. *The Pharma Innovation Journal* SP-11, 04–09.

Harika, G., Dhurua, S., Sreesandhya, N., Suresh, M. and Govind Rao, S. (2019) Biology of diamondback moth, *Plutella xylostella* (Lepidoptera: Plutellidae) of cauliflower under laboratory condition. *International Journal of Current Microbiology and Applied Sciences* 8, 866–873.

He, J.S., Wang, Z., Wang, X., Schmid, B., Zuo, W. *et al.* (2006) A test of the generality of leaf trait relationships on the Tibetan Plateau. *New Phytologist* 170, 835–848.

Higgins, S.I. and Richardson, D.M. (2014) Invasive plants have broader physiological niches. *Proceedings of the National Academy of Sciences USA* 111, 10610–10614.

Hull, R. (2014) Plant to plant movement. *Plant Virology* 5, 669–751.

Hutasoit, R.T., Triwidodo, H. and Anwar, R. (2017) Biology and demographic statistic of *Thrips Parvispinus* Karny (Thysanoptera: Thripidae) in chili pepper (*Capsicum annuum* Linnaeus). *Indonesian Journal of Entomology* 14, 107–116.

Influentialpoints.com (2023) *Pineus pini*: Pine woolly aphid, Scots pine adelgid. Available at: https://influentialpoints.com/Gallery/Pineus_pini_Pine_woolly_aphid_Scots_pine_adelgid.htm#identi (accessed 29 July 2023).

Jayarathnam, K. (1977) Studies on the population dynamics of the diamondback moth, *Plutella xylostella* (Linnaeus) (Lepidoptera: Yponomeutidae), and crop loss due to the pest in cabbage. Doctoral Thesis, University of Agricultural Sciences, GKVK, Bangalore, India.

Keifer, H.H. (1982) *An Illustrated Guide to Plant Abnormalities Caused by Eriophyid Mites in North America (no.573)*. Agricultural Research Service, United States Department of Agriculture, Washington DC, USA.

Körner, C. (2004) Mountain biodiversity, its causes and function. *AMBIO: A Journal of the Human Environment* 33, 11–17.

Kumari, S., Singh, B., Dhanda, S. and Dumra, N. (2021) Biological control of invasive pests in india. In: Kumar, V., Kumar, S., Kamboj, N., Payum, T. and Kumar, P. *et al* (eds) *Biological Diversity: Current Status and Conservation Policies*. Agro Environ Media, Haridwar, India, pp. 315–328.

Lal, K.B. and Singh, R.N. (1947) Seasonal history and field ecology of woolly aphids in Kumaon hills. *Indian Journal of Agricultural Sciences* 17, 211–218.

Lamsal, P., Kumar, L., Aryal, A. and Atreya, K. (2018) Invasive alien plant species dynamics in the Himalayan region under climate change. *Ambio* 47, 697–710.

Levy, R. and Habeck, D.H. (1976) Descriptions of the larvae of *Spodoptera sunia* and *S. latifascia* with a key to the mature *Spodoptera larvae* of the eastern United States (Lepidoptera: Noctuidae). *Annals of the Entomological Society of America* 69, 585–588.

Lim, W.H. (1989) Bunchy and malformed top of papaya cv. Eksotika caused by *Thrips parvispinus* and *Cladosporium oxysporum*. *Mardi Research Bulletin-Journal* 17, 200–207.

Lindquist, E.E. (1996) External anatomy and systematics 1.1.1. External anatomy and notation of structures. In: *World Crop Pests*, Vol. 6. Elsevier, Amsterdam, The Netherlands, pp. 3–31.

Lodaya, J.P., Suthar, M., Patel, H.C., Sisodiya, D.B., Acharya, R.R. *et al.* (2022) Status of invasive species of thrips, *Thrips parvispinus* (Karny) infesting chilli grown in Middle Gujarat. *The Pharma Innovation Journal* 11, 1298–1302.

Luginbill, P. (1928) *The fall army worm (Tech. Bull. No. 34)*. United States Department of Agriculture, Washington DC, USA, p. 34.

Ma, X., Xu, J., Luo, Y., Prasad Aggarwal, S. and Li, J. (2009) Response of hydrological processes to land-cover and climate changes in Kejie watershed, south-west China. *Hydrological Processes: An International Journal* 23, 1179–1191.
Mailu, A.M., Khamala, C.P.M. and Rose, D.J.W. (1978) Evaluation of pine woolly aphid damage to *Pinus patula* and its effect on yield in Kenya. *East African Agricultural and Forestry Journal* 43, 259–265.
Manson, D.C.M. and Oldfield, G.N. (1996) Biology and ecology 1.4.1. Life forms, deuterogyny, diapause and seasonal development. In: *World Crop Pests Vol. 6*. Elsevier, Amsterdam, The Netherlands, pp. 173–183.
Marri, D., Mensah, S.A., Kotey, D.A., Abraham, J., Billah, M.K. *et al.* (2023) Basic developmental characteristics of the fall armyworm, *Spodoptera frugiperda* (JE Smith) (Lepidoptera: Noctuidae), reared under laboratory conditions. *Psyche: A Journal of Entomology* 2023, 6917316.
Marullo, R. and Mound, L.A. (2002) Thrips and tospviruses. In: *Proceedings of the 7th International Symposium on Thysanoptera*, Australian National Insect Collection, Canberra, Australia.
Maurya, R.P. (2011) Natural enemies of apple woolly aphid, *Eriosoma lanigerum* Hausmann and their population dynamics in a mid hill apple orchard. *Indian Journal of Entomology* 73, 15–18.
McAuslane, H.J. and Smith, H.A. (2015) Sweetpotato whitefly B biotype, *Bemisia tabaci* (Gennadius) (Insecta: Hemiptera: Aleyrodidae). *IFAS Extension, University of Florida, Gainesville, Florida, USA* 1–9.
Molinari, F. (1986) Woolly apple aphid (*Eriosoma lanigerum* Hausm.). *Informatore Fitopatologico* 36, 35–37.
Monica, V.L., Kumar, A., Chand, H., Paswan, S. and Kumar, S. (2014) Population dynamics of *Tetranychus urticae* Koch on brinjal crop under north Bihar conditions. *Pest Management in Horticultural Ecosystems* 20, 47–49.
Myers, N. (1993) Environmental refugees in a globally warmed world. *Bioscience* 43, 752–761.
Navajas, M., de Moraes, G.J., Auger, P. and Migeon, A. (2012) Review of the invasion of *Tetranychus evansi*: Biology, colonization pathways, potential expansion and prospects for biological control. *Experimental and Applied Acarology* 59, 43–65.
Navik, O., Shylesha, A.N., Patil, J., Venkatesan, T., Lalitha, Y. *et al.* (2021) Damage, distribution and natural enemies of invasive fall armyworm *Spodoptera frugiperda* (JE smith) under rainfed maize in Karnataka, India. *Crop Protection* 143, 105536.
Nijssen, B., O'Donnell, G.M., Hamlet, A.F. and Lettenmaier, D.P. (2001) Hydrologic sensitivity of global rivers to climate change. *Climatic Change* 50, 143–175.
Osman, M.E., Abo Elnasr, A.A., Nawar, M.A. and Hefnawy, G.A. (2019). Myco-metabolites as biological control agents against the two-spotted spider mite, *Tetranychus urticae* Koch (Acari: Tetranychidae). *Egyptian Journal of Biological Pest Control*, 29(1), 64.
Parmesan, C. (2006) Ecological and evolutionary responses to recent climate change. *Annual Review of Ecology, Evolution, and Systematics* 37, 637–669.
Paschapur, A., Subbanna, A.R.N.S., Gupta, J., Parihar, M. and Mishra, K.K. (2022) Insect pest scenario in Uttarakhand Himalayas, India, under changing climatic conditions. *International Journal of Biometeorology* 66, 1445–1460.
Pathak, R., Negi, V.S., Rawal, R.S. and Bhatt, I.D. (2019) Alien plant invasion in the Indian Himalayan Region: State of knowledge and research priorities. *Biodiversity and Conservation* 28, 3073–3102.
Perring, T.M. and McMurtry, J.A. (1996) Other predatory arthropods. In: *World Crop Pests Vol. 6*. Elsevier, Amsterdam, the Netherlands, pp. 471–479.
Pimentel, D., McNair, S., Janecka, J., Wightman, J., Simmonds, C. *et al.* (2001) Economic and environmental threats of alien plant, animal, and microbe invasions. *Agriculture, Ecosystems & Environment* 84, 1–20.
Polat, B., Özpınar, A. and Şahin, A.K. (2016) Studies of selected biological parameters of tomato leafminer *Tuta absoluta* (Meyrick) (Lepidoptera: Gelechiidae) under natural conditions. *Phytoparasitica* 44, 195–202.
Prabaningrum, L., Moekasan, T.K., Udiarto, B.K., Den Belder, E. and Elings, A. (2008) Integrated pest management on sweet pepper in Indonesia: Biological control and control thresholds for thrips. *Acta Horticulturae* 767, 201–210.
Premalatha, K., Chinniah, C., Ravikumar, A. and Parthiban, P. (2016) Seasonal incidence and influence of weather factors on population dynamics of two spotted spider mite, tetranychus urticae koch in tomato eco-system. *Annals of Plant Protection Sciences* 24, 232–235.

Rahman, K.A. and Khan, A.W. (1941) Biology and control of woolly apple aphis, eriosoma lanigerum hausmann. *Indian Journal of Agricultural Sciences* 11, 265–278.

Rao, V.P., Ghani, M.A., Sankaran, T. and Mathur, K.C. (1971) *A review of the biological control of insects and other pests in South East Asia and the Pacific region*. Technical Communication No. 6, Commonwealth Institute of Biological Control, Trinidad, West Indies. Commonwealth Agricultural Bureaux Farnham Royal, England, UK.

Rasheed, V.A., Rao, S.K., Babu, T.R., Murali, T., Krishna, B.V. *et al*. (2018) Age specific fecundity life table of South American tomato leaf miner, *Tuta absoluta* (Meyrick) on tomato. *Journal of Entomology and Zoology Studies* 6, 546–549.

Rawat, U.S. and Pawar, A.D. (1991) Records of natural enemies of San Jose scale *Quadraspidiotus perniciousus* (Comstock) from Himachal Pradesh. *Journal of Biological Control* 5, 119–120.

Renault, D., Angulo, E., Cuthbert, R.N., Haubrock, P.J., Capinha, C. *et al*. (2022) The magnitude, diversity, and distribution of the economic costs of invasive terrestrial invertebrates worldwide. *Science of the Total Environment* 835, 155391.

Reshi, Z.A. and Khuroo, A.A. (2012) Alien plant invasions in India: Current status and management challenges. *Proceedings of the National Academy of Sciences, India Section B: Biological Sciences* 82, 305–312.

Riley, D.G., Joseph, S.V., Srinivasan, R. and Diffie, S. (2011) Thrips vectors of tospoviruses. *Journal of Integrated Pest Management* 2, I1–I10.

Rodgers, W.A. and Panwar, H.S. (1988) *Planning a wildlife protected areas network in India. Department of Environment, Forests, and Wildlife/Wildlife Institute of India report volume 1*. Wildlife Institute of India, Dehradun, India.

Rodgers, W.A., Panwar, H.S. and Mathur, V.B. (2002) *Wildlife Protected Area Network. India: A Review (Executive Summary)*. Wildlife Institute of India, Dehradun, India.

Ruiu, L. (2015). Insect pathogenic bacteria in integrated pest management. *Insects*, 6(2), 352–367.

Salick, J., Anderson, D., Woo, J., Sherman, R., Cili, N. *et al*. (2004) Tibetan ethnobotany and gradient analyses: Menri (medicine mountains), Eastern Himalaya. In: *Proceedings of Bridging scales and epistemologies: linking local knowledge and global science in multi-scale assessments*, The Millennium Ecosystem Assessment, Alexandria, Egypt.

Sandeepa, A.R., Pradeep, S. and Sridhara, S. (2017) Management of *Tetranychus urticae* Koch (Acarina: Tetranychidae) on carnation under polyhouse condition. *Journal of Global Biosciences* 6, 4771–4775.

Sandilyan, S. (2020) Invasive Species of India Report. National Biodiversity Authority, Ministry of Environment Forests and Climate Change Government of India. Available at: http://nbaindia.org/uploaded/pdf/Iaslist.pdf (accessed 4 July 2023).

Saxena, J.D., Rai, S., Srivastava, K.M. and Sinha, S.R. (1989) Resistance in the field populations of the diamondback moth to some commonly used synthetic pyrethroids. *Indian Journal of Entomology* 51, 265–268.

Sekar, K.C. (2012) Invasive alien plants of Indian Himalayan Region – Diversity and implication. *American Journal of Plant Sciences* 3, 177–184.

Sethy, S., Narayana, S., Twinkle, V.A. and Sunda, S. (2023) First report of invasive thrips, *Thrips parvispinus* (Karny) infestation on chilli from Eastern part of India. *Ecology, Environment, and Conservation* 29(January Suppl. Issue), S498–S503.

Sharma, A. and Pati, P.K. (2012) First record of the carmine spider mite, *Tetranychus urticae*, infesting *Withania somnifera* in India. *Journal of Insect Science* 12, 50.

Sharma, P.L. and Gavkare, O. (2017) New distributional record of invasive pest *Tuta absoluta* (Meyrick) in North-Western Himalayan Region of India. *National Academy Science Letters* 40, 217–220.

Shashank, P.R., Chandrashekar, K., Meshram, N.M. and Sreedevi, K. (2015) Occurrence of *Tuta absoluta* (Lepidoptera: Gelechiidae) an invasive pest from India. *Indian Journal of Entomology* 77, 323–329.

Simberloff, D. (2000) Global climate change and introduced species in United States forests. *Science of the Total Environment* 262, 253–261.

Singh, S.P. (1989) *Achievements of AICRP on Biological Control. Tech. Bull. 2*. Biological Control Centre, NCIPM, Faridabad.

Singh, S.S. (2003) Feeding potential of *Coccinella septempunctata* Linn. on woolly apple aphid, *Eriosoma lanigerum* Hausmann. *Journal of Entomological Research* 27, 213–216.

Singh, B., Mahla, M.K., Babu, R.S., Jain, D., Vyas, A.K. *et al*. (2023a) First report of the South American tomato pinworm, *Tuta absoluta* (Meyrick), as an invasive pest in Udaipur Region of Southern Rajasthan in India. *Bioinvasions Records* 12, 117–123. DOI: 10.3391/bir.2023.12.1.10.

Singh, S., Raghuraman, M., Keerthi, M.C., Das, A., Kar, S.K. *et al.* (2023b) Occurrence, distribution, damage potential, and farmers' perception on fall armyworm, *Spodoptera frugiperda* (JE smith): Evidence from the Eastern Himalayan Region. *Sustainability* 15, 5681.

Singh, S., Sharma, J.H., Udikeri, A. and Ansari, H. (2020) Invasive insects in India. In: El-Shafie, H. (ed.) *Invasive Species – Introduction Pathways, Economic Impact, and Possible Management Options*. IntechOpen, London, UK, pp. 49–64.

Skoracka, A., Smith, L., Oldfield, G., Cristofaro, M. and Amrine, J.W. (2010) Host-plant specificity and specialization in eriophyoid mites and their importance for the use of eriophyoid mites as biocontrol agents of weeds. *Experimental and Applied Acarology* 51, 93–113.

Sparks, A.N. (1979) A review of the biology of the fall armyworm. *Florida Entomologist* 62, 82–87.

Sparks Jr, A.S. and Riley, D.J. (2008) Spider mites. BugwoodWiki. Available at: https://wiki.bugwood.org/Spider_mites (accessed 29 July 2023).

Spear, D., Foxcroft, L.C., Bezuidenhout, H. and McGeoch, M.A. (2013) Human population density explains alien species richness in protected areas. *Biological Conservation* 159, 137–147.

Sreekumar, K.M., Srinivasa, N. and Sivamoorthy, R. (2019) First report of incidence of eriophyid mite *Aceria* sp. on Amaranthus. *Entomon* 44, 309–310.

Sridhar, V., Chakravarthy, A.K., Asokan, R., Vinesh, L.S., Rebijith, K.B. *et al.* (2014) New record of the invasive south american tomato leaf miner, *Tuta absoluta* (Meyrick) (Lepidoptera: Gelechiidae) in India. *Pest Management in Horticultural Ecosystems* 20, 148–154.

Sridhar, V., Chandana, P.S. and Rachana, R. (2021) Global status of *Thrips parvispinus*. *The Journal of Research PJTSAU* 49, 1–11.

Suby, S.B., Soujanya, P.L., Yadava, P., Patil, J., Subaharan, K. *et al.* (2020) Invasion of fall armyworm (*Spodoptera frugiperda*) in India: Nature, distribution, management and potential impact. *Current Science* 119, 44–51.

Tehri, K. (2014) A review on reproductive strategies in two spotted spider mite, *Tetranychus urticae* Koch 1836 (Acari: Tetranychidae). *Journal of Entomology and Zoology Studies* 2, 35–39.

Tewari, V.P., Verma, R.K. and Von Gadow, K. (2017) Climate change effects in the Western Himalayan ecosystems of India: Evidence and strategies. *Forest Ecosystems* 4, 1–9.

Thakur, J.R. (1970) Biology of woolly apple aphid, *Eriosoma lanigerum* (Hausmann) (Homoptera: Aphididae) and its control by the soil as well as foliar application with particular stress on timings. MSc Thesis, Panjab University, Chandigarh, India.

Thakur, J.R. and Dogra, G.S. (1980) Woolly apple aphid, *Eriosoma lanigerum*, research in India. *International Journal of Pest Management* 26, 8–12.

Tilman, D. (1999) The ecological consequences of changes in biodiversity: A search for general principles. *Ecology* 80, 1455–1474.

Timmanna, H., Baradevanal, G., Surpur, S., Raghavendra, D., Doddachowdappa, S. *et al.* (2022a) Diagnosis and potential invasion risk of *Thrips parvispinus* under current and future climate change scenarios. *PeerJ* 10, e13868.

Timmanna, H., Prashantha, C., Shashank, P.R., Nigam, V.D. and Birla, N. (2022b) Occurrence and spread of invasive thrips *Thrips parvispinus* (Karny) in North India. *Indian Journal of Entomology* e22965, 160–163.

TNAU (2015) Agritech Portal Crop Protection: Coconut Eriophyid: *Aceria guerreronis*. Available at: https://agritech.tnau.ac.in/crop_protection/coconut/crop_prot_crop_insect_oil_coconut_2.html (accessed 29 July 2023).

Tyagi, K., Kumar, V., Singha, D. and Chakraborty, R. (2015) Morphological and DNA barcoding evidence for invasive pest thrips, *Thrips parvispinus* (Thripidae: Thysanoptera), newly recorded from India. *Journal of Insect Science* 15, 105.

Westphal, E. and Manson, D.C.M. (1996) 1.4.6. feeding effects on host plants: Gall formation and other distortions. In: *World Crop Pests Vol. 6*. Elsevier, Amsterdam, the Netherlands, pp. 231–242.

Yurkonis, K.A., Meiners, S.J. and Wachholder, B.E. (2005) Invasion impacts diversity through altered community dynamics. *Journal of Ecology* 93, 1053–1061.

6 Invasive Ants in the Tropics

Arindam Kumar[1], Sudipa Das[2], Sourav Chakrabarty[2], Usha Nandini A.[3] and Akshay Kumar Chakravarthy[1*]
[1]Division of Crop Protection, ICAR–Indian Institute of Horticultural Research, Bengaluru, India; [2]National Pusa Collection, Division of Entomology, ICAR–Indian Agricultural Research Institute, New Delhi, India; [3]Society for Science and Technology Applications (SSTA), Bengaluru, India

Abstract
Invasive ant species represent a growing global threat to biodiversity, ecosystem stability, agriculture, infrastructure, and public health. Originating from localized regions, many of these ants, such as *Solenopsis invicta* (red imported fire ant) and *Anoplolepis gracilipes* (yellow crazy ant), have expanded rapidly due to anthropogenic activities, particularly global trade and transportation. The ecological consequences are extensive, including altered nutrient cycling, impaired pollination, collapse of native invertebrate populations, and degradation of symbiotic plant-insect networks. Additionally, invasive ants cause substantial economic damage through crop destruction, livestock injuries, and disruption of electrical systems. Despite existing control measures including chemical, biological, and regulatory the management remains fragmented and often ineffective due to logistical, financial, and legislative constraints. This chapter highlights the necessity of integrated pest management (IPM), early detection, international collaboration, and public awareness to combat ant invasions. Emphasis is placed on adopting multidisciplinary and transboundary approaches to mitigate their ecological and socio-economic impacts, thereby enhancing global biosecurity and sustainability efforts.

Keywords: Ants, invasions, displacement, impacts, mitigation

6.1 Introduction

Accelerated by climate change and the increased global movement of people and goods, biological invasions of insect pests, commonly defined as "the geographical expansion of a species into regions where it was not previously found" (Vermeij, 1996), are contributing to large-scale ecological changes. The movement of invasive species beyond their natural range poses significant risks to biodiversity, ecosystem function, agricultural sustainability, cropping systems, pesticide use, and both regional and national economies, as well as public health. The surge in global trade, coupled with advancements in agricultural practices and transportation, has led to a rise in accidental invasions (Perrings *et al.*, 2005; Mack *et al.*, 2000). Climate change may further exacerbate biological invasions in profound ways. Numerous studies have attempted to predict the identity, impact, and distribution of potential invasive species (Mooney and Cleland, 2001). Reflecting growing interest in the geographic spread of various insect species, many recent studies have employed bioclimatic and other modelling approaches

*Corresponding author: chakravarthyakshay@gmail.com

 Biological Invasions in the Tropics: Monitoring, Mechanism and Management (eds Akshay Kumar Chakravarthy and Pathour Rajendra Shashank)
DOI: 10.1079/9781836991977.0006

to: (a) identify suitable regions and forecast the potential spread of invasive species (De Meyer *et al.*, 2010) and (b) examine the possible effects of climate warming on the expansion of invasive species ranges (Gutierrez *et al.*, 2010). In addition, invasive species offer valuable models for studying adaptation processes, as well as the evolution of life-history, behavioural, and physiological traits in unfamiliar environments (Diamantidis *et al.*, 2011). Differences in life-history traits among populations of invasive species may indicate varying potential for invasion across these populations. However, the success of an invasion can also be influenced by intra- and interspecific competition, as well as geographic, climatic, and environmental barriers (Diamantidis *et al.*, 2011).

Invasive species face numerous challenges before they can successfully establish themselves in new environments. Successful biological invasion requires four distinct population processes: (a) arrival, (b) establishment, (c) naturalization, and (d) spread (Liebhold and Tobin, 2008; Seebens *et al.*, 2017). Arrival refers to the movement of individuals into an uninhabited area, with most invasive insect species' dispersal ability playing a central role. Human activities such as travel and trade facilitate the long-distance spread of these species. Establishment occurs when the immigrant population is able to maintain itself in the newly colonized area through local reproduction. The success of establishment depends on habitat suitability, the size of the founding population, the frequency of founder events, demographic traits, and interactions with the local biotic environment. There is a minimum population density required to ensure persistence in an area. The relationship between individual fitness and population size is described as the "Allee Effect", which is influenced by the rate at which propagules arrive (Liebhold and Tobin, 2008). Naturalization, often overlooked or included under establishment, is a protracted process where self-sustaining populations are formed and adapted to the new environment. This adaptation includes genetic changes driven by different selection pressures compared with the species' original habitat (Richardson *et al.*, 2000). Spread is the process by which an invasive species extends its range from its current habitat into new, unoccupied regions (Liebhold and Tobin, 2008). In this phase, the invasive species interacts with native species, and ecological factors such as host fruit quality and availability can drive further adaptation and evolutionary changes (Aluja *et al.*, 2014).

The dispersion of an established population can follow simple diffusion models or more complex models that incorporate environmental variables, such as topography and human movement. Consequently, successful invasion is a multifaceted biological process constrained by climatic, geographic, biological, and ecological barriers (Harmon *et al.*, 2009). Climate change can influence both abiotic and biotic factors, complicating the prediction of invasion patterns in a changing climate. Several forecast models have been employed to know the potential distribution of invasive species (Gallien *et al.*, 2010). However, accurately predicting the expansion of insect species' geographic ranges remains challenging. Phenomenological habitat suitability models are used to predict the potential distribution of non-native species, while mechanistic models aim to understand invasion dynamics after the species has been introduced. Habitat suitability models rely on native distribution patterns to identify suitable areas for establishment, assuming that underlying processes can be indirectly captured at large spatial scales. Mechanistic models focus on demographic responses, spread dynamics, and the effects of invasive species on native plant and animal communities. Hybrid models are considered to offer a more comprehensive approach to predicting and understanding invasion patterns (Gallien *et al.*, 2010).

Management activities are associated with each phase of the invasion process for notorious invasive insect species. To prevent arrival, international quarantines and thorough inspection protocols must be enforced (Liebhold and Tobin, 2008). Early detection networks and eradication programmes are implemented for incipient and established populations, while domestic quarantines and barrier zones are applied to manage populations in the final phase of spread. As naturalization is often overlooked, there are no specific management measures targeting this phase. The complex nature of invasion events makes decision making difficult, and managing populations is both risky and costly. Understanding the invasion biology of insect pests expanding their geographic range is crucial for effective pest management at national, regional, and international levels.

Ants, classified under the order Hymenoptera, are one of the most dominant and extensively studied insect groups in the terrestrial ecosystem (Siddiqui *et al.*, 2019). Historically, ants were not widespread across continents before human intervention. However, in the Anthropocene era, biological invasions, including those of ants (also known as tramp ants), have become a critical aspect of global environmental change. Invasive ant species represent a significant ecological threat as they aggressively outcompete and prey upon indigenous ant populations. Notable examples include the African big-headed ant (*Pheidole megacephala*), Argentine ant (*Linepithema humile*), and red imported fire ant (*Solenopsis invicta*), which have spread globally due to human activity. These species pose particular challenges for identification, as most measure less than 5 mm in length, exhibit minimal morphological differentiation, and frequently establish cryptic nesting sites rather than conspicuous mounds (FAO, 2018). A defining characteristic of many invasive ant species is the loss of intraspecific aggression, which facilitates the formation of super colonies. This social cohesion is considered one of the key factors behind their invasion success (Abbott, 2006). Through such behavioural adaptations, invasive ants are able to outcompete native species and establish themselves as dominant forces in ecosystems.

Various strategies have been employed to control the spread of exotic ants, including the use of chemical treatments and the introduction of natural predators (Hoffmann *et al.*, 2016). However, the effectiveness of these methods has been limited, highlighting the need for more comprehensive solutions. Researchers suggest that a coordinated, international approach is essential for protecting regional biodiversity (Paini *et al.*, 2016). Currently, countries have adopted varying strategies to address the issue. For example, China and Australia have heavily invested in monitoring and eradication efforts, utilizing a combination of chemical treatments and biological control methods (Wang *et al.*, 2015).

Despite these efforts, existing control measures often lack coordination and are limited in scope, complicating the fight against exotic ant invasions (Simberloff *et al.*, 2013). To effectively combat these invasions, it is imperative to develop integrated control strategies. This requires a multidisciplinary approach that fosters collaboration among researchers, policymakers, and local communities.

6.2 Dynamics of Invasion in the Context of Global Climate Change

Global climate change is an undeniable phenomenon, with its effects increasingly evident across the globe. The Fourth Assessment Report by the United Nations Intergovernmental Panel on Climate Change (IPCC) states that 11 out of the 12 years from 1995 to 2006 were among the warmest since 1850, with a global temperature rise of approximately 0.74°C over the past century. Projections indicate a rise in global temperatures by around 0.2°C per decade over the next two decades. This warming is primarily driven by the increased concentration of greenhouse gases (GHGs) and aerosols, which disrupt Earth's energy balance. Carbon dioxide (CO_2), the most significant anthropogenic greenhouse gas, has seen its emissions rise by approximately 80% between 1970 and 2004 (IPCC, 2007). The main contributors to the surge in CO_2 levels are the burning of fossil fuels and land-use changes, leading to concentrations of CO_2, which, at 379 ppm in 2005, are far above the natural range of 180–300 ppm seen over the last 650,000 years (IPCC, 2007). Understanding the impacts of shifting climate patterns on species' life history, phenology, and distribution is crucial for predicting how invasive species, particularly ants, may alter ecosystems. In a warming world, biological invasions are expected to become more pronounced. Climate change may result in (a) the expansion of ant species' geographic ranges into regions previously unsuitable for their establishment, (b) the shifting of ant species' ranges towards the poles (Kwon *et al.*, 2016), (c) an increase in the introduction of new ant species, and (d) population explosions in certain invasive ants, potentially causing significant ecological disruptions. Elevated temperatures are expected to impact all four stages of invasion. For example, higher temperatures may enhance the dispersal capabilities of invasive ants, as flight thresholds are reached earlier in the season (Parr and Bishop, 2022). Temperature increases could also affect key life-history traits such as development rates, lifespan, and fecundity, thereby improving the persistence of invasive ant populations during establishment and naturalization in new areas. Climate change will likely alter both abiotic and biotic constraints, complicating the prediction of invasion patterns and making management decisions more difficult, which could result in costly mistakes. Numerous modelling techniques have been employed to predict the potential spread of invasive species (Estay *et al.*, 2009; Ladanyi and Horvath, 2010). However, accurately forecasting the geographic expansion of invasive ant species remains challenging due to the complex interaction of social, biological, ecological, and climatic factors. Nonetheless, many models accounting for global warming suggest that temperate habitats will become more suitable for tropical ant species (Kwon *et al.*, 2016). Future research should focus on understanding the differential invasive potential among ant species, emphasizing biotypes with different life-history traits, and including data on life-history plasticity from wild populations to improve the accuracy of these predictions. While the spread of invasive ant species is not always solely dependent on climatic factors (Parr and Bishop, 2022), climate-driven range expansions are more likely for aggressive species. The ability of invasive ants to adapt to diverse environments, coupled with high levels of plasticity in their life-history traits, will support their geographic range expansion. Enhanced cold tolerance and thermal plasticity (the capacity to adjust thermal tolerance within a generation) may allow ants to colonize more marginal habitats, further aiding their global spread (Parr and Bishop, 2022). Invasive potential can vary significantly among biotypes of the same species (Diamantidis *et al.*, 2011), with populations adapted to marginal environments more likely to succeed in new habitats. The geographic expansion of invasive ants is driven by a complex interaction of climatic, social, and economic factors. Global warming is expected to ease several barriers to establishment, such as winter survival and host availability over extended periods. Even slight temperature increases can have a significant impact on the survival rates of tropical ant species in temperate regions. Moreover, climate change encourages the cultivation of tropical and subtropical fruits in cooler areas, extending the fruiting season and providing more opportunities for fruit-feeding ants. Increased human mobility and global trade, especially in fresh produce, will likely enhance the frequency of new tropical ant species reaching temperate zones. These factors, in conjunction with the ants' inherent adaptability and plasticity in life-history traits, will likely accelerate their spread, increasing the frequency and success of ant invasions in new environments.

6.3 Introduction and Establishment of Invasive Ants

With the increasing global population and expanding trade, opportunities for the transportation of species are expected to rise in the coming decades (Sardain *et al.*, 2019). Due to their small size and ability to thrive in close association with humans, many ant species are capable of "hitchhiking" in various commodities (Suhr *et al.*, 2019). Moreover, some ant species can establish a new population from just a single mated queen, or in certain cases, even from workers and brood, which increases the likelihood of successful establishment. Once introduced to a new area, an invasive ant species must first adapt to the local abiotic conditions to establish itself before it can spread. Species distribution models for 15 invasive ant species, based on 19 bioclimatic variables and incorporating six future climate change scenarios, predict a 6.3–35.8% increase in suitable habitats for five species, while eight species may experience a 2.6–64.3% decrease in suitable areas, with two species showing little net change (Bertelsmeier *et al.*, 2015). By 2080, it is anticipated that 37% of land in biodiversity hotspots will become suitable for one or more invasive ant species, compared with just 14.6% in other regions of the world (Bertelsmeier *et al.*, 2015). Additionally, reduced transit times or changes in abiotic filters could improve the survival rates of more species or populations, enabling alien species that are not yet invasive, or those confined to their native range, to become emerging priority invaders. While predictions about future distributions under climate change carry significant uncertainties, they collectively suggest that invasive ants will continue to pose a threat to biodiversity and human livelihoods (Table 6.1).

6.4 Features of Invasive Ants

Most invasive ant species belong to the subfamilies Myrmicinae, Formicinae, and Dolichoderinae. These ants are primarily native to Africa, Asia, and South America, and are typically omnivorous, with polymorphic worker castes. Interestingly, the climate of the regions into which these species have been introduced often resemble the environmental conditions of their native habitats. For example, *Linepithema humile* (Mayr, 1868), originally from the tropical and subtropical regions of Argentina, has successfully expanded into areas with mild temperate climates (Suarez *et al.*, 2000). Similarly, tropical species such as *Pheidole megacephala* (Fabricius, 1793) and *Anoplolepis gracilipes* Smith, 1857 have established themselves in new regions that mirror their native climatic conditions.

Invasive ants often display a high degree of phylogenetic diversity, especially in tropical regions. Within their colonies, worker ants may exhibit monomorphism, dimorphism, or polymorphism, depending on the species. In newly introduced habitats, many invasive ant populations tend to form unicolonial structures having large, cooperative networks of nests without territorial boundaries. Unicoloniality enables ants to achieve local dominance and high population densities (Hölldobler and Wilson, 1977). However, the social organization of some species, such as fire ants, is more complex (Holway *et al.*, 1998).

Another characteristic feature of invasive ants is their predominantly omnivorous diet. They feed on a wide range of food sources, including small carrion, invertebrates, and both insect and plant exudates (Hölldobler and Wilson, 1977). In tropical regions, invasive ants exhibit variable foraging behaviour influenced by local environmental conditions, availability of food resources, and the presence of other ant species in the area. These factors collectively shape their ecological impact and adaptability across different habitats.

6.5 Economic Impacts of Invasive Ant Species

Ants are diverse and economically significant insects, playing key roles in biological, behavioural, molecular, and evolutionary research (Ješovnik *et al.*, 2011; Thakur *et al.*, 2023). These insects are integral to ecosystem processes, particularly in seed dispersal, where they help in distributing and establishing plant species (Luo *et al.*, 2023). As ecosystem engineers, ants contribute to soil aeration through their burrowing activities, enhancing soil structure, and they participate in nutrient cycling by scavenging and decomposing organic matter (Folgarait, 1998; Perfecto and Philpott, 2023). In some cases, invasive ant species may even accelerate nutrient turnover. Many grass species in fire-prone grazing lands rely heavily on ants for survival (Fischer *et al.*, 2003). The introduction of invasive ants can significantly alter plant community dynamics and influence ecological succession.

In addition to their ecological roles, ants also function as natural biocontrol agents. Some invasive ant species prey on other pests, providing an effective form of biological control that helps to regulate harmful organisms (Rabitsch *et al.*, 2024). For example, the Asian weaver ant, *Oecophylla smaragdina* Fabricius has proven to be an effective biocontrol agent, controlling up to 50 different pest species (Peng and Christian, 2009; Offenberg, 2014). The interactions between ants, plants, and other organisms highlight the importance of ant diversity as an indicator of ecosystem health (Juárez-Juárez *et al.*, 2023). Ants fulfil multiple ecological functions, including scavenging, pollination, and herbivory, showcasing their complex and varied roles in nature (Costa *et al.*, 2023; Rabitsch *et al.*, 2024).

Despite the benefits that some invasive ants may offer, such as promoting nutrient cycling or acting as pest controllers, their negative impacts often outweigh these advantages. Invasive ant species, such as the yellow crazy ant and white-footed ant, can severely disrupt native plant reproduction by deterring legitimate pollinators, as seen in the Seychelles (Blüthgen *et al.*, 2017). The persistence of these ants in hindering pollination efforts underscores the

Table 6.1. Summary of major invasive ant species present globally. (Author's own table)

Genus and Species	Common Name	Origin	Invasive Areas	References
Subfamily: Dolichoderinae				
Linepithema humile (Mayr, 1868)	Argentine ant	Argentina, Brazil	Australia, New Zealand, Vanuatu, Hawaii, Japan, South Korea, Indonesia, Philippines, Malaysia, Vietnam	Clark, 1941; Don, 2007; Van Schagen *et al.*, 2021
Ochetellus glaber. (Mayr, 1862)	Black household ant	Australia	New Zealand, Solomon Islands, Hawaii, China, Japan, Borneo, Philippines	Don, 2007
Tapinoma melanocephalum (Fabricius, 1793)	Ghost ant	Indo-Malay	Australia, China, Japan, South Korea, India	Taylor, 1987; Wetterer, 2009a; Gruber *et al.*, 2017; Lee *et al.*, 2017
Technomyrmex albipes Smith, 1861	White-footed house ant	Indo-Pacific	Australia, Hawaii, China, Korea, Cambodia, Laos, Vietnam, Philippines, Malaysia	Taylor, 1987; Gruber *et al.*, 2017
Subfamily: Formicinae				
Anoplolepis gracilipes (F. Smith, 1857)	Yellow crazy ant	Indo-Malay	Australia, Hawaii, China, Japan, India	Taylor, 1987; Wetterer, 2005; Wetterer, 2009b; Gruber *et al.*, 2017
Paratrechina longicornis (Latrielle, 1802)	Black crazy ant	Indo-Malay	Australia, Hawaii, China, Japan, India	Taylor, 1987; Wetterer, 2008; Bharti *et al.*, 2016; Gruber *et al.*, 2017
Subfamily: Myrmicinae				
Monomorium floricola (Jerdon, 1851)	Bicoloured trailing ant	Indo-Malay	Australia, China, Japan, Southeast Asian countries	Taylor, 1987; Wetterer, 2008; Wetterer, 2010; Terayama *et al.*, 2014; Gruber *et al.*, 2017
Monomorium pharaonis (Linnaeus, 1758)	Pharaoh's ant	Indo-Malay, Africa, India	Australia, New Zealand, China, Japan, India	Taylor, 1987; Don, 2007; Terayama *et al.*, 2014; Gruber *et al.*, 2017
Pheidole megacephala (Fabricius, 1793)	African big-headed ant	Afrotropic	Australia, New Zealand, India	Taylor, 1987; Krushelnycky *et al.*, 2005; Don, 2007; Terayama *et al.*, 2014; Gruber *et al.*, 2017

Continued

Table 6.1. Continued

Genus and Species	Common Name	Origin	Invasive Areas	References
Solenopsis geminata (Fabricius, 1804)	Tropical fire ant	Neotropic	Australia, Hawaii, China, Japan, India	Hoffmann and O'Connor, 2004; Wetterer, 2011; Terayama *et al.*, 2014; Gruber *et al.*, 2017
Solenopsis invicta Buren, 1972	Red imported fire ant	Neotropic	Australia, China, Japan, South Korea	Wylie *et al.*, 2020
Tetramorium bicarinatum (Nylander, 1846)	Bicoloured pennant ant	Indo-Malay	Australia, New Zealand, China, Japan, India	Herben, 2005; Don, 2007; Terayama *et al.*, 2014; Gruber *et al.*, 2017
Trichomyrmex destructor (Jerdon, 1851)	Singapore ant	Palearctic	Australia, China, Japan, India	Herben, 2005; Terayama *et al.*, 2014; Gruber *et al.*, 2017
Wasmannia auropunctata (Roger, 1863)	Electric ant, little fire ant	Neotropic	Australia, American Samoa, French Polynesia, Guam, Federated States of Micronesia, New Caledonia, Papua New Guinea, Solomon Islands, Vanuatu, Wallis and Futuna, Hawaii	Wetterer and Porter, 2003

challenges faced in ecological restoration and the broader impacts on plant–pollinator relationships (Costa *et al.*, 2023).

Ants are one of the most studied insect groups and the most prevalent terrestrial insect species in ecosystems (Herben *et al.*, 2005; Siddiqui *et al.*, 2019). Prior to human intervention, ants were relatively limited in their spread across continents. However, through global trade and human activities, ants have dispersed far beyond their native ranges, becoming some of the most invasive species worldwide (Suarez *et al.*, 2005; Ingram *et al.*, 2006). Islands, in particular, are highly susceptible to ant invasions due to the lack of natural predators and environmental defences (Reimer, 2021). These invasions can have profound effects on native fauna, often leading to imbalances in local ecosystems. For example, a study by Rajesh *et al.* (2022) investigating ant communities in South Indian sacred groves found that urbanization significantly reduced native ant abundance while favouring invasive species like the yellow crazy ant (*Anoplolepis gracilipes*) and longhorn crazy ant (*Paratrechina longicornis*), though overall ant species richness remained unchanged across urbanization gradients; sacred groves with higher habitat diversity (Normalized Difference Vegetation Index – NDVI) showed lower *A. gracilipes* abundance, suggesting ecosystem integrity helps resist invasions, and these culturally protected forest fragments maintained distinct ant communities even in urbanized areas, highlighting their importance as biodiversity refuges despite invasive species pressures (Rajesh *et al.*, 2022).

The success of invasive ants can largely be attributed to their small size, polygyny (multiple queens), omnivorous diet, and unicolonial behaviour, which allows them to thrive in a variety of disturbed habitats (Passera, 2021; Masse *et al.*, 2023). Their adaptability enables them to survive in a wide range of environments, further contributing to their invasive potential.

As these ants establish large, competitive colonies in new areas, they cause significant ecological and economic damage. One of the most notable features of invasive ants is the transition from a multi-colonial to a unicolonial social structure, which enhances their foraging efficiency and facilitates their spread (Eyer *et al.*, 2021).

6.6 Potential Adverse Impacts of Invasive Ant Species

Invasive ant species pose significant threats to ecosystems, agriculture, and human health, leading to a variety of detrimental effects. This section examines the harmful impacts of invasive ants, focusing on their effects on local flora, fauna, human well-being, and the environment. One of the major concerns associated with invasive ants is their impact on local vegetation and essential crops. Native plant species are crucial for maintaining ecosystem balance, and the introduction of non-native ants can threaten their survival (Siddiqui *et al.*, 2021). These ants often damage plants by feeding on seeds or by burrowing into plant stems, which weakens their structure. They may also consume developing leaves and seedlings. For instance, the red imported fire ant (RIFA) has been known to cause substantial damage to crops by attacking their roots and leaves. Such aggressive feeding behaviour can result in reduced crop yields and poor-quality harvests. A study found that RIFA infestations led to a 35% decrease in sweet potato production (Adams, 2019). In addition to attacking crops, fire ants rely heavily on plant seeds for nutrition. In regions such as mulberry plantations in southern China, plant seeds made up nearly 50% of the ants' diet (Zhang *et al.*, 2015). RIFA has also been reported to damage over 50 different crops, including potatoes, citrus, soybeans, and sunflowers, posing a significant threat to agricultural economies worldwide (Wylie and Janssen-May, 2017).

The presence of invasive ants can also harm the local fauna, including a wide range of pets and livestock such as cattle, goats, horses, and poultry (Wylie and Janssen-May, 2017; Wang *et al.*, 2019). Research has shown that RIFA infestations have been a major concern for poultry facilities in regions such as Texas, where 94% of poultry farms reported issues with fire ants. The economic impact of these invasions is significant, particularly for sectors such as livestock farming. For example, the chicken industry of Australia, valued at $2.6 billion, faces considerable challenges from fire ants (Wylie and Janssen-May, 2017). Several studies have also shown that invasive ants harm native species. For example, RIFA has been found to negatively affect the density of small mammals, such as pygmy mice (*Baiomys* sp.) (Allen *et al.*, 2004). These ants can even harm birds and their young in both direct and indirect ways (Xiong *et al.*, 2019; Chen *et al.*, 2022). Additionally, fire ants disrupt honeybee colonies by stealing stored honey and larvae, affecting pollination and floral resource availability (Chen *et al.*, 2019; Wang *et al.*, 2020). Overall, invasive ants cause substantial ecological and economic damage, disrupting ecosystems and threatening biodiversity.

In addition to their environmental impacts, invasive ants pose serious risks to human health. Their stings can cause intense pain, allergic reactions, and even anaphylactic shock in susceptible individuals. In some cases, the stings of these ants can lead to blindness due to eye damage (Xu *et al.*, 2012; Wang *et al.*, 2020; Siddiqui *et al.*, 2021). Red fire ant stings, for instance, can cause swelling, nausea, and, in severe cases, chest pain, respiratory failure, and death (Siddiqui *et al.*, 2021). In the USA, studies have found that 55% of people stung by fire ants experienced multiple stings, and in some cases, these stings led to fatal allergic reactions (Klotz *et al.*, 2005; Bilo and Bonifazi, 2009). The risk of stings is widespread, with affected populations in infested areas encountering these ants both indoors and outdoors. Research has shown that 50–89% of residents in affected regions report being stung each year (de Shazo *et al.*, 2013). In southern China, a study found that 8.5% of people in certain villages had been stung, and some experienced severe allergic reactions (Zhang *et al.*, 2007). These findings highlight the significant health risks posed by invasive ants and the need for effective management strategies to minimize these dangers.

Invasive ants also cause damage to infrastructure, particularly electrical systems. These ants are known to gnaw on wires, which can lead to short circuits, fires, and other electrical malfunctions. In some regions, fire ants have been responsible for power outages and damage to communication equipment by nesting in transformers and power plants (Wang *et al.*, 2020). In areas such as North Carolina and Florida, fire ants have weakened asphalt roads, leading to the formation of potholes, which can be costly to repair (Banks *et al.*, 1990). In China's Guangdong province, RIFA infestations in power plants and transformer stations have caused significant damage to electrical equipment (Luo, 2005; Siddiqui *et al.*, 2021). These ants can disrupt power supplies by damaging critical infrastructure, resulting in both economic losses and potential safety hazards.

The success of invasive ants can be attributed to their ability to reproduce rapidly and adapt to various habitats. Each queen ant can lay a large number of eggs annually, and some species have multiple queens within a single colony, enabling exponential population growth (Tschinkel, 2019). The rapid spread of these ants is facilitated by their ability to be transported by birds, animals, and human activities. In regions like China, the absence of natural predators that regulate ant populations allows invasive species to proliferate unchecked (Chen and Fadamiro, 2018). Understanding the mechanisms behind their rapid spread and reproductive success is essential for developing effective control measures.

The environmental impact of invasive ants is profound, as they often outcompete and displace native ant populations and other invertebrates, leading to imbalances in ecosystem dynamics and biodiversity loss. Invasive ants can prey on native species and compete for resources, disrupting food webs and ecological processes (Moloney and Vanderwoude, 2002; Wylie and Janssen-May, 2017). In addition, native ants that share symbiotic relationships with butterfly species, such as the Illidge's ant-blue (*Acrodipsas illidgei*), face disruption due to fire ant invasions (Wylie and Janssen-May, 2017). Addressing the spread of invasive ants is crucial to protecting native species and maintaining ecosystem stability. Effective management strategies, including monitoring programmes and prevention measures, are essential to mitigate the environmental and societal risks associated with these pests.

For example, in the Indian Sunderbans, one of the largest mangrove ecosystems in the world, the spread of invasive ant species has posed a significant threat to local biodiversity and agricultural lands. A study by Bakra (2024) investigated the distribution of invasive ant species across five different locations within the region, including Sagar Island, Bakkhali, Patharpratima, Gosaba, and Hingalganj, during both pre-monsoon and post-monsoon seasons. Using honey baits, pitfall traps, and hand collection techniques, the study found that seven invasive ant species, comprising 22.847% of the total ant population, were present in the Sunderbans. The species identified included *Paratrechina longicornis*, *Tapinoma melanocephalum*, *Monomorium floricola*, *Technomyrmex albipes*, *Solenopsis geminata*, *Trichomyrmex destructor*, and *Anoplolepis gracilipes*, with *P. longicornis* being the most abundant. These invasive species outcompeted native ants, including *Camponotus compressus* and *Monomorium indicum*, particularly in agricultural areas such as Sagar Island, highlighting the ecological and economic risks posed by their presence. The study emphasizes the importance of early detection and rapid management of invasive species to protect both native biodiversity and agricultural productivity in such sensitive ecosystems (Bakra, 2024).

6.7 Challenges and Approaches in Managing Invasive Ant Species

The spread of invasive ant species poses significant challenges worldwide, affecting ecosystems, agriculture, and human health. Addressing these threats requires comprehensive strategies that consider a variety of constraints faced by decision makers, such as limited resources, public perception, technical difficulties, and the need for international collaboration.

A primary challenge in preventing ant invasions is lack of sufficient resources. The costs associated with research, education, prevention strategies, and the implementation of control measures are considerable. For instance, eradicating invasive ant species can be extremely expensive, with estimates suggesting that controlling some species in the USA could cost over a billion dollars (Xu *et al.*, 2022). Governments often face competing priorities, where the needs of citizens, such as healthcare and infrastructure, are given precedence over the resources allocated for controlling invasive species. Public opinion plays a significant role in shaping the approach to invasive ant management. Certain species may not be perceived as problematic by the public or policymakers, leading to a lack of resources dedicated to their control. Furthermore, legal restrictions or concerns over private land rights can hinder direct intervention, making it more difficult to implement control measures. Additionally, environmental policies protecting endangered species may limit the types of actions that can be taken to address invasive ants.

The technical challenges of controlling invasive ants are also significant. Their small size, adaptability, and ability to move over long distances make them difficult to track and eradicate. Invasive ants can establish large colonies that spread rapidly, and their complex nesting behaviours often make them hard to locate and treat. Moreover, the risk of pesticide resistance further complicates control efforts. Integrated pest management, involving collaboration across departments and research institutions, is critical to effectively managing these species and mitigating their impacts on agriculture, biodiversity, and human health. International cooperation is also crucial in addressing the spread of invasive ants, as these species frequently spread through international boundaries. The movement of goods and people facilitate the inadvertent transport of invasive species, making it challenging for any single nation to control their spread effectively (Hulme, 2009). Collaborative efforts between countries, particularly those with shared borders or trade routes, are necessary to develop and implement comprehensive strategies for preventing and managing ant invasions. Information sharing, joint surveillance efforts, and coordinated response plans are key to enhance the global capacity to combat these invasive species. Early detection is essential in preventing the spread of infestations. A well-coordinated surveillance system, employing trained personnel and advanced detection technologies, is vital for identifying and responding to ant invasions. Countries with extensive experience in managing invasive species, such as the USA, have developed specialized networks of entomologists who can rapidly identify and report ant infestations. In other regions, collaboration between universities, government agencies, and research institutions is also critical to develop sustainable control methods. Raising public awareness about the risks posed by invasive ants is another key component of their effective management. Educational campaigns can help communities understand the threats posed by these species and encourage proactive measures to prevent their spread. Providing resources and information to the public, through websites, fact sheets, and outreach programmes, is an essential part of any invasive species management strategy.

Regulation of international trade is another vital strategy in preventing the introduction of invasive ant species. Countries must enforce strict inspections of imported goods to minimize the risk of new species being introduced. This includes quarantine measures and thorough screening of goods at ports and borders. By regulating trade, governments can reduce the likelihood of new invasive species being introduced and protect native ecosystems and agriculture. Once infestations are detected, immediate and coordinated action is necessary to control and eradicate the ants. This typically involves a combination of biological control, chemical control, and integrated pest management strategies. Effective response efforts require financial resources, expertise, and collaboration between regional authorities, pest control teams, and research institutions. Collaboration among researchers, policymakers, and the public is essential to address the global threat posed by invasive ant species. By prioritizing research, improving surveillance systems, and fostering international cooperation, the spread of these harmful species can be mitigated, and their impact on ecosystems, agriculture, and human health can be minimized.

References

Abbott, K.L. (2006) Spatial dynamics of super colonies of the invasive yellow crazy ant, *Anoplolepis gracilipes*, on Christmas Island, Indian Ocean. *Diversity and Distributions* 12, 101–110. DOI: 10.1111/j.1366-9516.2006.00193.x.

Adams, C.T. (2019) *Agricultural and medical impact of the imported fire ants*. In: Lofgren, C.S. (ed.) *Fire Ants and Leaf-Cutting Ants*. Taylor and Francis, Boca Raton, Florida, USA, pp. 48–57. DOI: 10.1201/9780429038266-5.

Allen, C.R., Epperson, D.M. and Garmestani, A.S. (2004) Red imported fire ant impacts on wildlife: A decade of research. *American Midland Naturalist* 152, 88–103. DOI: 10.1674/0003-0031(2004)152[0088:RIFAIO]2.0.CO;2.

Aluja, M., Birke, A., Ceymann, M., Guillen, L., Arrigoni, E. *et al.* (2014) Agroecosystem resilience to an invasive insect species that could expand its geographical range in response to global climate change. *Agriculture, Ecosystems and Environment* 186, 54–63. DOI: 10.1016/j.agee.2014.01.017.

Bakra, D. (2024) Distribution of invasive ant species across various locations in and around the Indian Sunderbans. *Sustainable Chemical Insight in Biological Exploration* 36–44. DOI: 10.31674/book.2024scibe.004.

Banks, W.A., Adams, C.T. and Lofgren, C.S. (1990) Damage to North Carolina and Florida highways by red imported fire ants (Hymenoptera: Formicidae). *The Florida Entomologist* 73, 198–199. DOI: 10.2307/3495349.

Bertelsmeier, C., Luque, G.M., Hoffmann, B.D. and Courchamp, F. (2015) Worldwide ant invasions under climate change. *Biodiversity and Conservation* 24, 117–128. DOI: 10.1007/s10531-014-0794-3.

Bharti, H., Guénard, B., Bharti, M. and Economo, E.P. (2016) An updated checklist of the ants of India with their specific distributions in Indian states (Hymenoptera, Formicidae). *ZooKeys* 551, 1–83. DOI: 10.3897/zookeys.551.6767.

Bilo, M.B. and Bonifazi, F. (2009) The natural history and epidemiology of insect venom allergy: Clinical implications. *Clinical and Experimental Allergy* 39, 1467–1476. DOI: 10.1111/j.1365-2222.2009.03324.x.

Blüthgen, N., Kaiser-Bunbury, C. and Junker, R.R. (2017) 13 biological invasions. In: Oliveira, P.S. and Koptur, S. (eds) *Ant–Plant Interactions: Impacts of Humans on Terrestrial Ecosystems*. Cambridge University Press, UK, pp. 267–289. DOI: 10.1017/9781316671825.014.

Chen, L. and Fadamiro, H.Y. (2018) *Pseudacteon* phorid flies: Host specificity and impacts on *Solenopsis* fire ants. *Annual Review of Entomology* 63, 47–67. DOI: 10.1146/annurev-ento-020117-043049.

Chen, J., Rashid, T., Feng, G., Feng, Y., Zhang, A. *et al.* (2019) Insecticidal activity of methyl benzoate analogs against red imported fire ants, *Solenopsis invicta* (Hymenoptera: Formicidae). *Journal of Economic Entomology* 112, 691–698. DOI: 10.1093/jee/toy360.

Chen, S.Q., Zhao, Y., Lu, Y.Y., Ran, H. and Xu, Y.J. (2022) First record of the little fire ant, *Wasmannia auropunctata* (Hymenoptera: Formicidae), in Chinese mainland. *Journal of Integrative Agriculture* 21, 1825–1829. DOI: 10.1016/S2095-3119(22)63903-0.

Clark, J. (1941) Notes on the argentine ant and other exotic ants introduced into Australia. *Memoirs of the National Museum, Melbourne* 12, 59–70. DOI: 10.24199/j.mmv.1941.12.05.

Costa, A., Heleno, R., Freide, E.F., Dufrene, Y., Huckle, E. *et al.* (2023) Impacts of invasive ants on pollination of native plants are similar in invaded and restored plant communities. *Global Ecology and Conservation* 42, e02413. DOI: 10.1016/j.gecco.2023.e02413.

De Meyer, M., Robertson, M.P., Mansell, M.W., Ekesi, S., Tsuruta, K. *et al.* (2010) Ecological niche and potential geographic distribution of the invasive fruit fly *Bactrocera invadens* (Diptera, Tephritidae). *Bulletin of Entomological Research* 100, 35–48. DOI: 10.1017/S0007485309006713.

De Shazo, R.D., Williams, D.F., Goddard, J., Rockhold, R. and Kemp, S. (2013) Stings of imported fire ants: Clinical manifestations, diagnosis, and treatment. Available at: https://www.uptodate.com/contents/stings-of-imported-fire-ants-clinical-manifestations-diagnosis-and-treatment#topicContent

Diamantidis, A.D., Carey, J.R., Nakas, C.T. and Papadopoulos, N.T. (2011) Population-specific demography and invasion potential in medfly. *Ecology and Evolution* 1, 479–488. DOI: 10.1002/ece3.33.

Don, W. (2007) *Ants of New Zealand*, 1st edn. Otago University Press, Dunedin, New Zealand.

Estay, S.A., Lima, M. and Labra, F.A. (2009) Predicting insect pest status under climate change scenarios: Combining experimental data and population dynamics modelling. *Journal of Applied Entomology* 133, 491–499. DOI: 10.1111/j.1439-0418.2008.01380.x.

Eyer, P.A., Shults, P.T., Chura, M.R., Moran, M.N., Thompson, M.N. *et al.* (2021) Divide and conquer: Multicolonial structure, nestmate recognition, and antagonistic behaviors in dense populations of the invasive ant *Brachymyrmex patagonicus*. *Ecology and Evolution* 11, 4874–4886. DOI: 10.1002/ece3.7396.

FAO (2018). Available at: https://openknowledge.fao.org/server/api/core/bitstreams/707ef002-8052-4503-a7dd-cc58933050b7/content

Fischer, K., Bot, A.N., Brakefield, P.M. and Zwaan, B.J. (2003) Fitness consequences of temperature-mediated egg size plasticity in a butterfly. *Functional Ecology* 17, 803–810. DOI: 10.1111/j.1365-2435.2003.00798.x.

Folgarait, P.J. (1998) Ant biodiversity and its relationship to ecosystem functioning: A review. *Biodiversity and Conservation* 7, 1221–1244. DOI: 10.1023/a:1008891901953.

Gallien, L., Munkemuller, T., Albert, C.H., Boulangeat, I. and Thuiller, W. (2010) Predicting potential distributions of invasive species: Where to go from here? *Diversity and Distribution* 16, 331–342. DOI: 10.1111/j.1472-4642.2010.00652.x.

Gruber, M.A., Cooling, M. and Burne, A.R. (2017) An invasive ant distribution database to support biosecurity risk analysis in the Pacific. *Pacific Conservation Biology* 23(3), 258–261. DOI: 10.1071/PC17004.

Gutierrez, A.P., Ponti, L. and Gilioli, G. (2010) Climate change effects on plant-pest-natural enemy interactions. In: Hillel, D. and Rosenzweig, C. (eds) *Handbook of Climate Change and Agro Ecosystems. Impact, Adaptation, and Migration*. Imperial College Press, London, UK, pp. 209–237. DOI: 10.1142/9781848166561_0012.

Harmon, J.P., Moran, N.A. and Ives, A.R. (2009) Species response to environmental change: Impacts of food web interactions and evolution. *Science* 323, 1347–1350. DOI: 10.1126/science.1167396.

Herben, T. (2005) Species pool size and invasibility of island communities: A null model of sampling effects. *Ecology Letters* 8, 909–917. DOI: 10.1111/j.1461-0248.2005.00790.x.

Herben, T., Dostál, P., Březnová, M., Kozlíčková, V. and Kovář, P. (2005) Ant-induced soil modification and its effect on plant below-ground biomass. *Pedobiologia* 49, 127–137. DOI: 10.1016/j.pedobi.2004.09.004.

Hoffmann, B.D. and O'Connor, S. (2004) Eradication of two exotic ants from Kakadu National Park. *Ecological Management and Restoration* 5, 98–105. DOI: 10.1111/j.1442-8903.2004.00182.x.

Hoffmann, B.D., Luque, G.M., Bellard, C., Holmes, N.D. and Donlan, C.J. (2016) Improving invasive ant eradication as a conservation tool: A review. *Biological Conservation* 198, 37–49. DOI: 10.1016/j.biocon.2016.03.036.

Hölldobler, B. and Wilson, E.O. (1977) The number of queens: An important trait in ant evolution. *Naturwissenschaften* 64, 8–15. DOI: 10.1007/bf00439886.

Holway, D.A., Suarez, A.V. and Case, T.J. (1998) Loss of intraspecific aggression in the success of a widespread invasive social insect. *Science* 282, 949–952. DOI: 10.1126/science.282.5390.949.

Hulme, P.E. (2009) Trade, transport and trouble: Managing invasive species pathways in an era of globalization. *Journal of Applied Ecology* 46, 10–18. DOI: 10.1111/j.1365-2664.2008.01600.x.

Ingram, K.K., Bernardello, G., Cover, S. and Wilson, E.O. (2006) The ants of the Juan Fernández Islands: Genesis of an invasive Fauna. *Biological Invasions* 8, 383–387. DOI: 10.1007/s10530-004-3973-4.

IPCC (2007) Climate change 2007: Synthesis report. Available at: http://www.ipcc.ch (accessed 17 June 2025).

Ješovnik, A., Bujan, J. and Bračko, G. (2011) One genus and three species of ants (Hymenoptera: Formicidae) new for Croatia. *Entomologia Croatica* 15, 113–122.

Juárez-Juárez, B., Dáttilo, W. and Moreno, C.E. (2023) Synthesis and perspectives on the study of ant-plant interaction networks: A global overview. *Ecological Entomology* 48, 269–283. DOI: 10.1111/een.13227.

Klotz, J.H., deShazo, R.D., Pinnas, J.L., Frishman, A.M., Schmidt, J.O. *et al.* (2005) Adverse reactions to ants other than imported fire ants. *Annals of Allergy, Asthma and Immunology* 95, 418–425. DOI: 10.1016/S1081-1206(10)61165-9.

Krushelnycky, P.D., Loope, L.L. and Reimer, N.J. (2005) The ecology, policy, and management of ants in Hawaii. *Proceedings of Hawaiian Entomological Society* 37, 1–25.

Kwon, T.S., Li, F., Kim, S.S., Chun, J.H. and Park, Y.S. (2016) Modelling vulnerability and range shifts in ant communities responding to future global warming in temperate forests. *PLoS ONE* 11(8), e0159795. DOI: 10.1371/journal.pone.0159795.

Ladanyi, M. and Horvath, L. (2010) A review of the potential climate change impact on insect populations-general and agricultural aspects. *Applied Ecology and Environmental Research* 8, 143–152. DOI: 10.15666/aeer/0802_143151.

Lee, I.Y., Park, S.J., Seo, J.H., Sim, S., Kim, J.H. *et al.* (2017) Indoor occurrence of the ghost ant *Tapinoma melanocephalum* (Hymenoptera: Formicidae) in urban homes in Korea. *Korean Journal of Parasitology* 55, 225. DOI: 10.3347/kjp.2017.55.2.225.

Liebhold, A.M. and Tobin, P.C. (2008) Population ecology of insect invasions and their management. *Annual Review Entomology* 53, 387–408. DOI: 10.1146/annurev.ento.52.110405.091401.

Luo, Z. (2005) Harm of red imported fire ants to power facilities and their control. *Guangdong Electric Power* 18, 1–3.

Luo, B., Liao, J., Lu, M., Gu, H., Jing, X. *et al.* (2023) Warming-induced shifts on Tibetan Plateau: The overlooked ants and their ecological impacts. *Landscape Ecology* 38, 3999–4008. DOI: 10.1007/s10980-023-01770-3.

Mack, R.N., Simberloff, D., Mark Lonsdale, W., Evans, H., Clout, M. *et al.* (2000) Biotic invasions: Causes, epidemiology, global consequences, and control. *Ecological Applications* 689–710. DOI: 10.1890/1051-0761(2000)010[0689:BICEGC]2.0.CO;2.

Masse, P.S.M., Yede, M.N.P.A., Ebolo, M. and Titti, E. (2023) Interaction and resource exploitations between invasive and native ants in cocoa and surrounding farmlands in Southern Cameroon rainforest. *Acta Entomology and Zoology* 4, 126–132. DOI: 10.33545/27080013.2023.v4.i1b.105.

Moloney, S. and Vanderwoude, C. (2002) Red imported fire ants: A threat to eastern Australia's wildlife. *Ecological Management and Restoration* 3, 167–175. DOI: 10.1046/j.1442-8903.2002.t01-1-00109.x.

Mooney, H.A. and Cleland, E.E. (2001) The evolutionary impact of invasive species. In: *Proceedings of the National Academy of Sciences USA*, pp. 5446–5451 (Vol. 98). DOI: 10.1073/pnas.091093398.

Offenberg, J. (2014) The use of artificial nests by weaver ants: A preliminary field observation. *Asian Myrmecology* 6, 119–128.

Paini, D.R., Sheppard, A.W., Cook, D.C., Barro, P.J., Worner, S. *et al.* (2016) Global threat to agriculture from invasive species. In: *Proceedings of the National Academy of Sciences USA*, pp. 7575–7579 (Vol. 113). DOI: 10.1073/pnas.1602205113.

Parr, C.L. and Bishop, T.R. (2022) The response of ants to climate change. *Global Change Biology* 28(10), 3188–3205. DOI: 10.1111/gcb.16140.

Passera, L. (2021) Characteristics of tramp species. In: Williams, D.F. (ed.) *Exotic Ants*. Taylor and Francis, Boca Raton, Florida, USA, pp. 23–43.

Peng, R.K. and Christian, K. (2009) Determination and management of weaver ant, *Oecophylla smaragdina* (Fabricius) (Hymenoptera: Formicidae), marks on mango fruit in the Northern Territory of Australia. *International Journal of Pest Management* 55, 27–30. DOI: 10.1080/09670870802450227.

Perfecto, I. and Philpott, S.M. (2023) Ants (Hymenoptera: Formicidae) and ecosystem functions and services in urban areas: A reflection on a diverse literature. *Myrmecological News* 33, 103–122.

Perrings, C., Dehnen-Schmutz, K., Touza, J. and Williamson, M. (2005) How to manage biological invasions under globalization. *Trends of Ecology and Evolution* 20, 212–215.

Rabitsch, W., Werenkraut, V., Kenis, M., Roques, A. and Roy, H.E. (2024) Invasive alien insects and insect decline. In: Rodríguez, J., Pyšek, P. and Novoa, A. (eds) *Biological Invasions and Global Insect Decline*. Academic Press, pp. 143–171. DOI: 10.1016/B978-0-323-99918-2.00006-9.

Rajesh, T.P., Manoj, K., Ballullaya, U.P., Shibil, V.K., Asha, G. *et al.* (2022) Urban tropical forest islets as hotspots of ants in general and invasive ants in particular. *Scientific Reports* 12(1), 12003. DOI: 10.1038/s41598-022-16243-x.

Reimer, N.J. (2021) Distribution and impact of alien ants in vulnerable Hawaiian ecosystems. In: Reimer, N.J. (ed.) *Exotic Ants*. Taylor and Francis, Boca Raton, Florida, USA, pp. 11–22.

Richardson, D.M., Pyšek, P., Rejmanek, M., Barbour, M.G., Panetta, F.D. *et al.* (2000) Naturalization and invasion of alien plants: Concepts and definitions. *Diversity and Distribution* 6, 93–107. DOI: 10.1046/j.1472-4642.2000.00083.x.

Sardain, A., Sardain, E. and Leung, B. (2019) Global forecasts of shipping traffic and biological invasions to 2050. *Nature Sustainability* 2, 274–282. DOI: 10.1038/s41893-019-0245-y.

Seebens, H., Blackburn, T.M., Dyer, E.E., Genovesi, P., Hulme, P.E. *et al.* (2017) No saturation in the accumulation of alien species worldwide. *Nature Communications* 8, 14435. DOI: 10.1038/ncomms14435.

Siddiqui, J.A., Chen, Z., Li, Q., Deng, J., Lin, X. *et al.* (2019) DNA barcoding of aphid-associated ants (Hymenoptera, Formicidae) in a subtropical area of Southern China. *ZooKeys* 879, 117. DOI: 10.3897/zookeys.879.29705.

Siddiqui, J.A., Bamisile, B.S., Khan, M.M., Islam, W., Hafeez, M. *et al.* (2021) Impact of invasive ant species on native fauna across similar habitats under global environmental changes. *Environmental Science and Pollution Research* 28, 54362–54382. DOI: 10.1007/s11356-021-15961-5.

Simberloff, D., Martin, J.L., Genovesi, P., Maris, V., Wardle, D.A. *et al.* (2013) Impacts of biological invasions: What's what and the way forward. *Trends in Ecology and Evolution* 28, 58–66. DOI: 10.1016/j.tree.2012.07.013.

Suarez, A.V., Richmond, J.Q. and Case, T.J. (2000) Prey selection in horned lizards following the invasion of Argentine ants in southern California. *Ecological Applications* 10, 711–725. DOI: 10.1890/1051-0761(2000)010[0711:PSIHLF]2.0.CO;2.

Suarez, A.V., Holway, D.A. and Ward, P.S. (2005) The role of opportunity in the unintentional introduction of non-native ants. *Proceedings of the National Academy of Sciences USA* 102, 17032–17035. DOI: 10.1073/pnas.0506119102.

Suhr, E.L., O'Dowd, D.J., Suarez, A.V., Cassey, P., Wittmann, T.A. *et al.* (2019) Ant interceptions reveal roles of transport and commodity in identifying biosecurity risk pathways into Australia. *NeoBiota* 53, 1–24. DOI: 10.3897/neobiota.53.39463.

Taylor, R.W. (1987) A checklist of the ants of Australia, New Caledonia and New Zealand (Hymenoptera: Formicidae). Available at: https://www.antwiki.org/wiki/images/5/5c/Taylor_1987a.pdf

Terayama, M., Kubota, S. and Eguchi, K. (2014) *Encyclopaedia of Japanese Ants*. Asakura Shoten, Tokyo, p. 278.

Thakur, K., Jayakumar, P. and Sonawane, B.L. (2023) *In vitro* studies on the biology and predatory potential of commonly occurring ant species in a selected agro-ecosystem of the Himalayan Range, India. *Sociobiology* 70, e7698. DOI: 10.13102/sociobiology.v70i2.7698.

Tschinkel, W.R. (2019) The ecological nature of the fire ant: Some aspects of colony function and some unanswered questions. In: Lofgren, C.S. (ed.) *Fire Ants and Leaf-Cutting Ants*. Taylor and Francis, Boca Raton, Florida, USA, pp. 72–87.

Van Schagen, J.J., Davis, P.R. and Widmer, M.A. (2021) Ant pests of Western Australia, with particular reference to the Argentine ant (*Linepithema humile*). In: Williams, D.F. (ed.) *Exotic Ants*. Taylor and Francis, Boca Raton, Florida, pp. 174–180. DOI: 10.1201/9780429040795.

Vermeij, G.J. (1996) An agenda for invasion biology. *Biological Conservation* 78, 3–9. DOI: 10.1016/0006-3207(96)00013-4.

Wang, L., Zeng, L. and Chen, J. (2015) Impact of imidacloprid on new queens of imported fire ants, *Solenopsis invicta* (Hymenoptera: Formicidae). *Scientific Reports* 5, 17938.

Wang, L., Xu, Y., Zeng, L. and Lu, Y. (2019) Impact of the red imported fire ant *Solenopsis invicta* Buren on biodiversity in South China: A review. *Journal of Integrative Agriculture* 18, 788–796. DOI: 10.1016/S2095-3119(18)62014-3.

Wang, M., Liu, Y., Wen, T., Liu, W., Gao, Q. *et al.* (2020) Chromatin accessibility and transcriptome landscapes of *Monomorium pharaonis* brain. *Scientific Data* 7, 217. DOI: 10.6084/m9.figshare.12111210.

Wetterer, J.K. (2005) Worldwide distribution and potential spread of the long-legged ant, *Anoplolepis gracilipes* (hymenoptera: Formicidae). *Sociobiology* 45(1), 77–97.

Wetterer, J.K. (2008) Worldwide spread of the longhorn crazy ant, *Paratrechina longicornis* (Hymenoptera: Formicidae). *Myrmecological News* 11, 137–149.

Wetterer, J.K. (2009a) Worldwide spread of the ghost ant, *Tapinoma melanocephalum* (Hymenoptera: Formicidae). *Myrmecological News* 12, 23–33.

Wetterer, J.K. (2009b) Worldwide spread of the destroyer ant, *Monomorium destructor* (Hymenoptera: Formicidae). *Myrmecological News* 12, 97–108.

Wetterer, J.K. (2010) Worldwide spread of the flower ant, *Monomorium floricola* (Hymenoptera: Formicidae). *Myrmecological News* 13, 19–27.

Wetterer, J.K. (2011) Worldwide spread of the tropical fire ant, *Solenopsis geminata* (Hymenoptera: Formicidae). *Myrmecological News* 14, 21–35.

Wetterer, J.K. and Porter, S.D. (2003) The little fire ant, *Wasmannia auropunctata*: Distribution, impact and control. *Sociobiology* 41, 1–41.

Wylie, F.R. and Janssen-May, S. (2017) Red imported fire ant in Australia: What if we lose the war? *Ecological Management and Restoration* 18, 32–44. DOI: 10.1111/emr.12238.

Wylie, R., Yang, C.C.S. and Tsuji, K. (2020) Invader at the gate: The status of red imported fire ant in Australia and Asia. *Ecological Research* 35, 6–16. DOI: 10.1111/1440-1703.12076.

Xiong, T., Qiu, X.H., Ling, S.Q., Liu, J.L. and Zeng, X.N. (2019) Interaction of fipronil and the red imported fire ant (*Solenopsis invicta*): Toxicity differences and detoxification responses. *Journal of Insect Physiology* 115, 20–26. DOI: 10.1016/j.jinsphys.2019.03.001.

Xu, Y., Huang, J., Zhou, A. and Zeng, L. (2012) Prevalence of *Solenopsis invicta* (Hymenoptera: Formicidae) venom allergic reactions in mainland China. *Florida Entomologist* 95, 961–965. DOI: 10.1653/024.095.0421.

Xu, Y., Vargo, E.L., Tsuji, K. and Wylie, R. (2022) Exotic ants of the Asia-Pacific: Invasion, national response, and ongoing needs. *Annual Review of Entomology* 67, 27–42.

Zhang, B., Yu-Rong, H.E., Ting, C., Qi Guo-Jun, Q. and Li-Hua, L.U. (2015) Dietary composition of foragers of the red imported fire ant, *Solenopsis invicta* Buren (Hymenoptera: Formicidae) in two habitats, mulberry orchard and barren land, in South China. *Acta Entomologica Sinica* 58, 382–390. DOI: 10.16380/j.kcxb.2015.58.4.382390.

Zhang, R., Li, Y., Liu, N. and Porter, S.D. (2007) An overview of the red imported fire ant (Hymenoptera: Formicidae) in mainland China. *Florida Entomologist* 90, 723–731. DOI: 10.1653/0015-4040(2007)90[723:AOOTRI]2.0.CO;2.

7 Invasion and Management of South American Tomato Pinworm, *Tuta absoluta* (Meyrick) (Lepidoptera: Gelechiidae)

Satyapriya Singh[1,2*], Tanmaya Kumar Bhoi[3], Priyanka Nayak[4], M.N. Rudra Gouda[2] and Subham Gambhir[2]

[1]Central Horticultural Experiment Station (ICAR – Indian Institute of Horticultural Research), Bhubaneswar, Odisha, India; [2]Division of Entomology, ICAR – Indian Agricultural Research Institute, New Delhi, India; [3]ICFRE – Arid Forest Research Research Institute, Jodhpur, Rajasthan, India; [4]Institute of Agricultural Sciences, Siksha 'O' Anusandhan, Bhubaneswar, Odisha, India

Abstract

Tuta absoluta (Meyrick) (Lepidoptera: Gelechiidae) is a destructive invasive pest causing significant damage to tomato crops worldwide, leading to severe economic losses for farmers. Chemical constituents and phenology, coupled with favourable climatic conditions and habitat suitability, influence the damage pattern. The chapter discusses the factors that have contributed to the spread of *T. absoluta*, including globalization, international trade, anthropogenic factors, and climate change. The integrated management approach embracing all the available management practices suppressed this pest with lesser/no inimical effects on the environment and other living organisms. An overview of the management methods for *T. absoluta*, viz. crop rotation, the use of resistant varieties, biological control, and judicious use of pesticides are discussed.

Keywords: *Tuta absoluta*, Integrated Pest Management (IPM), biological control, chemical control

7.1 Introduction

Invasive species pose a significant threat to both natural and managed ecosystems, with the potential to cause severe economic, environmental, and ecological impacts (Pyšek *et al.*, 2020). One such invasive species is the South American tomato pinworm (TPW), *Tuta absoluta* (Meyrick, 1917) (Lepidoptera: Gelechidae), which has rapidly spread to new environments (Fand *et al.*, 2020). *T. absoluta* is considered one of the most damaging insect pests in the world, causing severe yield losses in solanaceous crops such as tomatoes, aubergines, and potatoes (Mahlangu *et al.*, 2022). The high reproductive potential and capacity to adapt to different climatic conditions have made *T. absoluta* a major phytosanitary concern and a threat to global food security (Desneux *et al.*, 2022). Since its first detection in South America in 2006, *T. absoluta* spread widely, to regions including the Mediterranean Basin, Europe, the Middle East, South Asia, and Africa (Guimapi *et al.*, 2020). The pest's

*Corresponding author: satyaiari05@gmail.com

 Biological Invasions in the Tropics: Monitoring, Mechanism and Management (eds Akshay Kumar Chakravarthy and Pathour Rajendra Shashank)
DOI: 10.1079/9781836991977.0007

rapid spread and establishment in new environments are attributed to strong baseline and plastic physiological tolerance and quick genetic adaptation alterations (Skendžić *et al.*, 2021). *T. absoluta* larvae feed on aerial plant parts, including fruits, leading to aesthetic harm, secondary infection, and significant yield losses, resulting in increased production costs, trade restrictions, and an overall rise in food and nutrition insecurity (Mansour and Biondi, 2021).

The management of *T. absoluta* is a challenging task for growers and researchers alike due to the pest's high adaptability, resistance to conventional insecticides, and limited available management options (Tarusikirwa *et al.*, 2020a). Integrated pest management (IPM) approaches combining cultural, physical, and biological control practices have been developed to manage *T. absoluta* populations effectively. However, the success of IPM operations is influenced by several factors, including environmental variability, host-plant resistance, and insecticide resistance (Walsh *et al.*, 2022). This book chapter aims to provide a comprehensive assessment of the ecology and management of *T. absoluta* by reviewing existing knowledge on the pest's bio-ecology, distribution, and management status worldwide. The chapter will discuss the potential ecological and economic impacts of *T. absoluta* on natural and managed systems and highlight the need for innovative and sustainable management strategies to address this growing phytosanitary challenge. By identifying current knowledge gaps and potential research needs, the chapter seeks to provide a platform for future research to develop effective and sustainable management strategies for *T. absoluta* and other invasive alien species.

7.2 Origin and Distribution of Tomato Pinworm

T. absoluta (Meyrick) is believed to have originated in Peru, where it was first reported in 1917 (Han *et al.*, 2019). The insect was initially considered a minor pest in South America, but in the last two decades, it has become a significant economic threat to tomato production worldwide (Roditakis *et al.*, 2015). It has a wide distribution range and has spread to many countries across the globe, primarily through the trade of infested plant material. The insect's distribution expanded rapidly in the past decade, and it is now present in over 100 countries all over the world, embracing Europe, Africa, Asia, and the Middle East (Verheggen and Fontus, 2019). The first record of TPW outside South America was in Spain in 2006, where it caused significant damage to tomato (Campos *et al.*, 2017). The insect quickly spread to other countries in Europe, including Italy, Portugal, Greece, and France. By 2008, *T. absoluta* reached North Africa, and by 2010, it was present in Israel and Jordan in the Middle East (Mansour *et al.*, 2018). In 2012, the insect was found in India, and by 2013, it established in Pakistan, Nepal, and Bangladesh. In Africa, *T. absoluta* was first reported in Morocco in 2008, and since then, it has spread across the continent. The insect is now present in most African countries, including Algeria, Tunisia, Libya, Egypt, Sudan, Ethiopia, Kenya, Tanzania, Uganda, Rwanda, Burundi, Zambia, Zimbabwe, Malawi, Mozambique, and South Africa (Shashank *et al.*, 2015; Santana *et al.*, 2019). The rapid spread of *T. absoluta* is attributed to high reproductive potential, absence of effective natural enemies, and its ability to adapt to a diverse set of environmental conditions (Mkonyi *et al.*, 2020). The insect has a short life cycle of 28–30 days under optimal conditions, and each female lays up to 300 eggs, making it a highly prolific pest (Shashank *et al.*, 2015). The larvae of *T. absoluta* are highly mobile and move from plant to plant quickly, allowing for rapid spread within a crop and to neighbouring fields (Guedes *et al.*, 2019). It is now considered a major phytosanitary factor affecting the global value chain of solanaceous crops, including tomato, potato, eggplant, and pepper. It has caused significant yield losses, reduced crop quality, and increased the use of pesticides, resulting in higher production costs and food prices. The pest has also caused trade restrictions on tomato seedlings and fruits, leading to food and nutrition insecurity (Ferracini *et al.*, 2019). Efforts are underway to control the spread and impact of *T. absoluta*.

7.3 Life Cycle of *Tuta Absoluta*

T. absoluta (Meyrick) undergoes complete metamorphosis consisting of egg, larva, pupa, and adult stages. The egg is oval, flattened, and measures approximately 0.3–0.5 mm in length (Bawin *et al.*, 2016). Initially translucent, the egg changes from white to yellow and finally to reddish-brown before hatching. Eggs are laid singly or in small clusters, typically on the underside of leaves near veins. Incubation lasts for 3–5 days under optimal conditions (Vivekanandhan *et al.*, 2024). The larva is the most destructive stage. Newly hatched larvae are small, whitish-yellow with a brown head and prothoracic shield. Mature larvae reach about 8 mm in length and display a light-and-dark banded body pattern. They feed internally on leaves, stems, flowers, and fruits, causing mines and galleries that significantly reduce marketability and yield. The larval period lasts about 12–14 days under favourable conditions (Vivekanandhan *et al.*, 2024). The pupa is a non-feeding, transitional stage. It is brownish and may occur within the leaf mines, soil, or leaf litter. Pupation lasts for 6–9 days depending on environmental conditions. The adult is a small, greyish-brown moth with a wingspan of 7–9 mm. Its narrow wings are fringed with long scales, giving it a powdery appearance. Adults are mostly nocturnal and strongly phototactic. Females begin oviposition soon after mating, laying up to 260 eggs during their lifespan. Adults live for 6–10 days under optimal temperature and humidity (Vivekanandhan *et al.*, 2024).

7.4 Host Range

T. absoluta (Meyrick) is a highly polyphagous pest with a host range that includes Solanaceae plants (Sylla *et al.*, 2019). The insect has been documented to feed on over 100 plants belonging to Solanaceae, including cultivated and wild plants (Biondi *et al.*, 2018). Tomato (*Solanum lycopersicum*) is the primary host plant of *T. absoluta* and is the most economically important crop damaged by this pest. The larvae of *T. absoluta* feed on all aerial parts of tomato (Cherif and Verheggen, 2019). They mine the leaf parenchyma tissue creating galleries and inflicting severe damage to the plant. In addition, the larvae bore into the fruit, causing further damage and making the crop unsuitable for sale or consumption. *T. absoluta* causes up to 100% yield loss in tomato if left uncontrolled (Idriss *et al.*, 2020). Other important cultivated crops attacked by *T. absoluta* include potato (*Solanum tuberosum*), aubergine (*Solanum melongena*), and pepper (*Capsicum annuum*). *T. absoluta* also feeds on wild Solanaceae species (Smith *et al.*, 2018). These include nightshade (*Solanum nigrum*), bittersweet nightshade (*Solanum dulcamara*), silverleaf nightshade (*Solanum elaeagnifolium*), black nightshade (*Solanum nigrum*), hairy nightshade (*Solanum sarrachoides*), and buffalo bur (*Solanum rostratum*). *T. absoluta* feeds on other plant families, although these hosts are not as important as the Solanaceae plants. These include tobacco (*Nicotiana tabacum*) and jimsonweed (*Datura stramonium*) (Bawin *et al.*, 2016).

7.5 Feeding and Damage Symptoms

The larvae of *T. absoluta* are the most damaging stage leading to significant yield losses (Larbat *et al.*, 2016). The larvae feed on all aerial parts of the tomato plant (Biondi *et al.*, 2018). Initially, the larvae create small mines on the leaves, visible as tiny spots on the upper sideof the leaf. These mines are generated by the larvae as they feed on the leaf mesophyll, resulting in a thin, transparent area on the leaf surface. As the larvae grow and become more mobile, they move on to other plant parts, such as buds, flowers, and fruits. The feeding damage to the buds and flowers may cause flower drop and, consequently, fruit abortion (Silva *et al.*, 2017). In severe infestations, the larvae feeding results in defoliation of the plant, leading to stunted growth and reduced yields (Bajracharya *et al.*, 2016). The larvae nibble on the fruit surface, creating characteristic mines or galleries, visible as brownish, irregular patches on the fruit surface (Tumuhaise *et al.*, 2016). The larvae also tunnel into the fruit, causing extensive internal damage, leading to fruit rot and secondary infections by microorganisms (Camargo *et al.*, 2016). Infested tomatoes are rejected in the market due to their poor appearance, and the growers suffer economic losses (Urbaneja *et al.*, 2012).

7.6 Behaviour and Ecology

The behaviour, ecology, and distribution of the tomato pinworm play a key part in invasion and persistence in new environments. Understanding the behavioural and ecological traits of *T. absoluta* is essential for developing effective management strategies (Guimapi *et al.*, 2016).

7.6.1 Behaviour

T. absoluta larvae are the primary cause of damage to the host plant. The larvae feed on the leaves, stems, and fruits of the plant, causing significant damage and yield losses. They produce silk threads that help them move around the plant and protect them from predators (Bawin *et al.*, 2015).

7.6.2 Ecology

T. absoluta is a polyphagous pest infesting a wide range of solanaceous crops, it has a rapid reproductive rate, with the potential to complete its life cycle in 19–27 days under optimal conditions. The pest can survive in a broad range of environmental conditions, including low humidity and high temperature. *T. absoluta* exhibits diapause, a state of arrested development in response to unfavourable environmental conditions, such as low temperature and short day-length. Diapause helps *T. absoluta* to survive during unfavourable conditions (Tarusikirwa *et al.*, 2020b).

7.6.3 Dispersal

T. absoluta has a high dispersal ability that is one of the reasons for rapid invasion of new environments. The pest disperses through natural means, such as wind, and human-mediated transport, such as the movement of infested plant materials. Studies have shown that *T. absoluta* disperses over long distances, up to several hundred kilometres, through wind-assisted migration. The pest has been found in regions far from its native range, indicating high dispersal ability (Cagnotti *et al.*, 2018). The behaviour, ecology, and dispersal of *T. absoluta* are closely interlinked and play a significant role in invasion and persistence in new environments. The pest's polyphagous nature, rapid reproductive rate, and capacity to survive in diverse environmental conditions make it a highly successful invader (Machekano *et al.*, 2018). The high dispersal ability of *T. absoluta* through natural and human-mediated means also contributes to successful invasion and spread. Effective management of *T. absoluta* requires an understanding of its behaviour, ecology, and dispersal.

7.7 Factors Influencing Invasion for TPW

The invasion of the tomato pinworm is influenced by several factors (Naselli *et al.*, 2017), including the following.

7.7.1 Cultivation practices

The management practices used in tomato cultivation significantly affect the invasion of the tomato pinworm. For instance, poor field sanitation, such as leaving crop residues in the field, increases the pest's survival and spread. Similarly, the use of poor-quality seedlings, poor crop nutrition, and inadequate water supply also favour TPW invasion. Good agricultural practices, such as crop rotation, selection of healthy seedlings, and IPM, assist in decreasing pest invasion (Li et al., 2020).

7.7.2 Climate

The tomato pinworm thrives in warm and humid conditions, with temperatures varying from 20–30°C. Therefore, regions with warm and humid climates are more susceptible to invasion. Additionally, TPW survival rate is high in regions with low winter temperatures since it can hibernate during winter and emerge in spring to continue its life cycle. Therefore, warmer regions are more vulnerable to continuous pest infestation (Shashank *et al.*, 2018).

7.7.3 Host plant availability

The tomato pinworm is a specialized pest that prefers to feed on Solanaceae, with tomato plants being its primary host. Therefore, the availability of tomato plants significantly affects the pest's invasion (Dhanapal *et al.*, 2021). Areas with large tomato plantations are more susceptible to pest invasion since there are more hosts available for the pest to feed on. TPW invasion can be exacerbated when host plant availability is coupled with conducive weather conditions, such as warm and humid environments (Silva *et al.*, 2021).

7.7.4 Human activities

Human activities such as trading of infested plant materials, movement of contaminated soil, and transportation of infected fruits, also influence the invasion. These activities facilitate the spread of TPW across regions and continents, leading to new infestations. Additionally, urbanization, deforestation, and land-use changes alter the ecosystem and can create new habitats and ecological niches that favour the pest's invasion (Fiaboe *et al.*, 2021).

7.7.5 Insecticide use

Over-reliance on insecticides to control pests also influences the invasion of the tomato pinworm. This is because the indiscriminate use of insecticides can lead to the development of resistance in the pest population (Santana *et al.*, 2019). As the pest becomes resistant to insecticides, controlling populations becomes more challenging, leading to continuous pest infestations. Therefore, application of insecticides should be combined with other management practices to reduce pest invasion (Zang *et al.*, 2022).

7.7.6 International trade

International trade also contributes to the invasion of the tomato pinworm. As tomato products are traded across regions, there is a high risk of infested products being introduced to new regions. Therefore, phytosanitary measures such as inspection of traded products, quarantine regulations, and proper disposal of infested materials are essential to further spread of the pest (Rostami *et al.*, 2020).

7.8 Pest Risk Analysis and Management of *Tuta absoluta*

Tuta absoluta (Meyrick), the tomato pinworm, is a highly invasive pest that poses a serious threat to tomato production globally. Its rapid spread, high reproductive capacity, and resistance to many conventional control measures classify it as a high-risk pest, necessitating a thorough Pest Risk Analysis (PRA) (Sankarganesh *et al.*, 2017). Accurate pest identification is critical in managing this species, and the use of both morphological characteristics and molecular diagnostic tools strengthens the accuracy of detection, thereby aiding in early and effective intervention (Tarusikirwa *et al.*, 2020a). Pathway analysis is an essential component of PRA, which involves identifying the potential routes through which *T. absoluta* can enter and establish in new areas (Pereira *et al.*, 2019). The most common pathway is international trade, where infested tomato fruits or seedlings facilitate the pest's introduction across borders. Additionally, human-mediated movement through contaminated clothing, packaging, or agricultural tools can support pest dispersal. The species is also capable of natural migration, spreading to nearby regions with the help of its flying ability and the presence of suitable environmental conditions.

The risk assessment of *T. absoluta* covers economic, environmental, and social aspects (Tonnang *et al.*, 2015). Economically, the pest can cause up to 100% yield loss in tomatoes under severe infestation, resulting in significant financial losses for farmers and disruption to local markets. Environmentally, its presence often leads to the overuse of insecticides, which can negatively impact non-target organisms, including natural predators, pollinators, and other beneficial insects. Socially, infestation by *T. absoluta* can lead to decreased agricultural productivity, loss of income, and job insecurity among farming communities. To address these challenges, an integrated risk management approach is vital. Quarantine measures, including strict phytosanitary inspections and the regulation of plant material movement, are crucial in preventing the pest's introduction through trade. Once established, biological control using natural enemies such as parasitoids (*Trichogramma* spp., *Necremnus* spp.), predators, and entomopathogenic fungi has shown promising results. Although chemical control is commonly practised and often effective, it should be used judiciously to avoid resistance development and minimize ecological disruption. Cultural control practices such as crop rotation, destruction of infested plant debris, use of healthy seedlings, and maintaining field hygiene play a significant role in reducing the pest population by interrupting its life cycle (Desneux *et al.*, 2022).

7.8.1 Detection and monitoring

TPW is a serious pest of tomato crops, and early detection and monitoring are crucial for its effective management. There are several techniques used for detecting and monitoring the pest, including visual inspection, pheromone traps, sticky traps, light traps, leaf sampling, molecular techniques, and integrated pest management programmes. Visual inspection involves examining tomato plants for signs of pest damage, while pheromone traps use synthetic chemicals that mimic the TPW sex pheromone to attract and trap male moths (Caparros Megido *et al.*, 2013). Sticky traps use coloured sticky cards placed at strategic locations to trap adult moths, while light traps use ultraviolet light to attract and trap them (Sadique *et al.*, 2025). Leaf sampling involves taking a sample of tomato leaves and examining them under a microscope for the presence of eggs or larvae, and molecular techniques use genetic markers to identify the pest's DNA in tomato plants or soil samples (Hadapad and Hire, 2019). IPM programmes combine a number of detection and monitoring methods with control measures such as cultural, chemical, and biological control to manage the pest effectively. These programmes aim to reduce the TPW impact on tomato production while minimizing the impact on non-target organisms and the environment. The choice of detection and monitoring method depends on the pest's population levels, the size of the tomato field, and the available resources. Early detection and monitoring can help growers take prompt action to prevent or control the pest's spread and reduce its impact on tomato crops (Desneux *et al.*, 2022).

7.8.2 Mass trapping and mating disruption

The application of synthetic female sex pheromone is an important technique for detection and mass trapping of *T. absoluta*, a pest that infests tomato crops (Attygalle *et al.*, 1995). Studies have shown that the use of homemade traps containing pheromones resulted in a greater reduction in leaf infestation compared with conventional insecticides in Argentinean tomato crops (Lobos *et al.*, 2013). Pheromone dispensers may be applied in conjunction with light sources and water traps to mass trap both adult genders simultaneously. Mating disruption, which involves the use of pheromones to disrupt the mating behaviour of the pest, is another effective control method for *T. absoluta* in tomato greenhouses (Caparros Megido *et al.*, 2013). The major sex pheromone component of *T. absoluta* females is (3E,8Z,11Z)−3,8,11-tetradecadien-1-yl acetate (TDTA), and the addition of minor pheromone components did not affect the attractiveness of TDTA in attracting *T. absoluta* males. Therefore, TDTA alone is adequate for attraction and can be utilized in the design of IPM programmes for *T. absoluta* (Ferrara *et al.*, 2001). Studies have confirmed the efficacy of synthetic TDTA under both laboratory and field conditions, making it an environmentally friendly and compatible method for modern cropping systems and organic agriculture (Michereff *et al.*, 2000).

7.9 Host Plant Resistance

Plants have various structural and biochemical traits that contribute to their resistance against pests. Among these traits, secondary compounds have been extensively studied. Secondary compounds are chemicals which are not directly involved in primary metabolic processes but play a key role in mediating interactions with other organisms. The three main categories of secondary compounds found in plants are terpenes, phenolics, and nitrogen-containing compounds. Terpenes are a diverse group of compounds which are accountable for the aromatic properties of plants. Phenolics are a group of compounds that play a vital role in plant defence against pests by acting as antioxidants and participating in the formation of structural barriers (Roumani *et al.*, 2022). Nitrogen-containing compounds, such as alkaloids, play a key part in the plant's defence against herbivores by acting as toxins (Hussain and AlJabr, 2022). The combination of these secondary compounds and other resistance-related traits in plants provides a comprehensive mechanism for resistance against pests (Singh *et al.*, 2022).

7.9.1 Trichomes

According to Glas *et al.* (2012), the *Solanum* genus has eight different types of trichomes, four of them being glandular and four non-glandular. Non-glandular trichomes have the potential to safeguard plants from herbivores by disrupting their feeding behaviour or movement. Glandular trichomes facilitate the production and storage of a variety of secondary plant compounds, including acyl-sugars, methyl ketones, and phenolics/oxidative enzymes (Dhankhar *et al.*, 2023). These compounds are believed to have a direct toxic effect on pests. In tomatoes, glandular trichomes with increased densities on newly emerging leaves are a crucial part of induced resistance (Peiffer *et al.*, 2009).

7.9.2 Acylsugars

Acylsugars are composed of ester bonds linking multiple acyl chains of varying lengths to glucose or sucrose backbones. These compounds are synthesized and accumulated in the glandular trichomes of tomatoes, mainly in type IV trichomes, and are a key constituent of the sticky exudate they produce (Resende *et al.*, 2006). Elevated levels of acylsugars are believed to confer protection against several insect pests.

7.9.3 Methyl ketones

Type VI trichomes in tomatoes are typically associated with methyl ketones, which include 2-tridecanone (2-TD) and 2-undecanone (2-UD). These ketones, derived from fatty acids with 7–15 carbons, are synthesized through the condensation of acetate units. 2-TD is particularly effective in deterring juvenile insects, leading to increased mortality and developmental issues, including pupal deformation in *T. absoluta* (de Almeida *et al.*, 2023).

7.9.4 Terpenes

Terpenes are a class of chemicals which are generated from acetyl-CoA through mevalonic acid and are composed of multiple isoprene units with a length of five carbons (Bottger *et al.*, 2018). In tomatoes, sesquiterpenes and their derivatives have been found to play a direct role in resisting herbivores, although they can also attract them. The metabolism of terpenoids in trichomes varies between wild and cultivated tomato species, and these differences may be responsible for variation in pest resistance (Wang *et al.*, 2020).

7.9.5 Phenolics

Phenolics are a group of secondary compounds that include flavonoids, tannins, lignin, and phenylpropanoids. The key phenolic compounds in tomatoes are chlorogenic acid and rutin, which are catecholic phenolics. These compounds can be involved in the generation of reactive oxygen species and can be converted to highly reactive quinones by polyphenol oxidase. The toxic and growth-reducing effects of these phenolics may be related to these processes (Vuolo *et al.*, 2019).

A study revealed that *Tuta absoluta* herbivory triggers a strong and localized metabolic response in tomato plants, marked by the massive accumulation of structurally diverse phenolamides, particularly in infested leaves. Notably, rare compounds such as spermine and dihydrocinnamic acid derivatives were identified. Transcriptomic analysis supported these findings by showing the upregulation of genes in both the phenylpropanoid and polyamine biosynthetic pathways. Two genes encoding putrescine hydroxycinnamoyl transferases were highly induced and functionally confirmed to catalyse the production of key phenolamides, including caffeoylputrescine (Roumani *et al.*, 2022).

7.10 Cultural Control

Incorporating traditional cultural control methods with chemical and organic techniques are beneficial for IPM targeting *T. absoluta*. These methods involve practices such as crop rotation, covering plants with insect-proof netting, selective removal and disposal of infested plant parts, and complete d(Chenestruction of plants after harvest. Additionally, the management of alternate wild hosts, such as black nightshade and devil's apple, surrounding the field or greenhouse can aid in controlling the distribution of *T. absoluta* (Desneux *et al.*, 2022). Creating habitats that offer food and shelter for natural antagonists of *T. absoluta* within and around tomato crops grown in open fields is also recommended (Biondi and Desneux, 2019).

7.11 Botanicals

Botanicals are extracts from plants that have insecticidal properties. Botanicals may be categorized based on their chemical constituents, such as essential oils, flavonoids, and alkaloids. These natural compounds have been successfully used to manage crop pests. They affect insects in various ways, including repelling, deterring feeding, acting as toxicants or growth retardants, chemo sterilizing, or attracting them (Singh *et al.*, 2021, 2022). The choice of parts of plants utilized depends on the targeted bioactive chemicals and their concentrations, with barks, leaves, roots, flowers, seeds, and stems being commonly used. Commonly used botanicals include neem, garlic, ginger, and pyrethrum. Botanicals are an attractive alternative for sustainable pest management due to their wide availability, application ease, and low operational costs. The extracts of neem and *Jatropha curcas* are effective against eggs and larvae of *T. absoluta*. In field studies, *A. sativum* and *A. indica* biopesticides resulted in 71.82 and 74.26% larval mortalities, respectively (Shiberu and Getu, 2018). Additionally, the ethanolic leaf extracts of *Piper amalago* var. *medium*, *P. glabratum*, and *P. mikanianum* elicited significant *T. absoluta* larval mortalities. *Piper* species contain amides such as piper amides that have neurotoxic and lipid metabolism effects, leading to knockdown, paralysis, and death of insects (De Brito *et al.*, 2015).

7.12 Entomopathogens Against TPW

Entomopathogenic microorganisms such as *Metarhizium anisopliae* var. *anisopliae* (Metsch.) Soroki, *Beauveria bassiana* (Balsamo) Vuillemin, and *Bacillus thuringiensis* (Berliner) have proven to be effective against *T. absoluta*. *Bacillus thuringiensis*, a soil-dwelling bacterium, is an important environmentally-friendly microorganism with entomopathogenic properties that can be widely applied in integrated approaches to pest management (Jalapathi *et al.*, 2020). Apart from Lepidoptera, *Bacillus thuringiensis* has also shown insecticidal effects on other insect orders, making it ideal for sustainable integrated pest management. *Bacillus thuringiensis* has already been widely used in the control of *T. absoluta*, and studies have revealed that third-instar larvae are the most susceptible to it. Additionally, the combination of *Beauveria bassiana* and *Bacillus thuringiensis* has been shown to have a synergistic effect for the management of *T. absoluta*. In addition, *Wolbachia* bacterial infection has also shown potential efficacy against *T. absoluta* by affecting reproduction. Spinosad, a fermentation product of *S. spinosa*, has been applied in *T. absoluta* control, but its continued application is threatened by resistance (Sabbour, 2014).

Entomopathogenic nematodes (EPNs) are a type of biological control agent that can eliminate insect pests by using their mutually beneficial intestinal bacteria. Recent research has demonstrated that the EPNs *Steinernema feltiae*, *S. carpocapsae*, and *Heterorhabditis bacteriophora* are effective against all developmental stages of *T. absoluta* larvae (Van Damme *et al.*, 2016). In addition, Kamali *et al.* (2018) found that *S. carpocapsae* and *H. bacteriophora* were highly effective against fourth-instar larvae of *T. absoluta*. These findings indicate that EPNs have potential for use in *T. absoluta* management and may be an important component of an integrated pest management strategy.

7.13 Predators

A diverse range of predators have been observed on *T. absoluta* in field. These include Araneidae, *Anthicus* sp., *Cycloneda sanguinea*, *Orius* sp., and *Xylocoris* sp. *Nesidiocoris tenuis*, *Macrolophus pygmaeus*, and *Dicyphus* spp., among the mirids, feed on *T. absoluta* eggs and young larvae, thus reducing and/or suppressing the pest population on tomato crop. *Xylocoris* sp., *C. sanguinea*, and members of Phlaeothripidae are identified as key predators of both egg and larval stages of *T. absoluta*. Another anthocorid, *Orius insidiosus*, is an important predator of *T. absoluta* eggs and larvae (Salehi *et al.*, 2016). The mite *Pyemotes* sp. has also been observed to feed on *T. absoluta* larvae, pupae, and adults, indicating its potential use in natural control of the pest. The predatory stinkbug, *Podisus nigrispinus* (Heteroptera: Pentatomidae), has been extensively studied for its life history, predatory behaviour, and dispersal in greenhouse environments, specifically in relation to *T. absoluta* (Vivan *et al.*, 2003). The green lacewing, *Chrysoperla externa*, has been proved to be an effective predator of *T. absoluta* larvae in laboratory (Strapasson *et al.*, 2014).

7.14 Parasitoids

The parasitic wasp *Trichogramma achaeae*, is one of the most effective (Cascone *et al.*, 2015) parasites in *T. absoluta* eggs, reducing the pest population in tomato crops. Other parasitic wasps, such as *Necremnus tutae* and *Pseudapanteles dignus*, have been proved to parasitize *T. absoluta* larvae, although their efficacy may vary depending on environmental conditions. The parasitic fly *Dacnusa sibirica* is another natural enemy of *T. absoluta*, with its larvae feeding on *T. absoluta* larvae.

7.15 Chemical Control

The introduction of the tomato pinworm, *T. absoluta*, to Neotropical America posed an immediate threat to tomato production, leading to widespread insecticide use in affected areas. However, due to the early colonization of tomato fields, insect attacks on many plant sections, and canopy protection, it is difficult to control this pest using insecticides alone. This has led to poor efficacy of insecticides against *T. absoluta*, resulting in repeated spraying and insecticide abuse. To mitigate, formulation adjuvants and improved spraying technology have been employed. However, the lack of control alternatives, particularly for open-field tomatoes, still necessitates multiple insecticide applications throughout the cropping cycle. Over time, the compounds used to combat *T. absoluta* have been modified, with organophosphates and pyrethroids being the only available insecticides. In tomato management as organophosphate use declined, the nereis toxin analogue cartap and avermectin abamectin were introduced and used in combination with pyrethroids. Later, oxadiazine indoxacarb and chitin biosynthesis inhibitors diflubenzuron, tefubenzuron, and trifumuron became commercially available (Silva *et al.*, 2011). Other chemical classes, such as pyrroles (chlorfenapyr), spinosyns (spinosad), and diamides (chlorantraniliprole and fubendiamide), have been employed for *T. absoluta* control (Silva *et al.*, 2019). Organic tomato cultivation relies on spinosad, azadirachtin, and *Bacillus thuringiensis* toxins (Bt toxins). Chemical pesticides remain an important component of insect pest management despite the development of other control methods. They play a key part in controlling *T. absoluta* infestation, with many insecticides proving effective against this pest in different parts of the world. For instance, dinotefuran, imidacloprid, fenoxycarb, phenthoate, and thiocyclam insecticides were found to have toxic effects on moths and 3rd instar larvae of *T. absoluta* under laboratory conditions. In addition, botanical extracts of neem (*Azadirachta indica*) and *Jatropha curcas* have insecticidal effects on *T. absoluta* eggs and larvae. However, *T. absoluta* is extremely difficult to control using chemical insecticides due to its ability to develop resistance and its larvae's ability to mine within plant tissue, providing protection from contact insecticides. Therefore, a system of IPM that includes selective chemicals, such as those that are beneficial to other insects, is necessary to control this pest in tomato crops.

7.16 Conclusion

T. absoluta is a serious pest of tomato crops, causing significant yield losses and economic damage to farmers worldwide. Its invasive potential and ability to rapidly develop resistance to pesticides make it a challenging pest to manage. However, integrated pest management strategies that combine cultural, biological, and chemical suppression techniques have shown positive results in reducing *T. absoluta* populations and minimizing crop damage. In the future, continued research and development of novel control methods, such as the use of pheromone-based traps and genetically modified crops, may prove effective in managing *T. absoluta*. Additionally, increased international cooperation and communication among researchers and growers can help to prevent the spread of this pest to new regions.

References

Attygalle, A.B., Jham, G.N., Svatoš, A., Frighetto, R.T., Meinwald, J. *et al.* (1995) Microscale, random reduction: Application to the characterization of (3E,8Z,11Z)-3,8,11-tetradecatrienyl acetate, a new lepidopteran sex pheromone. *Tetrahedron Letters* 36, 5471–5474.

Bajracharya, A.S., Mainali, R.P., Bhat, B., Bista, S., Shashank, P.R. *et al.* (2016) The first record of South American tomato leaf miner, *Tuta absoluta* (Meyrick 1917) (Lepidoptera: Gelechiidae) in Nepal. *Journal of Entomology and Zoology Studies* 4, 1359–1363.

Bawin, T., Dujeu, D., De Backer, L., Fauconnier, M.L., Lognay, G. *et al.* (2015) Could alternative solanaceous hosts act as refuges for the tomato leafminer, *Tuta absoluta*? *Arthropod-Plant Interactions* 9, 425–435. DOI: 10.1007/s11829-015-9383-y.

Bawin, T., Dujeu, D., De Backer, L., Francis, F. and Verheggen, F.J. (2016) Ability of *Tuta absoluta* (Lepidoptera: Gelechiidae) to develop on alternative host plant species. *The Canadian Entomologist* 148, 434–442. DOI: 10.4039/tce.2015.59.

Biondi, A. and Desneux, N. (2019) Special issue on *Tuta absoluta*: Recent advances in management methods against the background of an ongoing worldwide invasion. *Journal of Pest Science* 92, 1313–1315. DOI: 10.1007/s10340-019-01132-6.

Biondi, A., Guedes, R.N.C., Wan, F.H. and Desneux, N. (2018) Ecology, worldwide spread, and management of the invasive south american tomato pinworm, tuta absoluta: past, present, and future. *Annual Review of Entomology*. DOI: 10.1146/annurev-ento-031616-034933.

Böttger, A., Vothknecht, U., Bolle, C., Wolf, A., Böttger, A. *et al.* (2018) Terpenes and terpenoids. In: *Lessons on Caffeine, Cannabis & Co: Plant-Derived Drugs and Their Interaction with Human Receptors*. Springer, pp. 153–170.

Cagnotti, C.L., Riquelme Virgala, M., Botto, E.N. and López, S.N. (2018) Dispersion and persistence of *Trichogrammatoidea bactrae* (Nagaraja) over *Tuta absoluta* (Meyrick), in tomato greenhouses. *Neotropical Entomology* 47, 553–559. DOI: 10.1007/s13744-017-0573-4.

Camargo, R.A., Barbosa, G.O., Possignolo, I.P., Peres, L.E., Lam, E. *et al.* (2016) RNA interference as a gene silencing tool to control *Tuta absoluta* in tomato (*Solanum lycopersicum*). *PeerJ* 4, e2673. DOI: 10.7717/peerj.2673.

Campos, M.R., Biondi, A., Adiga, A., Guedes, R.N.C. and Desneux, N. (2017) From the western Palaearctic region to beyond: *Tuta absoluta* 10 years after invading Europe. *Journal of Pest Science* 90, 787–796.

Caparros Megido, R., Haubruge, E. and Verheggen, F. (2013) Pheromone-based management strategies to control the tomato leaf miner, *Tuta absoluta* (Lepidoptera: Gelechiidae). A review. *Biotechnology, Agronomy, Society and Environment* 17, 273–282.

Cascone, P., Carpenito, S., Slotsbo, S., Iodice, L., Sørensen, J.G. *et al.* (2015) Improving the efficiency of *Trichogramma achaeae* to control *Tuta absoluta*. *BioControl* 60, 761–771.

Cherif, A. and Verheggen, F. (2019) A review of *Tuta absoluta* (Lepidoptera: Gelechiidae) host plants and their impact on management strategies. *Biotechnology, Agronomy, Society and Environment* 23, 243–250.

de Almeida, K.C., de Resende, J.T.V. d., Hata, F.T., Oliveira, L.V.B. and Neto, J.G. (2023) Characterization of *Solanum* sp. *Lycopersicon* section for density and types of leaf trichomes and resistance to whitefly and tomato pinworm. *Scientia Horticulturae* 310, 111746.

De Brito, E.F., Baldin, E.L.L., de Carvalho Macedo Silva, R., do Prado Ribeiro, L. and Vendramim, J.D. (2015) Bioactivity of *Piper* extracts on *Tuta absoluta* (Lepidoptera: Gelechiidae) in tomato. *Pesquisa Agropecuária Brasileira* 50, 196–202.

Desneux, N., Han, P., Mansour, R., Arnó, J., Brévault, T. *et al.* (2022) Integrated pest management of *Tuta absoluta*: Practical implementations across different world regions. *Journal of Pest Science* 95, 1–23.

Dhanapal, R., Singh, R.N., Raghuraman, M., Mohan, M., Subaharan, K. *et al.* (2021) Evaluation of predatory potential and prey stage preference of mirid bug, *Nesidiocoris tenuis* on tomato pinworm, *Tuta absoluta*. *Biologia* 76, 2965–2971.

Dhankhar, R., Regmi, K., Kawatra, A. and Gulati, P. (2023) Trichomics: Trichomes as natural chemical factories. In: *Phytochemical Genomics: Plant Metabolomics and Medicinal Plant Genomics*. Springer Nature Singapore, Singapore, pp. 379–402.

Fand, B.B., Shashank, P.R., Suroshe, S.S., Chandrashekar, K., Meshram, N.M. *et al.* (2020) Invasion risk of the South American tomato pinworm *Tuta absoluta* (Meyrick) (Lepidoptera: Gelechiidae) in India: Predictions based on MaxEnt ecological niche modelling. *International Journal of Tropical Insect Science* 40, 561–571.

Ferracini, C., Bueno, V.H.P., Dindo, M.L., Ingegno, B.L., Luna, M.G. *et al.* (2019) Natural enemies of *Tuta absoluta* in the Mediterranean basin, Europe and South America. *Biocontrol Science and Technology* 29, 578–609.

Ferrara, F.A., Vilela, E.F., Jham, G.N., Eiras, A.E., Picanço, M.C. *et al.* (2001) Evaluation of the synthetic major component of the sex pheromone of *Tuta absoluta* (Meyrick) (Lepidoptera: Gelechiidae). *Journal of Chemical Ecology* 27, 907–917.

Fiaboe, K.R., Agboka, K., Agboyi, L.K., Koffi, D., Ofoe, R. *et al.* (2021) First report and distribution of the South American tomato pinworm, *Tuta absoluta* (Meyrick) (Lepidoptera: Gelechiidae) in Togo. *Phytoparasitica* 49, 167–177.

Glas, J.J., Schimmel, B.C., Alba, J.M., Escobar-Bravo, R., Schuurink, R.C. *et al.* (2012) Plant glandular trichomes as targets for breeding or engineering of resistance to herbivores. *International Journal of Molecular Sciences* 13, 17077–17103.

Guedes, R.N.C., Roditakis, E., Campos, M.R., Haddi, K., Bielza, P. *et al.* (2019) Insecticide resistance in the tomato pinworm *Tuta absoluta*: Patterns, spread, mechanisms, management and outlook. *Journal of Pest Science* 92, 1329–1342.

Guimapi, R.A., Srinivasan, R., Tonnang, H.E., Sotelo-Cardona, P. and Mohamed, A. (2020) Exploring the mechanisms of the spatiotemporal invasion of *Tuta absoluta* in Asia. *Agriculture* 10, 124.

Guimapi, R.Y.A., Mohamed, S.A., Okeyo, G.O., Ndjomatchoua, F.T., Ekesi, S. *et al.* (2016) Modeling the risk of invasion and spread of *Tuta absoluta* in Africa. *Ecological Complexity* 28, 77–93.

Hadapad, A.B. and Hire, R. (2019) Molecular characterisation of tomato leaf miner *Tuta absoluta* populations obtained from different geographical locations of India. *Journal of Biological Control* 33(2), 147–154.

Han, P., Bayram, Y., Shaltiel-Harpaz, L., Sohrabi, F., Saji, A. *et al.* (2019) *Tuta absoluta* continues to disperse in Asia: Damage, ongoing management and future challenges. *Journal of Pest Science* 92, 1317–1327.

Hussain, A. and AlJabr, A.M. (2022) Plant secondary metabolites: Emerging trends in agricultural pests control. In: *New and Future Development in Biopesticide Research: Biotechnological Exploration*. Springer, pp. 187–201.

Idriss, G.E.A., du Plessis, H., Khamis, F.M., Ekesi, S., Tanga, C.M. *et al.* (2020) Host range and effects of plant species on preference and fitness of *Tuta absoluta* (Lepidoptera: Gelechiidae). *Journal of Economic Entomology* 113, 1279–1289.

Jalapathi, S.K., Jayaraj, J., Shanthi, M., Theradimani, M., Venkatasamy, B. *et al.* (2020) Potential of Cry1Ac from *Bacillus thuringiensis* against the tomato pinworm, *Tuta absoluta* (Meyrick) (Gelechiidae: Lepidoptera). *Egyptian Journal of Biological Pest Control* 30, 1–4.

Kamali, S., Karimi, J. and Koppenhöfer, A.M. (2018) New insight into the management of the tomato leaf miner, *Tuta absoluta* (Lepidoptera: Gelechiidae) with entomopathogenic nematodes. *Journal of Economic Entomology* 111, 112–119.

Larbat, R., Adamowicz, S., Robin, C., Han, P., Desneux, N. *et al.* (2016) Interrelated responses of tomato plants and the leaf miner *Tuta absoluta* to nitrogen supply. *Plant Biology* 18, 495–504.

Lobos, E., Occhionero, M., Werenitzky, D., Fernandez, J., Gonzalez, L.M. *et al.* (2013) Optimization of a trap for *Tuta absoluta* Meyrick (Lepidoptera: Gelechiidae) and trials to determine the effectiveness of mass trapping. *Neotropical Entomology* 42, 448–457.

Machekano, H., Mutamiswa, R. and Nyamukondiwa, C. (2018) Evidence of rapid spread and establishment of *Tuta absoluta* (Meyrick) (Lepidoptera: Gelechiidae) in semi-arid Botswana. *Agriculture and Food Security* 7, 1–2.

Mahlangu, L., Sibisi, P., Nofemela, R.S., Ngmenzuma, T. and Ntushelo, K. (2022) The differential effects of *Tuta absoluta* infestations on the physiological processes and growth of tomato, potato, and eggplant. *Insects* 13, 754.

Mansour, R. and Biondi, A. (2021) Releasing natural enemies and applying microbial and botanical pesticides for managing *Tuta absoluta* in the MENA region. *Phytoparasitica* 49, 179–194.

Mansour, R., Brévault, T., Chailleux, A., Cherif, A., Grissa-Lebdi, K. *et al.* (2018) Occurrence, biology, natural enemies and management of *Tuta absoluta* in Africa. *Entomologiae Generalis* 38, 83–112.

Michereff Filho, M., Vilela, E.F., Jham, G.N., Attygalle, A., Svatos, A. *et al.* (2000) Initial studies of mating disruption of the tomato moth, *Tuta absoluta* (Lepidoptera: Gelechiidae) using synthetic sex pheromone. *Journal of the Brazilian Chemical Society* 11, 621–628.

Mkonyi, L., Rubanga, D., Richard, M., Zekeya, N., Sawahiko, S. *et al.* (2020) Early identification of *Tuta absoluta* in tomato plants using deep learning. *Scientific African* 10, e00590.

Naselli, M., Biondi, A., Tropea Garzia, G., Desneux, N., Russo, A. *et al.* (2017) Insights into food webs associated with the South American tomato pinworm. *Pest Management Science* 73, 1352–1357.

Peiffer, M., Tooker, J.F.LutheD.S.Felton, G.W. (2009) Plants on early alert: Glandular trichomes as sensors for insect herbivores. *New Phytologist* 184(3), 644–656.

Pereira, E., Farias, E., Ribeiro, A., Alvarenga, E., Aguiar, A. *et al.* (2019) Toxicity of piperine amide analogs toward the tomato pinworm *Tuta absoluta* (Lepidoptera: Gelechiidae) and risk assessment for two predators. *Horticulturae* 5, 70.

Pyšek, P., Hulme, P.E., Simberloff, D., Bacher, S., Blackburn, T.M. *et al.* (2020) Scientists' warning on invasive alien species. *Biological Reviews of the Cambridge Philosophical Society* 95, 1511–1534.

Resende, J.T.V.D., Maluf, W.R., Faria, M.V., Pfann, A.Z. and Nascimento, I.R. (2006) Acylsugars in tomato leaflets confer resistance to the South American tomato pinworm, *Tuta absoluta* Meyr. *Scientia Agricola* 63, 20–25.

Roditakis, E., Vasakis, E., Grispou, M., Stavrakaki, M., Nauen, R. *et al.* (2015) First report of *Tuta absoluta* resistance to diamide insecticides. *Journal of Pest Sciences* 88, 9–16.

Rostami, E., Madadi, H., Abbasipour, H., Allahyari, H. and Cuthbertson, A.G.S. (2020) Pest density influences on tomato pigment contents: The South American tomato pinworm scenario. *Entomologiae Generalis* 40, 195–205.

Roumani, M., Le Bot, J., Boisbrun, M., Magot, F., Péré, A. *et al.* (2022) Transcriptomics and metabolomics analyses reveal high induction of the phenolamide pathway in tomato plants attacked by the leafminer *Tuta absoluta*. *Metabolites* 12(6), 484.

Sabbour, M.M. (2014) Biocontrol of the tomato pinworm *Tuta absoluta* (Meyrick) (Lepidoptera: Gelechiidae) in Egypt. *Middle East Journal of Agriculture Research* 3, 499–503.

Sadique, M., Ishtiaq, M., Qayyum, M.A., Alkherb, W.A., Abbasi, A. *et al.* (2025) Comparison of pheromone lures and sticky pad color for capturing *Tuta absoluta* (Lepidoptera: Gelechiidae). *Insects* 16(1), 94.

Salehi, Z., Yarahmadi, F., Rasekh, A. and Sohani, N.Z. (2016) Functional responses of *Orius albidipennis* Reuter (Hemiptera, Anthocoridae) to *Tuta absoluta* Meyrick (Lepidoptera, Gelechiidae) on two tomato cultivars with different leaf morphological characteristics. *Entomologiae Generalis* 36, 127–136.

Sankarganesh, E., Firake, D.M., Sharma, B., Verma, V.K. and Behere, G.T. (2017) Invasion of the South American tomato pinworm, *Tuta absoluta*, in Northeastern India: A new challenge and biosecurity concerns. *Entomologiae Generalis* 36, 335–345.

Santana, P.A., Kumar, L., Da Silva, R.S. and Picanço, M.C. (2019) Global geographic distribution of *Tuta absoluta* as affected by climate change. *Journal of Pest Sciences* 92, 1373–1385.

Shashank, P.R., Chandrashekar, K., Meshram, N.M. and Sreedevi, K. (2015) Occurrence of *Tuta absoluta* (Lepidoptera: Gelechiidae) an invasive pest from India. *Indian Journal of Entomology* 77, 323–329.

Shashank, P.R., Twinkle, S., Chandrashekar, K., Meshram, N.M., Suroshe, S.S. *et al.* (2018) Genetic homogeneity in South American tomato pinworm, *Tuta absoluta*: A new invasive pest to oriental region. *3 Biotech*8350.

Shiberu, T. and Getu, E. (2018) Evaluation of bio-pesticides on integrated management of tomato leafminer, *Tuta absoluta* (Meyrick) (Gelechiidae: Lepidoptera) on tomato crops in Western Shewa of Central Ethiopia. *Entomol. Ornithol. Herpetol* 7(1), 10–14.

Silva, D.B., Weldegergis, B.T., Van Loon, J.J. and Bueno, V.H. (2017) Qualitative and quantitative differences in herbivore-induced plant volatile blends from tomato plants infested by either *Tuta absoluta* or *Bemisia tabaci*. *Journal of Chemical Ecology* 43, 53–65.

Silva, G.A., Picanço, M.C., Bacci, L., Crespo, A.L., Rosado, J.F. *et al.* (2011) Control failure likelihood and spatial dependence of insecticide resistance in the tomato pinworm, *Tuta absoluta*. *Pest Management Science* 67, 913–920.

Silva, G.A., Queiroz, E.A., Arcanjo, L.P., Lopes, M.C., Araújo, T.A. *et al.* (2021) Biological performance and oviposition preference of tomato pinworm *Tuta absoluta* when offered a range of solanaceous host plants. *Scientific Reports* 11, 1153.

Silva, J.E., Ribeiro, L.M., Vinasco, N., Guedes, R.N.C. and Siqueira, H.Á.A. (2019) Field-evolved resistance to chlorantraniliprole in the tomato pinworm *Tuta absoluta*: Inheritance, cross-resistance profile, and metabolism. *Journal of Pest Sciences* 92, 1421–1431.

Singh, S., Yadav, G.S., Das, A., Das, B., Devi, H.L. *et al.* (2021) Bioefficacy, environmental safety and synergistic impacts of biorational formulations against whitefly, leafhopper and blister beetle in organic okra ecosystem. *Journal of Agricultural Sciences* 159, 373–384.

Singh, S., Das, B., Devi, H.L., Sunani, S.K., Suklabaidya, A. *et al.* (2022) Evaluation of biorational pesticides against *Bemisia tabaci* Gennadius and *Amrasca biguttula biguttula* Ishida and its impact on natural enemies population in cowpea ecosystem. *Archives of Phytopathology Plant Protection* 55, 63–75.

Skendžić, S., Zovko, M., Živković, I.P., Lešić, V. and Lemić, D. (2021) The impact of climate change on agricultural insect pests. *Insects* 12, 440.

Smith, J.D., Dubois, T., Mallogo, R., Njau, E.F., Tua, S. *et al.* (2018) Hosts range of the invasive tomato pest *Tuta absoluta* Meyrick (Lepidoptera: Gelechiidae) on solanaceous crops and weeds in Tanzania. *The Florida Entomologist* 101, 573–579.

Strapasson, P., Pinto-Zevallos, D.M., Paudel, S., Rajotte, E.G., Felton, G.W. *et al.* (2014) Enhancing plant resistance at the seed stage: Low concentrations of methyl jasmonate reduce the performance of the leaf miner *Tuta absoluta* but do not alter the behavior of its predator *Chrysoperla externa*. *Journal of Chemical Ecology* 40, 1090–1098.

Sylla, S., Brévault, T., Monticelli, L.S., Diarra, K. and Desneux, N. (2019) Geographic variation of host preference by the invasive tomato leaf miner *Tuta absoluta*: Implications for host range expansion. *Journal of Pest Sciences* 92, 1387–1396.

Tarusikirwa, V.L., Machekano, H., Mutamiswa, R., Chidawanyika, F. and Nyamukondiwa, C. (2020a) *Tuta absoluta* (Meyrick) (Lepidoptera: Gelechiidae) on the "offensive" in Africa: Prospects for integrated management initiatives. *Insects* 11(11), 764.

Tarusikirwa, V.L., Mutamiswa, R., English, S., Chidawanyika, F. and Nyamukondiwa, C. (2020b) Thermal plasticity in the invasive South American tomato pinworm *Tuta absoluta* (Meyrick) (Lepidoptera: Gelechiidae). *Journal of Thermal Biology* 90, 102598.

Tonnang, H.E., Mohamed, S.F., Khamis, F. and Ekesi, S. (2015) Identification and risk assessment for worldwide invasion and spread of *Tuta absoluta* with a focus on Sub-Saharan Africa: Implications for phytosanitary measures and management. *PLoS ONE* 10, e0135283.

Tumuhaise, V., Khamis, F.M., Agona, A., Sseruwu, G. and Mohamed, S.A. (2016) First record of *Tuta absoluta* (Lepidoptera: Gelechiidae) in Uganda. *International Journal of Tropical Insect Science* 36(3), 135–139.

Urbaneja, A., González-Cabrera, J., Arno, J. and Gabarra, R. (2012) Prospects for the biological control of *Tuta absoluta* in tomatoes of the Mediterranean basin. *Pest Management Science* 68(9), 1215–1222.

Van Damme, V.M., Beck, B.K., Berckmoes, E., Moerkens, R., Wittemans, L. *et al.* (2016) Efficacy of entomopathogenic nematodes against larvae of *Tuta absoluta* in the laboratory. *Pest Management Science* 72(9), 1702–1709.

Verheggen, F. and Fontus, R.B. (2019) First record of *Tuta absoluta* in Haiti. *Entomologia Generalis* 38(4), 349–353.

Vivan, L.M., Torres, J.B. and Veiga, A.F.S.L. (2003) Development and reproduction of a predatory stinkbug, *Podisus nigrispinus* in relation to two different prey types and environmental conditions. *BioControl* 48(2), 155–168.

Vivekanandhan, P., Swathy, K., Sarayut, P. and Patcharin, K. (2024) Biology, classification, and entomopathogen-based management and their mode of action on *Tuta absoluta* (Meyrick) in Asia. *Frontiers in Microbiology* 15, 1429690.

Vuolo, M.M., Lima, V.S. and Junior, M.R.M. (2019) Phenolic compounds: Structure, classification, and antioxidant power. In: *Bioactive Compounds*. Woodhead Publishing, pp. 33–50.

Walsh, T.K., Heckel, D.G., Wu, Y., Downes, S., Gordon, K.H.J. *et al.* (2022) Determinants of insecticide resistance evolution: Comparative analysis among heliothines. *Annual Review of Entomology* 67, 387–406.

Wang, F., Park, Y.L. and Gutensohn, M. (2020) Glandular trichome-derived sesquiterpenes of wild tomato accessions (*Solanum habrochaites*) affect aphid performance and feeding behavior. *Phytochemistry* 180, 112532.

Zang, L.S., Akhtar, Z.R., Ali, A., Tariq, K. and Campos, M.R. (2022) Flubendiamide resistance and its mode of inheritance in tomato pinworm *Tuta absoluta* (Meyrick) (Lepidoptera: Gelechiidae). *Insects* 13(11), 1023.

8 Insects as Invasive Alien Species (IAS): A Threat to Agroforestry and their Management

Tanmaya Kumar Bhoi[1†], Varsha Giri[2†], Ipsita Samal[3], Deepak Kumar Mahanta[4], Atiraj Rathi[1], Komal J.[5], Prabhulinga T.[6*], Athulya R.[7], Sangeeta Singh[1] and Vipula Vyas[1]

[1]Forest Protection Division, ICFRE-Arid Forest Research Institute (ICFRE-AFRI), Jodhpur, India; [2]Silviculture and Forest management Division, Arid Forest Research Institute, Jodhpur, India; [3]ICAR-National Research Centre on Litchi, Mushahari, Ramna, Muzaffarpur, Bihar, India; [4]Forest Entomology Discipline, Forest Protection Division, Indian Council of Forestry Research and Education (ICFRE) – Forest Research Institute (ICFRE–FRI), Dehradun, Uttarakhand, India; [5]Basic Seed Multiplication and Training Centre, Central Silk Board, Pendari, Bilaspur, Chhattisgarh, India; [6]ICAR – National Bureau of Agricultural Insect Resources (ICAR–NBAIR), Bengaluru, Karnataka, India; [7]Forest Protection Division, Institute of Wood Science and Technology, Bengaluru, Karnataka, India

Abstract

Agroforestry boosts the environmental, economic, and social advantages of complex socio-ecological systems and delivers a variety of ecosystem services by breaking the chain reaction of deforestation, soil erosion, and degradation. As animals lose habitat, fewer ecosystem services are supplied, fewer resources become accessible, and biodiversity is reduced. These landscape alterations foster the introduction, spread, and abundance of non-native species, which modifies human population structures and threatens ecological equilibrium. Biological invasion is the introduction and spread of non-native species with great adaptation and dissemination. Due to inadvertent translocation, these thrive, reproduce, and become established. The International Union for the Conservation of Nature (IUCN) calls foreign organisms that colonize natural or semi-natural ecosystems invasive alien species (IAS). India has 173 foreign species of which 23 are insects and 47 have agricultural impact. Habitat destruction has contributed to the extinction of countless native species and the degradation of their natural habitats. Invasive species' hard-to-eliminate features give them an advantage. Invasive organisms grow quickly, reproduce easily, and have few natural adversaries. This chapter focuses on the routes by which pests enter the agroforestry system, the characteristics of IAS, compared with native pest species, the current status of IAS damaging agroforestry

*Corresponding author: prabhut07@gmail.com

†Tanmaya Kumar Bhoi and Vipula Vyas have equally contributed to this article, consider them as co-first author.

 Biological Invasions in the Tropics: Monitoring, Mechanism and Management (eds Akshay Kumar Chakravarthy and Pathour Rajendra Shashank)
DOI: 10.1079/9781836991977.0008

ecosystems, the effectiveness of integrated pest management (IPM) strategies for combating these pests, and the role played by regulatory organizations in eradicating invasives.

Keywords: Agroforestry, biological invasion, Invasive Alien Species (IAS), socio-economic impact

8.1 Introduction

Agroforestry, a sustainable land-use system that integrates trees with crops and/or livestock, plays a significant role in enhancing biodiversity, restoring soil fertility, improving microclimatic conditions, and strengthening livelihood resilience, especially in ecologically fragile regions like India (Roy *et al.*, 2017; Snyder, 2019). This practice not only boosts environmental sustainability but also delivers economic and social benefits by mitigating land degradation, halting deforestation, and reducing soil erosion (van Noordwijk, 2021). However, the intensification of agriculture and agroforestry into natural landscapes often leads to habitat fragmentation and ecological simplification, making these systems increasingly vulnerable to biological invasions (Phalan *et al.*, 2011). Landscape modifications reduce biodiversity, disrupt ecological functions, and create ecological niches that favour the establishment of Invasive Alien Species (IAS), especially insect pests that can rapidly adapt and colonize disturbed habitats (Pauchard and Alaback, 2004; Bateman *et al.*, 2013).

Insect IAS are one of the most pressing threats to agroforestry systems due to their capacity to outcompete native species, rapidly reproduce, and exploit new environments with minimal resistance. These invasive pests often lack natural enemies in the introduced regions, allowing them to proliferate unchecked and cause extensive ecological and economic damage (Ramos *et al.*, 2015). Traits such as high fecundity, broad host range, phenotypic plasticity, and mutualistic interactions enable them to dominate agroecosystems and outcompete native species (Simberloff and Holle, 1999). The establishment of monoculture plantations within agroforestry frameworks further facilitates pest invasions by eliminating natural competitors and increasing resource availability and light at potential invasion sites (Parendes and Jones, 2000). For instance, *Leptocybe invasa* and *Gonipterus* spp. have significantly impacted eucalyptus-based agroforestry systems, while *Spodoptera frugiperda* (fall armyworm) and *Phenacoccus solenopsis* (cotton mealybug) have disrupted multiple cropping systems integrated with agroforestry across India. The surge in international trade further exacerbates the problem. India, being one of the fastest-growing global economies, has witnessed increased movement of goods and plant materials, resulting in heightened risk of unintentional pest introductions. In 2014 alone, India exported to over 190 countries and imported goods worth $462.9 billion from approximately 140 nations, creating numerous pathways for invasive species to enter the country (Gupta *et al.*, 2019). Once introduced, these species often establish successfully due to favourable climatic conditions and ecological opportunities in disturbed or intensively managed agroforestry landscapes.

Biological invasions by insect pests are now recognized as the second most significant cause of biodiversity loss globally, after habitat destruction. In the context of agroforestry, their impact includes reduced crop and tree productivity, increased pest management costs, and ecological degradation through the displacement of native species and disruption of ecosystem services. These challenges highlight the urgent need for understanding the mechanisms and pathways of invasion, assessing the current status of invasive insect species in agroforestry systems, and evaluating the extent of economic and ecological losses they cause. To effectively manage these threats, a multifaceted strategy is required. This should include early detection and rapid response (EDRR) systems, strict quarantine and biosecurity regulations, development of pest-resistant agroforestry species, and the promotion of ecologically sound Integrated Pest Management (IPM) practices. Biological control using natural predators and parasitoids, along with active involvement of farming communities in surveillance and awareness, is critical. Regulatory agencies such as the Plant Protection Quarantine and Storage Division (PPQS), the National Bureau of Agricultural Insect Resources (NBAIR), and international bodies such as the International Plant Protection Convention (IPPC) must work collaboratively to track invasive pests and support research-based interventions. Our objective in this context was to assess the pest invasion pathways into the agroforestry system, status of invasives, economic and ecological losses incurred due to invasive pest species, and the role of regulatory organizations.

8.1.1 Introduced as contaminants

One of the most lucrative industries is thought to be the trade in logs. Although it generates a significant amount of foreign exchange, it also has the side effect of introducing alien species. These logs are a breeding ground for numerous insects and viruses due to their extended time of untreated storage in godowns. These logs' importation also spreads the resident alien species. Many different food items and seeds have been imported into India from other nations. When there was a lax legal restriction on the importation of food commodities in the past, it was demonstrated that mixing of pollutants with food lot was detrimental to the importing country (Piel *et al.*, 2008).

8.1.2 Living industry pathway

The term "living industry" refers to the trading of living things, such as horticulture, agriculture, aquaculture/aquariums, pets, and live seafood. These live things could very well accompany their respective pests as they are exported. For instance, it was speculated that the entry of apple rootstock from England may allow apple woolly aphids to spread in India (Gupta *et al.*, 2015).

8.1.3 Transportation related

Tropical countries such as India, known for their diverse landscapes, rich biodiversity, and deep cultural history, attract millions of domestic and international tourists each year. This surge in tourism, while economically beneficial, also poses significant ecological risks, particularly the increased likelihood of invasive alien species (IAS) introductions, a challenge that is often overlooked. As tourism expands across ecological zones, it unintentionally facilitates the spread of pests, pathogens, and other foreign organisms, posing threats to native biodiversity and agroforestry systems (Ramos *et al.*, 2015; Gupta *et al.*, 2019). One of the most common pathways is the inadvertent introduction of invasive organisms via tourists' personal belongings. Insects, fungal spores, or seeds can cling to clothing, footwear, luggage, or outdoor gear. When tourists travel between locations or countries, these stowaway organisms are transported into new habitats, often without being detected. Once introduced into favourable climates such as those in tropical regions, they can establish, reproduce rapidly, and outcompete native species, especially in disturbed or simplified landscapes such as agroforestry systems (Simberloff and Holle, 1999).

Tourism also results in the unregulated movement of biological materials across diverse ecological zones. Tourists commonly travel from forests to agricultural zones, national parks, and coastal areas, increasing the chance of organisms being accidentally transferred from one region to another. The landscape fragmentation and habitat modification that accompany the expansion of tourism infrastructure (roads, resorts, adventure sites) further create ecological openings that facilitate the establishment and proliferation of invasive species (Pauchard and Alaback, 2004; Bateman *et al.*, 2013). Cargo shipments linked to tourism and hospitality services including the transportation of luggage, outdoor equipment, furniture, ornamental plants, and food products serve as another critical invasion pathway. Insects can survive in wooden crates, bamboo artefacts, plant materials, and construction goods imported for the tourism industry. Increased movement through land, sea, and air routes provides multiple entry points for alien pests into new environments (Gupta *et al.*, 2019). The economic and biological costs of such invasions are often severe, as pests once established in agroforestry systems or natural ecosystems are difficult to eradicate and can cause long-term degradation (Ramos *et al.*, 2015).

Moreover, the exchange and transportation of fruits, vegetables, and food products between tourists and locals or across borders further intensifies this risk. Contaminated agricultural goods may carry insect eggs, larvae, or pathogens that survive transit and establish in new territories. For example, *Phenacoccus solenopsis* (cotton mealybug) and *Spodoptera frugiperda* (fall armyworm) have both spread rapidly in India due in part to increased trade and movement, including through unregulated plant material carried by travellers (Gupta *et al.*, 2019). Lastly, contaminated food waste left by tourists, especially fruits and vegetables, may create microhabitats for invasive pests in new environments. Discarded organic material in and around protected areas or agroforestry zones may act as breeding grounds for insect pests, which are then free to establish and spread due to a lack of natural predators and disturbed ecological balance (Simberloff and Holle, 1999). Monocultures and low-diversity systems common in agroforestry further exacerbate this, offering little ecological resistance to pest establishment (Parendes and Jones, 2000). The rise of tourism in tropical countries such as India, combined with increased international trade and inadequate biosecurity measures, contributes significantly to the spread of invasive insect species and other harmful organisms. These biological invasions threaten the stability and productivity of agroforestry systems, as well as the broader ecological integrity of the region. Therefore, an integrated approach involving regulatory agencies, researchers, and public awareness campaigns is urgently needed to prevent, detect, and manage these risks effectively (Ramos *et al.*, 2015; Roy *et al.*, 2017; Gupta *et al.*, 2019).

8.2 Criteria Adopted for Designating an Alien Species as Invasive

The National Biodiversity Authority, Centre for Biodiversity Policy and Law, and MoEFCC (Ministry of Environment, Forests, and Climate Change) of the Government of India gathered and carefully analysed the easily available ecosystem-wise literature on invasive alien species. Based on a straightforward process created by Centre for Biodiversity Policy and Law (CEBPOL), the list was largely examined to confirm the alien status and invasiveness of the species. The confirmed list was given to the National Biodiversity Authority (NBA) invasive species expert group for review after the initial confirmation. To determine the invasiveness of the species in India, the committee discussed the lists created by CEBPOL and NBA and advised adding the invasive features on a graded scale. The committee has recommended using crucial invasive features, such as invasiveness, effects, range of expansion, and others, to classify alien species as invasive in India after examining the literature (Table 8.1).

Table 8.1. Invasive attributes used to confirm the invasive status of the species reported in Indian ecosystems (Centre for Biodiversity Policy and Law (CEBPOL), 2019).

Status No.	Invasive attributes
Invasiveness	
1.	E– Invasive Elsewhere
2.	RMS – Rapid Multiplication and Spread in different ecosystems
3.	MMR – Multiple Modes of Reproduction
4.	MMD – Multiple Modes of Dispersion
Impacts	
1.	B1 – Affecting ecosystem functions and services
2.	B2 – Biodiversity loss
3.	B3 – Economic loss and health hazard
Invasion areas	
1.	RE – Range Extension

8.3 Invasion Process

Introduction, establishment, dissemination, and naturalization are the successive processes that make up the invasion of an alien species (Kolar and Lodge, 2002; Howard, 2008).

8.3.1 Introduction

The insect must first move from its current habitat or arrange to be moved in order to settle in a new area outside of its normal range. Passive transport, as this insect movement is termed, is caused by vectors. Humans or human consignments are currently the most prevalent vectors. The invasive species are often transported passively, which makes them challenging to control. Even after thorough quarantine inspections, it can be difficult to detect the ambiguous early stages of the bug that visitors may have carried in. A single gravid female bug can also wreak havoc in the newly introduced environment due to its profuse reproduction and lack of a natural enemy. One of the most typical examples of insect passive transportation is the presence of timber beetles (Cerambycidae) in Antarctica, which are anticipated to have arrived by transported timbers in one of the base stations (Osyczka *et al.*, 2012). Insects occasionally possess a built-in ability for long-distance migration. Therefore, they migrate away from a location once favourable conditions change. In India, *Spodoptera frugiperda*, a recent invasive insect pest, is known as the fall armyworm. Due to their exceptional flying skills, they are likely have emigrated from the African subcontinent to the Indian subcontinent (Bajracharya *et al.*, 2019), although the exact way the invasive pest arrived is still unknown. Another illustration is the migration of the North American-native Monarch butterfly, *Danaus plexippus* (Nymphalidae), which was discovered in Australia in the early 19th century (Nail *et al.*, 2019) (Appendix Table 1).

8.3.2 Establishment

Even while brief colonization is fairly common, an insect cannot be considered invasive unless it has successfully established itself in the new environment, which is only achievable once the invaded bug overcomes the environmental obstacles. Instead of local colonization, the development of an invasive species is based on the volume of propagules transported. It is believed that causing disturbances in the receiving environment aids in promoting the establishment. Because IAS are frequently adapted to disturbances, and so are poised for colonization in such habitats, or because native species are rarely suited to the timing of human-mediated disturbances, incidence and intensity favour establishment (Diez *et al.*, 2012). Insects are highly susceptible to invasiveness because of the rise in resource availability and the decline in biotic resistance.

8.3.3 Expansion/Spreading

The process of spreading is when a species that has already established itself extends to other places. Environmental elements such as weather, microclimate, and habitat quality influence how much something spreads (Bonte *et al.*, 2004). Interspecific rivalry and predators are a part of the community context for spreading. An individual insect's decision to spread is also influenced by either morphological or behavioural reactions, such as aggression or social behaviour (similar to how aphid wings develop). In addition, the established bug can spread via transportation facilitated by people. Since human-mediated transportation is swift, the latter preserves resources for the invasive insect, however the habitat porosity of the surroundings to which the infested insect has spread substantially affects the success rate of the spread (Renault *et al.*, 2018).

8.4 Characteristics of an Invasive Species

Invasive species are organisms that are introduced to new environments intentionally or unintentionally and establish, spread, and cause harm to native ecosystems, biodiversity, and even economies. Invasive insect pests, in particular, possess several biological and ecological traits that give them a competitive edge over native species. The following are key characteristics that define an invasive species:

Short life cycle: Invasive species typically have a short generation time, which means they can complete their life cycle from egg to adult in a matter of days or weeks. This rapid reproduction allows their populations to grow explosively in a short time span, increasing their chances of survival and colonization in new environments.

Broad host range: Many invasive pests are polyphagous, meaning they can feed on a wide variety of host plants or animals. This flexibility allows them to survive in diverse ecosystems and exploit multiple food sources, making them more adaptable than specialized native species that depend on a narrow range of hosts.

High fecundity: Invasive insects often lay large numbers of eggs during their lifespan. This high reproductive potential ensures rapid population build-up, even from a small founding population. High fecundity is a major reason why they can quickly dominate new habitats and outnumber native organisms.

Voracious feeders: Invasive species typically consume large amounts of resources, causing extensive damage to their host plants or prey. For example, invasive defoliating insects such as *Spodoptera frugiperda* (fall armyworm) can defoliate entire fields, drastically affecting crop and forest productivity.

Very resilient: They are often highly resilient to environmental stressors such as drought, temperature extremes, or poor food quality. This makes them well-suited to colonize disturbed or marginal habitats where native species may struggle to survive.

High dispersal ability: Invasive species usually possess excellent mobility either through flight, wind dispersal, water currents, or by hitchhiking on human-made transport (vehicles, cargo, packaging). Their ability to disperse over long distances allows them to colonize new regions rapidly and expand their range quickly.

Benefits from mutualist interaction: Some invasive species form mutualistic relationships with other organisms that help them establish and spread. For instance, invasive ants may protect sap-feeding pests such as aphids or mealybugs in return for honeydew, thereby helping both species to thrive and spread together.

Ability to endure a variety of environmental conditions: A hallmark of successful invasive species is their ecological plasticity; they can survive and reproduce under a wide range of temperatures, humidity levels, soil types, and altitudes. This adaptability allows them to invade and establish in different geographic regions, from tropical forests to arid landscapes.

8.5 Effects of Invading Pest Species on the Agricultural Ecosystem

The greatest risk to the agroecosystem is undoubtedly the effect of invading pest species. They take over the ecosystem and compete with local species because they lack natural enemies or controls in the new place. In exchange, the native species may gradually be replaced by non-native ones, which may ultimately destroy the ecosystem as a whole. They have detrimental effects on ecology, biodiversity, health, economy, other elements of human welfare, and agricultural yield. They can change the structure and species composition of ecosystems by eliminating native species or indirectly by altering how nutrients are cycled through the system (Diez *et al.*, 2012). Invasive species are thought to have cost the global economy $1.288 trillion over the previous 50 years.

The direct challenges posed by invasive species include beating native species in a race for food or other resources and spreading illness, which prevents native species from reproducing or results in the death of their young. The indirect risks posed by invasive species include: (1) Changing food webs: by displacing or eliminating native food sources, invasive species have the potential to alter the food web of an ecosystem. The invasive species may offer little or no food value to wildlife. (2) Decreased biodiversity: It has been noted that invasive species have altered the number and diversity of areas that are crucial habitat for native fauna. (3) Modifying ecological circumstances. Invasive species have the potential to alter ecosystem conditions, such as soil chemistry or the intensity of wildfires (Singh *et al.*, 2016).

8.6 Management of Invasive Species

The management of invasive species gets increasingly costly and challenging when populations of invasive insects, plants, and other pests grow and reach high densities, together with any diseases they may carry. The effectiveness of early intervention measures designed to either avoid the introduction of invasive species into an endangered area or to swiftly identify and address new species incursions has been supported by research over the past ten years (Leung *et al.*, 2014; Holden *et al.*, 2016). Proactive management practices, such as foresight, early identification, eradication, and other quick reactions, are more affordable and superior to ones that are reactive (Rout *et al.*, 2014; Epanchin-Niell and Wilen, 2015). Risk assessment, which forecasts the possibility and effects of an invasion, and pathway analysis are two methods for prediction. Regulations and quarantine procedures are just two examples of the many methods used to prevent diseases. In fact, according to Hulme (2009), pathway analysis and subsequent management of such pathways are the "frontline in the prevention of biological invasions" and represent efficient methods (Essl *et al.*, 2015; Tidbury *et al.*, 2016). Early detection relies heavily on surveillance, and eradication, containment, or suppression are the three main fast responses in the event that a target species is discovered (Beric and MacIsaac, 2015). Early intervention tactics frequently function at spatial sizes significantly larger than those used by the majority of land managers. Efficient cooperation between managers, regulators, and researchers at the local, sub-national, national, and global levels is essential. Many components of IPM tactics used to combat well-established pests are also present in early intervention measures for invasive species (Venette and Koch, 2009). IPM, in general, calls for the following: (1) systems to monitor presence and abundance of those species; (2) background knowledge of pestiferous species; (3) a clear statement of the system's goal; (4) a collection of complementary tools and tactics to affect the abundance; and (5) guidance on when management is worthwhile (e.g. economic thresholds). Years of in-depth study into the biology and management of single species in a variety of systems and habitats have resulted in the development of the current IPM programmes.

8.7 Effects of Invasive Species by Risk Analysis

The method of calculating the probability and potential severity for a specific unfavourable occurrence is known as risk assessment. Although the procedure seems straightforward and there is broad advice accessible (Venette, 2015), there are no necessary criteria for determining the threats associated with invasive species (Hulme, 2003). Identifying paths and mechanisms of movement and introduction, characterizing favourable environments and sensitive hosts, and analysing the potential effects of spread and establishment in formerly uninvaded areas are common methodologies to evaluate the dangers posed by invasives (Andersen *et al.*, 2004; Venette, 2015). However, information on the behaviour of alien species and their biology in unfamiliar environments is frequently lacking or nonexistent, which results in imprecise representations of danger that heavily rely on expert opinion (Landis, 2004). The outcomes of such investigations are qualitative, and a risk rating is ordinal (e.g. high, moderate, or low risk). Qualitative analyses could be sufficient to support managers in decisions like whether to permit import of specific goods for survey. When deciding whether to use qualitative or quantitative studies, Baker *et al.* (2015) provides a decision-support system. Discussions on difficult policy topics may be more focused with the aid of quantitative assessments of risk (Gray *et al.*, 1998). For such sophisticated models, numerical models that can evaluate triggers of invasion processes in real-world settings and handle massive geographic data sets are needed (Yemshanov *et al.*, 2009). In fact, multistate decision-support systems were named as crucial research gaps for enhanced risk assessments of invasive species (Andersen *et al.*, 2004) (Fig. 8.1).

8.8 Monitoring and Detection

For the management of invasives, and to create quantitative analyses of their ecological, economic, and social impacts, it is necessary to implement monitoring and inventory programmes and collect high-quality data of distribution. Strong monitoring and inventory systems offer insightful data that guide efforts to stop and contain invasive species. To create programmes it is necessary to know where to concentrate management efforts for educational outreach. There are a total of 71 plant quarantine stations spread out across major and minor ports in India (12 airports, 34 seaports, 11 foreign post offices, and 14 land frontiers), all of which are dedicated to limiting the import of any foreign contaminants. Early discovery of the invading insects is one of the key management

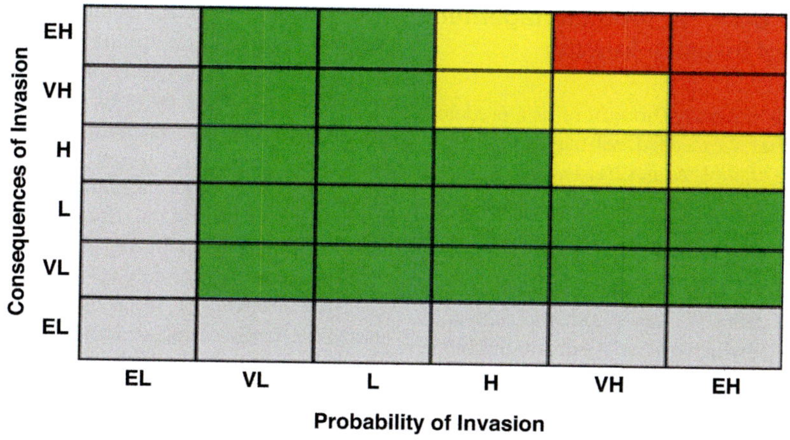

Fig. 8.1. Pest risk matrix: Overall risk (red = high risk; yellow = moderate risk; green = low risk; grey = effectively no risk) EL = Extremely low, VL = Very low, L = Low, H = High, VH = Very high, EH = extremely high. (Author's own image)

strategies. However, the precise identification of the insect is the issue (to species level). Expertise in insect taxonomy is essential for species identification, and this is a constraint in India (Sreedevi *et al.*, 2015). As a result, the Indian government has set up molecular diagnostic facilities throughout the quarantine stations to facilitate and speed up the detection of invasive insects (Station and Cargo, 2003). Monitoring can also be carried out physically using yellow sticky traps, light traps, and other devices, or chemically using pheromones (such as sex pheromones and attractants).

New developments in technologies and methods, such as remote sensing, GIS, GPS, etc., for quantifying the distribution and effects of invasive species, can have a strong impact on the management of IAS. Technologically advanced methods and tools have many benefits, and their application considerably improves the quality and amount of information. They are made to be coupled with other systems, such as monitoring and inventory, because the techniques are frequently used once a species of significance has been detected and a danger has been identified. They can also play a significant part in the creation of decision-support systems. The international movement of plants and animals is subject to legislative restrictions. The FAO, the United Nations, and its regional plant protection organizations signed the International Plant Protection Convention (IPPC) in 1951. (RPPOs). The Sanitary and Phytosanitary (SPS) Agreement under the WTO is primarily concerned with protecting from pests in international trade, and the IPPC provides an international framework for plant protection that includes the development of International Standards for Phytosanitary Measures (ISPMs) for safeguarding plant resources (guidelines for issuing Phytosanitary certificates (PSCs) for export purposes, PRA, Post Entry Quarantine Guidelines, etc.), information exchange on pest status and regulated pests in each country. The Directorate of Plant Protection, Quarantine and Storage (DPPQ&S), ICAR–NBPGR, ICAR–NBAIR, and other crop-specific institutes of ICAR, Ministry of Agriculture and Farmer's Welfare, Government of India, are in charge of monitoring invasive insects in India. The Destructive Insects and Pests Act, 1914 and amendments, the Plant Quarantine (Regulation of Import into India) Order, 2003 – Amendments, and international standards such as the International Plant Protection Convention (IPPC, 2005), the WTO-SPS Agreement, and the International Standards on Phytosanitary Measures (ISPMs) are adopted by DPPQ&S to prevent the entry of invasives into the nation through Phytosanitary Certificates (PSC) issued in compliance with the IPPC protocol for the export of agricultural goods.

8.9 Limit Concept Aids Management of Invasive Insects

A threshold is generally thought of as a predetermined level that, if exceeded, will trigger an action. According to Nichols *et al.* (2014), thresholds of three types, decision, utility, and ecological are pertinent to the management of invasive species in forests and rangelands and other natural resource management decisions. Varying techniques have been used to define ecological thresholds, e.g. "A point or zone at which [there is] a dramatic shift in the condition or of a biological system" (Nichols *et al.*, 2014). "Values of performance variables at which slight changes create considerable changes in the value of management outcomes" are known as utility thresholds. Utility thresholds are "driven" by human values, yet there may also be an ecological foundation. Values of system state variables that should trigger particular management operations are termed decision thresholds (Nichols *et al.*, 2014).

8.10 Goals and Objectives of Managing Invasives

Clear goals and objectives must be established for management efforts towards the invasive species under consideration. For instance, it might not be possible to eradicate a species or return an ecosystem or community to its pre-invasion species composition and functional status. If so, consideration of the detrimental effects of the invasive species, the significance of the invaded community, the effectiveness of any suggested interventions, the actions selected, and the desired outcome should be integrated into decision-making. After that, restoring the afflicted area to a usable, robust state may be the aim. Taking mass trapping, chemical or microbiological controls, and monitoring and detection surveys as examples, population management of *Lymantria dispar* (gypsy moth) can change biotic (e.g. population numbers and stand densities) and abiotic (e.g. climate) factors (Sharov *et al.*, 2002; Schweitzer *et al.*, 2014).

8.11 Integration of Input

Resource managers rarely work with a single-factor priority approach in practice. Instead, they need to integrate, placing special attention on the overarching management objective and prospective tools that might be deployed. Potential tools may be used individually or in groups. A site-specific, multi-tactic decision-making process called integrated pest management optimizes pest treatment in a way that is both financially and environmentally responsible. Strategies, including host vaccination, resistance, outreach, education and regulatory control, control of reproduction, biological control, chemical, cultural and physical control can be combined in a single programme.

8.12 Recent IAS Management Advancements

8.12.1 Improvements in surveys and traps for monitoring

Invasive insects are found in the early phases of invasion using intensive sampling techniques that were particularly used for that purpose. Terentev *et al.* (2022) used aerial photography in conjunction with hyperspectral sensors, multispectral imagery or satellite imagery (e.g. LiDAR, Landsat) to successfully identify invasive insect populations with distinctive physical structures. The semiochemical attractants of numerous insect species, such as the gypsy moth, have been recognized, and traps and artificial lures have been developed for early detection (Sharov *et al.*, 2002). Other approaches considered include employing sniffer dogs to find and identify *Anoplophora glabripennis* (Asian longhorned beetles) (Mankin *et al.*, 2008), using solitary ground nesting wasps for emerald ash borer biosurveillance (Careless *et al.*, 2014), and using acoustic detection for Asian longhorned beetles (Errico, 2013). Efforts to reduce the spotted lantern fly are hampered by the reliance on basic visual surveys for detection. Investigations are needed to identify pheromone behavioural cues and adult and nymphal capturing techniques. The identification and detection of invading plant pathogens, insects, animal and plant diseases, and aquatic creatures are improved by recent developments in molecular diagnostic technologies. To facilitate quick identification of invasives, DNA barcodes are developed (Ball and Armstrong, 2006; Hollingsworth *et al.*,

2011). The capacity to detect morphologically unclear insect developmental stages, like eggs and larvae, that would be challenging to identify using conventional approaches, is one of the main benefits of DNA-based identification. To discriminate between morphologically identical species that are found in cryptic species complexes, molecular approaches are helpful (Lopez and Hoddle, 2014; Cooperband *et al.*, 2016).

8.13 Conclusions

Globalization has benefited humans in many ways, but at a cost. The increased connection by globalization has led to the introduction of several invasive insect pests into the tropics, which have not only destroyed many agricultural crops but also resulted in significant financial loss. Over the 20th century, the scope and overall impacts of biological invasions grew quickly. Each decade has seen an increase in the number of invasive species, the irreversible alteration of more ecosystems, and the impact of invasive alien species on a wider range of functions and processes. These data are offered to help with identification, monitoring, and management. Rapid monitoring methods are required to identify potential areas for the entry of new species. Understanding species interactions and their effects on regional ecosystems is essential. It is necessary to analyse how a specific new species becomes an invasive, what determines invasion rates, and how preventative measures might be established, on regional and global scales. It is important to adopt awareness-raising tactics, such as enlisting the aid of neighbourhood groups, local schools, or volunteers to manage IAS in certain neighbourhoods. The utilization of numerous beneficial data sets and instruments is possible while doing a pest risk analysis. It is crucial to emphasize that pest risk studies are not exhaustive. To make it as quick and simple as possible for individuals who are new to pest risk analysis to understand the discipline, good pest risk analysis practice needs to be summarized in a type of handbook. The construction of pest risk analyses might be better understood if they were published more frequently.

References

Andersen, M.C., Adams, H., Hope, B. and Powell, M. (2004) Risk analysis for invasive species: General framework and research needs. *Risk Analysis* 24(4), 893–900. DOI: 10.1111/j.0272-4332.2004.00487.x.

Bajracharya, A.S.R., Bhat, B., Sharma, P., Shashank, P.R., Meshram, N.M. *et al.* (2019) First record of fall army worm *Spodoptera frugiperda* (J.E. Smith) from Nepal. *Indian Journal of Entomology* 81(4), 635–639. DOI: 10.5958/0974-8172.2019.00137.8.

Baker, R., Eyre, D., Brunel, S., Dupin, M., Reynaud, P. *et al.* (2015) Mapping endangered areas for pest risk analysis. *Pest Risk Modelling and Mapping for Invasive Alien Species* 7, 18–34.

Ball, S.L. and Armstrong, K.F. (2006) DNA bar codes for insect pest identification: A test case with tussock moths (Lepidoptera: *Lymantriidae*). *Canadian Journal of Forest Research* 36(2), 337–350. DOI: 10.1139/x05-276.

Bateman, I.J., Harwood, A.R., Mace, G.M., Watson, R.T., Abson, D.J. *et al.* (2013) Bringing ecosystem services into economic decision-making: Land use in the United Kingdom. *Science* 341(6141), 45–50. DOI: 10.1126/science.1234379.

Beric, B. and MacIsaac, H.J. (2015) Determinants of rapid response success for alien invasive species in aquatic ecosystems. *Biological Invasions* 17(11), 3327–3335. DOI: 10.1007/s10530-015-0959-3.

Bonte, D., Baert, L., Lens, L. and Maelfait, J.P. (2004) Effects of aerial dispersal, habitat specialisation, and landscape structure on spider distribution across fragmented grey dunes. *Ecography* 27(3), 343–349. DOI: 10.1111/j.0906-7590.2004.03844.x.

Careless, P., Marshall, S.A. and Gill, B.D. (2014) The use of *Cerceris fumipennis* (Hymenoptera: *Crabronidae*) for surveying and monitoring emerald ash borer (Coleoptera: *Buprestidae*) infestations in eastern North America. *The Canadian Entomologist* 146(1), 90–105. DOI: 10.4039/tce.2013.53.

Centre for Biodiversity Policy and Law (CEBPOL) (2019) Invasive alien species of India. Available at: http://nbaindia.org/cebpol/pub/iaslist.pdf

Chakravarthy, A.K., Kumar, K.P., Sridhar, V., Prasannakumar, N.R., Nitin, K.S., Nagaraju, D.K. and Reddy, P. V. (2017). Incidence, hosts and potential areas for invasion by rugose spiraling whitefly, *Aleurodicus rugioperculatus* Martin (Hemiptera: Aleyrodidae) in India. *Pest Management in Horticultural Ecosystems*, 23(1), 41–49.

Cooperband, M.F., Stouthamer, R., Carrillo, D., Eskalen, A., Thibault, T., Cossé, A. A., *et al.* (2016). Biology of two members of the *Euwallacea fornicatus* species complex (Coleoptera: Curculionidae: Scolytinae), recently invasive in the U.S.A., reared on an ambrosia beetle artificial diet. *Agricultural and Forest Entomology*, 18, 223–237. DOI: 10.1111/afe.12155.

Diez, J.M., D'Antonio, C.M., Dukes, J.S., Grosholz, E.D. and Olden, J.D. (2012) Will extreme climatic events facilitate biological invasions. *Frontiers in Ecology and the Environment* 10(5), 249–257. DOI: 10.1890/110137.

Epanchin-Niell, R.S. and Wilen, J.E. (2015) Individual and cooperative management of invasive species in human-mediated landscapes. *American Journal of Agricultural Economics* 97(1), 180–198. DOI: 10.1093/ajae/aau058.

Errico, M. (2013) Asian longhorned beetle detector dog pilot project. In: McManus, K.A. and Gottschalk, K.W. (eds) *Proceedings, 23rd U.S. Department of Agriculture Interagency Research Forum on Invasive Species 2012*, Gen. Tech. Rep. NRS-P-114, United States Department of Agriculture, Forest Service, Northern Research Station, Newtown Square, p. 18.

Essl, F., Bacher, S., Blackburn, T.M., Booy, O., Brundu, G. *et al.* (2015) Crossing frontiers in tackling pathways of biological invasions. *BioScience* 65(8), 769–782. DOI: 10.1093/biosci/biv082.

Faizal, M.H., Prathapan, K.D., Anith, K.N., Mary, C.A., Lekha, M. and Rini, C.R. (2006) Erythrina gall wasp *Quadrastichus erythrinae*, yet another invasive pest new to India. *Current Science*, 90(8), 1082.

Gray, G.M., Allen, J.C., Burmaster, D.E., Gage, S.H., Hammitt, J.K. *et al.* (1998) Principles for conduct of pest risk analyses: Report of an expert workshop. *Risk Analysis* 18(6), 773–780. DOI: 10.1023/B:RIAN.0000005922.65955.69.

Gupta, N., Verma, S.C., Sharma, P.L., Thakur, M., Sharma, P. *et al.* (2019) Status of invasive insect pests of India and their natural enemies. *Journal of Entomology and Zoology Studies* 7(1), 482–489.

Gupta, R., Tara, J.S. and Sharma, R. (2015) Management and comparative efficacy of various treatments against *Eriosoma lanigerum* Hausmann on apple trees (*Malus domestica* Borkh.) in Jammu province of Jammu and Kashmir state. *International Journal of Advanced Biological Research* 5(4), 319–321.

Holden, M.H., Nyrop, J.P. and Ellner, S.P. (2016) The economic benefit of time-varying surveillance effort for invasive species management. *Journal of Applied Ecology* 53(3), 712–721. DOI: 10.1111/1365-2664.12617.

Hollingsworth, P.M., Graham, S.W. and Little, D.P. (2011) Choosing and using a plant DNA barcode. *PLoS ONE* 6(5), e19254. DOI: 10.1371/journal.pone.0019254.

Howard, F.W., Pemberton, R.R. and Liu, H. (2008) Erythrina gall wasp, *Quadrastichus erythrinae* (Hymenoptera: Eulophidae) in Florida, and susceptibility of *Erythrina herbacea* (Fabaceae). *Proceedings of the Florida State Horticultural Society* 121, 1.

Hulme, P.E. (2003) Biological invasions: Winning the science battles but losing the conservation war? *Oryx* 37(2), 178–193. DOI: 10.1017/S003060530300036X.

Hulme, P.E. (2009) Trade, transport and trouble: Managing invasive species pathways in an era of globalization. *Journal of Applied Ecology* 46(1), 10–18. DOI: 10.1111/j.1365-2664.2008.01600.x.

IPPC (2005) Identification of risks and management of invasive alien species using the IPPC framework. In: *Proceedings of the workshop on invasive alien species and the International Plant Protection Convention*, 22–26 September 2003, Braunschweig, Germany.

Jalali, S.K. and Singh, S.P. (1989) Release and recovery of an exotic coccinellid predator, *Curinus coeruleus* (Muls.) on subabul psyllid, *Heteropsylla cubana* Crawf. in India. *Journal of Insect Science* 2(2), 158–159.

Jhala, R.C., Bharpoda, T.M. and Patel, M.G. (2008) *Phenacoccus solenopsis* Tinsley (Hemiptera: Pseudococcidae), the mealy bug species recorded first time on cotton and its alternate host plants in Gujarat, India. *Uttar Pradesh Journal of Zoology* 28(3), 403–406.

Josephrajkumar, A., Mohan, C., Babu, M., Krishna, A., Krishnakumar, V. *et al.* (2019) First record of the invasive Bondar's nesting whitefly, *Paraleyrodes bondari* Peracchi on coconut from India. *Phytoparasitica* 47(3), 333–339. DOI: 10.1007/s12600-019-00741-2.

Kolar, C.S. and Lodge, D.M. (2002) Ecological predictions and risk assessment for alien fishes in North America. *Science* PMID(5596), 1233–1236. DOI: 10.1126/science.1075753.

Landis, W.G. (2004) Ecological risk assessment conceptual model formulation for nonindigenous species. *Risk Analysis* 24(4), 847–858. DOI: 10.1111/j.0272-4332.2004.00483.x.

Leung, B., Springborn, M.R., Turner, J.A. and Brockerhoff, E.G. (2014) Pathway-level risk analysis: The net present value of an invasive species policy in the US. *Frontiers in Ecology and the Environment* 12(5), 273–279. DOI: 10.1890/130311.

Lopez, V.M. and Hoddle, M.S. (2014) Effects of body size, diet, and mating on the fecundity and longevity of the goldspotted oak borer (Coleoptera: Buprestidae). *Ann Entomol Soc Am* 107(2), 539–548

Mankin, R.W., Smith, M.T., Tropp, J.M., Atkinson, E.B. and Jong, D.Y. (2008) Detection of *Anoplophora glabripennis* (Coleoptera: Cerambycidae) larvae in different host trees and tissues by automated analyses of sound-impulse frequency and temporal patterns. *Journal of Economic Entomology* PMID(3), 838–849. DOI: 10.1603/0022-0493(2008)101[838:doagcc]2.0.co;2.

Mishra, S.C. (1920) The American blight or wooly apple aphid *E. lanigerum* (Hausmann). *Agricultural Research Journal* 15, 627.

Muniappan, R. and Viraktamath, C.A. (1986) Status of biological control of the weed, *Lantana camara* in India. *International Journal of Pest Management* 32(1), 40–42. DOI: 10.1080/09670878609371025.

Nagrare, V.S., Kranthi, S., Biradar, V.K., Zade, N.N., Sangode, V. *et al*. (2009) Widespread infestation of the exotic mealybug species, *Phenacoccus solenopsis* (Tinsley) (Hemiptera: Pseudococcidae), on cotton in India. *Bulletin of Entomological Research* 99(5), 537–541. DOI: 10.1017/S0007485308006573.

Nail, K.R., Drizd, L. and Voorhies, K.J. (2019) Butterflies across the globe: A synthesis of the current status and characteristics of monarch (*Danaus plexippus*) populations worldwide. *Frontiers in Ecology and Evolution* 7, 362. DOI: 10.3389/fevo.2019.00362.

Nichols, J.D., Eaton, M.J. and Martin, J. (2014) Thresholds for conservation and management: Structured decision-making as a conceptual framework. In: Guntensperger, G.R. (ed.) *Application of Threshold Concepts in Natural Resource Decision-Making*. Springer Science+Business Media, NY, pp. 9–28.

Osyczka, P., Mleczko, P., Karasiński, D. and Chlebicki, A. (2012) Timber transported to Antarctica: A potential and undesirable carrier for alien fungi and insects. *Biological Invasions* 14(1), 15–20. DOI: 10.1007/s10530-011-9991-0.

Palaniswami, M.S., Pillai, K.S., Nair, R.R. and Mohandas, C. (1995) A new cassava pest in India. *Cassava News* 19(1), 6–7.

Parendes, L.A. and Jones, J.A. (2000) Role of light availability and dispersal in exotic plant invasion along roads and streams in the H.J. Andrews Experimental Forest. *Conservation Biology* 14, 64–75.

Pauchard, A. and Alaback, P.B. (2004) Influence of elevation, land use, and landscape context on patterns of alien plant invasions along roadsides in protected areas of South-Central Chile. *Conservation Biology* 18(1), 238–248. DOI: 10.1111/j.1523-1739.2004.00300.x.

Phalan, B., Onial, M., Balmford, A. and Green, R.E. (2011) Reconciling food production and biodiversity conservation: Land sharing and land sparing compared. *Science* 333(6047), 1289–1291. DOI: 10.1126/science.1208742.

Piel, F., Gilbert, M., De Cannière, C. and Grégoire, J.C. (2008) Coniferous round wood imports from Russia and Baltic countries to Belgium. A pathway analysis for assessing risks of exotic pest insect introductions. *Diversity and Distributions* 14(2), 318–328. DOI: 10.1111/j.1472-4642.2007.00390.x.

Ramos, N.C., Gastauer, M., de Almeida Campos Cordeiro, A. and Meira-Neto, J.A.A. (2015) Environmental filtering of agroforestry systems reduces the risk of biological invasion. *Agroforestry Systems* 89(2), 279–289. DOI: 10.1007/s10457-014-9765-7.

Renault, D., Laparie, M., McCauley, S.J. and Bonte, D. (2018) Environmental adaptations, ecological filtering and dispersal central to insect invasions. *Annual Review of Entomology* 63, 345–368. DOI: 10.1146/annurev-ento-020117-043315.

Rout, T.M., Moore, J.L. and McCarthy, M.A. (2014) Prevent, search or destroy? A partially observable model for invasive species management. *Journal of Applied Ecology* 51(3), 804–813. DOI: 10.1111/1365-2664.12234.

Roy, S., Roy, M.M., Saxena, P. and Bano, R. (2017) Role of entomology in sustaining agroforestry productivity. In: *Agroforestry*. Springer, pp. 691–704. DOI: 10.1007/978-981-10-7650-3_27.

Schweitzer, C., Clark, S.L., Gottschalk, K.W., Stringer, J. and Sitzlar, R. (2014) Proactive restoration: Planning, implementation, and early results of silvicultural strategies for increasing resilience against *Lymantria dispar* infestation in upland oak forests on the Daniel Boone National Forest, Kentucky. *Journal of Forestry* 112(5), 401–411. DOI: 10.5849/jof.13-085.

Selvaraj, K., Sundararaj, R. and Sumalatha, B.V. (2019) Invasion of the palm infesting whitefly, *Aleurotrachelus atratus* Hempel (Hemiptera: Aleyrodidae) in the Oriental region. *Phytoparasitica* 47(3), 327–332. DOI: 10.1007/s12600-019-00742-1.

Sharov, A.A., Leonard, D., Liebhold, A.M., Roberts, E.A. and Dickerson, W. (2002) Slow the Spread': A national program to contain the gypsy moth. *Journal of Forestry* 100(5), 30–36. DOI: 10.1093/jof/100.5.30.

Shashank, P.R., Chandrashekar, K., Meshram, N.M. and Sreedevi, K. (2015) Occurrence of *Tuta absoluta* (Lepidoptera: Gelechiidae) an invasive pest from India. *Indian Journal of Entomology* 77(4), 323–329.

Shashank, P.R., Suroshe, S.S., Singh, P.K., Chandrashekar, K., Nebapure, S.M. *et al.* (2016) Report of invasive tomato leaf miner, *Tuta absoluta* (Lepidoptera: Gelechiidae) from Northern India. *Indian Journal of Agricultural Sciences* 86(12), 1635–1636.

Shylesha, A.N., Jalali, S.K., Gupta, A.N., Varshney, R.I., Venkatesan, T. *et al.* (2018) Studies on new invasive pest *Spodoptera frugiperda* (J.E. Smith) (Lepidoptera: Noctuidae) and its natural enemies. *Journal of Biological Control* 32(3), 1–7. DOI: 10.18311/jbc/2018/21707.

Simberloff, D. and Von Holle, B.V. (1999) Positive interactions of nonindigenous species: Invasional meltdown? *Biological Invasions* 1(1), 21–32. DOI: 10.1023/A:1010086329619.

Singh, S.P. (2004) *Some Success Stories in Classical Biological Control of Agricultural Pests in India*. APAARI publication, Bangkok, Thailand, p. 73.

Singh, S., Pandey, A.K., Singh, B.K. and Rajak, D.C. (2016) Impact of exotic pests on agro-biodiversity and their management: A review. *Biological Bulletin* 2(1), 58–73.

Snyder, W.E. (2019) Give predators a complement: Conserving natural enemy biodiversity to improve biocontrol. *Biological Control* 135, 73–82. DOI: 10.1016/j.biocontrol.2019.04.017.

Sreedevi, K., Meshram, N. and Shashank, P.R. (2015) Insect taxonomy – basics to bar coding. In: *New Horizons in Insect Science: Towards Sustainable Pest Management*. Springer, New Delhi, pp. 3–12. DOI: 10.1007/978-81-322-2089-3_1.

Srinivasan, T., Saravanan, P.A., Josephrajkumar, A., Rajamanickam, K., Sridharan, S. *et al.* (2016) Invasion of the rugose spiralling whitefly, *Aleurodicus rugioperculatus* Martin (Hemiptera: Aleyrodidae) in Pollachi tract of Tamil Nadu, India. *Madras Agricultural Journal* 103(10–12), 349–353. DOI: 10.29321/MAJ.10.001047.

Station, A.B. and Cargo, A. (2003) *Plant Quarantine Including Internal Quarantine Strategies in View of Onslaught of Diseases and Insect Pests*. NAAS, New Delhi, India, p. 1067.

Sujithra, M., Prathibha, V.H.R., Hegde, V. and Poorani, J. (2019) Occurrence of nesting whitefly *Paraleyrodes minei* Iaccarino (Hemiptera: Aleyrodidae) in India. *Indian Journal of Entomology* 81(3), 507–510. DOI: 10.5958/0974-8172.2019.00109.3.

Sundararaj, R., Selvaraj, K., Swamy, C.M.K., Ranjith, M. and Sumalatha, B.V. (2020) First record of the invasive woolly whitefly *Aleurothrixus floccosus* (Maskell) from India. *Indian Journal of Entomology* 82(1), 88–91. DOI: 10.5958/0974-8172.2020.00022.X.

Terentev, A., Dolzhenko, V., Fedotov, A. and Eremenko, D. (2022) Current state of hyperspectral remote sensing for early plant disease detection: A review. *Sensors (Basel)* 22(3), 757. DOI: 10.3390/s22030757.

Tidbury, H.J., Taylor, N.G.H., Copp, G.H., Garnacho, E. and Stebbing, P.D. (2016) Predicting and mapping the risk of introduction of marine non-indigenous species into Great Britain and Ireland. *Biological Invasions* 18(11), 3277–3292. DOI: 10.1007/s10530-016-1219-x.

van Noordwijk, M. (2021) Agroforestry-based ecosystem services: Reconciling values of humans and nature in sustainable development. *Land* 10(7), 699. DOI: 10.3390/land10070699.

Varaprasad, K.S., Balasubramanian, S., Diwakar, B.J. and Rao, C.R. (1994) First report of an entomogenous nematode, *Panagrolaimus* sp. from coffee-berry borer, *Hypothenemus hampei* (Ferrari) from Karnataka, India. *Plant Protection Bulletin* 46(2/3), 34.

Venette, R.C. (2015) *Pest risk modelling and mapping for invasive alien species*. CABI International, Wallingford, p. 268.

Venette, R.C. and Koch, R.L. (2009) IPM for invasive species. In: Radcliffe, E.B., Hutchison, W.D. and Cancelado, R.E. (eds) *Integrated Pest Management*. Cambridge University Press, Cambridge, p. 424.

Viraktamath, C.A., Tewari, G.C., Srinivasan, K. and Gupta, M. (1993) American serpentine leaf miner is a new threat to crops. *Indian Farming* 43, 10–12.

Yemshanov, D., McKenney, D.W., Pedlar, J.H., Koch, F.H. and Cook, D. (2009) Towards an integrated approach to modelling the risk and impacts of invasive forest species. *Environmental Reviews* 17, 163–178. DOI: 10.1139/A09-007.

Appendix

Table 1 Status of invasive pests in India, their host range, damage symptoms, and their natural enemy (Author's own table).

SL. No.	Invasive insect (Order: Family)	Introduced from	Entry to India	Host range	Damage symptoms	Natural enemy	Reference
1	Woolly aphid, *Eriosoma lanigerum* (Hemiptera: Aphididae)	China, 1889	1889, Coonoor, (Tamil Nadu)/ 1909, Uttar Pradesh	Apple, pear	Nymphs and adults are sap suckers, roots are damaged by the formation of swellings, and plants look unhealthy even leading to death	Parasitoids, *Aphelinus mali*	Mishra, 1920
2	San Jose scale, *Quadraspidiotus perniciosus* (Hemiptera: Diaspididae)	China, 1911	Kashmir, 1921	*Populus* spp., *Salix* spp., Celtis spp., *Fagus* spp., *Morus* spp.	Nymphs and adults suck the sap usually from the twigs and branches, even fruits and leaves, depending on severity. Heavy infestation on bark causes gumming which will be fatal to plant. Pink discolouration around the parts infested by scales	Parasitoids, *Encarsia perniciosi*	Singh, 2004

Continued

Table 1 Continued

SL. No.	Invasive insect (Order: Family)	Introduced from	Entry to India	Host range	Damage symptoms	Natural enemy	Reference
3	Lantana bug, *Orthezia insignis* (Hemiptera: Orthezidae)	Sri Lanka/ West Indies, 1915	NA	Lantana, coffee, jacaranda, citrus, sweet potato, gumwood, brinjal, rose	Reproduction is parthenogenetic. Bug feeds by sucking the sap from the phloem. As a by-product, honeydew is excreted which leads to attraction of ants and development of sooty mould resulting in foul odour and decrease of market value of fruits	Predator, *Hyperaspis pantherina*	Muniappan and Viraktamath, 1986
4	Cottony cushion scale, *Icerya purchase* (Hemiptera: Margarodidae)	Australia, 1921	Tamil Nadu, 1920	*Acacia decurrens, Acacia dealbata*, and also a wide range of forest trees and agriculture are affected by this insect	Insect is a sap sucker. Abundant amount of honeydew is excreted leading to the sooty mould development eventually affecting photosynthesis and yield	Predator, *Rodolia cardinalis*	Singh, 2004

5	Pine woolly aphid, *Pineus pini* (Hemiptera: Adelgidae)	Western & Central Europe, 1970	Nilgiris (Tamil Nadu), 1970	*Pinus* spp., *Pinus patula*	Aphid feeds on shoots of plants causing dieback from the tip	Predator, *Cheilomenes aurora*	Singh, 2004
6	Subabul psyllid, *Heteropsylla cubana* (Hemiptera: Psyllidae)	Central America, 1988	Tamil Nadu & Bangalore, 1988	*Leucaena* sp.	These insects are sap suckers from the young shoots, leaves and inflorescences leading to complete deformation of young shoots	Predator, *Curinus coeruleus*	Jalali and Singh, 1989
7	Spiralling whitefly, *Aleurodicus dispersus* (Hemiptera: Aleyrodidae)	Caribbean region, Central America, 1993	Kerala, 1993	Multiple host range	Nymphs and adults suck sap from host plants. Premature leaf drop. Nymphs secrete white, waxy, and flocculent material. Honeydew is produced which develops sooty mould affecting photosynthesis	Parasitoids, *Encarsia perniciosi, Encarsia guadeloupe*, Predator, *Oecophylla smaragdina*	Palaniswami *et al.*, 1995

Continued

Table 1 Continued

SL. No.	Invasive insect (Order: Family)	Introduced from	Entry to India	Host range	Damage symptoms	Natural enemy	Reference
8	Silver leaf whitefly, *Bemisia argentifolii* (Hemiptera:Aleyrodidae)	California, Texas, 1999	NA	Tomato, squash, poinsettia, cucumber, eggplants, okra, beans, and cotton	Four types of damages can be seen. Disorder symptoms (irregular ripening), sooty mould development due to honeydew secretion, viral disease (tomato leaf curl virus transmitted by adults), and cosmetic damage due to the presence of adults and nymphs on leaves	Parasitoids, *Encarsia formosa*, *Eretmocerus eremicus*	Singh, 2004
9	Papaya mealybug, *Paracoccusmarginatus* (Hemiptera:Pseudococcidae)	Central America, 2007	Coimbatore (Tamil Nadu), 2007	Mulberry, tapioca, Jatropha, cotton and several fruits, flowers, and plantation crops	Stylets are inserted into leaves and skin of the fruits and suck the sap. Injection of toxic substance leads to chlorosis, stunting, distortion, early leaf and fruit fall. Sooty mould is developed disturbing photosynthesis	Parasitoid, *Acerophagus papayae*, Predator, *Cryptolaemus montrouzieri*	Jhala *et al.*, 2008

10	Cotton mealybug, *Phenacoccus Solenopsis* (Hemiptera: Pseudococcidae)	USA, 2005	Gujarat, 2006	Cotton, brinjal, okra, tomato, sesame, sunflower, rose	Sap suckers Infestation of this mealybug on cotton causes: stunting, yellowing, distortion and premature drop of leaves and fruits	Parasitoid, *Aenasius bambawalei*	Nagrare *et al*., 2009
11	Rugose spiralling whitefly, *Aleurodicus rugioperculatus* (Hemiptera: Aleyrodidae)	Central America, 2016	Tamil Nadu, 2016	Coconut, guava, banana, mango, drumstick, jackfruit	Young ones and adults suck the sap from ventral surface of the leaves	Parasitoid, *Encarsia* sp.	Srinivasan *et al*., 2016
12	Nesting whitefly, *Paraleyrodes minei* (Hemiptera: Aleyrodidae)	Syria, 2018	Kerela, 2018	Coconut	Young ones and adults suck the sap from ventral surface of the leaves	Parasitoid, *Encarsia* sp.	Sujithra *et al*., 2019
13	Bondar's nesting whitefly, *Paraleyrodes bondari* (Hemiptera: Aleyrodidae)	Central America, 2018	Kerela, 2018	Coconut	Young ones and adults suck the sap from ventral surface of the leaves	Parasitoid, *Encarsia* sp.	Josephrajkumar *et al*., 2019
14	Neotropical Whitefly, *Aleurotrachelus atratus* (Hemiptera:Aleyrodidae)	Brazil, 2019	Mandya (Karnataka), 2019	Coconut	Young ones and adults suck the sap from ventral surface of the leaves	Parasitoid, *Encarsia* sp.	Selvaraj *et al*., 2019
15	Woolly whitefly, *Aleurothrixus floccosus* (Hemiptera:Aleyrodidae)	Neotropical region, 2019	Kerela, 2019	Guava, citrus	Young ones and adults suck the sap from ventral surface of the leaves	Parasitoid, *Cales noacki*	Sundararaj *et al*., 2020

Continued

Table 1 Continued

SL. No.	Invasive insect (Order: Family)	Introduced from	Entry to India	Host range	Damage symptoms	Natural enemy	Reference
16	Coffee berry borer, *Hypothenemus hampei* (Coleoptera:Scolytidae)	Northeast Africa, 1990	Gudalur- Tamil Nadu, 1990	Coffee	The fertilized female bores an entrance hole at the terminal pore or in the calyx ridge	Parasitoids, *Prorops nasuta, Phymastichus coffea*	Varaprasad *et al*., 1994
17	Blue gum chalcid, *Leptocybe invasa* (Hymenoptera: Eulophidae)	Australia	Karnataka/Tamil Nadu, 2001	Eucalyptus	Severely affected trees show leaf fall, a twisted appearance, loss of vigour, stunted growth and dieback, which eventually leads to the death of the tree	Parasitoids, *Quadrastichus mendeli*	Singh, 2004
18	Erythrina gall wasp, *Quadrastichus erythrinae* (Hymenoptera: Eulophidae)	Tanzania, East Africa	Kerela, 2006	*Erythrina* sp., black pepper vanilla	Leaves become swollen and deformed, and look unlike the leaves of healthy trees	Parasitoids, *Eurytoma erythrinae*	Faizal *et al*., 2006

19	South American tomato leaf miner, *Tuta absoluta* (Lepidoptera: Gelechiidae)	South America, 2014	Pune, 2014	Tomato, potato, pepper, brinjal	Larvae attack leaves, buds, stalks and fruits. Feeding results in blotches visible from both sides, galleries on leaves and pinholes on fruits from stalk end stuffed with frass	Parasitoid, *Trichogramma achaea*, Predator, *Nesidiocoris tenuis*	Shashank *et al.*, 2015; Shashank *et al.*, 2016
20	Fall armyworm, *Spodopterafrugiperda* (Lepidoptera:Noctuidae)	America, 2018	Karnataka, 2018	Maize, millet, sorghum, sugarcane, rice, wheat, cowpea, groundnut, potato, soybean, cotton	Hatched larvae feed on leaves by scraping and skeletonizing leaving a silvery transparent membrane. Feeds unopened leaves in whorls and stuffs with frass. Older larvae feed on primordial shoot and tassel leading to dead heart symptom	Parasitoids, *Telenomus* sp., *Trichogramma* sp.	Shylesha *et al.*, 2018
21	Serpentine leaf miner, *Liriomyza trifolii* (Diptera: Agromyzidae)	Florida, 1990	Telangana, 1991	Tomato, castor, pea, cucurbits	Irregular or serpentine pale grey lines appear on both sides of the leaf blades as the larvae feed	Parasitoids, *Hemiptarsenus varicornis*	Viraktamath *et al.*, 1993

Continued

Table 1 Continued

SL. No.	Invasive insect (Order: Family)	Introduced from	Entry to India	Host range	Damage symptoms	Natural enemy	Reference
22	The coconut eriophyid mite, *Aceria guerreronis*(Archnida: Eriophyidae)	Mexico	Kerela, 1997	Coconut	Nuts are discoloured resulting in the reduction of market value	Predator *Neoseiulus baraki*, Fungal pathogen, *Hirsutella thompsonii*	Singh, 2004
23	Rugose Spiralling Whitefly, *Aleurodicus rugioperculatus* (Hemiptera: Aleyrodidae)	Mexico	Karnataka, 2017	Coconut	Formation of spiral-shaped white waxy egg spirals on the underside of leaves, yellowing and curling of leaves, sticky honeydew excretion leading to sooty mould development, reduced photosynthesis, premature leaf drop, and overall decline in plant health, vigour, and fruit yield	Wasps from families such as Braconidae, Ichneumonidae, and Encyrtidae	Chakravarthy *et al*. (2017).

9 Major Invasive Insect Pests on Vegetable Crops in the Tropics

Nadur Lakkanna Naveena[1], Peketi Krishna Reddy[1] and Nagalapura Ramakrishnappa Prasannakumar[2]*
[1]Department of Entomology, College of Agricultural Sciences, Keladi Shivappa Nayaka University of Agriculture and Horticultural Sciences, Iruvakki, Shivamogga, Karnataka, India; [2]Division of Crop Protection, ICAR – Indian Institute of Horticultural Research, Hessarghatta Lake, Bengaluru, India

Abstract
Vegetable crops are vital in tropical regions, providing essential nutrients, supporting food security and the economic well-being of communities. However, their productivity and nutritional quality are increasingly threatened by invasive insect pests. This chapter provides an outline of invasive species in vegetable crops in the tropics. Further, it highlights their impact on crop production, outlines the factors contributing to their introduction and spread, and discusses their key biological characteristics. Additionally, current management strategies aimed at mitigating their effects are examined.

Keywords: Invasive insect pests, natural enemies, vegetable crops, outbreaks

9.1 Introduction

Food security is a major concern for the growing population in the world, particularly in the tropics and sub-tropical countries. Sustainable crop production is required to meet food demands, bring economic stability, and reduce a country's poverty (George, 2011). The global vegetable production volume has increased significantly, from 682 million metric tons in 2000 to more than 1.18 billion metric tons in 2023 and >90% of the area in vegetable production is in Asia and Africa (FAO Stat, 2025). China was the leading producer with a production volume of over 616 million metric tons, followed by India with approximately 145 million metric tons of fresh vegetables in 2022 (Statista, 2025). Vegetables supply vital nutrients to the human diet. Vegetable legumes improve soil fertility. Further, these crops help in raising small-scale farmers in the tropics from poverty (Srinivasan, 2016). However, vegetable production is often constrained by severe biotic and abiotic factors that reduce yields and profits. Insect pests and diseases are one of the major production constraints in tropical vegetable production (Srinivasan, 2016). Globally, 20–40% of loss in crop yield is due to pests, and with annual costs of more than US$220 billion to the global economy, and invasive insects such as desert locusts and red palm weevils at least US$70 billion (IPPC, 2025).

*Corresponding author: prasannaent@gmail.com

 Biological Invasions in the Tropics: Monitoring, Mechanism and Management (eds Akshay Kumar Chakravarthy and Pathour Rajendra Shashank)
DOI: 10.1079/9781836991977.0009

Invasive alien species (IAS), one of the drivers of biodiversity loss and species extinctions (IUCN, 2025), can affect the economic status of a country (Pimentel *et al.*, 2001). Globally, IAS cause multitrillion-dollar economic impacts (Cuthbert *et al.*, 2025). The rate of new introductions is growing, as evidenced in the IUCN red list, where 1 in 10 species are affected by invasive species (IUCN, 2025), due to advances in trade and travel globally and also to changes in climate which makes some areas more hospitable to new species. Invasive insects have often been accidentally introduced to countries, some becoming very well established and becoming serious pests, while others are expanding their host ranges rapidly. Invasive pests threaten local biodiversity by competing for food and shelter and have been known to impact crop production adversely (Naveena *et al.*, 2020; Ramdwar *et al.*, 2023). Due to increased globalization, insect pests are moving from one region to another through human agencies, intentionally or accidentally (Raghuteja *et al.*, 2022). Unintentional entry of exotic pests increased due to an increase in trade (Export/Import) between countries across the world. These pests may grow rapidly, establish and destroy crops. In order to protect plants from pests globally, the International Plant Protection Convention (IPPC) was established in 1951 under the Food and Agriculture Organization of the United Nations (FAO). The IPPC is the only international convention and reference organization dedicated to developing plant health standards and is recognized under the World Trade Organization (WTO)'s Agreement on the Application of Sanitary and Phytosanitary Measures (SPS Agreement). Further, it is ratified by 185 contracting parties that collaborate to develop, adopt and promote the application of International Standards for Phytosanitary Measures (ISPMs) as a main tool to safeguard global food security, facilitate safe trade and protect the environment. These provide the basis for countries to make national legislation, guidelines and measures to protect their plant resources from harmful pests. They also allow the development of standards related to storage places, packaging, conveyances, containers, soil and any other organism, object or material capable of harbouring or spreading plant pests, particularly where international transportation is involved. At present there are 46 ISPMs (IPPC, 2025).

In tropical countries invasive pests have been restricted by implementing the laws. In India entry of invasive species is regulated through The Destructive Insect and Pest Act of 1914, and this huge task is presently managed by the Directorate of Plant Protection, Quarantine, and Storage, Ministry of Agriculture and farmers welfare, Government of India by implementing the Plant Quarantine (Regulation of Import into India) Order, 2003 through its quarantine centres located at entry points across the land border, airports and seaports. This order aims to prevent entry, establishment and spread of exotic pests.

9.2 Features of Invasive Pests

Invasive pests can reproduce and grow rapidly, they can spread fast and over large areas due to a strong potential to compete with native species. They can survive in diverse environmental and habitat conditions due to favourable weather conditions and absence of natural enemies. They can live on diverse food items, outcompeting the native species for their food and shelter. Invasive pests usually have association with humans: as a hitch-hiker (Atree, 2025).

The traits of an organism play a role in its introduction and establishment in non-native areas. Further, environment, phenotypic plasticity (rapid changes according to surrounding biotic and abiotic conditions through regulation of epigenetic mechanisms, e.g. aphids, develop wings (alates) due to changes in the environment) and local selective pressures help the invasive species/pests to establish very quickly in non-native areas (Sinclair *et al.*, 2020; Cuthbert *et al.*, 2025).

9.3 Steps of Invasion

Invasion by alien organisms can be categorized in four steps: 1. Introduction; 2. Establishment; 3. Spread; and 4. Naturalization (Fig. 9.1) (Naveena *et al.*, 2020; AVRDC, 2025). The potential pathways for invasion into new areas include cross-border movements and

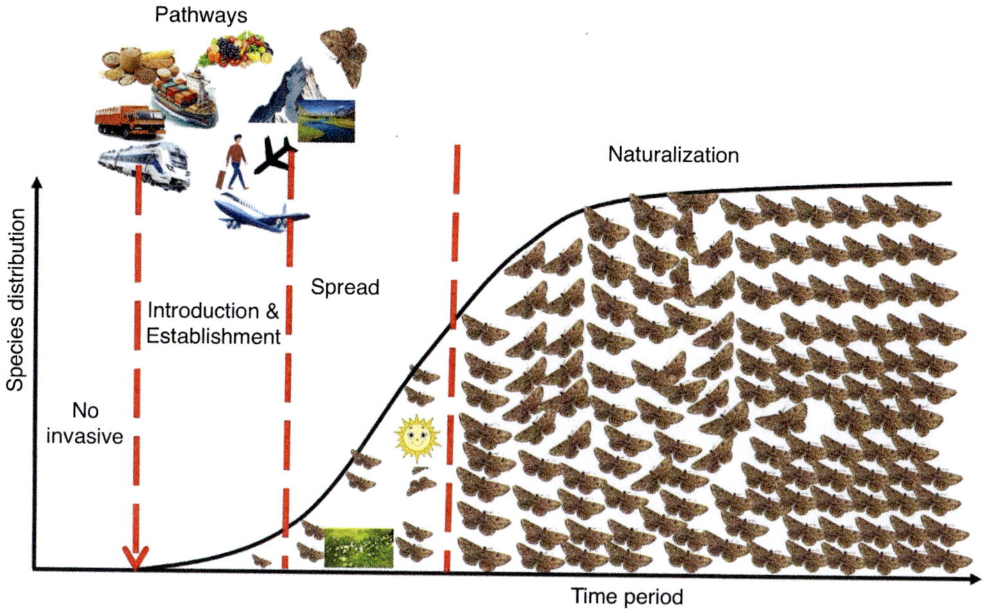

Fig. 9.1. Steps and pathways of invasion by invasive species. (Author's own image)

transportation/trade movements from infested areas, as well as the movement of containers, packaging and vehicles that can carry seeds or seedlings/plants (Fig. 9.1) (AVRDC, 2025). Further, range expansion and human-mediated long-distance transfer, with widening of the pest's trophic niche lead to arrival of a pest to a new territory and its further naturalization (Musolin *et al.*, 2022).

9.4 Effects of Invasive Pests

Sustainable Development Goal 15 of UN depicts that biological invasions put human health, food security and livelihoods at risk (IUCN, 2025). India has experienced >28 invasions to date, out of which 12 invasive insect pests have been recorded on vegetable crops, negatively impacting vegetable crop productivity. These invasions led to significant ecological and economic losses and caused negative social consequences. Invasive species threaten native biota, impacting both the environment and human welfare (Simberloff *et al.*, 2013). They have effects on numerous ecological processes and ecosystem services (Cuthbert *et al.*, 2025), including altering food webs. For wildlife, the invasive species may provide little to no food, decreasing biodiversity.

9.5 Invasions in Vegetables

9.5.1 Diamondback moth, *Plutella xylostella* (Linnaeus) (Lepidoptera: Plutellidae)

The diamondback moth (DBM) is a highly invasive, migratory and destructive pest which feeds on Brassica crops such as cabbage, broccoli, cauliflower and canola, and also on various cruciferous weeds (CABI, 2004). *P. xylostella* has spread from its origins in Europe, America, Southeast Asia, Australia, New Zealand, South Africa and East Asia to nearly every region where cruciferous crops are grown (Keerthika and Jayakumar, 2024). The estimated annual global cost of DBM management, crop damage and yield losses is estimated at US$4–5 billion (Zalucki *et al.*, 2012). Depending on the crop and the extent of infestation, *P. xylostella* causes significant economic yield losses of up to 92% in cabbage, 75% in broccoli and 30% in cauliflower (Farias *et al.*, 2020).

9.5.1.1 Biology of the DBM

The adult is a small, slender, greyish-brown moth with pronounced antennae. It is about 6 mm long, and marked with a broad cream or light brown band along the back. The band is sometimes constricted to form one or more light-coloured diamonds on the back, which forms the basis for its common name, diamondback moth. When viewed from the side, the tips of the wings can turn upward slightly. Adult males and females live about 12 and 16 days, respectively, and females deposit eggs for about 10 days. Freshly laid eggs are oval in shape, pale yellowish in colour and laid mostly singly or in small groups on the lower surface of the leaves near the midrib or petiole of the plant. The incubation period varies from 2 to 4 days. The caterpillar is light green and sometimes with irregular dark markings on the head, which turn light yellow when mature. Newly hatched caterpillars feed as leafminers inside the leaf tissue while the older ones feed on all plant parts. The larvae undergo three moults with four instars during their larval period of 7–12 days. Yellow-whitish to green pupae are found in flimsy cocoons on the underside of the leaves and the stage lasts for 3–5 days. The entire lifespan of *P. xylostella*, depending on environment conditions, varies from 13 to 22 days for the completion of one generation (CABI, 2004; Harika *et al.*, 2019).

9.5.1.2 Nature of damage and symptoms

The caterpillars feed on the underside of leaves causing window-like holes that break the leaf as it grows. In cases of severe infestation entire leaves are damaged leaving only the veins and some portions of the crop will be contaminated with the pest excrement, which makes the produce unsuitable for consumption and sale. The larval feeding can cause significant yield losses ranging from 30–92% depending on the crop (CABI Compendium, 2022).

9.5.1.3 Management

Clean cultivation, early planting, crop rotation with non-cruciferous vegetable crops, intercrops, trap crops, host plant resistance, regular pest monitoring, utilization of effective natural enemies (parasitoids, predators, entomopathogens), behavioural management and application of insecticides with different modes of action to avoid resistance are amongst the methods available for the management of DBM (Philips *et al.*, 2014). The available management strategies when used alone do not work to their full potential and lead to other serious concerns like host plant resistance breakdown, resistance to insecticides and so on. In combination, these tactics can complement each other and lead to improved pest suppression, reduced insecticide resistance and a more sustainable pest management programme.

9.5.2 Lantana bug, *Orthezia insignis* (Browne) (Homoptera: Ortheziidae)

The lantana bug, *Orthezia insignis*, is a scale insect native to the Neotropical region and the tropics and now has spread elsewhere in the world where the lantana weed, *Lantana camara* has been introduced (Wittenberg and Cock, 2001). *O. insignis* was first observed in Sri Lanka in 1893 and subsequently in India in 1915 (Beeson and Chatterjee, 1939). *O. insignis* is polyphagous, usually preferring woody shrubs, sugarcane, citrus, potato, tomato, chrysanthemum, jacaranda, casuarina and other host plants (CABI, 2021).

9.5.2.1 Biology of the lantana bug

The lantana bug reproduces parthenogenetically and the life cycle involves nymphs hatching from eggs within a white, waxy ovisac, which the female carries, and these nymphs then move to feed on the plant's phloem. Adult females are brownish olive green, oval, approximately 1.5 mm long and 1.3 mm wide, with parallel, brittle white wax plates that curve slightly upwards, giving an elegant appearance. The adult males are single-winged, greyish with powdery wax, about 2.0 mm in length, covered with short hair, and terminating in a tuft of white wax filaments.

9.5.2.2 Nature of damage and symptoms

O. insignis extracts large quantities of sap, causing general host debilitation, but not death.

Build-up of sticky honeydew deposits occurs on nearby surfaces, which may attract attendant ants. Unsightly sooty moulds grow on the sugary deposits, and badly fouled leaves may be dropped prematurely and the quality of fruits may be reduced. The older females are easy to see on young stems, especially when they walk about and the movement of the white ovisacs catches the light. They also cause indirect damage to the plants by excreting (or more correctly eliminating) honeydew, a growth medium for sooty moulds, which produce a black coating over the leaf surface. This coating interferes with photosynthesis and may cause poor growth, a reduction in fruit size and generally give an unsightly appearance to the crop. *O. insignis* is ideally suited to be transported by water, wind and animal agents including domestic animals and humans for its dissemination. Accidental introductions to new countries apparently occur on infested planting material.

9.5.2.3 Management of lantana bug

This ensign scale can be effectively controlled by a predatory coccinellid, *Hyperaspis pantherina* (GISD, 2025) along with common cultural practices and in severe infestations systemic insecticides such as imidacloprid and thiamethoxam keep the pest under control.

9.5.3 Potato tuber moth, *Phthorimaea operculella* (Lepidoptera: Gelechiidae)

Phthorimaea operculella is a cosmopolitan pest, especially in warm temperate and tropical regions where crops are grown. The principal host is potato, but other Solanaceae plants are also attacked, especially tomato, tobacco and chilli. There are many related wild hosts; 60 plant species are listed as hosts by Das and Raman (1994).

9.5.3.1 Biology of the potato tuber moth

The adult females lay up to 300 eggs singly or in groups on leaf surface or exposed tubers with an incubation period of 3–15 days at warm temperatures in tropical countries (20–30°C) and are very sensitive to cold/lower temperatures. The newly hatched larvae bore into leaf veins, petioles, tender shoots or tubers and feed for about 33 days then move for pupation in the soil. Adults will emerge between 6–29 days from pupae and live for about 10 days (Carter, 1984; Trivedi and Rajagopal, 1992; CABI, 2025).

9.5.3.2 Nature of damage and symptoms

On growing plants, leaf mines betray the occurrence of larvae, and in addition, the stem is weakened or broken. On tubers, detection is more difficult without cutting open some tubers, when galleries and larvae will be apparent within the potatoes. Larvae feed on the entire foliage, leaving leaf veins. This affects seedlings, and disrupts formation of head in cabbage, broccoli and cauliflower.

9.5.3.3 Management

Potato tuber moth can be effectively controlled by various parasites such as *Trichogramma* sp., *Bracon* sp., *Chelonus* sp., *Apanteles* sp., predators such as coccinellids (*Brumoides suturalis*), *Chrysoperla* sp. and pathogens such as *Bacillus thuringiensis* (CABI, 2025).

9.5.4 American serpentine leaf miner, *Liriomyza trifolii* (Burgess) (Diptera: Agromyzidae)

Liriomyza trifolii originated in North America and spread to other parts of the world in the 1960s–1980s. It has been recorded in several countries, presumably associated with the global trade of ornamental plants. It entered India through *Chrysanthemum* cuttings in 1990.

9.5.4.1 Biology of L. trifolii

Adults emerge during morning hours, females are larger than males and males live for 10–15 days, while females live for 15–20 days. Females have multiple matings within 24 hours of emergence and lay individual eggs on upper/lower surface of leaves by ovipositional punctures. Eggs are whitish and hatch in 2–3 days. The larvae, which look like maggots, feed by mining in the mesophyll, having four instars and black sclerotized exuvia can be found inside the leaf mines which indicate the number of moults. Before pupation the larvae make a semicircular cut on the leaf surface at the end of the leaf mine and complete larval life in 3–4 days; pupation takes place in soil for about 8–10 days (Rai *et al.*, 2023).

9.5.4.2 Nature of damage and symptoms

Damage to the plant is caused by the stippling made by the female, and internal mining by the larvae, allowing pathogenic fungi to enter the leaf via the feeding punctures and mechanical transmission of plant viruses. Leaf mining and stippling greatly reduce photosynthesis in the plant, thus affecting plant growth. Punctures made by females while feeding and oviposition result in a stippled appearance on leaves, especially at the leaf tip and on leaf margins. Feeding punctures appear as white speckles 0.13–0.15 mm in diameter.

9.5.4.3 Management

Cultural practices include field sanitation, weed removal, use of parasitoids belonging to Pteromalidae, Eulophidae and Braconidae, use of neem oil, and insecticides such as emamectin benzoate and flufenoxuron (Rai *et al.*, 2023).

9.5.5 Spiralling whitefly *Aleurodicus dispersus* (Russell) (Hemiptera: Aleyrodidae)

Aleurodicus dispersus was first described from Florida in 1965, and now it is present worldwide on ~500 host plants in 53 families (Boopathi, 2017; Pradhan and Borthakur, 2020). These include agricultural and horticultural plants/crops including vegetables, fruits, ornamentals, etc. In India it was noticed for the first time in 1993 on tapioca in Kerala and by the year, 2000 it spread to the entire country (Pradhan and Borthakur, 2020).

9.5.5.1 Biology

Spiralling whitefly can be seen throughout the year with peak population in March to June at average temperatures of about 30–35°C; a slight increase (>35°C) leads to mortality of nymphs and adults. Further, lower temperatures with intermittent rainfall also decrease the whitefly populations. Adults live for 15 days, their forewings have dark spots and eyes are dark reddish-brown. Females lay single stalked yellowish-white eggs along with deposits of waxy material in a spiralling fashion on the under surface of leaf veins. Each female lays up to 64 eggs with an incubation period of 5–8 days. The first instar larvae are mobile, travelling in search of food for a short distance, and the third instar nymphs are static; all the stages have a white waxy covering on the body, with the last stage having more of this waxy covering. The larval/nymphal duration completes in 12–14 days and 2–3 days of pupal period. The total life cycle completes in about 30 days (Pradhan and Borthakur, 2020).

9.5.5.2 Nature of damage and symptoms

Both nymphs and adults pierce leaves and suck the sap, leading to drop off and early death when populations are high. Damage is also caused due to sooty mould development, and also by transmission of viruses (Berlinger, 1986). They cause >40 diseases of vegetable and fibre crops worldwide with a yield loss of 20–100% (Kessing and Mau, 1993).

9.5.5.3 Management

Control of whitefly is a difficult task when it occurs in the field. However, reduction of their damage level can be achieved by adopting integrated approaches such as removal of alternate weed hosts in the cropped area, use of yellow sticky traps, use of parasitoids such as *Encarsia haitiensis* and *E. guadeloupae* and coccinellid predators, pathogens such as *Paecilomyces farinosus*, use of neem seed kernel extract (NSKE at 3%), neem oil (2 ml/litre) and chemicals such as triazophos (at 0.03%).

9.5.6 Silverleaf whitefly, *Bemisia argentifolii* Bellows & Perring (Hemiptera: Aleyrodidae)

Bemisia argentifolii is a race (B-Biotype) of *B. tabaci* (De Barro *et al.*, 2005). Distributed throughout the world on various host plants it is a potential vector of *Ipomovirus lenisbatatae* (SPMMVO) and leaf curl virus (LCV) and a known vector of Begomovirus in various crops (EPPO, 2025). The insect causes damage by sucking the sap from (phloem) leaves through its piercing and sucking mouthparts, and also due to the accumulation of honeydew, leading to the growth of fungi.

9.5.6.1 *Biology of* B. argentifolii

Adults are soft bodied and light-yellow with a light dusting of wax. Females are larger in size than the males. Females have arrhenotokous parthenogenesis where they lay haploid eggs which give rise to males and mated females can produce both haploid and diploid oval-shaped eggs with tapering towards the distal end on leaves. Females live for 10–24 days and each female lays 66–300 eggs depending on host and temperature with an incubation period of 6–7 days. The nymphs are flattened oval, greenish yellow in colour, having four nymphal stages and first nymphal stage is active, mobile, whereas remaining stages are static and they become mature in 10–15 days. All the nymphal stages secrete white waxy material on their body surface.

9.5.6.2 *Nature of damage and symptoms*

Both nymphs and adults suck the sap, heavy infestations lead to death of seedlings, or reduced plant vigour and reduction in yield. They also cause damage by excreting honeydew, which further attracts sooty mould and affects photosynthesis.

9.5.6.3 *Management*

Silverleaf whitefly can be effectively managed by field sanitation, plastic mulching, use of UV-absorbing greenhouse plastic films, use of aphelinid parasitoids, and general predators such as *Chrysoperla* sp., minute pirate bugs and bigeyed bugs, *Geocoris* sp. (McAuslane, 2009).

9.5.7 Papaya mealybug, *Paracoccus marginatus* Milliams and Granara de Willink (Hemiptera: Pseudococcidae)

Paracoccus marginatus is a polyphagous pest having a wide host range. It originated in Mexico and spread across the world and first reported in India in 2008 from Tamil Nadu (Muniappan *et al.*, 2009).

9.5.7.1 *Biology*

Adult female is apterous, yellowish body covered with white mealy matter, bigger than winged males. Females have three instars and males have four instars. In the male, first and second instars are yellowish and remaining stages are pinkish in colour. Females lay 100–600 eggs in an ovisac and incubation period is 10 days. The newly hatched nymphs are active and they crawl for short distances in search of food/feeding places, the remaining instars are not mobile and feed by sucking the plant sap. All the stages secrete waxy material. Males complete their life cycle in 27–30 days, while females complete theirs in 24–26 days (Makale *et al.*, 2023).

9.5.7.2 *Nature of damage and symptoms*

The insects suck the sap by inserting their stylets into the epidermis of the leaves, fruit and stem. It injects a toxic substance into the leaves/stem/fruit while feeding, which leads to chlorosis, yellowing, plant stunting, leaf deformation or crinkling, early leaf and fruit drop, and death of plants. On the infested plant parts, cotton-like masses can be observed.

9.5.7.3 *Management*

It can be effectively managed by classical biological control using exotic parasitoids such as *Acerophagus papayae*, and Indian apefly, *Spalgis epius* (Westwood) (Lycaenidae: Lepidoptera), a native parasitoid in the north-eastern region of India.

9.5.8 Cotton mealybug, *Phenacoccus solenopsis* Tinsley (Hemiptera: Pseudococcidae)

Phenacoccus solenopsis is a polyphagous pest that originated in the USA and its spread is more in tropical and subtropical conditions and limited in temperate regions of the world. It feeds on >200 plants of about 55 families, and can cause up to 60% yield loss in various agricultural and horticultural crops.

9.5.8.1 *Biology of* P. solenopsis

Cotton mealybugs exhibit sexual dimorphism, where males are winged and females are apterous. They have both sexual and ovoviviparous mode of reproduction. Only mated females lay 150–600 viable eggs in a white waxy ovisac at temperatures between 25–32°C, and they will hatch in times varying from a few minutes to 2 hours. Females have three nymphal instars and males have four instars including prepupal stage and have longer duration than females. The insect can complete 12–15 generations per year.

9.5.8.2 *Nature of damage and symptoms*

Cotton mealybugs suck the sap from leaves, stems and tender shoots, which leads to distorted, crinkled leaves, stunted growth and may further lead to wilting and drying. White waxy cottony masses are visible on leaves, shoots and tender parts of the plants along with egg sacs.

9.5.8.3 *Management*

P. solenopsis can be effectively managed by various parasites and predators; successful classical biological control is by using Australian ladybird beetle *C. montrouzieri*, which is effective in tropical countries. Many chemicals are recommended (quinolphos, methyl demeton, triazophos, profenophos, dichlorvos, monocrotophos, acephate, cypermethrin, imidacloprid, thiamethoxam, buprofezin, alphamethrin, malathion, chlorpyrifos) for use in management with careful consideration of natural enemies, although these chemicals are less effective on this insect due to its cryptic habit and waxy cover over the body (Fand and Suroshe, 2015).

9.5.9 South American tomato leaf miner, *Tuta absoluta* (Meyrick) (Lepidoptera: Gelechiidae)

Tuta absoluta, a severe pest in both field and protected cultivation was reported in 1917 in Peru, by 1990 in South America, in 2006 it was found in Spain and within three years the entire Mediterranean basin, and in 2011 had spread to Europe. In Asia in 2009, it spread from Israel, Jordan and Turkey and then to Iran in 2010 and India in 2014, and to Bangladesh and China in 2016 and 2017, respectively (AVRDC, 2025). It mainly occurs on tomato, solanaceous vegetables (eggplant, beans, potato, pepper, etc.), solanaceous weeds such as *Solanum nigrum* and *Datura stramonium* (Halder *et al.*, 2025).

9.5.9.1 *Biology of* T. absoluta

Tuta absoluta is active during night time and early morning hours. They are small insects (7 mm), with forewings with a black spot, females live longer (10–15 days) than males (6–7 days). Each female lays 250–300 whitish to yellow-coloured eggs in single or rarely in groups on undersurface of leaves with an incubation period of 7 days. The larvae have four instars in 11 days, fully fed last instar larvae fall on the ground and form a silken cocoon for pupation in soil, and dried leaves, rarely. Pupation also happens on stem/in the mines itself without formation of cocoon. Pupa turns from greenish to dark brown in 5 days. *T. absoluta* completes one life cycle in 30–35 days and 10–12 generations in a year under greenhouse conditions with an optimum temperature of 30°C (Pandey *et al.*, 2023).

9.5.9.2 *Nature of damage and symptoms*

It causes damage on leaves, buds, calyces, young green/ripened red fruits. Infested fruits show minute holes. The larvae mine in the mesophyll tissue, forming irregular blotches or bigger zig-zag papery gallery than that of *Liriomyza trifolii* (Halder *et al.*, 2025). Causes up to 100% crop damage in tomato (Acharya *et al.*, 2023).

9.5.9.3 *Management*

In order to avoid the spread, frequent survey and surveillance, quarantine and effective phytosanitary measures are very important (Acharya *et al.*, 2023). Use of bio-insecticides, *Bacillus thuringiensis* (Bt) and azadirachtin (neem based), entomopathogenic microbes, *Metarhizium anisopliae* and *Beauveria bassiana*, *Phthorimaea operculella* granulovirus (PhopGV), natural enemies such as *Nesidiocoris tenuis* Reuter (Hemiptera: Miridae), *Chrysoperla carnea* (Stephens) and *Dicyphus errans* (Wolff), *Trichogramma pretiosum* (Riley), an egg parasitoid, novel bio-pesticide, such as Prev-Am® (orange oil, salt borax and biodegradable surfactants) causes reduction in *T. absoluta* population by up to 75% (Buragohain *et al.*, 2021; Pandey *et al.*, 2023) in addition to use of chlorantraniliprole.

9.5.10 Black thrips *Thrips parvispinus* (Karny) (Thysanoptera: Thripidae)

Thrips parvispinus is a cosmopolitan pest native to Southeast Asia and is also reported from Indonesia, France, Greece, Hawaii, Mauritius, Reunion, Spain, Tanzania, the Netherlands and India (Tyagi *et al.*, 2015). This polyphagous pest affects various crops including beans, eggplant, papaya, pepper, potato, shallot, ornamental flowers and strawberry, causing substantial damage (Sridhar *et al.*, 2021). Alhough it is a pollen feeder and colonizes on flowers, during outbreaks it damages plants by rasping and sucking of the sap by both larvae and adults. A recent severe outbreak of *T. parvispinus* occurred in major pepper-growing regions of India in 2021, resulting in a substantial yield loss of 70–80% across more than 1 million acres (Sridhar *et al.*, 2021; Prasannakumar *et al.*, 2021; Barik *et al.*, 2025).

9.5.10.1 Biology of T. parvispinus

The biology of thrips is dependent on temperature, they are very active between 25–30°C and females lay up to 69 eggs, with an incubation period of ~5 days. There are two nymphal instars, the prepupa completes in ~6 days and pupa in 2 days, and it completes its life cycle in 13–14 days. Adult females live longer (~9 days) than males (6 days). Further, it shows a type III survivorship curve with intrinsic rate of increase of 0.15 individuals per day per female (Hutasoit *et al.*, 2017; Thorat *et al.*, 2022).

9.5.10.2 Nature of damage and symptoms

Crinkles and upward curling of infested leaves, elongated petiole, and buds become brittle and drop down. Further, they significantly reduce productivity by causing widespread flower loss, fruit deformity and fruit drop in chilli.

9.5.10.3 Management

This pest can be effectively managed by adopting various cultural methods such as early planting, avoiding excessive use of N fertilizers, mechanical methods such as removal of weed hosts, use of white coloured sticky traps instead of blue or yellow traps. However, these can be used for mass collection. Botanicals such as neem oil, Pongamia oil and eucalyptus oil (2.0%) gave up to 30% control during vegetative stage. Biological control methods such as use of the predatory ladybird beetle, *Menochilus sexmaculatus*, entomopathogenic fungi, *Lecanicillium lecanii*, bacteria, *Pseudomonas* sp., biopesticides, viz. neem oil (azadirachtin) and insecticides such as fipronil 80 WG, 0.2 g/litre; cyantraniliprole 10.26 OD, 1.25 ml/litre; acetamiprid 20 SP, 0.2 g/litre; spirotetramat 150 OD, 0.8 ml/litre; or spinosad 45 SC, 0.3 ml/litre water as sequential sprays at weekly intervals were found effective for the management of *T. parvispinus* (Thorat *et al.*, 2022).

9.5.11 Madeira mealy bug, *Phenacoccus madeirensis* Green (Hemiptera: Pseudococcidae)

The highly invasive Madeira mealybug *Phenacoccus madeirensis* Green (Hemiptera: Pseudococcidae) is a polyphagous pest that seriously infests fruits, vegetables and ornamental plants. A few economically important crops include cassava, pineapple, citrus and potatoes. *P. madeirensis* was first reported in

1920 in Madeira, Portugal, and in 1993 was first reported in Asia (Japan) and had spread across Asia including India by 2012 (GBIF, 2018; Wang *et al.*, 2019). It is prevalent in all of Africa, America, Asia and in European countries.

9.5.11.1 Biology

Females are slightly flattened and oval; pale green to grey in colour; reddish legs; on the dorsal surface dusted with white mealy wax; with transverse dark bands on the intersegmental areas of thorax and abdomen, forming in many samples one pair of dark longitudinal lines on dorsum; with 18 pairs of lateral wax filaments, caudal pair of wax filaments longer than lateral filaments; ovisac covering entire dorsum, white, irregular form. The adult male is reddish in colour, with a single pair of wings. Male pupae are covered with a loose white, silky cocoon (Badr and Moharum, 2017).

9.5.11.2 Nature of damage and symptoms

Madeira mealybugs suck the sap from plant parts, which leads to yellowing of leaves, reduced plant vigour and stunted growth. Further, excretion of honeydew attracts the growth of sooty mould, which hinders photosynthesis.

9.5.11.3 Management

Madeira mealybugs can be effectively managed by using parasitoids such as *Allotropa* sp., *Anagyrus amnestos* Ramkumar and Poorani, *Anagyrus quadrii* (Hayat, Alam & Agarwal) and *Anagyrus loeckii* Noyes & Menezes (Mani and Shylesha, 2019).

9.6 Conclusion

Globalization led to the invasive spread of insects, which is a serious threat to local flora and fauna and even led to the extinction of several species. While the globalization of agriculture has boosted worldwide trade, movement of plant materials has raised the potential of invasive pest introductions into the tropics (Table 9.1). These pests multiply greatly and harm economically significant crops and other plants if they are not accompanied by the natural enemies that keep them under control in their native habitat. By having a fundamental understanding of invasive pest species and cooperating internationally through the sharing of information, the entry of a new pest can be minimized.

Accurate identification of invasive species/pests through DNA barcoding (Dicky *et al.*, 2015; Musolin *et al.*, 2022), identification, conservation and augmentation of natural enemies of invasive pests, modification in crop cultivation approaches/cropping patterns, use of insecticides with different modes of action, and phytosanitary regulations, help to prevent entry, establishment and spread of invasive species in a country. Understanding and quantification of rapid changes in specific traits (such as mobility, food choice and life history strategies) will also help in developing robust and effective management strategies for invasive species (Garvey *et al.*, 2020; Cuthbert *et al.*, 2025). Plant quarantine laboratories should be upgraded for the quick identification of invasive pests at the entry points and implementing the effective quarantine procedures in international trade is a key means to reduce the risk of spread of invasive species across the borders in the tropics and also at a global level.

Table 9.1. List of invasive insects on vegetable crops in tropics (modified from Naveena *et al.*, 2020).

Sl. No	Common name	Scientific name	Origin	Year	Host plants
01	Diamondback moth	*Plutella xylostella* (Linnaeus) (Plutellidae: Lepidoptera)	Europe and East Asia	1914	Cruciferous vegetables
02	Lantana bug	*Orthezia insignis* (Browne) (Ortheziidae; Hemiptera)	Sri Lanka/ West Indies	1915	Lantana, coffee, jacaranda, gumwood, eggplant
03	Potato tuber moth	*Phthorimaea operculella* (Zeller) (Gelechiidae: Lepidoptera)	Italy	1937	Stored potatoes, tomato, eggplant
04	Serpentine leafminer	*Liriomyza trifolii* (Burgess) (Agromyzidae: Diptera)	Florida (USA)	1990	Pea, tomato, cucurbits, ornamental plants
05	Spiralling whitefly	*Aleurodicus disperses* (Russell) (Aleyrodidae: Hemiptera)	Central America	1993	Host range of 481 plants
06	Silverleaf whitefly	*Bemisia argentifolii* (Bellows) (Aleyrodidae: Hemiptera)	USA	1999	Tomato
07	Papaya mealybug	*Paracoccus marginatus* (Williams & Granarade Willink) (Pseudococcidae: Hemiptera)	Mexico	2005	Eggplant
08	Cotton mealybug	*Phenococcus solenopsis* (Tinsley) Pseudococcidae: Hemiptera)	North America	2006	Eggplant, okra, tomato
09	Madeira mealybug	*Phenacoccus madeirensis* Green (Pseudococcidae: Hemiptera)	Portugal	2012	Tomato, potato, eggplant
10	Tomato pinworm	*Tuta absoluta* (Meyrick) (Gelechidae: Lepidoptera)	South America	2014	Tomato, potato, eggplant, tobacco
11	Solanum whitefly	*Aleurothrixus trachoides* (Back) (Aleyrodidae: Hemiptera)	Neotropical region	2014	Solanaceae plants like eggplant, chilli and tomato
12	Western flower thrips	*Frankliniella occidentalis* (Pergande)	USA	2015	Tomato, capsicum and cucumber
13	Black thrips	*Thrips parvispinus* (Karny) (Thripidae: Thysanoptera)	South-East Asia	2015	Papaya, dahlia and chilli
14	Leek moth	*Acrolepiopsis assectella* (Zeller) (Acrolepiidae: Lepidoptera)	Europe	2020	Onion, leek, garlic

References

Acharya, R., Barman, A.K., Sharma, S.R., Kafle, L., Kim, S. *et al.* (2023) Biology, distribution, and management of invasive South American tomato leafminer, *Tuta absoluta* (Meyrick) (Lepidoptera; Gelechiidae), in Asia. *Archives of Insect Biochemistry and Physiology* 114, e22056. DOI: 10.1002/arch.22056.

ATREE (2025). Available at: http://miap.atree.org/ias.php (accessed 9 June 2025).

AVRDC (2025) Asian Vegetable Research and Development Center. Available at: https://avrdc.org/wp-content/uploads/2024/05/WorldVeg-2023-Annual-Report.pdf (accessed 26 May 2025).

Badr, S.A. and Moharum, F.A. (2017) Scientific note: The Madeira *Phenacoccus madeirensis* Green (Hemiptera: Pseudococcidae) as a new record of mealybug in Egypt. *Alexandria Journal of Agricultural Science* 62(3), 329.

Barik, S., Mahankuda, B., Sahu, N., Mohapatra, S., Sridhar, V. *et al.* (2025) Thrips and mites in peppers: Challenges and resistance strategies. *Euphytica* 221, 82. DOI: 10.1007/s10681-025-03515-6.

Beeson, C.F.C. and Chatterjee, N.C. (1939) Possibilities of control of lantana (*Lantana aculeata* L.) by indigenous insect pests. *Indian Forest Records (New Series)* 6, 41–84.

Berlinger, M.J. (1986) Host plant resistance to *Bemisia tabaci*. *Agriculture Ecosystems & Environment* 17, 69–82. DOI: 10.1016/0167-8809(86)90028-9.

Boopathi, T. (2017) CABI compendium. DOI: 10.1079/cabicompendium.4141. (accessed 29 May 2025).

Buragohain, P., Saikia, D.K., Sotelo-Cardona, P. and Srinivasan, R. (2021) Evaluation of Bio-Pesticides against the South American Tomato Leaf Miner, *Tuta absoluta* Meyrick (Lepidoptera: Gelechiidae) in India. *Horticulturae* 7(9), 325.

CABI (2004) Crop Protection Compendium. Available at: www.cabi.org (accessed 28 May 2025).

CABI (2021) CABI Compendium 37938. Available at: https://doi.org/10.1079/cabicompendium.37938 (accessed 17 June 2025).

CABI (2022) CABI Compendium 42318. Available at: https://www.cabidigitallibrary.org/doi/full/10.1079/cabicompendium.42318 (accessed 17 June 2025).

CABI (2025) CABI Compendium. Available at: https://doi.org/10.1079/cabicompendium.37938. https://www.cabidigitallibrary.org/doi/full/10.1079/cabicompendium.40686#core-ref-23-1 (accessed 28 May 2025).

Carter, D.J. (1984) *Pest Lepidoptera of Europe: With Special Reference to the British Isles*, Vol. 31. Springer Science & Business Media.

Cuthbert, R.N., Bodey, T.W., Briski, E., Capellini, I., Dick, J.T.A. *et al.* (2025) Harnessing traits to predict economic impacts from biological invasions. *Trends in Ecology and Evolution (In Press)*. DOI: 10.1016/j.tree.2025.03.016.

Das, G.P. and Raman, K.V. (1994) Alternate hosts of the potato tuber moth, *Phthorimaea operculella* (Zeller). *Crop Protection* 13(2), 83–86. DOI: 10.1016/0261-2194(94)90155-4.

De Barro, P.J., Trueman, J.W. and Frohlich, D.R. (2005) *Bemisia argentifolii* is a race of *B. tabaci* (Hemiptera: Aleyrodidae): The molecular genetic differentiation of *B. tabaci* populations around the world. *Bulletin of Entomological Research* 95(3), 193–203. DOI: 10.1079/ber2004351.

Dickey, A.M., Kumar, V., Hoddle, M.S., Funderburk, J.E., Morgan, J.K. *et al.* (2015) The *Scirtothrips dorsalis* species complex: Endemism and invasion in a global pest. *PLoS ONE* 10(4), e0123747. DOI: 10.1371/journal.pone.0123747.

EPPO (2025). Available at: https://gd.eppo.int/taxon/BEMITA/vectorof (accessed 29 May 2025).

Fand, B.B. and Suroshe, S.S. (2015) The invasive mealybug *Phenacoccus solenopsis* Tinsley, a threat to tropical and subtropical agricultural and horticultural production systems – a review. *Crop Protection* 69, 34–43. DOI: 10.1016/j.cropro.2014.12.001.

FAO Stat (2025) FAO statistics. Available at: https://www.fao.org/faostat/en/#data/QCL (accessed 26 May 2025).

Farias, E.S., Santos, A.A., Ribeiro, A.V., Carmo, D.G., Paes, J.S. *et al.* (2020) Climate and host plants mediating seasonal dynamics and within-plant distribution of the diamondback moth (*Plutella xylostella*). *Crop Protection* 134, 105172. DOI: 10.1016/j.cropro.2020.105172.

Garvey, P.M., Banks, P.B., Suraci, J.P., Bodey, T.W., Glen, A.S. *et al.* (2020) Leveraging motivations, personality, and sensory cues for vertebrate pest management. *Trends in Ecology and Evolution* 35(1), 990–1000. DOI: 10.1016/j.tree.2020.07.007.

GBIF (2018) *Phenacoccus madeirensis* Green, 1923. Available at: https://www.gbif.org/

George, R.A.T. (2011) *Tropical Vegetable Production*. CABI Digital Library. DOI: 10.1079/9781845937539.0001.

GISD (Global Invasive Species Database) (2025) Species profile: *Orthezia insignis*. Available at: http://www.iucngisd.org/gisd/species.php?sc=1462 (accessed 15 June 2025).

Halder, J., Rai, A.B. and Divekar, P.A. (2025) Invasive and emerging insect pests problems in vegetables: Current status and way forward. In: Singh, A.N., Pandey, K.K., Srivastava, K., Halder, J. and Gowda, M.T. (eds) *Innovative Approaches for the Biotic Stress Management in Vegetable Crops*. ICAR-IIVR, Training Manual No. 117. ICAR-Indian Institute of Vegetable Research, Varanasi, Uttar Pradesh, India, pp. 7–22.

Harika, G., Dhurua, S., Sreesandhya, N., Suresh, M. and Govinda Rao, S. (2019) Biology of diamond-back moth, *Plutellae xylostella* (Lepidoptera: Plutellidae) of cauliflower under laboratory condition. *International Journal of Current Microbiology and Applied Sciences* 8(1), 866–873. DOI: 10.20546/ijcmas.2019.801.094.

Hutasoit, R.T., Triwidodo, H. and Anwar, R. (2017) Biology and demographic statistics of *Thrips parvispinus* Karny (Thysanoptera: Thripidae) in chili pepper (*Capsicum annuum* Linnaeus). *Indonesian Journal of Entomology* 14(3), 107–116.

IPPC (2025) International Plant Protection Convention. Available at: https://www.ippc.int/en/about/overview/ (accessed 26 May 2025).

IUCN (2025). Available at: https://iucn.org/our-work/topic/invasive-alien-species (accessed 9 June 2025).

Keerthika, K. and Jayakumar, M. (2024) Forecasting the future geographic distribution of diamondback moth, *Plutella xylostella* (L.) in changing climatic conditions using machine learning models. In: *National Conference on Emerging trends in Biological Sciences (NCETBS-2024)*, SRM Institute of Science and Technology, Ramapuram, Chennai, September 2024.

Kessing, J.L.M. and Mau, R.F.L. (1993) *Aleurodicus dispersus* (Russell). Available at: http://www.extento.hawaii.edu/kbase/crop/type/a_disper.htm (accessed 29 May 2025).

Makale, F., Kansiime, M.K. and Rwomushana, I. (2023) *Papaya Mealybug in Kenya: Identification, Management, and Future Outlook*. CABI Plant Health Cases. DOI: 10.1079/planthealthcases.2023.0018.

Mani, M. and Shylesha, A.N. (2019) Invasive mealybugs and their management in India: Past, present & future. *Hexapoda* 28, 64–77.

McAuslane, J.H. (2009) EENY-129 (IN286), one of a series of Featured Creatures from the Entomology and Nematology Department, Florida Cooperative Extension Service. Institute of Food and Agricultural Sciences, University of Florida. Available at: http://entomology.ifas.ufl.edu/creatures

Muniappan, R., Shepard, B.M., Watson, G.W., Carner, G.R., Rauf, A. *et al.* (2009) New records of invasive insects (Hemiptera: Sternorrhyncha) in Southeast Asia and West Africa. *Journal of Agricultural and Urban Entomology* 26(4), 167–174. DOI: 10.3954/1523-5475-26.4.167.

Musolin, D.L., Kirichenko, N.I., Karpun, N.N., Aksenenko, E.V., Golub, V.B. *et al.* (2022) Invasive insect pests of forests and urban trees in Russia: Origin, pathways, damage, and management. *Forests* 13, 521. DOI: 10.3390/f13040521.

Naveena, N.L., Shashank, R.P., Raghavendra, D. and Jeer, M. (2020) Invasive insect pests in India: Current scenario and future perspective. *Indian Entomologist* 1, 23–31.

Pandey, M., Bhattarai, N., Pandey, P., Chaudhary, P., Katuwal, D.R. *et al.* (2023) A review on biology and possible management strategies of tomato leaf miner, *Tuta absoluta* (Meyrick), Lepidoptera: Gelechiidae in Nepal. *Heliyon* 9(6), e16474. DOI: 10.1016/j.heliyon.2023.e16474.

Philips, C.R., Fu, Z., Kuhar, T.P., Shelton, A.M. and Cordero, R.J. (2014) Natural history, ecology, and management of diamondback moth (Lepidoptera: Plutellidae), with emphasis on the United States. *Journal of Integrated Pest Management* 5(3), D1–D11.

Pimentel, D., McNair, S., Janecka, J., Wightman, J., Simmonds, C. *et al.* (2001) Economic and environmental threats of alien plant, animal, and microbe invasions. *Agriculture, Ecosystems and Environment* 84(1), 1–20. DOI: 10.1016/S0167-8809(00)00178-X.

Pradhan, P.P. and Borthakur, A.B. (2020) Menace of spiralling whitefly, *Aleurodicus dispersus* Russell on the agrarian community. *International Journal of Current Microbiology and Applied Sciences* 9(12), 2756–2772. DOI: 10.20546/ijcmas.2020.912.329.

Prasannakumar, N.R., Venkataravanappa, V., Rachana, R.R., Sridhar, V., Govindappa, M.R. *et al.* (2021) Status of the outbreak of *Thrips parvispinus* (Karny) on chilli in Karnataka. *Pest Management in Horticultural Ecosystems* 27(2), 286–290.

Raghuteja, P.V., Chalapathi Rao, N.B.V. and Padma, E. (2022) A review on invasive pests of horticultural crop ecosystems. *Biological Forum – An International Journal* 14(3), 1209–1217.

Rai, D., Nautiyal, P., Raghav, D.K. and Dubey, A.K. (2023) American Serpentine leaf miner (*Liriomyza trifolii*, Burgess), a serious threat of horticultural crops. In: Teodor, R. (ed.) *Emerging Issues in Agricultural Sciences*. BP International, India, pp. 136–146. DOI: 10.9734/bpi/eias/v4/5641B.

Ramdwar, M.N.A., Harripersad, J. and Badal, S. (2023) Examining the predatory relationship between the indigenous firefly (*Aspisoma ignitum*) and the invasive giant African snail (*Achatina fulica*) in Trinidad, West Indies. *Journal of Agricultural Science* 15(5), 47–56.

Simberloff, D., Martin, J.L., Genovesi, P., Maris, V., Wardle, D.A. *et al.* (2013) Impacts of biological invasions: What's what and the way forward. *Trends in Ecology and Evolution* 28(1), 58–66.

Sinclair, J.S., Lockwood, J.L., Hasnain, S., Cassey, P. and Arnott, S.E. (2020) A framework for predicting which non-native individuals and species will enter, survive, and exit human mediated transport. *Biological Invasions* 22, 217–231. DOI: 10.1007/s10530-019-02086-7.

Sridhar, V., Chandana, S.P. and Rachana, R.R. (2021) Global status of *Thrips parvispinus* (Karny 1922), an invasive pest. *Journal of Research* PJTSAU 49(4), 1–11.

Srinivasan, R. (2016) Integrated pest management in tropical vegetable crops. In: Abrol, D.P. (ed.) *Integrated Pest Management in the Tropics. New India Publishing Agency*. New Delhi (India), pp. 219–247.

Statista (2025) Statista. Available at: https://www.statista.com/statistics/264662/top-producers-of-fresh-vegetables-worldwide/ (accessed 26 May 2025).

Thorat, S.S., Sisodiya, D.B. and Gangwar, R.K. (2022) Invasive thrips, *Thrips parvispinus* (Karny) an invasive threat: A review. *Environment and Ecology* 40(4A), 2170–2175.

Trivedi, T.P. and Rajagopal, D. (1992) Distribution, biology, ecology and management of potato tuber moth, *Phthorimaea operculella* (Zeller) (Lepidoptera: Gelechiidae): A review. *International Journal of Pest Management* 38(3), 279–285.

Tyagi, K., Kumar, V., Singha, D. and Chakraborty, R. (2015) Morphological and DNA barcoding evidence for invasive pest thrips, *Thrips parvispinus* (Thripidae: Thysanoptera), newly recorded from India. *Journal of Insect Science* 15(1), 105. DOI: 10.1093/jisesa/iev087.

Wang, Y.S., Dai, T.M., Tian, H., Wan, F.H. and Zhang, G.F. (2019) *Phenacoccus madeirensis* Green (Hemiptera: Pseudococcidae): New geographic records and rapid identification using a species-specific PCR assay. *Crop Protection* 116, 68–76.

Wittenberg, R. and Cock, M.J.W. (eds) (2001) *Invasive Alien Species: A Toolkit of Best Prevention and Management Practices*. CAB International, Wallingford, Oxon, UK, pp. xvii–228.

Zalucki, M.P., Shabbir, A., Silva, R., Adamson, D., Shu-Sheng, L. *et al.* (2012) Estimating the economic cost of one of the world's major insect pests, *Plutella xylostella* (Lepidoptera: Plutellidae): Just how long is a piece of string? *Journal of Economic Entomology* 105, 1115–1129.

10 Status of Invasive Mealybugs in the Indian Tropics

Arunkumara C.G.[1]*, Dhyan Chowdary D.[2], Anil Kumar G.[2] and Keerthi M.C.[3]
[1]CRRI – Central Rainfed Upland Rice Research Station, Hazaribag, Jharkhand, India; [2]Department of Entomology, Agricultural College, Bapatla, ANGRAU, Andhra Pradesh, India; [3]ICAR – Indian Institute of Horticultural Research, Bengaluru, Karnataka, India

Abstract

Mealybugs are soft-bodied, sap-sucking insects found worldwide, with approximately 2000 identified species. They are named for the mealy wax secretion that covers their bodies, and over 160 species have been recognized as destructive pests affecting crops globally. Known for their rapid reproduction and widespread distribution, mealybugs pose a significant threat as invasive pests. In India, several species of mealybugs, including the cotton mealybug, *Phenacoccus solenopsis* and the papaya mealybug, *Paracoccus marginatus*, have caused considerable economic losses. Additionally, the cassava mealybug, *Phenacoccus manihoti* has recently become a major problem in Tamil Nadu and Kerala. The hidden nature of mealybugs, as they often reside within plant structures, makes chemical control difficult, as pesticides frequently do not reach the targeted pests. As a result, natural enemies play a vital role in reducing mealybug populations and mitigating the associated damage. This chapter provides a comprehensive overview of invasive mealybug species in tropical regions, covering their host range, distribution, biology, damage they cause, natural enemies, and management strategies.

Keywords: Biology, distribution, inflicted damage, invasive pests, host range, mealybugs, natural enemies

10.1 Introduction

Biological invasions, often referred to as biological pollution, lead to significant losses in biodiversity. Invasive species pose risks to agricultural diversity, human well-being, and animal health due to their non-native and exotic nature, which gives them remarkable abilities to disperse and adapt. Several species of mealybugs have inadvertently entered countries such as India, Malaysia, Vietnam, Brazil, Sri Lanka, and Bangladesh, resulting in substantial economic losses in cultivated ecosystems (Atwal, 1976). In 1879, India encountered its first invasive pest, the San Jose scale, *Comstockaspis perniciosa* (Comstock, 1881) (Hemiptera: Coccomorpha: Diaspididae), which originated from China and has since been documented in a total of 32 species. Invasive pests have a significant impact on agro-ecosystems, leading to considerable economic losses (Shivakumara *et al.*, 2024a). The severity of crop losses is worsened by the limited availability of natural enemies that can effectively manage these pests. This results in potential economic and environmental damage, as well as harm to human health.

*Corresponding author: arunkumarabruce93@gmail.com

 Biological Invasions in the Tropics: Monitoring, Mechanism and Management (eds Akshay Kumar Chakravarthy and Pathour Rajendra Shashank)
DOI: 10.1079/9781836991977.0010

Mealybugs are classified within the insect category known as scale insects. These insects have soft, segmented oval bodies but lack an external shell. The name "mealybug" aptly describes these insects, as their bodies are covered in a white, powdery substance that resembles cornmeal (Kosztarab and Kozár, 1988; Mani and Shylesha, 2021). Due to their visual characteristics, mealybugs are often mistaken for cushion scale insects or woolly aphids. Unlike their close relatives, scale insects, mealybugs retain their legs throughout their life cycle. While mealybug species may appear similar externally, each possesses unique biological and morphological characteristics. Identification of mealybugs is mainly based on adult females. This group constitutes the second-largest family within the Coccoidea, with over 2000 described species and approximately 290 genera (Downie and Gullan, 2005; Ben-Dov, 2006). As phloem feeders, mealybugs extract sap from a variety of plants, including angiosperms, gymnosperms, and ferns. Beyond their feeding habits, they pose a threat by transmitting certain plant diseases, resulting in significant economic losses. Mealybugs consume large quantities of plant fluids and excrete a sugary liquid waste known as honeydew, which fosters the growth of a black fungus called sooty mould. Consequently, a severe mealybug infestation can create a messy, sticky, and black microenvironment. In recent years, mealybugs, once considered minor pests, have now escalated to the status of major pests, particularly affecting commercial crops, vegetables, and fruits. The primary method of control for mealybugs is chemical intervention. However, their cryptic behaviour, distinctive waxy covering, and clustered distribution create challenges that render the effectiveness of many insecticides questionable. The repeated use of insecticides has negatively impacted the natural enemies of mealybugs. Insecticide resistance has made certain chemicals unsustainable for continued use, and there is increasing restriction of insecticides due to their toxicity and lack of selectivity. Some substances have been phased out, while others are targeted for reduction as part of programmes and regulations that promote the sustainable use of insecticides, guided by risk or hazard assessments (Franco *et al.*, 2004; Charles *et al.*, 2006; Walton *et al.*, 2006).

Moreover, because mealybugs often inhabit concealed parts of plants, chemicals frequently fail to reach the intended targets, making chemical control largely ineffective. Mealybugs typically proliferate during the fruiting phase of plants, necessitating repeated insecticide applications for effective management. This frequent use results in residue issues in agricultural produce, rendering them unfit for export and posing a risk in the domestic market.

10.2 Biology of Mealybugs

Most mealybug species reproduce through parthenogenesis, although certain species, such as *Maconellicoccus hirsutus* (Green) (Pseudococcidae) (pink hibiscus mealybug), exhibit biparental reproduction. Mature females lay eggs within a white wax egg sac, typically found in clusters on twigs, branches, or the bark of the host plant, and occasionally on leaves or terminal ends. Each egg sac can contain up to 600 eggs, with the majority developing into female offspring, leading to rapid population growth. Some species, such as *Dysmicoccus brevipes* Cockerell (Pseudococcidae) (pink pineapple mealybug), are ovoviviparous, meaning their eggs hatch internally. The tiny eggs measure between 0.3–0.4 mm in length, with development taking 3–9 days. After hatching, the mobile nymphs, called crawlers, resemble female adults. Female mealybugs go through three nymphal instars, while males have four, with duration of 22–25 days. In the final instar stage, males enter an inactive phase with wing buds that are enclosed in a cocoon made of mealy wax. Under typical conditions, mealybugs may take up to 30 days to complete all nymphal stages and can produce as many as 15 generations per year (Tanwar *et al.*, 2007).

10.3 Mode of Transport

Plants that are not infected are at risk of infection from those that are already infested, as juvenile mealybugs can crawl from one plant to another. These small crawlers can be transported through various means, including wind, birds, clothing, and vehicles, often settling in crevices, especially on newly planted specimens. The wax coating on each egg aids in transport via equipment, animals, or humans. Notably, female mealybugs are inactive and unable to fly, relying on external factors to move. Humans inadvertently contribute significantly to the spread of mealybugs. As infestations develop, mealybug colonies often migrate from shoot tips to twigs and branches, eventually moving down the trunk. The potential for long-distance movement increases when infested plant materials and fresh produce are transported across regions or even within a farm. Ants attracted to the honeydew secreted by mealybugs have been observed transporting them between plants (Tanwar *et al.*, 2007).

10.4 Association of Ants and Mealybugs

Mealybugs engage in a mutually beneficial relationship with ants by attracting them with their sweet secretion, known as honeydew. In return, ants help spread mealybugs and provide protection against predators such as ladybird beetles and parasites. Additionally, ants play a vital role in maintaining their colony's cleanliness by removing detritus that accumulates in the honeydew, which could potentially harm the ant community. Ant species such as *Oecophylla smaragdina* (Fabricius) (Hymenoptera: Formicidae), *Crematogaster* sp., and *Anoplolepis gracilipes* F. Smith (Hymenoptera: Formicidae) have been observed tending to mealybugs, particularly *Planococcus citri* Antoine Risso (Hemiptera: Pseudococcidae) (citrus mealybug), while feeding on honeydew produced on hibiscus plants (Tanwar *et al.*, 2007).

10.5 Nature of Damage

Mealybug infestations can have various detrimental effects on host plants, with the severity often depending on the plant's susceptibility. Infested growth points may exhibit stunted development, and dense clusters of mealybugs form thick, cottony mats beneath leaf surfaces due to their waxy secretions. These pests excrete large amounts of honeydew, which not only attracts ants but also promotes the growth of black sooty mould, significantly reducing the plant's ability to photosynthesize. Both nymphs and adults of mealybugs consume sap from leaves, leading to yellowing, withering, and even premature fruit drop. In severe cases, heavy infestations can result in defoliation and plant death. Moreover, mealybugs adversely affect the formation of flowers and stems, particularly in succulents with fleshy tissues. Infested fruits may become entirely coated in white wax, resulting in premature drop or shrinkage on the plant, making them unmarketable (Shivakumara *et al.*, 2024b).

Like many pests, mealybugs can originate from various sources, including endemics, immigrants, and mutants. Invasive mealybugs can quickly become significant pests upon successfully colonizing new regions, negatively impacting crop yields as they infest both native and non-native plants. *Planococcus citri* (Hemiptera: Pseudococcidae), an introduced pest in citrus-growing regions globally, weakens citrus saplings; although its impact on mature fruit-bearing trees is relatively minor. Mealybug outbreaks are attributed to three main factors: (1) the introduction of a foreign mealybug species, (2) the use of broad-spectrum pesticides that disrupt ecological balances, and (3) environmental factors that influence the tri-trophic interactions among host plants, mealybugs, and their natural enemies. The invasive mealybug species in India, along with their origins, introduction years, and host dynamics, are documented in Table 10.1 and illustrated in Fig. 10.1.

Invasive and economically significant mealybug species, including their distribution, host range, key diagnostic characteristics, biology, damage potential, and the natural enemies associated with them are described here under.

10.6 Pest Descriptions

10.6.1 Striped mealybug, *Ferrisia virgata* (*Cockerell, 1893*)

The striped mealybug is a small insect that poses challenges to various crops. Its global spread has been facilitated by international trade. This pest has an elongated, oval body that is dark grey in colour, with brown legs, and it is covered in white, mealy wax. It is distinguished by a pair of dark dorsal stripes and measures 4–5 mm in length, featuring two long tails. The body is

Table 10.1. Invasive mealybugs in India with origin and host dynamics. (Author's own table)

Common name	Scientific name	Origin	Introduced year	Host dynamics	Reference
Striped mealybug	*Ferrisia virgata*	Western hemisphere	1894	Guava, avocado, banana, betel vine, black pepper, cassava, cashew, cauliflower, citrus, cocoa, coffee, cotton, custard apple, grapevine, litchi, mango, oil palm, pigeon pea, pineapple, soyabean and tomato	Newstead (1894)
Fruit tree mealybug	*Rastrococcus invadens*	Oriental region	1986	Mango, citrus, breadfruit banana	Williams (1986)
Citrus mealybug	*Planococcus citri*	China	1986	Primary – citrus, Other hosts – curry leaf, Annona, coffee, cotton, banana, carambola, cocoa, ginger and mango	Prakasan (1987)
Pineapple mealybug	*Dysmicoccus brevipes*	Central and South America	1987	Primary host – pineapple Other hosts – avocado, banana, celery, citrus, clover, cocoa, coconut, coffee, cotton, custard apple, figs, ginger, guava, maize, mango, oil palm, orchids, groundnut, peppers, plantain, potato and sugarcane	Mani and Thontadarya (1987)
Sugarcane mealybug	*Saccharicoccus sacchari*		1933	Sugarcane	Isaac and Misra (1933)
Papaya mealybug	*Paracoccus marginatus*	Mexico	2005	Primary – papaya Other hosts – guava, cassava, sweet potato and eggplant	Muniappan *et al.* (2008)
Cotton mealybug	*Phenacoccus solenopsis*	North America	2006	Eggplant, okra, tomato and China rose (hibiscus)	Jhala *et al.* (2008)
Madeira mealybug	*Phenacoccus madeirensis*	Neotropical origin	2012	Cassava, pineapple, citrus, potato and ornamentals	Shylesha and Joshi (2012)
Jack Beardsley mealybug	*Pseudococcus jackbeardsleyi*	Neotropical origin	2012	Citrus, papaya, capsicum, banana, tomato, orchids, pepper and hibiscus plants	Mani *et al.* (2013)
Lantana mealybug	*Phenacoccus parvus*	Neotropical origin	2012	Lantana, tomato, potato, eggplant, China aster	Sridhar *et al.* (2012)
Cassava mealybug	*Phenacoccus manihoti*	South America	2020	Cassava	Sampath *et al.* (2021)

adorned with long, thin, glossy threads that resemble crystals and extend in all directions, lacking lateral filaments. Typically, this species is ovoviviparous (Mani and Shivaraju, 2016).

10.6.1.1 Distribution and host range

Originally native to America, the genus *Ferrisia* Fullaway, 1923 has now expanded worldwide, spreading across tropical and temperate regions (Williams and Watson, 1988). The striped mealybug, one of the two *Ferrisia* species that transitioned to the Old World, has become widely distributed (Williams and Watson, 1988; Williams and Granara de Willink, 1992; Kosztarab, 1996; Williams, 2004). This species has economic significance as a pest affecting agricultural products, including cotton and cocoa (Kosztarab, 1996; Ameyaw *et al.*, 2014; Oliveira *et al.*, 2014). The striped mealybug is polyphagous, with common hosts such as citrus, croton, hibiscus, and copperleaf (Garcia Morales *et al.*, 2015). Additionally, it significantly impacts crops such as Amaranthus, cassava, Ficus, magnolia, Plumeria, silk tree,

Fig. 10.1. Different species of mealybugs. (A) Adult *Paracoccus marginatus*. (B) Adult *Phenacoccus solenopsis*. (C) *Ferrisia virgata* infestation on custard in the IIHR field. (Author's own image)

sugar-apple, cocoa, and cotton (Ameyaw *et al.*, 2014; Oliveira *et al.*, 2014).

10.6.1.2 Biology of the striped mealybug

The striped mealybug completes development through three nymphal instars, with females maturing into adults after the third instar. Males begin constructing cocoons at the beginning of the second instar, where they continue to feed and develop (Highland, 1956). During the third instar, males undergo a pre-pupa stage before forming a pupa. After moulting, males emerge as winged adults and reach sexual maturity. The reproductive process of the striped mealybug involves sexual reproduction, unlike the closely related *Ferrisia malvastra*, which is parthenogenic (Kaydan and Gullan, 2012). Instead of producing ovisacs, female striped mealybugs create "pads" from waxy filaments. Nymphs of the first instar hatch from eggs laid on these pads within a few hours (Awadallah *et al.*, 1979). According to various studies (Highland, 1956; Ammar *et al.*, 1979; Awadallah *et al.*, 1979), there are typically 3–5 generations of striped mealybugs each year.

10.6.1.3 Damage

In recent years, notable infestations of *F. virgata* have been observed on ornamental plants, raising concerns due to the species' classification as a significant pest, particularly affecting cotton. They are typically found on foliage above ground, causing sooty mould-type damage. Under severe weather conditions, especially in Africa, they may move to the crown and roots of their host plants. An unusual characteristic of the striped mealybug is its role as a vector for plant diseases, transmitting viruses such as citrus tristeza, piper yellow mottle, and cocoa swollen shoot viruses (Thorold, 1975; Bhat *et al.*, 2003; Ameyaw *et al.*, 2014).

10.6.1.4 Natural enemies

Important natural enemies of *F. virgata* include *Acerophagus texanus*, *Anagyrus* sp., *Scymnus coccivora*, *Chrysopa* sp., *Cryptolaemus montrouzieri*, *Spalgis epius*, and *Chrysoperla carnea* (Sergio et al., 2022).

10.6.2 Citrus mealybug, *Planococcus citri Risso, 1813*

Mealybugs are significant pests of fruit crops and can reach epidemic levels in affected areas. *Planococcus citri*, commonly known as the citrus mealybug, is native to Asia but has spread globally due to accidental introductions. This pest infests a wide variety of crop plants, ornamentals, and wild plants. In India, losses attributed to *P. citri* are particularly severe in the northeastern region and Peninsular India (Mani and Krishnamoorthy, 1996). For instance, it has been identified as a major pest for specific fruit varieties, such as kinnow in Punjab (Arora, 1999) and Khasi mandarin in Meghalaya (Shylesha and Pathak, 1997; Pathak *et al.*, 1999). In Nagpur, Central India, the prevalence of this pest in mandarin orchards varies from low to medium (Kalidas and Shivankar, 1994).

10.6.2.1 Host range

The citrus mealybug frequently affects curry leaf plants, *Murraya koenigii*, poinsettias, and other flowering plants. It also targets several other crops, including Annona, arabica and robusta coffee, citrus fruits, cotton, bananas, carambola, cocoa, ginger, and mangoes (Jayma *et al.*, 1992).

10.6.2.2 Biology of the citrus mealybug

The body of the citrus mealybug exhibits a colour variation ranging from yellow to pink, adorned with moderately sized, slightly curved waxy filaments, and lacks a pyramid shape. Adult mealybugs display a single dorsal median stripe on their backs and have a clear body fluid. Fully developed females are wingless, with a robust and flattened body, and have short waxy filaments that are covered with white mealy powder along their edges. The presence of these pests is most notable from January to April, and they descend to the ground between April and May. Females deposit yellow eggs within an ovisac, with a single mass containing anywhere from 300 to 800 eggs. After 10–20 days, the eggs hatch, releasing amber-coloured nymphs that begin feeding by piercing the lower leaf surfaces with their mouthparts. As nymphs mature, they develop filaments and a waxy coating. Within a few weeks, after undergoing three moults, a female nymph reaches full maturity. The winged males, resembling grey midges, have long antennae, lack mouthparts, and do not feed. Three weeks post-hatching, male nymphs create cotton-like cocoons, undergo pupation, and experience four moults to become winged adults. This life cycle results in three overlapping generations each year.

10.6.2.3 Damage

Citrus plants at various growth stages consistently show signs of attack, particularly at leaf joints or between branches of mature trees. The mealybugs prefer to target split bark injuries along the stems of declining trees. Both females and nymphs predominantly feed on the undersides of leaves, causing considerable damage to both mature plants and seedlings. In severe infestations, citrus flowers may fail to develop into fruit, leading to the discovery of dead young berries before they mature. The feeding activity of mealybugs causes leaves to wrinkle, and significant horticultural features, such as variegation, may increase pest prevalence (Bindra, 1970; Jadhav and Pujari, 1999).

10.6.2.4 Natural enemies

Anagyrus sp., *Blepyrus insularis*, *Diversinervus* sp., *Tetrastichus* sp., *Microterys* sp., *Cryptochaetum* sp., *S. coccivora*, *Pulluspallidicollis*, *Nephus* sp., *Chrysopa* sp., *Micraspis cardoni*, *Pseudaspidimerus uttami*, *C. montrouzieri*, and *S. epius* were recorded in Kodagu, Karnataka (Singh, 1990). *Coccidoxenoides peregrines* (Krishnamoorthy and Mani, 1989), *Mallada boninensis*, *Plesiachrysa lacciperda*, *Anisochrysa basalis*, and *C. carnea* (Krishnamoorthy and Mani, 1989) were recorded in Karnataka. In Assam, *C. montrouzieri* and *Entomophthora fumosa* were recorded on *P. citri* (Chowdhury and Majid, 1954).

10.6.3 Papaya mealybug, *Paracoccus marginatus* williams & Granara de Willink, 1992

Papaya mealybug, *P. marginatus*, is a small, polyphagous sucking insect that poses a significant threat to various plants, particularly economically important tropical fruits, vegetables, and ornamental plants. First reported in Coimbatore, Tamil Nadu, in 2007, this pest has caused extensive damage to agricultural and horticultural crops, escalating to major pest status by 2009. As a result, farmers in Coimbatore, Erode, Tirupur, and Salem have faced substantial losses (Tanwar *et al.*, 2010).

10.6.3.1 Distribution and host range

The papaya mealybug originated in Mexico and Central America, where it was first collected in 1955 and officially described in the Neotropical region in 1992, encompassing Belize, Costa Rica, Guatemala, and Mexico. It gained pest status after invading the Caribbean, with records found in 14 Caribbean countries since 1994. *P. marginatus* was observed in South and Southeast Asia between 2008 and 2009. In India, it was first documented in July 2007 in Coimbatore and has since spread to nearby districts, as well as to Karnataka, Andhra Pradesh, and Kerala. This mealybug exhibits a high level of polyphagy, affecting plants across 136 genera within 49 families, as reported by Garcia Morales *et al.* (2019). Notable host plants include papaya, hibiscus, avocado, citrus, cotton, tomato, eggplant, peppers, beans, and peas, along with sweet potato, mango, cherry, and pomegranate.

10.6.3.2 Biology

Adult female *P. marginatus* are yellow and covered with a white waxy coating, featuring short waxy filaments around the margins. Adult males initially exhibit a yellow colour that changes to pink during the pre-pupal and pupal stages. A single female can lay approximately 230–400 eggs within an ovisac. Males undergo four instars, with first-instar nymphs, known as crawlers, having developmental durations of 6.5, 6.6, 2.4, and 41 days for the first, second, third, and fourth instars, respectively, at 25°C. The fourth instar is enclosed in a cocoon. Female nymphs have shorter durations for the first, second, and third instars, lasting 6.5, 5.5, and 5.2 days, respectively, at the same temperature. This species reproduces both sexually and through parthenogenesis.

10.6.3.3 Damage

In Tamil Nadu, India, agricultural losses attributed to the papaya mealybug are estimated to range from 10–60%, depending on the specific crop, with some instances exceeding these figures (Shylesha *et al.*, 2010; Selvaraj, 2012).

10.6.3.4 Natural enemies

Several indigenous hymenopteran parasitoids from the Encyrtidae family have been identified as natural enemies of the mealybug, including *Anagyrus loecki*, *Pseudophycus* sp., *Acerophagus papayae*, *Pseudleptomastix mexicana*, and *Apoanagyrus californicus*. To enhance biological control efforts, three exotic encyrtid parasitoids – *A. loecki*, *A. papayae*, and *P. mexicana* – were introduced to India by the USDA – Animal and Plant Health Inspection Service (APHIS) from Puerto Rico on 15 July 2010. This initiative was part of a collaborative effort with the ICAR – National Bureau of Agricultural Insect Resources (NBAIR) in Bangalore, India. Among these parasitoids, *A. papayae* demonstrated a higher parasitization rate, ranging from 75.6–81.7%.

10.6.4 Cotton mealybug, *Phenacoccus solenopsis Tinsley, 1898*

Phenacoccus solenopsis was first identified by Tinsley in 1898. Since its arrival in India in 2005, this mealybug species has caused significant damage to cotton crops. It has been found on various plants, including food crops, fibre crops, fruits, ornamental plants, plantation species, vegetables, and weeds (Kumar *et al.*, 2024). Due to its invasiveness, rapid spread, and morphological and biological variations, *P. solenopsis* has become a key subject of research among entomologists worldwide. The frequent outbreaks have resulted in severe losses for both Bt and non-Bt cotton, leading growers to apply large quantities of pesticides for control.

10.6.4.1 Distribution and host range

Originally from Central America, *P. solenopsis* is now widespread, affecting various agro-ecological zones across more than 24 countries. It is recognized as one of the most destructive mealybug species in the developing world, causing notable crop losses. The pest has a broad host range, affecting 154 plant species, including 20 field crops, 64 weeds, 45 ornamental plants, and 25 shrubs and trees, spanning 53 plant families (Arif *et al.*, 2009). Among the 47 cultivated crops reported, severe infestations have been documented in cotton, okra (Malvaceae), tomato, chilli, eggplant, potato (Solanaceae), cluster bean, green gram (Fabaceae), papaya (Caricaceae), and sunflower (Asteraceae).

10.6.4.2 Biology

The body of *Phenacoccus solenopsis* is covered in a white waxy substance; female adults measure approximately 3 mm wide and 5 mm long. The dorsum features distinct black bands, stout legs with moderate hair, and brown, nine-segmented antennae (Tinsley, 1898). This species reproduces via ovo-viviparity, requiring mating for females to reproduce. An adult female can lay around 500 eggs in an ovisac; these eggs are minute, oval-shaped, and light yellow. The smooth, translucent eggs are retained within the female's body until they are ready to hatch. The female produces a sac with a cottony covering that protrudes from the anal end. The incubation period for the eggs is approximately 6.6 days. Female nymphs undergo three moults, while males undergo four. The neonate nymphs are oblong. The average duration of the first, second, and third instar nymphs are 4.8, 5.6, and 6.4 days, respectively, with a total nymphal duration of 16.8 days for females. The first and second instar nymphs are pale yellow and oblong, while a white waxy substance covers the dorsal surface of the third instar. The adult female, which is oblong and wingless, ranges from light to dark yellow, and has pre-oviposition, oviposition, and post-oviposition periods lasting 4.3, 8.0, and 2.7 days, respectively. The female adult survives for 15.5 days, leading to an overall lifespan of approximately 31 days. Dark spots on the thorax and abdomen form two longitudinal stripes.

10.6.4.3 Damage

Since the introduction of Bt cotton in India, there has been a marked increase in the prevalence of cotton mealybugs. These mealybugs feed on various parts of the cotton plant, including leaves, petioles, squares, and bolls. Severe infestations can lead to extensive damage, resulting in stunted plant growth, reduced boll size, and a decline in yield. Significant outbreaks of *P. solenopsis* were reported in the Punjab and Sindh regions of Pakistan in 2005 (Arif *et al.*, 2009), and by 2006, the infestation spread to numerous cotton-growing districts, severely compromising cotton yields. From 2005 to 2009, cotton yield losses ranging from 30 to 60% were documented in both Pakistan and India, particularly affecting crops during 2006 and 2007 (Dutt, 2007).

10.6.4.4 Natural enemies

In India, a solitary endoparasitoid, *Aenasius* sp., was reported to attack *P. solenopsis* (Sharma, 2007). Hayat (2009) described a new species of parasitoid, *A. bambawalei* (previously known as *A. arizonensis*), associated with *P. solenopsis*, which shows promising biocontrol potential. This parasitoid was found to effectively parasitize *P. solenopsis* on several plants, achieving parasitization rates up to 72%. An even higher parasitism rate of 89% was recorded in okra (Sharma, 2007). The predator *C. montrouzieri* has also been effectively utilized to control various mealybug species. In India, two coccinellid species, *Brumoides suturalis* and *Hyperaspis maindroni*, were identified as being associated with *P. solenopsis* (Patel *et al.*, 2009). Other potential predators, such as *C. carnea*, have been observed consuming approximately 30 mealybug eggs daily in laboratory tests (Rabinder *et al.*, 2008).

10.6.5 Cassava mealybug, *Phenacoccus manihoti* Matile-Ferrero, 1976

Phenacoccus manihoti poses a significant threat to cassava on a global scale. Originally from South America, the first documented presence of this pest was in the Democratic Republic of Congo in 1973, from which it spread throughout sub-Saharan Africa (Hahn and Williams, 1973). It was not until 2008 that *P. manihoti* was identified in Thailand, indicating its first appearance in Asia. Since then, it has rapidly expanded across cassava-growing regions in Thailand and neighbouring countries, extending to Indonesia, with yield losses of up to 100% reported in affected areas (Herren, 1981).

10.6.5.1 Distribution and host range

The *P. manihoti* is native to South America, particularly Argentina, Guyana, and Paraguay. After its unintentional introduction to the Republic of the Congo in 1973 (Herren and Neuenschwander, 1991), the mealybug has proliferated across Africa, reaching nearly every cassava-producing nation from West to East Africa, and extending to South Africa's eastern border. According to reports by Winotai *et al.* (2010), Muniappan *et al.* (2011), Parsa *et al.* (2012), and Sartiami *et al.* (2015), *P. manihoti* has also been found in Asia, including Thailand (2008), Indonesia (2010), Cambodia and Vietnam (2012), Laos (2014), and Malaysia (2015). Currently, this pest is present in tropical regions worldwide, with the exception of the Nearctic region (Garcia Morales *et al.*, 2016). In India, it was recorded on cassava in Thrissur, Kerala, in April 2020 (Sampath *et al.*, 2021). *P. manihoti* primarily targets cassava and other Manihot species, with rare instances of infestations on other crops or wild hosts. Research and laboratory tests have shown a preference for poinsettia, croton, and *Talinum triangulare*. There is no conclusive evidence that *P. manihoti* can sustain multiple generations on plants other than cassava and possibly other Euphorbiaceae (Herren and Neuenschwander, 1991).

10.6.5.2 Biology of cassava mealybug

Female mealybugs are ovoid in shape, characterized by a distinctive rose-pink colour and a coating of white powdery wax. *P. manihoti* can have nine generations per year, with populations starting to grow in February. Its life cycle includes an egg stage and four instars, with the adult stage being the fourth. Due to its parthenogenic reproduction, *P. manihoti* only produces female offspring, meaning that a single immature or adult mealybug can initiate an outbreak. The average development time from egg to adult is about 33 days. Adults lay 200–600 eggs in ovisacs on the undersides of leaves and around apical and lateral buds under favourable conditions. These ovisacs are sticky, aiding in the mealybug's spread over great distances. The first instars, or crawlers, hatch from the eggs and go through three moults before reaching maturity.

10.6.5.3 Damage

P. manihoti causes significant damage by inhibiting the growth points of cassava plants, sometimes leading to complete defoliation. Reported losses in storage root yield have reached as high as 84% (Nwanze *et al.*, 1982). The mealybug prefers the lower surfaces of fully expanded leaves (Nwanze, 1978) and gradually moves to the stems and shoot tips. As the crawlers mature, they induce leaf curling at the growing tip, prompting the formation of adventitious buds on nearly all internodes and resulting in 'bunchy tops'.

10.6.5.4 Natural enemies

Anagyrus lopezi, a parasitoid native to Central America, has been successfully used to control *P. manihoti* in Africa and other Asian nations. *A. lopezi* is host-specific and environmentally adaptable, demonstrating parasitism levels ranging from 10–57%. Various predators, such as coccinellids, *Exochomus flavipes*, *Hyperaspis delicatula*, and the dipteran predator *Dicrodiplosis manthe*, as well as lepidopterans such as *Spalgis lemolea*, also contribute to the integrated management of *P. manihoti*.

10.7 Management of Mealybugs

10.7.1 Cultural control

Early detection is crucial for controlling mealybugs, as they are most susceptible during their immature stages (Dreistadt, 2001; Franco *et al.*, 2009). When acquiring plants, inspect them carefully for signs of mealybugs or cottony wax masses, particularly on sensitive species. Mealybugs can be difficult to identify with the naked eye because their bodies are covered in a waxy substance and they often hide in concealed areas, such as the undersides of leaves or bends in branches. Signs of mealybug infestation, including sooty mould and ant activity, may be present even if the mealybugs themselves are not visible. Isolate, remove, and dispose of any infested plants promptly to prevent further spread of the infestation. In commercial greenhouses, plants with heavy infestations are often destroyed due to the difficulty of managing such outbreaks.

10.7.2 Mechanical control

Integrated management of mealybugs combines cultural and mechanical practices aimed at reducing pest populations and preventing their spread. During winter, pruning affected shoots and opening the tree canopy can enhance light penetration, helping to suppress pest development. It is also essential to eliminate ant colonies in orchards, as they play a significant role in transporting mealybugs to feeding sites. In the summer, raking the soil around the main stem helps to dry out the eggs and exposes pests to their natural enemies. Applying sticky bands (7–8 cm wide) around the main stem, approximately 50 cm above the ground in mid-December, effectively traps ascending nymphs; these bands should be replaced as needed. Other containment measures include debarking and directly eliminating pest populations (Shivankar and Singh, 2000). Additional integrated approaches may involve applying adhesive materials to the main stem to deter mealybug movement (Michelakis and Hamid, 1995).

10.7.3 Pheromones

Males have shown positive responses to synthetic pheromones, specifically [(1R-cis)−2, 2-dimethyl −3-(1-methyl phenyl) cyclobutyl methyl acetate], on filter paper discs in Petri dishes, even at rates of 19 mm g/108 cm^3. However, this response diminished at higher doses. According to Hefetz and Auber (1990), the synthetic sex pheromone of *P. citrii* in the field has a half-life of approximately 2 weeks, with doses ranging from 400 to 700 mg resulting in the highest number of male captures. Regarding Insect Growth Regulators (IGRs), Mendel *et al.* (1991) reported that the IGR buprofezin demonstrated strong ovicidal activity, leading to a 91–99% mortality rate in nymphs with over 80% inhibition of egg hatch.

10.7.4 Biological control

Biological control is one of the most effective and sustainable methods for managing invasive mealybug species in tropical regions. This approach involves the introduction or conservation of natural enemies, primarily parasitoid wasps such as *Cheiloneurus* sp., *Aenasius arizonensis*, *Pseudleptomastix mexicana*, *Chartocerus* sp., *Marietta leopardine*, *Aprostocetus* sp. and *Homalotylus turkmenicus*. *Anagyrus lopezi* and *Acerophagus papaya* (Fig. 10.2) and predatory insects, mainly ladybird beetles, such as *Cryptolaemus montrouzieri*, *Coccinella septempunctata*, *Cheilomenes sexmaculata*, *Coccinella transversali*, *Brumoides suturalis*, *Scymnus nubilu*, *Harmonia octomaculata*, and *Cheilomenes propinqua* (Fig. 10.3) which naturally suppress mealybug populations (Dreistadt, 2001; Franco *et al.*, 2009). One notable success is the classical biological control of the cassava mealybug, *P. manihoti*, in Africa, using *Anagyrus lopezi*, which has significantly reduced crop losses and minimized pesticide use. Similarly, *Acerophagus papayae* has been utilized against the papaya mealybug, *P. marginatus*, in several Asian and Pacific countries with promising results. The effectiveness of biological control depends on the careful selection of agents, understanding host specificity, and ongoing monitoring. When integrated with other pest management strategies, biological control offers an environmentally friendly and economically viable solution to invasive mealybug outbreaks in tropical agro-ecosystems.

10.7.5 Chemical control

Controlling mealybugs with pesticide sprays can be challenging due to their waxy and hydrophobic covering (Dreistadt, 2001; Franco *et al.*, 2009). Contact insecticides are most effective when targeting the younger stages, as mealybugs in their first and second instars lack the protective wax layer that develops later. It is recommended to apply insect growth regulators (IGRs), horticultural oils, and insecticidal soaps directly to the plants and mealybugs for treating early instar mealybugs. Systemic insecticides, when sprayed on the soil and absorbed by the host plant through its roots, have also shown effectiveness. However, because of the overlapping generations of mealybugs, repeated and frequent applications are often necessary, which can lead to the accelerated emergence of pesticide resistance.

10.7.6 Classical biological control of invasive mealybug species in India

Several species of mealybugs have invaded India over the years, notably the cotton mealybug (*Phenacoccus solenopsis*) and the papaya mealybug (*Paracoccus marginatus*), both of which have caused severe economic losses for farmers. More recently, the cassava mealybug (*Phenacoccus manihoti*) has led to significant losses for cassava growers in Tamil Nadu and Kerala. Alongside the invasive cotton mealybug, an exotic parasitoid, *Acerophagus bambawalei*, was also introduced

Fig. 10.2. Hymenopteran parasitoids associated with *Phenacoccus solenopsis* and *Paracoccus marginatus*. (A) *Cheiloneurus* sp. (B) *Aenasius arizonensis*. (C) *Pseudleptomastix mexicana*. (D) *Chartocerus* sp. (E) *Marietta leopardine*. (F) *Aprostocetus* sp. and (G) *Homalotylus turkmenicus*. (Author's own images)

Fig. 10.3. Predatory fauna of mealybugs. (A) *Coccinella septempunctata*. (B) *Cheilomenes sexmaculata*. (C) *Coccinella transversali*. (D) *Brumoides suturalis*. (E) *Scymnus nubilis*. (F) *Harmonia octomaculata*. (G) *Cheilomenes propinqua*. (Author's own images)

and has since established naturally. Classical biological control has been implemented for two mealybug species in India: the papaya mealybug and the cassava mealybug. In 2006, farmers in Tamil Nadu began reporting damage to their papaya crops inflicted by a pest. Despite numerous applications of insecticides, crop losses continued to escalate, as documented by Regupathy and Ayyasamy (2010). In July 2008, Dr R. Muniappan from the Integrated Pest Management Innovation Laboratory (IPM IL) collected specimens of the pest infesting a papaya tree at Tamil Nadu Agricultural University (Muniappan, 2009). These specimens were sent to Gillian Watson at the California Department of Food and Agriculture, who confirmed the pest to be *P. marginatus*. The pest's effects extended to various crops, including papaya, mulberry, Deccan hemp, physic nut, guava, pomegranate, eggplant, potato, tomato, and teak, resulting in losses ranging from 10 to 60% depending on the crop (Shylesha *et al.*, 2010; Selvaraj, 2012; Srinivasan *et al.*, 2024). In February 2009, *P. marginatus* was identified in mulberry seedlings in the Erode and Tiruppur districts (USAID, 2012).

The impact of *P. marginatus* on crop yield was substantial due to its ability to move from one plant to another, injecting saliva and feeding on plant sap by inserting stylets into the vascular system (Shylesha *et al.*, 2010). This pest caused noticeable damage, including sooty mould, dried leaves, and stunted growth (Kumar and Shekhar, 2010; Sharma and Muniappan, 2021). Insecticides proved ineffective against *P. marginatus*, as the insect's waxy coating prevents chemicals from penetrating its body (Kumar and Shekhar, 2010). In July 2010, three parasitoids – *Acerophagus papayae*, *Pseudleptomastix mexicana*, and *Anagyrus loecki* – were imported from Puerto Rico (Regupathy and Ayyasamy, 2011). These parasitoids specifically targeted the papaya mealybug and did not affect other mealybug species in India.

The imported parasitoids laid their eggs on the second and third instar nymphs of the papaya mealybug. Upon hatching, the parasitoid larvae consumed the mealybug's body contents, ultimately leading to the pest's death. The larvae then pupated inside, transforming the mealybug's body into a mummy (Muniappan *et al.*, 2006). Following approval from the Plant Protection Advisor to the Government of India, mass production of parasitoids was carried out at 57 locations in Tamil Nadu, and they were subsequently released into farmers' fields free of charge. Only *A. papayae* was cultured, and starting in October 2010, 250 parasitoids per hectare were released in affected areas. The other two parasitoids were not field-released. Remarkably, parasitoids released in one hectare expanded their presence to 50 hectares within 3 months (Shylesha *et al.*, 2010). Approximately 5.7 million parasitoids were released across Tamil Nadu in multiple phases. These parasitoids not only established themselves but also extended their impact beyond papaya to other crops affected by the papaya mealybug (Shylesha *et al.*, 2010). Within a month of release, pest populations began to decline, and by the fifth month, effective control was achieved throughout the affected area. This included a 97% decline in the mealybug population on papaya and mulberry, and complete control in cassava (Kalyanasundaram *et al.*, 2010).

The introduction of parasitoids led to a decrease in pesticide use, which resulted in increased production and income for farmers. The mealybug problem, which could have posed a significant threat to crops across South Asia, was effectively addressed at a local cost of $200,000 in the first year (USAID, 2012) and an additional $100,000 per year for the subsequent 3 years. This solution proved cost-effective, preventing potential devastation of crops in the region. Cassava is a vital root crop and the primary food source for approximately 300 million people in the tropics. Tapioca is commonly used in food preparation in Kerala. The invasion of the cassava mealybug (*P. manihoti*) resulted in massive crop damage and losses for farmers in Tamil Nadu and Kerala. The cassava mealybug was first detected in India at Thrissur in 2020 by the ICAR-National Bureau of Agricultural Insect Resources (NBAIR) in Bengaluru. To protect the cassava crop from mealybug infestation, ICAR-NBAIR imported a highly host-specific parasitoid wasp, *Anagyrus lopezi*, from the International Institute of Tropical Agriculture (IITA) in Benin, West Africa, in August 2021, following Government of India guidelines (Sampathkumar *et al.*, 2020).

In the previous year, the Bureau conducted several field release initiatives across the districts of Tamil Nadu, where cassava is the dominant crop. The classical biological control agent demonstrated exceptional effectiveness at all release sites in Tamil Nadu, successfully reducing the cassava mealybug population in a short period. Recognizing the positive impact of the parasitoid on cassava mealybug management in Tamil Nadu, and given the current severity of the cassava mealybug issue in Kerala, ICAR-NBAIR collaborated with ICAR-CTCRI, Thiruvananthapuram, AICRP-BC, and KAU Thrissur Centre to organize a field release programme for the parasitoid *Anagyrus lopezi* in Madakkathara Panchayat, Thrissur (Sampathkumar *et al.*, 2020). Invasive mealybugs continue to pose a significant threat to tropical agriculture, biodiversity, and economic stability. Their rapid spread, polyphagous nature, and ability to develop resistance to conventional control methods complicate management efforts. Key invasive species such as *Phenacoccus manihoti*, *Paracoccus marginatus*, and *Planococcus lilacinus* have established themselves in various tropical regions, often causing severe damage to staple and commercial crops.

Despite advancements in monitoring, quarantine, and biological control strategies, gaps remain in early detection, regional cooperation, and the implementation of integrated pest management (IPM) at the farmer level. Climate change further complicates the challenge by expanding suitable habitats and altering pest dynamics. To mitigate the impact of mealybugs in tropical ecosystems, a coordinated approach that includes robust surveillance, sustainable management practices, community engagement, and ongoing research is crucial. Strengthening international collaboration and investing in local capacity-building will be essential for developing long-term, adaptive solutions to this persistent pest threat.

References

Ameyaw, G., Dzahini-Obiatey, H. and Domfeh, O. (2014) Perspectives on cocoa swollen shoot virus disease (CSSVD) management in Ghana. *Crop Protection* 65, 64–70.

Ammar, E., Awadallah, K. and Rashad, A. (1979) Ecological studies on Ferrisia virgata on *Acalypha shrubs* in Dokki, Giza (Homoptera, Pseudococcidae). *Deutsche Entomologische Zeitschrift* 26, 207–213.

Arif, M.I., Rafiq, M. and Ghaffar, A. (2009) Host plants of cotton mealybug (*Phenacoccus solenopsis*): A new menace to cotton agroecosystem of Punjab, Pakistan. *International Journal of Agriculture and Biology* 11, 163–167.

Arora, P.K. (1999) *Proceedings International Symposium on Citriculture Held at Nagpur during Nov. 23–27th, 1999* 931–934.

Atwal, A.S. (1976) *Agricultural Pests of India and South East Asia*. Kalyani Publishers, Ludhiana, pp. 195–213.

Awadallah, K., Ammar, E., Tawfik, M. and Rashad, A. (1979) Life-history of the white mealybug *Ferrisia virgata* (Homoptera, Pseudococcidae). *Deutsche Entomologische Zeitschrift* 26, 101–110.

Ben-Dov, Y. (2006) Scales in a family/genus query. *Family Pseudococcidae & Genus*. Available at: http://www.sel.barc.usda.gov/calecgi/chklist.exe

Bhat, A., Devasahayam, S., Sarma, Y. and Pant, R. (2003) Association of a Badnavirus in black pepper (*Piper nigrum* L.) transmitted by mealybug (*Ferrisia virgata*) in India. *Current Science* 84, 1547–1550.

Bindra, O.S. (1970) *Citrus Decline in India*. Ludhiana, Punjab: Joint Publication of PAU/OSU/USAID.

Charles, J.G., Cohen, D., Walker, J.T.S., Forgie, S.A., Bell, V.A. *et al*. (2006) A review of grapevine leafroll associated virus type 3 (GLRaV-3) for the New Zealand wine industry. *The Horticulture and Food Research Institute of New Zealand, Auckland*.

Chowdhury, S. and Majid, S. (1954) *Handbook of Plant Protection*. Shillong: Department of Agriculture, Assam Publication, p. 177.

Cockerell, T.D.A. (1893) Note on the genus *Pseudococcus* Westwood. *Entomological News* 4(10), 317–318.

Comstock, J.H. (1881) Report of the entomologist. Part II Report on scale insects. Report of the Entomologist of the United States Department of Agriculture for the year 1880, pp. 276–349.

Downie, D.A. and Gullan, P.J. (2005) Phylogenetic congruence of mealybugs and their primary endosymbionts. *Journal of Evolutionary Biology* 18, 315–324.

Dreistadt, S.H. (2001) *Integrated Pest Management for Floriculture and Nurseries*. UCANR Publications.

Dutt, U. (2007) *Mealy bug infestation in Punjab: Bt cotton falls flat*. Available at: https://www.countercurrents.org/dutt210807.htm

Franco, J.C., Suma, P., Silva, E.B., Blumberg, D. and Mendel, Z. (2004) Management strategies of mealybug pests of citrus in Mediterranean countries. *Phytoparasitica* 32, 507–522.

Franco, J.C., Zada, A. and Mendel, Z. (2009) Novel approaches for the management of mealybug pests. In: *Biorational Control of Arthropod Pests*. pp. 233–278.

Garcia Morales, M., Denno, B., Miller, D.R., Miller, G.L. and Ben-Dov, Y. (2015) *ScaleNet: A literature-based model of scale insect biology and systematics*. Available at: http://scalenet.info (accessed 22 December 2016).

Garcia Morales, M., Denno, B.D., Miller, D.R., Miller, G.L., Ben-Dov, N.B. *et al*. (2016) *ScaleNet: A literature-based model of scale insect biology and systematics*. Available at: http://scalenet.info/

Garcia Morales, M., Denno, B.D., Miller, D.R., Miller, G.L., Ben-Dov, Y. *et al*. (2019) *ScaleNet: A literature-based model of scale insect biology and systematics*. Available at: http://scalenet.info/

Hahn, S.K. and Williams, R.J. (1973) *Enquête sur le manioc en République du Zaïre, 12–20 March 1973*. Report to the Minister of Agriculture of the Republic of Zaire.

Hayat, M. (2009) Description of a new species of *Aenasius walker* (Hymenoptera: Encyrtidae), parasitoid of the mealybug, *Phenacoccus Solenopsis* Tinsley (Homoptera: Pseudococcidae) in India. *Biosystematica* 3, 21–26.

Hefetz, A. and Auber, O. (1990) Male response to the synthetic sex pheromone of *Planococcus citri* (Risso) (Hom., Pseudococcidae) and its application for population monitoring. *Journal of Applied Entomology* 109, 502–506.

Herren, H.R. (1981) Biological control of the cassava mealybug. In: Terry, E.R., Oduro, K.A. and Caveness, F. (eds) *Tropical Root Crops: Research Strategies for the 1980s, Proceedings of the First Triennial*

Root Crops Symposium, 8–12 September 1980, Ibadan, Nigeria: International Institute of Tropical Agriculture, pp. 79–80.

Herren, H.R. and Neuenschwander, P. (1991) Biological control of cassava pests in Africa. *Annual Review of Entomology* 36, 257–283.

Highland, H.A. (1956) The biology of *Ferrisia virgata*, a pest of azaleas. *Journal of Economic Entomology* 49, 276–277.

Isaac, P.V. and Misra, C.S. (1933) The chief insect pests of sugarcane and methods for their control. *Agriculture and Livestock in India* 3, 315–324.

Jadhav, V.B. and Pujari, C.V. (1999) *Proceedings of the International Symposium on Citriculture Held at Nagpur during Nov. 23–27th, 1999* 23–27th.

Jayma, L., Martin, R.F.L. and Mau, R. (1992) Citrus mealybug, *Planococcus citri* (Risso). In: *Crop Knowledge Master*. Available at: http://www.extento.hawaii.edu/kbase/crop/type/p_citri.htm

Jhala, R.C., Bharpoda, T.M. and Patel, M.G. (2008) *Phenacoccus solenopsis* Tinsley (Hemiptera: Pseudococcidae), the mealybug species recorded first time on cotton and its alternate host plants in Gujarat. *Uttar Pradesh Journal of Zoology* 28, 403–406.

Kalidas, P. and Shivankar, V.J. (1994) *Final Report of the Project "Studies on Chemical Control of Insect Pests of Nagpur Mandarin with Special Reference to Citrus Blackfly, Psylla and Leaf Miner"*. Nagpur: NRCC, p. 35.

Kalyanasundaram, M., Karuppuchamy, P., Divya, S., Sakthivel, P., Rabindra, R.J. *et al.* (2010) Impact of release of the imported parasitoid *Acerophagus papayae* on the management of papaya mealybug *Paracoccus marginatus* in Tamil Nadu. In: Shylesha, A.N., Joshi, S., Rabindra, R.J. and Bhumannavar, B.S. (eds) *Classical Biological Control of Papaya Mealybug (Paracoccus marginatus) in India, Proceedings of the National Consultation Meeting on Strategies for Deployment and Impact of the Imported Parasitoids of Papaya Mealybug*, pp. 68–72.

Kaydan, M. and Gullan, P. (2012) A taxonomic revision of the mealybug genus *Ferrisia* Fullaway (Hemiptera: Pseudococcidae), with descriptions of eight new species and a new genus. *Zootaxa* 3543, 1–65.

Kosztarab, M. (1996) *Scale Insects of Northeastern North America: Identification, Biology, and Distribution*. Virginia Museum of Natural History.

Kosztarab, M. and Kozár, F. (1988) *Scale Insects of Central Europe*. p. 456.

Krishnamoorthy, A. and Mani, M. (1989) Records of green lacewings preying on mealybugs in India. *Current Science* 58, 155–156.

Kumar, J.B.N. and Shekhar, M.A. (2010) *Outbreak of a new invasive pest, papaya mealybug (Paracoccus marginatus), in South India: A serious threat to sericulture industry*. Silkworm. Available at: http://silkwormmori.blogspot.com/2010/10/outbreak-of-new-invasive-pest-papaya.html (accessed 30 July 2012).

Kumar, G.N.K., Suroshe, S.S., Gupta, A., Keerthi, M.C., Srinivas, K. *et al.* (2024) Morphological and molecular diagnosis of platygastrid parasitoid, *Allotropa* sp. near Phenacocca (Hymenoptera) from India and its performance on cotton mealybug, *Phenacoccus solenopsis*. *Phytoparasitica* 52(1), 1.

Mani, M. and Thontadarya, T.S. (1987) Record of mealybug species on grapevine in Karnataka. *Current Science* 56, 1192.

Mani, M. and Krishnamoorthy, A. (1996) Mealybug problem on fruit crops. *Indian Horticulture* 41, 43–45.

Mani, M. and Shivaraju, C. (2016) Mealybugs and their management in agricultural and horticultural crops. DOI: 10.1007/978-81-322-2677-2.

Mani, M. and Shylesha, A. (2021) Invasive mealybugs and their management in India: Past, present & future. *Hexapoda, Insecta Indica* 28(1 & 2), 64–77. DOI: 10.55446/hexa.2021.43.

Mani, M., Joshi, S., Kalyanasundaram, M., Shivaraju, C., Krishnamoorthy, A. *et al.* (2013) A new invasive jack beardsley mealybug, *Pseudococcus jackbeardsleyi* (Hemiptera: Pseudococcidae) on papaya in India. *Florida Entomologist* 96, 242–245.

Matile-Ferrero, D. (1976) La fauneterrestre de l'ile de sainte-hélène 7. coccoidea. In: *Annales, Musée Royal de l'Afrique Central, Tervuren, Serie in 8°, Sciences Zoologiques 215*. pp. 292–317.

Mendel, Z., Blumberg, D. and Ishaaya, I. (1991) Effect of buprofezin on Icerya purchase and *Planococcus citri*. *Phytoparasitica* 19, 103–112.

Michelakis, S. and Hamid, H.A. (1995) Integrated control methods of the citrus mealybug, *Planococcus citri* (Risso) in Crete, Greece. *Israel Journal of Entomology* 29, 277–284.

Muniappan, R. (2009) Papaya mealybug, new invader in Asia. *IAPPS Newsletter 1*,

Muniappan, R., Meyerdirk, D.E., Sengebau, F.M., Berringer, D.D. and Reddy, G.V.P. (2006) Classical biological control of the papaya mealybug, *Paracoccus marginatus* (Hemiptera: Pseudococcidae) in the Republic of Palau. *Florida Entomologist* 89, 212–217.

Muniappan, R., Shepard, B.M., Watson, G.W., Carner, G.R., Sartiami, D. *et al.* (2008) First report of the papaya mealybug, *Paracoccus marginatus* (Hemiptera: Pseudococcidae), in Indonesia and India 25, 37–40.

Muniappan, R., Watson, G.W., Vaughan, L., Gilbertson, R. and Noussourou, M. (2011) New records of mealybugs, scale insects, and whiteflies (Hemiptera: Sternorrhyncha) from Mali and Senegal. *Journal of Agricultural and Urban Entomology* 28, 1–7.

Newstead, R. (1894) Scale insects in Madras. *Indian Museum Notes* 3, 21–32.

Nwanze, K.F. (1978) Biology of the cassava mealybug, *Phenacoccus manihoti* Mat.-Ferr. in the Republic of Zaire. In: Nwanze, K.F. and Leuschner, K. (eds) *Proceedings of the International Workshop on the Cassava Mealybug Phenacoccus manihoti Mat.-Ferr. (Pseudococcidae) held at INERA-M'vuazi, Bas-Zaire, Zaire, June 26–29, 1977*, Ibadania, Niger: International Institute of Tropical Agriculture, pp. 20–28.

Nwanze, K.F., Leuschner, K. and Ezumah, H.C. (1982) The cassava mealybug, *Phenacoccus* sp. in the Republic of Zaire. *PANS* 25, 125–130.

Oliveira, M.D., Silva-Torres, C.S., Torres, J.B. and Oliveira, J.E.M. (2014) Population growth and within-plant distribution of the striped mealybug *Ferrisia virgata* (Cockerell) (Hemiptera: Pseudococcidae) on cotton. *Revista Brasileira de Entomologia* 58, 71–76.

Parsa, S., Kondo, T. and Winotai, A. (2012) The cassava mealybug (*Phenacoccus manihoti*) in Asia: First records, potential distribution, and an identification key. *PLoS ONE* 7(10), e47675. Available at: https://doi.org/10.1371/journal.pone.0047675

Patel, H.P., Patel, A.D. and Bhatt, N.A. (2009) Record of coccinellids predating on mealy bug, *Phenacoccus solenopsis* Tinsley (Homoptera: Pseudococcidae) in Gujarat. *Insect Environment* 14, 179–181.

Pathak, K.A., Rao, K.R., Mishra, M., Gupta, S.G. and Yadav, D.S. (1999) Integrated management of citrus mealy bug, *Planococcus citri* (Risso) in micro-propagated plantlets in Meghalaya. In: *Proceedings of International Symposium on Citriculture, held at Nagpur During November 23–27*, pp. 956–960.

Prakasan, C.B. (1987) Biological control of coffee pests. *Journal of Coffee Research* 17, 114–117.

Rabinder, K., Ramandeep, K. and Brar, K.S. (2008) Development and predation efficacy of *Chrysoperla carnea* (Stephens) on mealy bug, *Phenacoccus solenopsis* (Tinsley) under laboratory conditions. *Journal of Insect Science (Ludhiana)* 21, 93–95.

Regupathy, A. and Ayyasamy, R. (2011) Impact of withdrawal of pesticide application against papaya mealybug, *Paracoccus marginatus* Williams and Granara de Willink, on the biodiversity of natural enemy complex in small-scale papaya farming system in Tamil Nadu. In: *National Symposium on Harnessing Biodiversity for Biological Control of Crop Pests*, p. 2 (Vol. 25).

Risso, P.A. (1813) Sur l'histoire naturelle des Oranges, Bigaradiers, Limettiers, Cédratiers, Limoniersou Citroiniers, cultivés dans le département des Alpes-Maritimes. *Annales Du Museum National d'Histoire Naturelle* 20, 401–431.

Regupathy, A. and Ayyasamy, R. (2010) Impact of papaya mealybug on papain industry. In: Shylesha, A.N., Joshi, S., Rabindra, R.J. and Bhumannavar, B.S. (eds) *Classical Biological Control of Papaya Mealybug (Paracoccus marginatus) in India, Proceedings of the National Consultation Meeting on Strategies for Deployment and Impact of the Imported Parasitoids of Papaya Mealybug*, pp. 57–59.

Sampath, K.M., Mohan, M., Shylesha, A.N., Joshi, S., Venkatesan, T. *et al.* (2021) Occurrence of cassava mealybug, *Phenacoccus manihoti* Matile-Ferrero (Pseudococcidae: Hemiptera), a new invasive pest on cassava in India and prospects for its classical biological control. *Current Science* 120, 432–435.

Sampathkumar, M., Mohan, M., Shylesha, A.N., Ankita, G., Sunil, J. *et al.* (2020) Invasive cassava mealybug (CMB) *Phenacoccus manihoti* a potential threat for cassava cultivation in India. ICAR-NBAIR Publication:31. Available at: https://nbair.res.in/taxonomy/term/103

Sartiami, D., Watson, G.W., Roff, M.N.M., Hanifah, M.Y. and Abd Ghani, I. (2015) First record of cassava mealybug, *Phenacoccus manihoti* (Hemiptera: Pseudococcidae), in Malaysia. *Zootaxa* 3957, 235–238.

Selvaraj, K.N. (2012) Assessing the economic impact of parasitoids of papaya mealy bug. *PowerPoint Presentation with Unpublished Data from Entomologists at Tamil Nadu Agricultural University, Coimbatore, India.*

Sergio, A., Mejia-Ortiz, J.S., Noyes, F., Infante, F., Cisneros-Hernandez, J. *et al.* (2022) Biology of *Acerophagus texanus* (Howard, 1898) (Hymenoptera: Encyrtidae), a parasitoid of *Ferrisia virgata* Cockerell (Hemiptera: Pseudococcidae). *International Journal of Tropical Insect Science* 3, 2523–2529.

Sharma, S.S. (2007) *Aenasius* sp. nov. effective parasitoid of mealy bug (*Phenacoccus solenopsis*) on okra. *Haryana Journal of Horticultural Sciences* 36, 412.

Sharma, A. and Muniappan, R. (2021) Ecology and management of *Paracoccus marginatus* (papaya mealybug) (Hemiptera: Pseudococcidae) in the Indian Subcontinent – achievements, and lessons. *Indian Journal of Entomology* 84(2), 475–482. DOI: 10.55446/IJE.2021.43.

Shivakumara, K.T., Keerthi, M.C., Polaiah, A.C., Gupta, A. and Joshi, S. (2024a) Incidence of invasive mealybugs, *Phenacoccus solenopsis* Tinsley and *Paracoccus marginatus* Williams & Granara de Willink (Hemiptera: Pseudococcidae) on Gymnema sylvestre (R. Br) and its natural enemies in the semi-arid region of India. *International Journal of Pest Management* 70(4), 1126–1140.

Shivakumara, K.T., Chinapolaiah, A., Keerthi, M.C., Ramya, R.S. and Gotyal, B.S. (2024b) Identification and characterization of novel resistant genotypes of *Gymnema sylvestre* (Retz.) R. Br. ex Sm. against invasive mealybug species, *Phenacoccus solenopsis* Tinsley and *Paracoccus marginatus* Williams and Granara de Willink for sustainable pest management. *Journal of Applied Research on Medicinal and Aromatic Plants* 39, 100534.

Shivankar, V.J. and Singh, S. (2000) *Citrus Insect Pests*. Nagpur: NRCC Publication, p. 252.

Shylesha, A.N. and Pathak, K.A. (1997) Integrated management of major insect-pests of Khasi mandarin in Meghalaya. In: *Proceedings of national symposium on citriculture, November 17–19, 1997, at National Research Centre for Citrus, Nagpur*, pp. 332–334.

Shylesha, A.N. and Joshi, S. (2012) Occurrence of Madeira mealybug, *Phenacoccus madeirensis* Green (Hemiptera: Pseudococcidae) on cotton in India and record of associated parasitoids. *Journal of Biological Control* 26, 272–273.

Shylesha, A.N., Joshi, S., Rabindra, R.J. and Bhumannavar, B.S. (2010) Classical biological control of the papaya mealybug. In: Shylesha, A.N., Joshi, S., Rabindra, R.J. and Bhumannavar, B.S. (Eds.), *Classical Biological Control of Papaya (Paracoccus marginatus) in India*, Proceedings of the National Consultation Meeting on Strategies for Deployment and Impact of the Imported Parasitoids of Papaya Mealybug, pp. 73–78.

Singh, S.P. (1990) In: *Proceedings of the Indo-USSR Joint Workshop on Problems and Potentials of Biocontrol of Pests and Diseases, Bangalore, June 26–28, 1990*, p. 294.

Sridhar, V., Joshi, S., Kumar, R. and Jhansirani, G. (2012) First record of lantana mealybug, *Phenacoccus parvus* Morrison (Hemiptera: Pseudococcidae), as a pest on China aster from South India. *Journal of Horticultural Science* 7, 108–109.

Srinivasan, G., Shanthi, M. and Naveena, K. (2024) Efficacy of some IPM strategies against cotton mealybug *Phenacoccus solenopsis* Tinsley. *Indian Journal of Entomology* 86(4), 1188–1192. DOI: 10.55446/IJE.2023.1514.

Tanwar, R.K., Jeyakumar, P. and Monga, D. (2007) *Mealybugs and Their Management*, Technical Bulletin 19, September 2007. National Centre for Integrated Pest Management, LBS Building, Pusa Campus, New Delhi–110012.

Tanwar, R.K., Jeyakumar, P. and Vennila, S. (2010) Papaya mealybug and its management strategies. *Technical Bulletin* 22. National Centre for Integrated Pest Management, New Delhi, pp.22.

Thorold, C.A. (1975) *Diseases of Cocoa*. Clarendon Press, Oxford.

Tinsley, J. (1898) An ants'-nest coccid from New Mexico. *Canadian Entomologist* 30, 47–48.

USAID (2012) Partnership for success: Controlling the papaya mealybug in India. Available at: http://transition.usaid.gov/in/newsroom/pdfs/ss_papaya_bug.pdf (accessed 8 January 2013).

Walton, V.M., Daane, K.M., Bentley, W.J., Millar, J.G., Larsen, T.E. *et al.* (2006) Pheromone-based mating disruption of *Planococcus ficus* (Hemiptera: Pseudococcidae) in California vineyards. *Journal of Economic Entomology* 99, 1280–1290.

Williams, D.J. (1986) *Rastrococcus invadens* sp. n. (Hemiptera: Pseudococcidae) introduced from the Oriental Region to West Africa and causing damage to mango, citrus and other trees. *Bulletin of Entomological Research* 76, 695–699.

Williams, D.J. (2004) *Mealybugs of Southern Asia*. The Natural History Museum, London, and Southdene SDN BHD, Kuala Lumpur, p. 896.

Williams, D.J. and Watson, G.W. (1988) *The Scale Insects of the Tropical South Pacific Region. Pt. 2: The Mealybugs (Pseudococcidae)*. London: CAB International Institute of Entomology, p. 260.

Williams, D.J. and Granara de Willink, M.C. (1992) *Mealybugs of Central and South America*. CAB International.

Winotai, W., Goergen, G., Tamò, M. and Neuenschwander, P. (2010) Cassava mealybug has reached Asia. *Biocontrol News and Information* 31, 10–11.

11 Invasive Whiteflies in Tropics and Sub-tropics: Their Impact on Agro-ecosystems

Sainath G., Sampath Reddy, Saipooja Nelluri and Kariyanna B.*
Division of Fluoro-Agrochemicals, CSIR – Indian Institute of Chemical Technology, Tarnaka, Hyderabad, India

Abstract
Whiteflies are a significant pest group in tropical and subtropical regions, posing a threat to agro-ecosystems and agricultural productivity. The important six whiteflies, viz. silverleaf whitefly (*Bemisia tabaci* Gennadius), giant whitefly (*Aleurodicus dugesii* Cockerell), Bondari nesting whitefly (*Paraleyrodes bondari* Peracchi), spiralling whitefly (*Aleurodicus dispersus* Russell), rugose spiralling whitefly (*Aleurodicus rugioperculatus* Martin) and woolly whitefly (*Aleurothrixus floccosus* Maskell) pose a threat to the biodiversity and ecology after their invasion in a new location. This chapter emphasizes their impact on agro-ecosystems and the need for effective management strategies. Whiteflies cause yield losses, stunted plant growth and reduced crop quality through sap depletion and virus transmission. Integrated pest management (IPM) approaches have proved effective in managing whitefly infestations. Collaboration among stakeholders is crucial for promoting awareness and adoption of sustainable practices. By implementing integrated management strategies, the negative effects of whiteflies can be mitigated, ensuring the long-term sustainability of agro-ecosystems.

Keywords: Whiteflies, invasive, impact, management, tropical areas, agroecosystem

11.1 Introduction

The proliferation of invasives, especially those in relation to plants such as scale insects and whiteflies, has become a global concern due to their significant impact on native ecosystems, biodiversity and economies. The increased volume, diversity and speed of product movement worldwide have facilitated the spread of invasives, including insects and mites that act as vectors for plant diseases (Wosula *et al.*, 2018). In India alone, over 110 exotic insects were reported, with whiteflies and mealybugs comprising a substantial portion (Mandal, 2011). Whiteflies are considered some of the most problematic invasive pests worldwide (Naveena *et al.*, 2020).

The accidental introduction of alien whitefly species occurs frequently when they hitchhike on host plants, enabling their dispersion across countries through plant trade. Their small size, cryptic nature and attachment to host plants during their immature stages contribute to their successful establishment in new geographic areas (Simala *et al.*, 2015). In addition, exotic whitefly species are known

*Corresponding author: kariyanna.iict@csir.res.in; kariyannabento@gmail.com

 Biological Invasions in the Tropics: Monitoring, Mechanism and Management (eds Akshay Kumar Chakravarthy and Pathour Rajendra Shashank)
DOI: 10.1079/9781836991977.0011

for their rapid reproduction rates, high levels of phenotypic plasticity and their potential to outcompete native species, leading to significant damage to economically valuable crops. This invasive behaviour results in challenges for crop protection and agricultural productivity (Byrne and Bellows, 1991; Naranjo and Ellsworth, 2009). In India, there are 469 whitefly species belonging to 71 genera feeding on agricultural, horticultural and forestry crops, including eight invasive species (David and Regu, 1995).

Whiteflies from the Aleurodicinae and Aleyrodinae subfamilies cause significant harm to host plants by feeding on them, which results in the loss of vital nutrients and water. This feeding behaviour leads to reduced plant growth and early leaf shedding, severely impacting the health and yield of the affected plants (Oliveira *et al.*, 2001; Stansly and Naranjo, 2010). They also induce indirect damage by secreting wax and sticky honeydew, which encourages the development of black sooty mould on affected plants and thus reduces photosynthetic capacity. Some whitefly species serve as vectors for plant viruses, such as begomoviruses, which are transmitted by solanum whitefly *Aleurotrachelus trachoides* Back (Brown *et al.*, 1995; Jones, 2003; Chandrashekar *et al.*, 2020).

The organization of alien populations carries significant implications in the development of early warning systems and effective control strategies. Timely identification of invasives and the immediate implementation of biological approaches have the potential to mitigate economic losses (Simberloff and Rejmánek, 2011). Furthermore, it is crucial to continuously monitor introduced species to understand their status, observe temporal trends, track variations in species composition, monitor the geographic range and host plants and assess their invasiveness and effects on crops (Gilbert *et al.*, 2004; Goolsby *et al.*, 2005; Pimentel *et al.*, 2005; Liebhold *et al.*, 2012; Simberloff *et al.*, 2013). This chapter aims to comprehensively document and evaluate the invasion patterns, expansion of host plants and geographic spread, co-occurrence patterns, extent of damage caused and natural predators of invasive whitefly species to develop a sustainable management strategy.

11.2 History and Importance of Invasive Whitefly

A. dispersus was the first non-native species of whitefly reported in India. The insect was first recognized in the Western Ghats of South India in 1995 (David and Regu, 1995). This introduction marked the beginning of the invasion of whiteflies in the country. In subsequent years, several other invasive whitefly species have been identified in different regions of India. Some notable discoveries are listed in Table 11.1.

The success of an invasive is influenced by favourable environments and the availability of suitable hosts (Duan *et al.*, 2015). Features influencing their successful invasion are unclear, but the separation of non-natives from natural

Table 11.1. Invasive whiteflies in India

Common name	Scientific name	Year of invasion	Authors
Acacia whitefly	*Tetraleurodes acaciae* (Quaintance)	2017	Sundararaj and Vimala, 2018
Solanum whitefly	*Aleurotrachelus trachoides*	2015	Dubey and Sundararaj, 2015
Rugose spiralling whitefly	*Aleurodicus rugioperculatus*	2016	Sundararaj and Selvaraj, 2017
Bondar's nesting whitefly	*Paraleyrodes bondari* Peracchi	2018	Josephrajkumar *et al.*, 2019
Palm-infesting whitefly	*Aleurotrachelus atratus* Hempel	2019	Selvaraj *et al.*, 2019a
Nesting whitefly	*Paraleyrodes minei* Iaccarino	2018	Mohan *et al.*, 2019
Woolly whitefly	*Aleurothrixus floccosus*	2019	Sundararaj *et al.* (2020a)

enemies and the ongoing dispersal of exotic species under changing global climates play a role (Diez *et al.*, 2012). Studies predict an increase in invasive species in certain regions, including north-western Europe, India, north-eastern USA and China (Bellard *et al.*, 2013). Further, effective management options include prevention, early diagnosing, eradication and classical biocontrol programmes. Implementing successful biocontrol strategies has saved millions of rupees by managing the adverse impacts of invasive whiteflies on agriculture. Accurate identification and timely research on bioecology and population dynamics are crucial for effective management strategies (Duan *et al.*, 2015).

11.3 Invasive Whiteflies Across the Tropics and Subtropics

11.3.1 *Bemisia tabaci* Gennadius

B. tabaci, the silverleaf whitefly, originated in tropical regions but has spread to many parts of the world. It is known for its ability to infest a wide range of plants, including fruits, vegetables, field crops and ornamental plants (Costa *et al.*, 2006). *B. tabaci* is a pest that feeds exclusively on phloem, but its feeding habits differ notably from aphids. While aphids extensively probe and relocate during feeding, whitefly nymphs remain stationary and feed continuously for more than 28 days throughout their nymphal stage (Gill, 1990; Byrne and Bellows, 1991; Johnson and Walker, 1999; Freeman *et al.*, 2001). This prolonged feeding period creates a close relationship between whiteflies and their host plants, possibly leading to distinct defence responses compared with aphids. Furthermore, the salivary components of whiteflies and aphids are likely different, which can trigger varied host responses (Walling, 2000). *B. tabaci* cause direct damage by feeding on plant sap, leading to stunted growth, leaf yellowing and reduced crop yield (Legg and Brown, 2006). They excrete a sticky substance called honeydew, promoting the growth of sooty mould, further affecting the plant's health. In addition, they transmit viral diseases to plants, compounding the damage they cause (Byrne and Bellows, 1991).

11.3.2 *Aleurodicus dugesii* Cockerell

A. dugesii, the giant whitefly, is native to the Caribbean and has become invasive in parts of Asia, the Americas and the Africa (Perring and Cooper, 2010). It primarily attacks woody plants, such as fruit and ornamental trees and palms. The feeding activities of *A. dugesii* cause similar damage as silverleaf whiteflies, including leaf yellowing, stunted growth and honeydew secretion. This species also transmits certain plant viruses (Ghanim *et al.*, 1998).

The adult and nymph stages of *A. dugesii* become plant pests by extracting plant fluids and vital nutrients, leading to leaf shedding (defoliation), stunting and even plant death (Bellows and Hoddle, 2010a). This species also secretes smooth, elongated white wax filaments resembling a beard, which can completely coat leaves. These wax filaments are easily dispersed by the wind and can create nuisances on outdoor surfaces such as furniture, cars and swimming pools, and ultimately detracting from the aesthetic appeal of ornamental plants (Gill, 1992).

11.3.3 *Paraleyrodes bondari* Peracchi

P. bondari, Bondar's nesting whitefly, was first discovered on citrus trees (*Citrus* spp.) in Brazil in 1971 (Peracchi, 1971). It originated in the Neotropical region and spread to different areas including Belize, Honduras, Puerto Rico, Hawaii and Florida in the USA (Stocks, 2012). The pest's impact on coconut production has raised concerns among farmers and researchers alike (Dinesh *et al.*, 2020). It was initially recorded in India on coconut palms in Kerala in 2018 (Josephrajkumar *et al.*, 2019), and subsequently reported in Karnataka and later in the Andaman as well as Nicobar Islands (Vidya *et al.*, 2019). More recently, its presence has been observed in the Lakshadweep Islands, as well as in various places in Andhra Pradesh and Tamil Nadu, South India (Bhat *et al.*, 2020; Kumar *et al.*, 2021). Effective management strategies, including cultural practices, biological control methods and targeted insecticide applications, are essential for controlling *P. bondari* populations and minimizing crop damage (Serra *et al.*, 2014). Adults of *P. bondari* are smaller compared with *A. dispersus*. They lay eggs covered in waxy threads, giving the impression of eggs within abundant woolly and glassy wax threads. Female *P. bondari* typically live in circular nests, which led to the name Bondar's nesting whitefly (Iaccarino *et al.*, 1990).

11.3.4 *Aleurodicus dispersus* Russell

A. dispersus, the spiralling whitefly, is the first alien whitefly documented in India; it was first recorded in the Western Ghats and spread to different parts of India including the islands. Its presence in India has raised issues due to its impact on coconut cultivation and other host plants (Karthikeyan *et al.*, 2019; Vasanthakumar *et al.*, 2020). Russell (1965) described this pest based on samples collected from coconut in West Florida, USA. Its rapid reproduction and adaptability have contributed to its persistent nature (Bansal *et al.*, 2021). In India, the spiralling whitefly was first reported in the South Indian Western Ghats and has since spread across the country, including the Lakshadweep, Nicobar and Andaman Islands. Its presence poses a significant threat to agricultural production and ecosystem health (Deshpande *et al.*, 2014). Effective management strategies, including biological control methods and targeted insecticide applications, are essential for controlling its populations and minimizing crop damage (Vasanthakumar *et al.*, 2020). The direct damage inflicted by *A. dispersus* results from the piercing and sucking of sap from foliage by both young ones and adult stages. The most significant damage occurs during the initial three nymphal stages, often leading to premature leaf shedding. Despite heavy infestations, the direct feeding damage typically does not reach levels that would fatally harm plants (Waterhouse and Norris, 1989).

11.3.5 *Aleurodicus rugioperculatus* Martin

A. rugioperculatus, the rugose spiralling whitefly, was first discovered on coconut trees in Belize and later in 2009, this was reported in Miami-Dade County, South Florida on gumbo limbo (*Bursera simaruba* (L.) Sarg) as a pest (Martin, 2004). *A. rugioperculatus* originated in Central America and is currently found in limited regions such as Mexico and Florida (Evans, 2008). Its establishment in these regions is attributed to factors such as climate adaptability and the availability of suitable host plants (Bansal *et al.*, 2019). In India, it was first documented on coconut trees and various other crops in 2016 in Pollachi, Tamil Nadu (Sundararaj and Selvaraj, 2017). Afterwards, it was found in many areas in South India, West Bengal and Assam (Selvaraj *et al.*, 2017, 2019b). More recently, it was recorded in the Telangana, Gujarat, Maharashtra, Lakshadweep Islands, Odisha, Chhattisgarh, and a few districts in Meghalaya (Nandan *et al.*, 2021; Singh *et al.*, 2022; Varma *et al.*, 2023). *A. rugioperculatus* adults are identifiable due to their significant size and the existence of two irregular light brown bands spanning across their wings (Stocks and Hodges, 2012). Piercing and sucking of sap from foliage is carried out by both immature and adult stages. The most significant damage occurs during the initial three nymphal stages, often leading to premature leaf shedding. Despite heavy infestations, the direct feeding damage typically does not reach levels that would fatally harm plants (Waterhouse and Norris, 1989).

11.3.6 *Aleurothrixus floccosus* Maskell

A. floccosus, the woolly whitefly, was originally described on *Citrus* sp. in Cuba (Maskell, 1896) and is naturally found in Neotropical regions (Malumphy and Treseder, 2011). Its presence in India was initially detected on guava in Kerala in 2019 (Sundararaj *et al.*, 2020b). Since then, infestations have been documented on guava in Karnataka, Tamil Nadu, in particular found in Salem, Dharmapuri, Tirupathur, Coimbatore, Krishnagiri, Karur and Namakkal districts (Bhat *et al.*, 2021; Ramesh *et al.*, 2022). More recently, infestations of whitefly have been noted on guava in the Lakshadweep Islands (Rani *et al.*, 2023). *A. floccosus* has a wide host range, including mulberry, citrus, mango and other fruit trees, as well as ornamental plants. Its presence poses a significant threat to agricultural production and ecosystem health (Thangavel *et al.*, 2015). *A. floccosus* lays eggs and feeds on the lower surfaces of young citrus leaves, coating them with a woolly substance made from threads and honeydew generated by nymphs that do not move (Tello *et al.*, 2014). The woolly whitefly selects young leaves that are yellowish-green as the preferred location for depositing its eggs (Walker and Zareh, 1990).

11.4 Economic Damage

Invasive whiteflies have resulted in significant economic damage across various regions. Their effects on agriculture and horticulture lead to substantial economic losses. Key points regarding the economic losses incurred by invasive whiteflies include the following.

Crop Yield Reduction: Invasive whiteflies feed on the sap of host plants, leading to weakening and stunting of growth. This directly affects crop yields and quality (Schuster *et al.*, 2010). Crops such as cotton, vegetables (e.g. tomato, brinjal/eggplant), and other commercially important plants suffer reduced productivity and resulting economic losses (Kumar *et al.*, 2018a).

Quality Decline: Whitefly infestations often result in a decline in the quality of agricultural produce (Heinz and Parrella, 1990; Navas-Castillo and Moriones, 2017). Damaged crops may exhibit signs of wilting, discolouration, premature leaf drop and reduced market value (Cohen *et al.*, 1988). This negatively impacts farmers' income and market competitiveness ((Navas-Castillo and Moriones, 2017; Kumar *et al.*, 2021).

Increased Production Costs: Managing and controlling invasive whitefly populations require the implementation of pest management strategies (Simmons and Shaaban, 2011). Farmers incur additional expenses on insecticides, biological control agents, monitoring systems and labour for pest control measures. These costs contribute to the economic burden associated with whitefly infestations (Cuthbertson *et al.*, 2012).

Trade Restrictions and Market Access: Invasive whitefly infestations can lead to trade restrictions and bans on agricultural commodities (Navas-Castillo *et al.*, 2011). Importing nations can make phytosanitary regulations mandatory to prevent introduction of whiteflies. This can limit market access for affected regions, affecting farmers' incomes and export opportunities (Krause-Sakate *et al.*, 2020).

Environmental Impact: The ecological consequences of invasive whitefly infestations extend beyond economic damage. Whiteflies disrupt the balance of ecosystems, negatively affecting beneficial insect populations, natural pollination processes and overall biodiversity. These environmental impacts can have long-term consequences on agricultural sustainability (Heinz and Parrella, 1990).

11.5 Interaction Between Whiteflies and Plants

The relationship between whiteflies and host plants involves intricate dynamics. Whiteflies have specialized mouthparts, called stylets, which they use to effectively feed on the phloem sap of plants (Jiang *et al.*, 2019). In response to whitefly infestation, plants produce chemical compounds toxic to the pests and release specific volatile organic compounds (VOCs) to attract natural enemies of the whiteflies (Dicke and van Loon, 2000). Whiteflies, on their part, have developed mechanisms to defend themselves against plant-induced defences. They activate detoxification systems to handle the plant's secondary defence metabolites and secrete salivary proteins that regulate plant defence responses, including protease, proteinase inhibitor and catalase (Li *et al.*, 2020). Additionally, when plants are infected with begomoviruses, the plants' jasmonic acid signalling pathway, which is involved in defence responses, can be suppressed, benefiting the survival of whiteflies on these viruliferous plants (Wang *et al.*, 2017).

11.6 Host Range and Biology

11.6.1 Host range

The host range of alien whiteflies is expanding (Table 11.2). In India, *A. dispersus* feeds on over 320 plant species, including coconut, banana and mango (Husain and Nath, 2018). *A. rugioperculatus* feeds on about 120 plant species, including coconut, banana and guava (Dinesh and Parvatha, 2021). *A. trachoides* breeds on 24 host plants (Maharjan and Ghimire, 2020)

Table 11.2. Host range of major invasive whiteflies in India. (Author's own table)

Whitefly species	Primary host plants	Number of host plants	References
Bemisia tabaci	Cucurbits, *Solanum lycopersicum*, *S. melongena*, *Vigna radiata*, *Abelmoschus esculentus*	1000 plant species in 74 plant families	Abd-Rabou and Simmons, 2010; Singh and Aggarwal, 2025
Aleurodicus dispersus	Coconut, banana, mango, guava, hibiscus	>320	Husain and Nath, 2018; Vasanthakumar *et al.*, 2020
Aleurodicus rugioperculatus	Coconut, banana, guava, oil palm	~120	Dinesh and Parvatha, 2021; Singh *et al.*, 2022
Paraleyrodes bondari	Coconut, citrus, guava, ornamental plants	>25	Ajesh and Alok, 2019; Josephrajkumar *et al.*, 2019
Aleurothrixus floccosus	*Fraxinus excelsior* L., *Punica granatum* L., *Citrus limon* Burm., *C. sinensis* Osbeck, *C. aurantium* L., *Poncirus trifoliata* Raf and *C. reticulata* Blanco	>7	Ulusoy *et al.*, 2003
Aleurodicus dugesii	Dominant host species belongs to family Fabaceae, Euphorbiaceae and Solanaceae	>200 species from 35 plant families	Bellows and Hoddle, 2010b; Hidayat *et al.*, 2020
Aleurothrixus trachoides	Duranta, potato, tomato, bell pepper	24	Chandrashekar *et al.*, 2020; Maharjan and Ghimire, 2020

and *P. bondari* feeds on more than 25 host plants (Ajesh and Alok, 2019). These whiteflies coexist on various host plants, including coconut and guava. Natural enemies such as parasitoids and predators help control these pests (Bellows and Hoddle, 2010b).

11.6.2 Biology

The life cycle of invasive whiteflies generally includes the egg, nymph and adult stages. The length of each stage varies based on the specific whitefly species and environmental factors (Barbedo, 2014). Below is a summary of the life stages. The adult female whitefly lays eggs on the undersides of plant leaves, usually in circular clusters. The eggs are oval or elongated in shape and are usually pale yellow or transparent (Jones, 2003). After the incubation period, the eggs hatch, and small, wingless nymphs emerge. These nymphs actively feed on plant sap and undergo several moulting stages, or instars, as they grow and mature (Bellows and Perring, 1994). The final nymphal stage transitions into the adult form of the whitefly. Adult whiteflies are tiny, winged insects characterized by a white, powdery coating. They use their piercing-sucking mouthparts to feed on plant sap, with females typically being larger than males (Liu and Stansly, 1996). Adult whiteflies reproduce through sexual mating, with females laying eggs that allow the life cycle to continue. Due to their rapid reproductive rate, whitefly populations can increase quickly, resulting in significant infestations on host plants (Fereres and Moreno, 2009).

11.6.3 Expanded host associations of invasive whiteflies

Recent studies have documented a rapid expansion in host plant range among key invasive whiteflies. *Aleurodicus dispersus* is known to feed on more than 320 host species, including economically important crops such as coconut, mango, banana and guava (Husain and Nath, 2018). Similarly, *Aleurodicus rugioperculatus* has been observed on over 120 species, notably guava and oil palm (Dinesh and Parvatha, 2021). This adaptability in host preference contributes to their successful colonization and makes control efforts more challenging. Understanding these associations is essential for predicting future outbreaks and formulating plant protection strategies.

Table 11.3. Natural enemies of invasive whiteflies and their role in biological control. (Author's own table)

Natural enemy species	Target whitefly species	Biological control role	References
Encarsia guadeloupae	*Aleurodicus dispersus*	Nymphal parasitoid	Selvaraj *et al.*, 2016
Delphastus catalinae	*Bemisia tabaci, A. rugioperculatus*	Predate on eggs and nymphs	Bellows and Hoddle, 2010a
Chrysoperla carnea	Multiple whitefly species	Generalist predator of immature stages	Naranjo, 2011
Isaria fumosorosea (fungus)	*Aleurodicus rugioperculatus*	Entomopathogenic fungus (EPF) effective in humid areas	Kumar *et al.*, 2018a

11.6.4 Natural enemies and biocontrol strategies

Biological control agents play a crucial role in managing whitefly populations in tropical and subtropical ecosystems (Table 11.3). *Encarsia guadeloupae*, a parasitic wasp, effectively parasitizes *Aleurodicus dispersus* (Selvaraj *et al.*, 2016). Predatory insects such as *Delphastus catalinae* and *Chrysoperla carnea* have been deployed against *Bemisia tabaci* and *A. rugioperculatus*, reducing nymphal densities (Bellows and Hoddle, 2010b; Naranjo, 2011). In addition, entomopathogenic fungi such as *Isaria fumosorosea* offer promising eco-friendly alternatives for whitefly suppression, particularly under humid conditions (Kumar *et al.*, 2018b). Incorporating such biocontrol strategies into integrated pest management (IPM) frameworks is key to sustainable pest regulation.

11.6.5 Role of climate change in whitefly invasiveness

Climate change significantly influences the spread, survival and population dynamics of invasive whiteflies. Rising temperatures and altered precipitation patterns create favourable environments for whitefly proliferation, especially in tropical and subtropical regions (Diez *et al.*, 2012). Extended warm seasons facilitate more generations per year, while reduced cold stress expands their potential geographic range. Moreover, climate-induced stress on host plants can increase susceptibility to whitefly infestation. Studies suggest that under global warming scenarios, whitefly invasions may intensify in previously unaffected areas, posing new challenges to crop protection and biosecurity (Naranjo and Ellsworth, 2009; Bellard *et al.*, 2013). Therefore, understanding the implications of climate change is essential for anticipating whitefly outbreaks and adapting pest management strategies accordingly.

11.7 Techniques for Managing Whiteflies

Combatting invasive whiteflies requires the use of advanced tools and strategies to effectively manage their populations. Genetic techniques, such as RNA interference (RNAi) and gene editing, offer potential solutions for controlling whitefly populations (Karunker *et al.*, 2008). RNAi can be used to silence specific genes in whiteflies, inhibiting their vital biological processes and reducing their ability to survive and reproduce (Xu *et al.*, 2020). Recent advanced digital technologies, including remote sensing, drones, and machine learning algorithms, can aid in the detection and monitoring of whitefly populations over large areas. These technologies enable timely interventions and more precise targeting of control measures (Ferriol *et al.*, 2014).

11.8 Integrated Pest Management (IPM) for Whiteflies

Integrated Pest Management tools play a crucial role in the management of invasive whiteflies. They aim to reduce chemical pesticides and instead emphasize a combination of biological, cultural, physical and chemical control methods (Simmons and Shaaban, 2008; Peshin and Dhawan, 2009). By reducing reliance on chemical pesticides, IPM tools help minimize the negative impact on the environment. (Pretty *et al.*, 2010). Monitoring and surveillance techniques are employed to detect the presence and abundance of whiteflies, enabling timely intervention when populations reach economic thresholds (Byrne and Bellows, 1991). Invasive whiteflies have the potential to develop resistance to chemical pesticides over time. IPM tools utilize multiple approaches, such as rotating different control methods, to mitigate the risk of resistance development and preserve the efficacy of available control options and also reduce the health risk (Horowitz *et al.*, 2005; Peshin and Dhawan, 2009). They help optimize pest management practices, leading to improved crop yields and reduced economic losses caused by invasive whiteflies. By employing cost-effective and efficient control measures, farmers can achieve better economic outcomes (Liu and Stansly, 1996).

Breeding and selecting host plant varieties that exhibit resistance or tolerance to whiteflies can be an effective long-term strategy. Plant resistance can involve traits such as reduced attractiveness to whiteflies, the ability to inhibit whitefly feeding or reproduction, or the production of toxic compounds against whiteflies (Firdaus *et al.*, 2013). Traps and monitoring systems, such as yellow sticky traps or pheromone traps, can be used to monitor whitefly populations. They help in detecting early infestations and provide valuable information for making informed management decisions (Byrne and Bellows, 1991). A sterile insect technique (SIT) consists of mass rearing and sterilizing male whiteflies using techniques such as irradiation. These sterile males are released into the target area, where they mate with wild females, leading to infertile eggs and a reduction in the population over time (Hendrichs *et al.*, 2002). Use of biopesticides derived from naturally-occurring organisms, such as fungi, bacteria or viruses, can be effective in controlling whitefly populations. These biopesticides specifically target whiteflies while minimizing harm to beneficial insects and the environment (Vega *et al.*, 2009). Natural enemies to suppress whitefly populations include predatory insects such as ladybugs (e.g. *Delphastus* spp.), lacewings (e.g. *Chrysoperla* spp.) and parasitic wasps (e.g. *Encarsia* spp.) which can be introduced to control whitefly infestations. These natural enemies feed on whiteflies or their eggs, helping to reduce their numbers (Naranjo, 2011).

11.9 Conclusion

Whiteflies are a threat to agricultural systems by directly damaging plants through their feeding activities, leading to reduced plant vigour, stunted growth and yield losses. Additionally, they act as vectors for numerous plant viruses, contributing to the spread of diseases among crops. Their presence can disrupt natural enemy populations and alter plant–insect interactions, leading to imbalances in the ecosystem. The use of chemical pesticides to control whiteflies can have unintended consequences, harming beneficial insects, contaminating the environment and contributing to pesticide resistance. To effectively manage whiteflies in tropical regions, IPM strategies that combine various control methods are essential. Timely monitoring, accurate identification and knowledge of whitefly biology and behaviour are critical for implementing effective management practices. Understanding the impact of whiteflies on agro-ecosystems in the tropics and subtropics provides valuable insights for designing and implementing appropriate management approaches. By integrating ecological principles, scientific knowledge and practical solutions, it is possible to mitigate the negative effects of whiteflies and foster sustainable agricultural systems.

Acknowledgement

The authors would like to thank the Director CSIR–IICT. "(IICT/Pubs./2024/376)"

References

Abd-Rabou, S. and Simmons, A.M. (2010) Survey of reproductive host plants of *Bemisia tabaci* (Hemiptera: Aleyrodidae) in Egypt, including new host records. *Entomological News* 121(5), 456–465.

Ajesh, K. and Alok, S. (2019) Distribution, host range and management of *Paraleyrodes bondari* in India. *Asian Journal of Agriculture and Life Sciences* 92, 140–146.

Bansal, R., Singh, B. and Sharma, S. (2019) Ecological factors influencing the spread of *Aleurodicus rugioperculatus* in North America and Central America. *International Journal of Pest Management* 65, 135–142.

Bansal, R., Singh, B. and Sharma, S. (2021) Invasive whiteflies and their management strategies: A review. *International Journal of Pest Management* 67, 14–23.

Barbedo, J.G.A. (2014) Using digital image processing for counting whiteflies on soybean leaves. *Journal of Asia-Pacific Entomology* 17, 685–694.

Bellard, C., Thuiller, W., Leroy, B., Genovesi, P., Bakkenes, M. *et al.* (2013) Will climate change promote future invasions? *Global Change Biology* 19, 3740–3748.

Bellows, T. and Hoddle, M. (2010a) *Giant whitefly: Aleurodicus dugesii*. University of California, Riverside, California (US). Available at: http://cisr.ucr.edu/giant_whitefly.html

Bellows, T.S. and Hoddle, M.S. (2010b) Biological control of whiteflies: An update on the efficacy of natural enemies. *California Agriculture* 54, 24–30.

Bellows, T.S. and Perring, T.M. (1994) Whitefly biology. *Annual Review of Entomology* 39, 331–357.

Bhat, S.H., Shankarnarayan, K. and Shivananda, T. (2020) First report of *Paraleyrodes bondari* Peracchi on coconut in Lakshadweep Islands, India. *Journal of Entomology and Zoology Studies* 8, 903–905.

Bhat, S.H., Shankarnarayan, K. and Shivananda, T. (2021) Spread of *Aleurothrixus floccosus* in guava orchards of Karnataka, India. *Indian Journal of Entomology* 83, 489–492.

Brown, J.K., Frohlich, D.R. and Rosell, R.C. (1995) The sweet potato or silverleaf whiteflies: Biotypes of *Bemisia tabaci* or a species complex. *Annual Review of Entomology* 40, 511–534.

Byrne, D.N. and Bellows, T.S. (1991) Whitefly biology. *Annual Review of Entomology* 36, 431–457.

Chandrashekar, K., Rao, A., Gorane, A., Verma, R. and Tripathi, S. (2020) *Aleurothrixus trachoides* (Back) can transmit begomovirus from Duranta to potato, tomato and bell pepper. *Journal of Biosciences* 45, 1–8.

Cohen, S., Kern, J., Harpaz, I. and Ben-Joseph, R. (1988) Epidemiological studies of the tomato yellow leaf curl virus (TYLCV) in the jordan valley, israel. *Phytoparasitica* 16(3), 259–270. DOI: 10.1007/BF02979527.

Costa, H.S., Ullman, D.E. and Johnson, M.W. (2006) Associations between whiteflies and plants. In: Byrne, J.N. and Bellows, D.M. (eds) *Whiteflies: Their Bionomics, Pest Status and Management*. pp. 233–243.

Cuthbertson, L., Kuhlmann, U. and Tuck, R. (2012) Cost implications of invasive species management: A case study of the whitefly. *Pest Management Science* 68, 993–999.

David, B.V. and Regu, K. (1995) *Aleurodicus dispersus* Russell (Aleyrodidae: Homoptera), a whitefly pest new to India. *Pestology* 19, 5–7.

Deshpande, A.A., Ramasamy, M., Kalleshwaraswamy, C.M. and Asokan, R. (2014) Occurrence and management of spiralling whitefly *Aleurodicus dispersus* Russell (Hemiptera: Aleyrodidae) on coconut in India. *Journal of Plantation Crops* 42(2), 227–231.

Dicke, M. and van Loon, J.J.A. (2000) Multitrophic effects of herbivore-induced plant volatiles in host-plant selection by herbivores and their natural enemies. *Entomologia Experimentalis et Applicata* 97, 237–249.

Dinesh, S. and Parvatha, R.P. (2021) Host range and potential impact of *Aleurodicus rugioperculatus* on agriculture. *Journal of Agricultural and Environmental Research* 21, 45–52.

Diez, J.M., D'Antonio, C.M., Dukes, J.S., Grosholz, E.D., Olden, J.D. *et al.* (2012) Will extreme climatic events facilitate biological invasions? *Frontiers in Ecology and the Environment* 10, 249–257.

Dinesh, S., Reddy, K.R. and Prakash, A. (2020) The impact of *Paraleyrodes bondari* on coconut production in India. *Journal of Coconut Research* 32, 100–105.

Duan, J.J., Leah, S.B., Kristopher, J.A., Michael, D.U. and Roy, G.V.D. (2015) Population dynamics of invasive forest insects and associated natural enemies in the aftermath of invasion: Implications for biological control. *Journal of Applied Ecology* 52, 1246–1254.

Dubey, A.K. and Sundararaj, R. (2015) A new combination and first record of genus *Aleurothrixus* Quaintance and Baker (Hemiptera: Aleyrodidae) from India. *Biosystematica* 9, 23–28.

Evans, G.A. (2008) *The Whiteflies (Hemiptera: Aleyrodidae) of the World and Their Host Plants and Natural Enemies*. USDA Agricultural Research Service, Publication.

Fereres, A. and Moreno, A. (2009) Behavioural aspects influencing plant virus transmission by homopteran insects. *Virus Research* 141, 158–168.

Ferriol, P., Escalera, A. and González-Cabrera, J. (2014) Remote sensing and digital technology in agricultural pest control: The case of the whitefly. *Precision Agriculture* 15, 610–628.

Firdaus, S., Vosman, B., Hidayati, N., Supena, E.D.J., Visser, R.G. *et al*. (2013) The *Bemisia tabaci* species complex: Additions from different tomato varieties under controlled and field conditions. *Bulletin of Entomological Research* 103, 395–406.

Freeman, T.P., Gullberg, G.T. and Schmitt, T.M. (2001) Effect of silverleaf whitefly (*Bemisia argentifolii*) feeding on plant nutrition in cotton. *Entomologia Experimentalis Applicata* 100, 207–213.

Ghanim, M., Morin, S., Zeidan, M. and Czosnek, H. (1998) Evidence for transovarial transmission of tomato yellow leaf curl virus by its vector, the whitefly *Bemisia tabaci*. *Virology* 240(2), 295–303.

Gilbert, M., Grégoire, J.C., Freise, J.F. and Heitland, W. (2004) Long-distance dispersal and human population density allow the invasion of the alien leaf miner *Cameraria ohridella*. *Journal of Animal Ecology* 73, 1150–1158.

Gill, R.J. (1990) The morphology of whiteflies. In: *Bemisia Tabaci: A Literature Survey on the Cotton Whitefly with an Annotated Bibliography*. California Department of Food and Agriculture.

Gill, R.J. (1992) The morphology of whiteflies: *Aleurodicus dugesii* and its waxy secretions. In: *Proceedings of the First International Symposium on Whiteflies in Agricultural Systems*, pp. 133–148.

Goolsby, J.A., DeBarro, P.J., Kirk, A.A., Sutherst, R.W., Canas, L. *et al*. (2005) Post-release evaluation of the biological control of *Bemisia tabaci* biotype "B" in the USA and the development of predictive tools to guide introductions for other countries. *Biological Control* 32, 70–77.

Heinz, K.M. and Parrella, M.P. (1990) Impact of the sweet potato whitefly on the environment: A review. *Environmental Entomology* 19, 1038–1043.

Hendrichs, J., Vreysen, M.J.B., Enkerlin, W.R. and Cayol, J.P. (2002) Strategic options in using sterile insects for area-wide integrated pest management. In: *Sterile Insect Technique*. CRC Press, pp. 3–24.

Hidayat, P., Maryana, E., Kusumah, Y.M. and Nurulaila, L. (2020) Host range and population density of the giant whitefly *Aleurodicus dugesii* Cockerell (Hemiptera: Aleyrodidae) on horticultural crops in Cipanas-Cianjur, West Java, Indonesia. In: *IOP Conference Series: Earth and Environmental Science*, Vol. 468. IOP Publishing, p. 012008.

Horowitz, A.R., Kontsedalov, S., Khasdan, V. and Ishaaya, I. (2005) Biotypes B and Q of *Bemisia tabaci* and their relevance to neonicotinoid and pyriproxyfen resistance. *Archives of Insect Biochemistry and Physiology* 58, 216–225.

Husain, M.A. and Nath, D.C. (2018) Host range of the spiraling whitefly *Aleurodicus dispersus* Russell in India. *Indian Journal of Entomology* 80, 1135–1140.

Iaccarino, F., Vaccaro, L. and Stefani, R. (1990) Morphology and behavior of Bondar's nesting whitefly (*Paraleyrodes bondari*). *Bollettino Di Zoologia Agraria e Di Bachicoltura* 22, 37–46.

Jiang, Z., Wang, Y. and Zhang, Y. (2019) Feeding behavior of whiteflies and their impact on plant physiology. *Journal of Insect Physiology* 115, 103869.

Johnson, M.W. and Walker, G.P. (1999) Intracellular punctures by the whitefly, *Bemisia argentifolii*: A novel mechanism of plant tissue damage. *Annals of the Entomological Society of America* 92, 291–297.

Jones, D.R. (2003) Plant viruses transmitted by whiteflies. *European Journal of Plant Pathology* 109, 195–219.

Josephrajkumar, A., Mohan, C. and Babu, M. (2019) First record of the invasive Bondar's nesting whitefly, *Paraleyrodes bondari* Peracchi on coconut from India. *Phytoparasitica* 47, 333–339.

Karthikeyan, K., Suresh, S., Samuthiravelu, P. and Prakash, A. (2019) Incidence, spread and management of spiralling whitefly *Aleurodicus dispersus* Russell (Hemiptera: Aleyrodidae) in coconut ecosystems of India. *Journal of Entomology and Zoology Studies* 7(3), 1016–1020.

Karunker, I., Morou, E., Nikou, D., Nauen, R., Sertchook, R. *et al*. (2008) Structural model and functional characterization of the *Bemisia tabaci* CYP6CM1 protein, a cytochrome P450 that confers resistance to neonicotinoid insecticides. *Insect Biochemistry and Molecular Biology* 38, 920–930.

Krause-Sakate, R., Córdova, J. and Araujo, R. (2020) Economic impact of invasive whiteflies on agricultural trade. *Agricultural Economics* 51, 511–520.

Kumar, A., Gupta, M. and Kumar, P. (2021) Distribution and impact of invasive whiteflies in South India: A case study of *Paraleyrodes bondari*. *Journal of Agricultural Science and Technology* 23, 129–138.

Kumar, P., Yadav, A. and Poonia, A. (2018a) Effects of whitefly infestations on crop yield: A case study of tomato and brinjal. *Journal of Agricultural Research* 56, 145–157.

Kumar, V., Francis, A., Avery, P.B., McKenzie, C.L. and Osborne, L.S. (2018b) Assessing compatibility of *Isaria fumosorosea* and buprofezin for mitigation of *Aleurodicus rugioperculatus* (Hemiptera: Aleyrodidae): An invasive pest in the Florida landscape. *Journal of Economic Entomology* 111, 1069–1079.

Legg, J.P. and Brown, J.K. (2006) Cassava mosaic and cotton leaf curl geminiviruses: A comparison of the whitefly-transmitted geminiviruses that cause major losses to cassava and cotton in Africa and India. *African Journal of Biotechnology* 5, 1710–1719.

Liu, T.X. and Stansly, P.A. (1996) Oviposition by *Bemisia argentifolii* (Homoptera: Aleyrodidae) on tomato and cucumber: Effects of leaf factors and insecticide residues. *Journal of Economic Entomology* 89, 664–670.

Li, Y., Sun, Y. and Liu, X. (2020) Whitefly responses to plant defense mechanisms: A review. *Pest Management Science* 76, 1951–1962.

Liebhold, A.M., Brockerhoff, E.G., Garrett, L.J., Parke, J.L. and Britton, K.O. (2012) Live plant imports: The major pathway for forest insect and pathogen invasions of the US. *Frontiers in Ecology and the Environment* 10, 135–143.

Maharjan, M. and Ghimire, B. (2020) Host plant records of *Aleurothrixus trachoides* in South Asia. *International Journal of Tropical Insect Science* 40, 21–28.

Malumphy, C. and Treseder, K. (2011) Palm-infesting whitefly *Aleurotrachelus atratus* (Hempel) (Hemiptera: Aleyrodidae) established in England at a botanical garden. *Entomologist's Monthly Magazine* 147, 23–31.

Mandal, F.B. (2011) The management of alien species in India. *International Journal of Biodiversity and Conservation* 3, 467–473.

Martin, J.H. (2004) The whitefly fauna of *Belize* (Hemiptera: Aleyrodidae) with the description of three new species. *Zootaxa* 722, 1–35.

Maskell, W.M. (1896) Contributions towards a monograph of the *Aleurodidae*, a family of Hemiptera – Homoptera. *Transactions and Proceedings of the New Zealand Institute* 28, 411–449.

Mohan, C., Josephrajkumar, A. and Babu, M. (2019) Non-native Neotropical nesting whitefly, *Paraleyrodes minei* laccarino on coconut palms in India and its coexistence with Bondar's nesting whitefly *Paraleyrodes bondari* Peracchi. *Current Science* 117, 515–519.

Nandan, D., Rao, G.K. and Patil, R.P. (2021) First report of *Aleurodicus rugioperculatus* in Lakshadweep Islands, India. *Journal of Insect Science* 21, 85–88.

Naranjo, S.E. (2011) Impacts of biotic and abiotic factors on whitefly-natural enemy interactions in cotton. *Environmental Entomology* 40, 789–800.

Naranjo, S.E. and Ellsworth, P.C. (2009) Fifty years of the integrated control concept: Moving the model and implementation forward in Arizona. *Pest Management Science* 65, 1267–1286.

Navas-Castillo, J. and Moriones, E. (2017) Whitefly transmission of plant viruses: Mechanisms and epidemiology. *European Journal of Plant Pathology* 147, 1–15.

Navas-Castillo, J., Luque, A. and Moriones, E. (2011) Trade and agriculture: The whitefly problem. *Crop Protection* 30, 615–628.

Naveena, N.L., Shashank, P.R., Devaramane, R. and Malikarjuna, J. (2020) Invasive insect pests in India: Current scenario and future perspective. *Indian Entomologist* 1, 23–31.

Oliveira, M.R.V., Henneberry, T.J. and Anderson, P. (2001) History, current status, and collaborative research projects for *Bemisia tabaci*. *Crop Protection* 20, 709–723.

Peracchi, A.L. (1971) Dois aleirodideos pragas de Citrus no Brasil (Homoptera, Aleyrodidae). *Arquivos Do Museu Nacional, Rio de Janeiro* 54, 145–151.

Perring, T.M. and Cooper, A.D. (2010) The biology and management of the giant whitefly (*Aleurodicus dugesii*). In: Byrne, J.N. and Bellows, D.M. (eds) *Whiteflies: Their Bionomics, Pest Status and Management*. pp. 433–449.

Peshin, R. and Dhawan, A.K. (2009) *Integrated Pest Management: Innovation-Development Process*, Vol. 1. Springer Science & Business Media.

Pimentel, D., Zuniga, R. and Morrison, D. (2005) Update on the environmental and economic costs associated with alien-invasive species in the United States. *Ecological Economics* 52, 273–288.

Pretty, J., Toulmin, C. and Williams, S. (2010) Sustainable intensification in African agriculture. *International Journal of Agricultural Sustainability* 9, 5–24.

Ramesh, K., Kumar, A. and Sharma, S. (2022) Distribution of *Aleurothrixus floccosus* in South India: New records and implications for guava cultivation. *Journal of Agricultural Science* 14, 302–310.

Rani, M., Prakash, A. and Rao, P. (2023) Recent infestations of whiteflies on guava in the Lakshadweep Islands. *Journal of Entomology and Zoology Studies* 11, 509–514.

Russell, L.M. (1965) A new species of *Aleurodicus* Douglas and two close relatives. *Florida Entomologist* 48, 47–55.

Schuster, D.J., Capinera, J.L. and Braman, S.K. (2010) Impact of invasive whiteflies on crop health: A review. *Florida Entomologist* 93, 395–400.

Selvaraj, K., Venkatesan, T., Sumalatha, B.V. and Kiran, C.M. (2016) Potential natural enemies of the invasive rugose spiraling whitefly, *Aleurodicus rugioperculatus* Martin in India. *Journal of Biological Control* 30, 236–239.

Selvaraj, K., Gupta, A., Venkatesan, T., Jalali, S.K., Sundararaj, R. *et al.* (2017) First record of invasive rugose spiraling whitefly *Aleurodicus rugioperculatus* Martin (Hemiptera: Aleyrodidae) along with parasitoids in Karnataka. *Journal of Biological Control* 31, 74–78.

Selvaraj, K., Sundararaj, R. and Sumalatha, B.V. (2019a) Invasion of the palm infesting whitefly, *Aleurotrachelus atratus* Hempel (Hemiptera: Aleyrodidae) in the Oriental region. *Phytoparasitica* 47, 327–332.

Selvaraj, K., Venkatesan, T., Sumalatha, B.V. and Kiran, C.M. (2019b) Invasive rugose spiraling whitefly *Aleurodicus rugioperculatus* Martin, a serious pest of oil palm *Elaeis guineensis* in India. *Journal of Oil Palm Research* 31, 651–656.

Serra, G., Lentini, A., Delrio, G. and Cocco, A. (2014) Integrated pest management strategies for whiteflies (*Aleurodicus* spp.) in Mediterranean crops. *Journal of Pest Science* 87, 45–53.

Simala, M., Milek, T.M. and Pintar, M. (2015) Alien whiteflies (Hemiptera: Aleyrodidae) of Europe recorded in Croatia. *Zbornik Predavanj in Referatov* 12, 3–4.

Simberloff, D. and Rejmánek, M. (2011) *Encyclopedia of Biological Invasions*. University of California Press.

Simberloff, D., Martin, J.L., Genovesi, P., Maris, V., Wardle, D.A. *et al.* (2013) Impacts of biological invasions: What's what and the way forward. *Trends in Ecology and Evolution* 28, 58–66.

Simmons, A.M. and Shaaban, A.M. (2008) Cultural control of *Bemisia tabaci* using row covers and plant spacing in squash. *Journal of Insect Science* 8, 45.

Simmons, A.M. and Shaaban, M. (2011) Pest management strategies for invasive species: A case study of whitefly. *Journal of Integrated Pest Management* 2, 1–8.

Singh, G. and Aggarwal, N. (2025) Alternate hosts of *Bemisia tabaci* (Gennadius) and their role in its carry-over to cotton in North-Western India. *International Journal of Pest Management* 71(2), 170–182.

Singh, J., Gupta, R. and Sharma, V. (2022) New records of *Aleurodicus rugioperculatus* in Maharashtra and Gujarat. *Indian Journal of Entomology* 84, 564–570.

Stansly, P.A. and Naranjo, S.E. (2010) Integrated systems for managing *Bemisia tabaci* in protected and open field agriculture. *Crop Protection* 29, 822–831.

Stocks, I.C. and Hodges, G.S. (2012) *Identification of the Rugose Spiraling Whitefly, Aleurodicus rugioperculatus Martin*. Florida Department of Agriculture and Consumer Services, Pest Alert.

Sundararaj, R. and Selvaraj, K. (2017) Invasion of rugose spiraling whitefly, *Aleurodicus rugioperculatus* Martin (Hemiptera: Aleyrodidae): A potential threat to coconut in India. *Phytoparasitica* 45, 71–74.

Sundararaj, R. and Vimala, D. (2018) New record of the legume feeding whitefly *Tetraleurodes acaciae* (Quaintance) (Hemiptera: Aleyrodidae) from India. *Indian Journal of Entomology* 80, 1168–1169.

Sundararaj, R., Selvaraj, K. and Kalleshwaraswamy, C.M. (2020a) First record of the invasive woolly whitefly *Aleurothrixus floccosus* (Maskell) from India. *Indian Journal of Entomology* 82, 88–91.

Sundararaj, R., Selvaraj, K., Vimala, D. and Venkatesan, T. (2020b) Whiteflies (Hemiptera: Aleyrodidae) of India. *Biodiversity Survey Report*.

Tello, J., Aguilar, M. and González, P. (2014) Biological aspects and control strategies for citrus woolly whitefly (*Aleurothrixus floccosus*). *Spanish Journal of Agricultural Research* 12, 347–356.

Thangavel, P., Gopal, R. and Kumar, S. (2015) Host range and potential impact of *Aleurothrixus floccosus* on agriculture. *Journal of Agricultural and Environmental Research* 21, 123–132.

Ulusoy, M.R., Vatansever, G., Erkılıç, L. and Uygun, N. (2003) *Aleurothrixus floccosus* (Maskell) (Homoptera: Aleyrodidae) and its parasitoid, Cales noacki Howard (Hymenoptera, Aphelinidae) in the East Mediterranean Region of Turkey. *Journal of Pest Science* 76, 163–169.

Varma, S., Kumar, A. and Bhat, P. (2023) Distribution of *Aleurodicus rugioperculatus* in Telangana and Chhattisgarh: Implications for pest management. *Journal of Plant Protection Research* 63, 195–202.

Vasanthakumar, R., Ravi, K. and Sathia Raj, D. (2020) Distribution and management of spiraling whitefly, *Aleurodicus dispersus* in the Western Ghats of India. *Journal of Biological Control* 34, 24–29.

Vega, F.E., Meyling, N.V., Luangsa-ard, J.J. and Blackwell, M. (2009) Fungal entomopathogens. In: *Insect Pathology*. Elsevier, pp. 171–220.

Vidya, C.V., Sundararaj, R., Dubey, A.K., Bhaskar, H., Chellappan, M. *et al.* (2019) Invasion and establishment of Bondar's nesting whitefly, *Paraleyrodes bondari* Peracchi (Hemiptera: Aleyrodidae) in Indian mainland and Andaman and Nicobar Islands. *Entomon* 44, 149–154.

Walker, M. and Zareh, N. (1990) The behavior and ecology of *Aleurothrixus floccosus* in citrus orchards. *Entomological Bulletin* 18, 233–242.

Walling, L.L. (2000) The myriad plant responses to herbivores. *Journal of Plant Growth Regulation* 19, 195–216.

Wang, X., Li, W. and Liu, S.S. (2017) Whitefly interactions with plants. *Current Opinion in Insect Science* 19, 70–75.

Waterhouse, D.F. and Norris, K.R. (1989) *Biological Control: Pacific Prospects*. Inkata Press, Melbourne.

Wosula, E.N., Evans, G.A., Issa, K.A. and Legg, J.P. (2018) Two new invasive whiteflies (Hemiptera: Aleyrodidae) to Tanzania. *African Entomology* 26, 259–264.

Xu, Z., Zhong, M., Li, Z. and Su, Q. (2020) The potential use of RNA interference (RNAi) in whitefly control. *Insects* 11, 301.

12 Rugose Spiralling Whitefly, *Aleurodicus rugioperculatus* Martin (Hemiptera: Aleyrodidae): From Invasion to Management

Tanmaya Kumar Bhoi[1], Vipula Vyas[1], Ipsita Samal[2], Deepak Kumar Mahanta[3], Ifrah Khan[1], Komal J.[4], Prabhulinga T.[5]*, Athulya R.[6], Sangeeta Singh[1] and Varsha Giri[7]

[1]Forest Protection Division, ICFRE – Arid Forest Research Institute (ICFRE–AFRI), Jodhpur, India; [2]ICAR – National Research Centre on Litchi, Mushahari, Ramna, Muzaffarpur, Bihar, India; [3]Forest Entomology Discipline, Forest Protection Division, Indian Council of Forestry Research and Education (ICFRE) – Forest Research Institute (ICFRE–FRI), Dehradun, Uttarakhand, India; [4]Basic Seed Multiplication & Training Centre, Central Silk Board, Pendari, Bilaspur, Chhattisgarh; [5]ICAR – National Bureau of Agricultural Insect Resources (ICAR–NBAIR), Bengaluru, Karnataka; [6]Forest Protection Division, Institute of Wood Science and Technology, Bengaluru, Karnataka, India; [7]Silviculture and Forest Management Division, Arid Forest Research Institute, Jodhpur, India

Abstract

Alien species are exotic or non-native organisms that escaped their habitat and dispersed away to new habitats/regions. When non-native/invasive species are unintentionally or purposefully introduced to new areas, they quickly become invasive and threaten local biodiversity. Without natural predators in natural environments, such organisms become invasive and harm economically significant crops. Tropical regions with a warm and humid climate are subjected to exotic pests. Global anthropogenic climate change, rising temperatures, limited rainfall, and human migrations have all contributed to the proliferation and establishment of invasive species in the tropics. Rugose spiralling whitefly (RSW), *Aleurodicus rugioperculatus* Martin, is one such new invasive and highly polyphagous species, initially described from Belize, Florida in 2004 from coconut, but suspected from Central America as place of origin. Understanding the biology, host range, and climatic factors affecting abundance of RSW and management can help in mitigating the invasive RSW damage.

Keywords: Exotic pests, rugose spiralling whitefly (RSW), management options, host range

*Corresponding author: prabhut07@gmail.com

DOI: 10.1079/9781836991977.0012

12.1 Introduction

Many plant, insect, and pathogen species have been introduced to new areas because of free commerce and travel of plant resources across political and geographic borders. These alien species are referred to as non-native species. Exotic or non-native organisms that appear outside their natural environment and dispersal capabilities are considered alien species. Numerous alien species provide significant assistance to farming and forestry systems. Some alien species, however, turn into invasives, mistakenly or purposely transported outside natural habitats into new areas where they demonstrate the ability to invade, establish, and dominate the native species (Raghubanshi *et al.*, 2005). Such species multiply greatly and harm economically significant crops and other plants, rendering them invasive, since they are not surrounded by the natural predators that keeps them in check in their naturally occurring areas. Among the invasive species, exotic whiteflies constitute the greatest number of invasives in India recently (Bakthavatsalam *et al.*, 2021). Exotic invasive whiteflies in India are responsible for both direct and indirect yield losses in agri-horticulture crops. Rugose spiralling whitefly (RSW), *Aleurodicus rugioperculatus* Martin, is one of them. It is a polyphagous, sap-sucking, tiny, phloem-feeding insect of the order Hemiptera, that was documented from Belize in 2004. By inserting their pointed stylets, the nymphs and adults suck the sap from the underside of the leaflets of palm trees. Due to this pest's feeding habits, a lot of honeydew is excreted, which then accumulates on the upper layer of the leaves and below as well as on other understorey crops. Egg spirals appear in cases of severe attack on leaves, petioles, and tender coconuts. Since honeydew excrement is sweet and liquid, ants are drawn to it and sooty mould quickly develops, disrupting typical leaf physiology and increasing its invasive potential. In March 2009, Miami-Dade County, Florida, received a report of this exotic whitefly pest (Stocks and Hodges, 2012). Also, RSW was discovered in India during July 2016 on coconuts in several sites in Coimbatore, Tamil Nadu, and the Palakkad (Kerala) (Selvaraj *et al.*, 2016; Srinivasan *et al.*, 2016; Sundararaj and Selvaraj, 2017), and in other regions (Chandrika Mohan *et al.*, 2016, 2017; Rao *et al.*, 2018). RSW feeds on a variety of host plants, such as fruit trees, woody ornamentals, and palms (Mannion, 2010; Elango *et al.*, 2019; Alagar *et al.*, 2020). Since RSW is polyphagous in nature and has the ability to migrate to new areas, the current incidence of the pest in India is worrying, and a systematic strategy is needed to stop its infestation (Selvaraj *et al.*, 2017). Biological control techniques should be included within IPM programmes to manage such groups of insects (Skinner *et al.*, 2014).

12.2 Origin and Distribution

Gumbo limbo, *Bursera simaruba* (L.) Sarg, a tree, was the source of the first reports of rugose spiralling whiteflies in Miami Dade County, Florida, in 2009. However, it was first documented from Belize in 2004 on coconut (Martin, 2004), where it was noted to have a natural population. The USA, Belize, Mexico, and Guatemala are the only countries in North and Central America where this pest is distributed. It is thought that RSW originated in Central America (Evans, 2008). Since the first confirmed population of RSW in Florida in 2009, its spread has risen significantly within the state and it has dispersed to 22 other regions in Central and South America (Evans, 2013). In the Oriental region, RSW has only been introduced to India. RSW was initially noticed in multiple coconut fields in the Pollachi area of the Coimbatore district, Tamil Nadu, during July 2016 and was then first documented in Kottayam from Kerala (Sundararaj and Selvaraj, 2017). RSW was accidentally introduced to coconut palms, mango trees, and guava trees in Changanassery, Kottayam District, Kerala, India. Additionally, between October and November 2016, it was observed in nurseries in Kadiyapulanka, Coastal Andhra Pradesh. Coconut saplings from nurseries in South India may be the potential entrance point into Andhra Pradesh. Chakravarthy *et al.* (2017) recorded hosts and potential areas for invasion by RSW in India.

12.3 Host Range

RSW is a polyphagous pest recorded on host plants, such as palms, woody ornamentals, and fruits (Mannion, 2010). Records from the Florida Department of Agriculture and Consumer Services (FDACS), Division of Plant Industry (DPI), indicate that rugose spiralling whiteflies

Table 12.1. Biology and morphometrics of Rugose spiralling whitefly.

SL. no.	Life stages	Period (days)	Length (mm)	Width (mm)
1	Egg	8.47 ± 0.26	0.31 ± 0.01	0.11 ± 0.02
2	First nymphal stage	5.80 ± 0.19	0.34 ± 0.01	0.18 ± 0.01
3	Second nymphal stage	5.03 ± 0.34	0.53 ± 0.01	0.79 ± 0.02
4	Third nymphal stage	6.63 ± 0.20	0.94 ± 0.01	0.82 ± 0.01
5	Fourth or Pseudopupal stage	10.30 ± 0.29	1.23 ± 0.01	1.00 ± 0.01
6	Egg – Adult emergence	36.23 ± 1.51	****	****
7	Adult longevity	20.00 ± 1.00	****	****
8	Total development period	56.23 ± 2.20	****	****

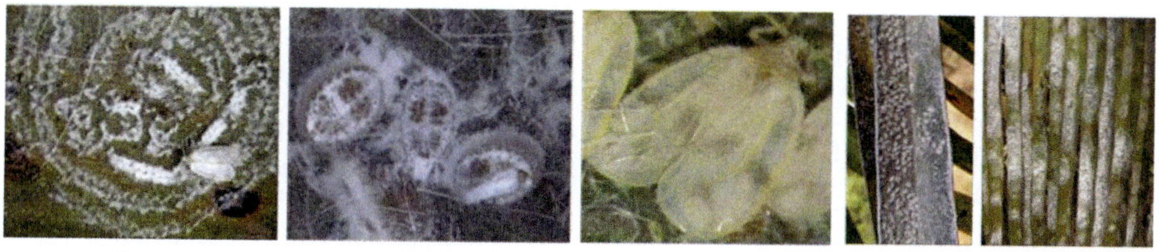

Fig. 12.1. Rugose spiralling whitefly, eggs laid in spiralling manner, eggs and nymphs covered with dense wax, severe sooty mould, sluggish adult, largest whitefly. (Authors' own photos).

were found on at least 118 plant species between 2009 and 2015, including a mix of native, edible, ornamental, palm, invasive, and weed species (Stocks, 2012). The plants infested by RSW between 2009 and 2012 in Florida included 16% gumbo limbo, 22% of palm species, 9% avocado, 10% *Calophyllum* spp., 3% of mango spp., and 4% black olive (Francis *et al.*, 2016). Forty-four per cent of host data for members of the palm family Arecaceae came from coconuts. These species are regarded as preferable hosts for RSW based on incidence statistics. All plant species, meanwhile, if not confirmed as the pest's actual hosts, need to be managed. In Kerala, 17 plants of 11 families were identified as favoured hosts of RSW (Shanas *et al.*, 2016).

12.4 Description and Biology

Martin originally identified RSW in 2004 using samples from Belize obtained on coconut palm leaves (Martin, 2004). RSW have four distinct stages in their stage of development, i.e. egg, nymph, pupa, and adult. The details of RSW biology described below and its morphometrics are shown in Table 12.1.

Adults: Rugose spiralling whiteflies are naturally sluggish and nearly thrice (about 2.5 mm) as long compared with the typically encountered whiteflies. Two irregularly shaped light brown stripes on the adult RSW wings make them stand out in addition to their size (Stocks and Hodges, 2012). At the base of their abdomens, males have lengthy features that resemble pincers.

Eggs: Females place their eggs on the underside of the leaves in concentric circles or spirals, covering them with a white waxy substance. Eggs range in shape from ellipsoid to creamy white to darker yellow (Fig. 12.1).

Nymph: Rugose spiralling whiteflies go through five developmental stages before becoming adults. The first instar, sometimes referred to as the crawler phase since it is the solely moving juvenile stage, hatches from the egg and seeks a site to start feeding by sucking plant sap with its needle-like mouth parts. Crawlers moult into immature stages which are initially stationary, round, and flat but eventually become more convex as their life cycle progresses (Mannion, 2010). The nymphs are light to golden yellow and produce long, thin waxy filaments that gradually become heavier as well as a dense, cottony wax (Stocks and Hodges, 2012).

Puparium: This species is distinguished by a broad cordate vasiform aperture, an operculum that is characteristically rugose on the dorsum and has two ventromedian fine setae, a dorsally typically rugose operculum, and a lingula head that protrudes beyond the orifice and is apically acute and finely spinulose with four setae that are proximate to the apex (Martin, 2004; Sundararaj and Selvaraj, 2017). A single submedian pair of pro-, meso-, and meta-throacic setae as well as the eighth abdominal setae located fully anterior to the vasiform orifice as well as the opposing anterior corners of the operculum are present. The anterior marginal, cephalic, and first abdominal setae are lacking. Two pairs of much smaller pores are present on abdominal segments VII and VIII, as well as four separate pairs of compound abdominal pores in the cephalic and anterior regions.

12.5 Nature of Damage

By withdrawing nutrients and water, RSW can stress the plant. In addition, whiteflies release honeydew, a viscous, sticky liquid that sooty moulds can grow on and turn into a dark liquid (Mani, 2000a). The sooty mould dries to develop thick coats on the host leaf and other parts of a plant. The accumulation of sooty mould on leaves may interfere with photosynthesis, causing physiological problems in the host. Additionally, honeydew draws wasps and ants, which defend the whiteflies against their natural predators (Stocks and Hodges, 2012). Rugose spiralling whiteflies not only harm their hosts, but they also cause an annoyance where they are present. On non-plant surfaces including cars, patios, and furniture as well as understorey plants, sticky wax, honeydew, sooty mould, and the bodies of the dead whiteflies fall and accumulate. Additionally, dried sooty mould is difficult to remove and may need professional cleaning or pressure washing (Mayer *et al.*, 2010). Several symptoms that show the presence of whiteflies include egg spirals beneath the leaves, sticky honeydew, dense white waxy material, black sooty mould growth, leaf damage, and early leaf fall.

12.6 Identification

Identification of whiteflies and their natural enemies uses morphology and molecular methods. Whitefly identification is mostly dependent on the fourth larval instar, or puparium. According to Dubey and David (2012), infested host plant leaves containing puparium were obtained and wrapped in paper, and mature whiteflies were collected and stored in 70% ethanol with the appropriate collecting information for molecular and morphological identification, respectively. To check the emergence of parasitoids, a portion of the collected host plant leaves/parts contaminated with puparium and immature stages were placed in a rearing jar (21 × 10 cm) for up to 10–15 days. The parasitoids that emerged were collected with an aspirator and kept for later identification in tubes containing 70% ethanol. On the basis of proportion of parasitized pupae to unparasitized in host leaves, parasitism was evaluated. At the NBAIR (National Bureau of Agricultural Insect Resources), Bengaluru, the species was identified. The puparium's permanent mounts were created (Martin, 1987). Its host plants were discovered, and taxonomy of its natural enemies was validated. Molecular analysis of the adult RSW's partial mitochondrial cytochrome c oxidase I (COI) gene (658 bp) was done using samples from five different Karnataka locales. Following industry standard procedures, the extracts were subjected to PCR amplification of the COI gene's 5' terminus.

12.7 Co-existence with Other Whiteflies

Four Neotropical non-native, exotic whiteflies, including the Bondar's nesting whitefly (*Paraleyrodes bondari* Peracchi), the nesting whitefly (*P. minei* Iaccarino), the rugose spiralling whitefly, and the palm infesting whitefly (*Aleurotrachelus atratus*), were reported from

Table 12.2. Damage rating scale for the infestation of *A. rugioperculatus*.

No. of egg spirals	Grade	Category	Infestation index
No egg spirals and sooty mould encrustation noticed	0	Nil	0
Fewer than 10 egg spirals per leaflet; presence of sooty mould encrustation in 5–6 lowermost fronds	1	Low	0.01–1.00
Ten to 20 egg spirals per leaflet; presence of sooty mould encrustation in 10–12 fronds	2	Medium	1.01–2.0
More than 20 egg spirals per leaflet; presence of sooty mould encrustation in more than 12 fronds	3	High	2.01–3

India over a 2-year period (Josephrajkumar *et al.*, 2020). The cohabitation of these whiteflies in the same palms or leaves suggests a novel method for the coexistence of these competing insect species. Except for the fact that they are demographically relatively similar and require in-depth investigation, the synchrony of coexistence and mutual existence of these competing insect species may be caused by the marked time partitioning of resource usage among the species (Mani, 2000b). Invasives are shown to coexist in situations where one species uses another species' feeding and breeding niche when the weather conditions are ideal and attempts to gradually supplant one or more of its rivals, causing temporal fluctuation.

12.8 Assessment of Infestation

The damage rating scale created by Srinivasan *et al.* (2016) can be used to determine the extent of infestation. A diagonal selection of at least 20 palm trees was made in a garden, and the severity of the damage was determined. With the aid of the method below, an infestation index was worked out to classify the gardens as low, medium, or high intensity (Table 12.2).

$$\text{Intensity of damage (\%)} = \frac{(\text{No of fronds infested}/\textit{Tree})}{(\text{Total no of fronds observed}/\textit{tree})} \times 100$$

$$\text{Infestation Index} = \frac{(\text{No.of palms under Scale } 0 \times 0) + (\text{No.of palms under Scale } 1 \times 1) + \cdots. + (\text{No.of palms under Scale } 3 \times 3)}{(\text{Total no. of palms observed})}$$

12.9 Climatic Factors

The population, survival, growth, outbreaks, ability to reproduce, and activity of pests as well as predators and parasites are all greatly influenced by the climate, directly or indirectly (Arif *et al.*, 2007). Information on population dynamics in relation to meteorological variables is crucial for constructing a weather-based pest forecasting system. The model must have an understanding of the interactions between the crop development stage and meteorological criteria dynamics (Zia *et al.*, 2015). Activity of the pest will help to create effective management methods that will boost productivity (Damos and Savopoulou-Soultani, 2010). An integral component of the integrated pest management module is population dynamics (Fakhri and Jamal, 2012; Taravati, 2013). To understand population dynamics and their relationship to meteorological conditions, a study was conducted. Maximum and minimum temperatures had a considerable impact on the population of RSW and its natural enemies, although rainfall and RH had no significant impact. However, in coconut and guava gardens, the population of RSW was highly significant and inversely linked with populations of natural enemies, suggesting that an increase in coccinellids and spiders may lead to a decline in the RSW population (Sushmitha *et al.*, 2020). Arthropods exhibit varied patterns in their occurrence as well as in the type and degree of damage they cause to a crop due to change in the agro-climatic parameters of different places. The distribution and abundance of the new exotic whitefly species *A. rugioperculatus* (RSW) and their corresponding indigenous enemies were assessed on five-year-old Chowghat Orange Dwarf coconut palms at Tamil Nadu Agricultural University's Coconut Farm. The study found RSW on coconut all year long, and data showed that the population of *A. rugioperculatus* increased from 130.8 nymphs per leaf per frond during the first week of July 2018 to 161.0 nymph/leaf/frond in the first week of October before declining in April. *Encarsia guadeloupae* parasitized RSW at a rate ranging from 31.60% (August 2018) to 57.60% (December 2018). The relationship between abiotic and biotic factors and the population of *A. rugioperculatus* was negatively correlated with *Cryptolaemus montrouzieri* and *E. guadeloupae*. The relative humidity, maximum and minimum temperatures, showed a strong positive association. However, there was no connection between the population of *A. rugioperculatus* and rainfall (Elango and Nelson, 2020). Chavan and colleagues (2022) carried out an experiment from December 2018 to December 2019 in the Department of Horticulture, Dr Balasaheb Sawant Konkan Krishi Vidyapeeth, Dapoli. RSW occurred throughout the year on coconut palms, but their density ranged from 4.24 to 102.87 adults/frond per palm in various standard meteorological weeks (SMW). The investigations on the pest population began during the 50th SMW (10–16 December), recording 34.31 RSW per frond per palm. It peaked during the 15th SMW (9–15 April), reaching 102.87 RSW per frond per palm, and then stayed rather consistent until the 21st SMW (21–27 May). The population of RSW showed a statistically highly significant link with maximum temperature, brilliant sunshine hours, and evaporation, which was reported as r=0.574, 0.473, and 0.708, respectively, according to a correlation study between pest incidence and meteorological factors. The multiple regression analysis R^2 value showed that weather parameters accounted for 56.0% of the overall variation in the RSW population on coconuts.

12.10 Monitoring and Surveillance

The fundamental instruments for creating an efficient IPM are monitoring and surveillance (Singh and Gupta, 2016). Numerous researchers have emphasized the advantages of adopting various trapping techniques for monitoring. Entomologists take advantage of the fact that insects are variably drawn to different colours for the purpose of collecting Coleoptera, Hemiptera, Hymenoptera, and Thysanoptera (Elango *et al.*, 2017). RSW is an invasive pest, therefore understanding its activities and flight patterns is crucial (Taravati *et al.*, 2014). According to Elango and Jeyarajan Nelson (2020), RSW infested 20 other plants in Tamil Nadu, with banana and coconut being the most popular hosts.

12.11 Cultural and Physical Control

Even though leaf removal may be an environmentally favourable method, the insect is not entirely eradicated; rather, the plant's whitefly population is only reduced. Even a small infestation can easily infest additional plants. The pupae and nymphs of whiteflies can be eliminated from dense foliage by removing leaves. Additionally, as whiteflies are drawn to yellow colours, sticky yellow traps are employed to catch adults (Barbedo, 2014). A trap made of paper strips and sticky materials like petroleum is set up in a greenhouse. As the insects fly, they are captured. The disadvantage of this strategy is that it can only capture flying specimens. Because they lack the ability to fly, it is typically useless for insects that are still in their initial stages (Barbedo, 2014). This strategy tries to lower the whitefly population rather than completely eradicate it. Hence it is not a foolproof control scheme for farmers (Nakamura *et al.*, 2007).

12.12 Biological Control

A biological control may be traditional, augmentative, or conservation-oriented. *Encarsia guadeloupae* Viggiani (Hymenoptera: Aphelinidae) was found to parasitize *A. rugioperculatus*, while *Encarsia dispersa* Polaszek was found to do the same by Poorani and Thanigairaj (2017) in surveys made in Tamil Nadu. Banana was heavily parasitized by *E. guadeloupae*, between 40–70% (Poorani and Thanigairaj, 2017). Selvaraj *et al.* (2016) found that in South India, parasitism of *A. rugioperculatus* by *E. guadeloupae* on coconuts ranged from 20–60%. The extent of parasitism was introduced and nomenclature and diagnostic characteristics of *Encarsia dispersa* are given in this study along with remarks on the predators of the rugose whitefly. Of the two parasitoids, *E. guadeloupae* was more prevalent and responsible for 60–70% of the overall parasitism.

12.13 Chemical Control

Only use insecticides when absolutely required, and always read and follow all label directions. Chemical treatment is currently considered the major technique of control for this pest. Numerous insecticides are available to control whitefly, although the mode of application and location of use will differ depending on the label (Mannion, 2010). Systemic application (soil or trunk) pesticides can provide effective control. Systemic insecticides can be applied to the soil (drenching, granular formulations, burying pellets, or soil injection), the trunk (basal bark spray and trunk injection), or the foliage by licensed professionals; however, soil and trunk applications benefit from the systemic properties of these products and provide longer-term control. Most contact insecticides can be applied to the foliage and can provide quick knockdown of the whiteflies, but will typically provide only a few weeks of control. Pesticides can kill the biological control agents and should be used only as a last resort.

12.14 Bio-intensive Pest Management

The current infestation of RSW in India is of concern in view of it having polyphagous characteristics and propensity to distribute to new places, and a planned strategy is needed to stop such incursion (Selvaraj *et al.*, 2017). Biological control techniques must be combined with integrated pest management programmes to manage pests (Skinner *et al.*, 2014). It is known that a large variety of entomopathogenic fungi (EPF) naturally infect whitefies throughout the world. *Isaria fumosorosea* Wize, formerly known as *Paecilomyces fumosoroseus* (Hypocreales: Cordycipitaceae), is one of these and is a successful bio-control agent for whiteflies. Under normal field conditions, it results in epizootics (Luangsa-Ard *et al.*, 2005). In Florida, *A. rugoperculatus*, *P. bondari*, *Singhiella* simplex, and coconut infestations are all managed by *I. fumosorosea* (Ali *et al.*, 2015; Kumar *et al.*, 2018). *I. fumosorosea* (ICAR-NBAIR Pfu-5) blastospore formulation has demonstrated promise for treating RSW (Sumalatha *et al.*, 2020). While several conventional insecticides, biopesticides, and insect growth regulators were effective over RSW (Pradhan *et al.*, 2020), when applied in conjunction with the proper insecticides at lower rate for efficient pest management, *I. fumosorosea* (ICAR-NBAIR Pfu-5) can increase its effectiveness in the field. This method could be an alternative for farmers to manage *A. rugioperculatus* because it uses fewer pesticides, which reduces environmental risk and resistance. The findings revealed that the combination of *I. fumosorosea* at 5 ml/l + profenofos 50 EC at 2 ml/L spray drastically decreased RSW infestation percentage to 82.97%, with the intensity percent of RSW dropping to 80.49%, and the mean number of RSW colonies to 79.68%. The combination of *I. fumosorosea* at 5 ml/l + profenofos 50 EC at 2 ml/L spray twice with a 15-day gap between each application successfully reduced the RSW infestation upon coconuts (Sandeep *et al.*, 2022). The group of insecticides known as organophosphates includes profenofos 50 EC, paralysing and killing the pest (Elbert *et al.*, 2008). *A. rugioperculatus* responded significantly to profenofos four days after treatment, as seen by the darkening of the eggs' colour (Pradhan *et al.*, 2020). The first selective insect growth regulator, Buprofezin 25 SC, was created to control *B. tabaci* on cotton (Ishaaya and Horowitz, 1992). The chemical is an inhibitor for production of chitin that causes pest mortality by preventing cuticle formation in the early stages of development (Palumbo *et al.*, 2001). During the 2017–2018 academic year, laboratory evaluation was made to determine the effectiveness of three neem formulations based on azadirachtin (NeemAzal), mineral oil (2%), Neem seed kernel extract (NSKE, 5%), fish oil rosin soap (FORS, 2%), Econeem (3%) and entomopathogens *Metarhizium anisopliae*, *Beauveria bassiana* following NeemAzal F (AI 5% azadirachtin) at 0.05%. The data showed that NeemAzal T (AI 1% azadirachtin) at 0.2% caused 63.33% death of nymphs after 72 hours of treatment (HAT). Neem oil 3% and

NeemAzal (AI azadirachtin 0.15%) at 0.05% and 0.05%, respectively, resulted in 63.33% and 46.67% mortality at 72 HAT. Adults that received NeemAzal F saw a mortality rate of 70.0% at 72 HAT, followed by those who received NeemAzal T (56.67%) and NeemAzal (53.33%). The use of *Lecanicillium lecanii* (National Bureau of Agriculture Insect Resources [VL 15]) at a rate of 1×10^8 conidia/ml at 10 days after treatment resulted in mortality of 50.0%. This proved to be better than other entomopathogen treatments (Elango and Nelson, 2020).

12.15 Conclusion

In conclusion, the invasive rugose spiralling whitefly (RSW) is an important pest in the tropics and it has shown significant economic damage in India. The population of RSW persists throughout the year, making this pest difficult to manage. There are several areas which need further study, viz. (a) The economic yield loss brought on by RSW on a range of hosts needs to be studied. (b) It is necessary to identify and confirm other probable hosts of RSW. (c) It is necessary to improve mass multiplication procedures for parasitoids and predators in a lab setting and test them in the field for effectiveness against RSW. (d) Finding eco-friendlier and more potential bio-pesticides is necessary to combat RSW more efficiently. (e) To create the forecasting models, it is necessary to record how the meteorological parameters affect RSW. (f) Vigilance is required to stop the RSW from spreading more quickly by identifying hot spots and conducting routine surveys and monitoring to that end.

References

Alagar, M., Rajamanikam, K., Chinnadurai, S., Yasmin, A. and Maheswarappa, H.P. (2020) Surveillance, assessment of infestation, biology, host range of an invasive rugose spiraling whitefly, *Aleurodicus rugioperculatus* (Martin) and status of its natural enemies in Tamil Nadu. *Journal of Entomology and Zoology Studies* 8(3), 2041–2047.

Ali, A.D., Harlow, J.L., Avery, P.B. and Kumar, V. (2015) Investigating the role of fungal entomopathogens in whitefly landscape IPM programs. *Journal of Entomological Science* 50(3), 254–261. DOI: 10.18474/JES14-35.1.

Arif, M.J., Abbas, G. and Saeed, S. (2007) Cotton in Danger, p. 4. *Dawn*, the Internet Edition. [cited Mar 24, 2007]. Available at: http://DAWN.com

Bakthavatsalam, N., Selvaraj, K., Sundararaj, R. and Sumalatha, B.V. (2021) Biosecurity and economic concerns on invasion by exotic whiteflies in India and their biocontrol strategies. *Hexapoda, Insecta Indica* 28(1 & 2), 78–87. DOI: 10.55446/hexa.2021.44.

Barbedo, J.G.A. (2014) Using digital image processing for counting whiteflies on soybean leaves. *Journal of Asia-Pacific Entomology* 17(4), 685–694. DOI: 10.1016/j.aspen.2014.06.014.

Beevi, S.P. and Lyla, K.R. (2001) Host plant mediated natural parasitism by *Encarsia* spp. (Aphelinidae: Hymenoptera) on spiralling whitefly, *Aleurodicus dispersus* Russell (Aleyrodidae: Homoptera). In: Verghese, A. and Reddy, P.P. (eds) *Proceedings, 2nd National Symposium on Integrated Pest Management in Horticultural Crops: New Molecules, Biopesticides & Environment*, Association for Advancement of Pest Management in Horticultural Ecosystems, Bangalore, India, October 17–19, 2001, pp. 208–210.

Chakravarthy, A.K., Kumar, K.P., Sridhar, V., Prasannakumar, N.R., Nitin, K.S. *et al.* (2017) Incidence, hosts and potential areas for invasion by rugose spiraling whitefly, *Aleurodicus rugioperculatus* Martin (Hemiptera: Aleyrodidae) in India. *Pest Management in Horticultural Ecosystems*, 23(1), 41–49.

Chandrika Mohan, J.A., Vinayaka Hegde, K.V., Renjith, P.B., Anjali, A.S. and Chowdappa, P. (2016) Gradient outbreak and bio-suppression of spiralling whitefly in coconut gardens in South India. *Indian Coconut Journal* 59(8), 9–12.

Chandrika Mohan, J.A., Merin Babu, P.P., Krishnakumar, S. and Vinayaka Hegde, C.P. (2017) Invasive rugose spiralling whitefly on coconut. In: *Technical Bulletin No. 117, Centenary Series 60*, Vol. 16. ICAR-CPCRI, Kasaragod, p. 117.

Chavan, S.S., Narangalkar, A.L. and Sapkal, S.D. (2022) Population dynamics of rugose spiralling whitefly, *Aleurodicus rugioperculatus* (Martin) on coconut. *The Pharma Innovation Journal* 11(3), 864–867.

Damos, P.T. and Savopoulou-Soultani, M.S. (2010) Development and statistical evaluation of models in forecasting moth phenology of major lepidopterous peach pest complex for integrated pest management programs. *Crop Protection* 29(10), 1190–1199. DOI: 10.1016/j.cropro.2010.06.022.

Dubey, A.K. and David, B.V. (2012) Collection, preservation and preparation of specimens for taxonomic study of whiteflies (Hemiptera: Aleyrodidae). In: David, B.V. (ed.) *The Whiteflies or Mealy Winged Bugs: Biology, Host Specificity and Management*. Lambert Academic Publishing, Germany, pp. 1–19.

Elango, K. and Jeyarajan Nelson, S.J. (2020) Population dynamics of exotic rugose spiralling whitefly, *Aleurodicus rugioperculatus* Martin (Hemiptera: Aleyrodidae) on coconut as influenced by weather factors and natural enemies. *Journal of Plantation Crops* 48(2), 120–125. DOI: 10.25081/jpc.2020.v48.i2.6370.

Elango, K., Sridharan, S., Saravanan, P.A. and Balakrishnan, S. (2017) Relative performance of different colour laden sticky traps on the attraction of sucking pests in pomegranate. *International Journal of Current Microbiology and Applied Sciences* 6(11), 2997–3004. DOI: 10.20546/ijcmas.2017.611.350.

Elango, K., Jeyarajan Nelson, S., Sridharan, S., Paranidharan, V. and Balakrishnan, S. (2019) Biology, distribution and host range of new invasive pest of India coconut rugose spiralling whitefly *Aleurodicus rugioperculatus* (Martin) in Tamil Nadu and the status of its natural enemies. *International Journal of Agricultural Sciences* 11(9), 8423–8426.

Elbert, A., Haas, M., Springer, B., Thielert, W. and Nauen, R. (2008) Applied aspects of neonicotinoid uses in crop protection. *Pest Management Science* 64(11), 1099–1105. DOI: 10.1002/ps.1616.

Evans, G.A. (2008). The whiteflies (Hemiptera: Aleyrodidae) of the world and their host plants and natural enemies. *USDA-APHIS*, 31, 1–2013.

Evans, G.A. (2013) The whiteflies (Hemiptera: Aleyrodidae) of the world and their host plants and natural enemies. USDA-APHIS 31, 2008.

Fakhri, M.S.A. and Jamal, K. (2012) Population dynamics of major insect-pests of cotton in relation to abiotic factors. *International Journal of Advanced Biological Research* 2(3), 500–505.

Francis, A.W., Stocks, I.C., Smith, T.R., Boughton, A.J., Mannion, C.M. *et al.* (2016) Host plants and natural enemies of rugose spiraling whitefly (Hemiptera: Aleyrodidae) in Florida. *Florida Entomologist* 99(1), 150–153. DOI: 10.1653/024.099.0134.

Ishaaya, I. and Horowitz, A.R. (1992) Novel phenoxy juvenile hormone analog (pyriproxyfen) suppresses embryogenesis and adult emergence of sweetpotato whitefly (Homoptera: Aleyrodidae). *Journal of Economic Entomology* 85(6), 2113–2117. DOI: 10.1093/jee/85.6.2113.

Josephrajkumar, A., Mohan, C., Babu, M., Prathibha, P.S., Hegde, V. *et al.* (2020) Diagnosis of invasive whitefly species co-occurring on coconut. *Current Science* 119(7), 1101–1105. DOI: 10.18520/cs/v119/i7/1101-1105.

Joshi, M. (1999) *Green Revolution and its Impacts*. A.P.H. Publishing Corporation, New Delhi, p. 10.

Kumar, V., Francis, A., Avery, P.B., McKenzie, C.L. and Osborne, L.S. (2018) Assessing compatibility of *Isaria fumosorosea* and buprofezin for mitigation of *Aleurodicus rugioperculatus* (Hemiptera: Aleyrodidae): An invasive pest in the Florida landscape. *Journal of Economic Entomology* 111(3), 1069–1079. DOI: 10.1093/jee/toy056.

Luangsa-Ard, J.J., Hywel-Jones, N.L., Manoch, L. and Samson, R.A. (2005) On the relationships of Paecilomyces sect. Isarioidea species. *Mycological Research* 109(5), 581–589. DOI: 10.1017/s0953756205002741.

Madhusudhan, L. (2015) Agriculture role on Indian economy. *Business and Economics Journal* 6(4), 176. DOI: 10.4172/2151-6219.1000176.

Mani, M., Krishnamoorthy, A. and Dinesh, M.S. (2000a) Biological control studies on the spiralling whitefly, *Aleurodicus dispersus* russell (Homoptera: Aleyrodidae). In: *Abstracts, Entomocongress 2000: Perspectives for the New Millennium*, Trivandrum, India. Association for Advancement of Entomology, November 5–8, , pp. 37–38. (abstract SI-0-40).

Mani, M., Dinesh, M.S. and Krishnamoorthy, A. (2000b) Presence of *Encarsia* spp. on spiralling whitefly *Aleurodicus dispersus* (Russell) in peninsular India. *Insect Environment* 6, 100.

Mannion, C. (2010). *Rugose Spiraling Whitefly – A New Whitefly in South Florida*. University of Florida, IFAS Extension Publication. Available at: http://trec.ifas.ufl.edu/mannion/pdfs/Rugose/spiraling/whitefly.pdf

Martin, J.H. (1987) An identification guide to common whitefly pest species of the world (Homoptera: Aleyrodidae). *Tropical Pest Management* 33, 298–322.

Martin, J.H. (2004) The whiteflies of Belize (Hemiptera: Aleyrodidae) Part 1. *Zootaxa* 681(1), 1–119. DOI: 10.11646/zootaxa.681.1.1.

Mayer, H., McLaughlin, J., Hunsberger, A., Vasquez, L., Olcyzk, T. *et al.* (2010) Common questions about the gumbo limbo spiraling whitefly (*Aleurodicus rugioperculatus*). *The Miami-Dade Cooperative Extension.*

Nakamura, S., Inoue, M., Fujimoto, H. and Kasamatsu, K. (2007) The efficacy of yellow tape formulation of pyriproxyfen against the sweet potato whitefly, *Bemisia tabaci* (Gennadius) (Homoptera: Aleyrodidae). *Journal of Asia-Pacific Entomology* 10(1), 75–79. DOI: 10.1016/S1226-8615(08)60334-9.

Palumbo, J.C., Horowitz, A.R. and Prabhaker, N. (2001) Insecticidal control and resistance management for *Bemisia tabaci*. *Crop Protection* 20(9), 739–765. DOI: 10.1016/S0261-2194(01)00117-X.

Poorani, J. and Thanigairaj, R. (2017) First report of *Encarsia dispersa* Polaszek (Hymenoptera: Aphelinidae) as a parasitoid of rugose spiralling whitefly, *Aleurodicus rugioperculatus* Martin (Hemiptera: Aleyrodidae), a recent invasive pest in India, with notes on its predators. *Journal of Biological Control* 31(1), 1–4. DOI: 10.18311/jbc/2017/16263.

Pradhan, S.K., Shylesha, A.N., Selvaraj, K. and Sumalatha, V.B. (2020) Comparative biology of invasive rugose spiralling whitefly *Aleurodicus rugioperculatus* Martin on three host plants. *Indian Journal of Entomology* 82(3), 498–503. DOI: 10.5958/0974-8172.2020.00125.X.

Raghubanshi, A.S., Raj, L.C., Gaur, J.P. and Singh, J.S. (2005) Invasive alien species and biodiversity in India. *Current Science* 88, 539–540.

Ramani, S. (2000) Fortuitous introduction of an aphelinid parasitoid of the spiralling whitefly, *Aleurodicus dispersus* Russell (Homoptera: Aleyrodidae) into the Lakshadweep Islands, with notes on host plants and other natural enemies. *Journal of Biological Control* 14, 55–60.

Rao, N.B.V.C., Roshan, D.R., Rao, G.K. and Ramanandam, G. (2018) A review on rugose spiralling white-fly, *Aleurodicus rugioperculatus* Martin (Hemiptera: Aleyrodidae) in India. *Journal of Pharmacognosy and Phytochemistry* 7(5), 948–953.

Sandeep, A., Selvaraj, K., Kalleshwaraswamy, C.M., Hanumanthaswamy, B.C. and Mallikarjuna, H.B. (2022) Field efficacy of *Isaria fumosorosea* alone and in combination with insecticides against *Aleurodicus rugioperculatus* on coconut. *Egyptian Journal of Biological Pest Control* 32(1), 1–8.

Selvaraj, K., Sundararaj, R., Venkatesan, T., Ballal, C.R., Jalali, S.K. *et al.* (2016) Potential natural enemies of the invasive rugose spiraling whitefly, *Aleurodicus rugioperculatus* Martin in India. *Journal of Biological Control* 30(4), 236–239. DOI: 10.18311/jbc/0/15598.

Selvaraj, K., Gupta, A., Venkatesan, T., Jalali, S.K., Sundararaj, R. *et al.* (2017) First record of invasive rugose spiraling whitefly *Aleurodicus rugioperculatus* Martin (Hemiptera: Aleyrodidae) along with parasitoids in Karnataka. *Journal of Biological Control* 31(2), 74–78. DOI: 10.18311/jbc/2017/16015.

Shanas, S., Job, J., Joseph, T. and Krishnan, G.A. (2016) First report of the invasive rugose spiralling whitefly, *Aleurodicus rugioperculatus* Martin (Hemiptera: Aleyrodidae) from the Old World. *Entomon* 41(4), 365–368.

Singh, N. and Gupta, N. (2016) ICT based decision support systems for Integrated Pest Management (IPM) in India: A review. *Agricultural Reviews* 37(4), 309–316. DOI: 10.18805/ag.v37i4.6461.

Skinner, M., Parker, B.L. and Kim, J.S. (2014) Role of entomopathogenic fungi in integrated pest management. In: Abrol, D.P. (ed.) *Integrated Pest Management, Current Concepts and Ecological Perspective*. Academic Press, London, pp. 169–191. DOI: 10.1016/B978-0-12-398529-3.00011-7.

Srinivasan, T., Saravanan, P.A., Josephrajkumar, A., Rajamanickam, K., Sridharan, S. *et al.* (2016) The madras. *Agricultural Journal* 103(10–12), 349–353.

Stocks, I. (2012) Rugose spiraling whitefly host plants. In: *Florida Department of Agriculture and Consumer Services*. Division of Plant Industry, p. 6.

Stocks, I. and Hodges, G. (2012) Pest alert – DACS-p-01745. In: *Florida Department of Agriculture and Consumer Services, Division of Plant Industry*. p. 6.

Sumalatha, B.V., Selvaraj, K., Poornesha, B. and Ramanujam, B. (2020) Pathogenicity of entomopatho-genic fungus *Isaria fumosorosea* on rugose spiralling whitefly *Aleurodicus rugioperculatus* and its effect on parasitoid *Encarsia guadeloupae*. *Biocontrol Science and Technology* 30(10), 1150–1161. DOI: 10.1080/09583157.2020.1797632.

Sundararaj, R. and Selvaraj, K. (2017) Invasion of rugose spiralling whitefly, *Aleurodicus rugioperculatus* Martin (Hemiptera: Aleyrodidae): A potential threat to coconut in India. *Phytoparasitica* 45(1), 71–74. DOI: 10.1007/s12600-017-0567-0.

Sushmitha, S., Sujatha, A., Emmanuel, N., Krishna, K.U. and Suneetha, D.R.S. (2020) Incidence, intensity and host dynamics of rugose spiralling whitefly, *Aleurodicus rugioperculatus* Martin (Hemiptera:

Aleyrodidae) in Andhra Pradesh. *Pest Management in Horticultural Ecosystems* 26(1), 109–120. DOI: 10.5958/0974-4541.2020.00018.1.

Taravati, S., Glenn, H. and Mannion, C. (2014). Daily flight activity of the rugose spiraling whitefly (Hemiptera: Aleyrodidae). *Florida Entomologist*, 97(4), 1842–1844.

Taravati, S., Mannion, C., Glenn, H. and Osborne, L. (2013) Natural enemies of rugose spiraling whitefly, *Aleurodicus rugioperculatus* Martin (Insecta: Hemiptera: *Aleyrodidae*) in the south Florida landscape. *ENY-870*. In: University of Florida, IFAS Extension. Available at: http://edis.ifas.ufl.edu/pdffiles/IN/IN100400.pdf

Zia, K., Fareed, M.S., Arshad, M., Hafeez, F. and Khan, R.R. (2015) Impact of abiotic factors on population fluctuation of cotton whitefly (*Bemisia tabaci*) on transgenic and nontransgenic cotton cultivars in Faisalabad. *Pakistan Entomologist* 37(2), 127–131.

13 Invasion and Management of *Thrips Parvispinus* (Karny) (Thysanoptera: Thripidae) in Tropics and Sub-tropics

Basavaraj K.[1]*, Saleemali Kannihalli[1], Abhinaya[2] and Eurekha Savadatti[3]

[1]*Department of Entomology, College of Agriculture, Gangavathi, Karnataka, India;*
[2]*Department of Entomology, College of Agriculture, Raichur, Karnataka, India;*
[3]*Department of Entomology, College of Agriculture, Dharwad, Karnataka, India*

Abstract

Thrips parvispinus (Karny), a member of the family Thripidae, has emerged as a significant invasive pest across tropical and subtropical agroecosystems. Native to Southeast Asia, this polyphagous species has rapidly expanded its geographic range, affecting horticultural and ornamental crops in South and Southeast Asia, Africa, Europe and the Americas. In India and other tropical countries, severe infestations have been reported on economically important crops such as chilli (*Capsicum annuum*), papaya (*Carica papaya*), eggplant (*Solanum melongena*) and various ornamental plants. Its damage is primarily caused by direct feeding on plant tissues, resulting in leaf bronzing, flower and fruit deformation and significant yield losses. The invasion and persistence of *T. parvispinus* are closely linked to factors such as climate change, intensive monocropping and the indiscriminate use of insecticides. Effective management of this pest requires an integrated approach combining regular monitoring, cultural practices, use of selective insecticides and biological control agents. This chapter highlights the invasive spread, ecological impact and sustainable management strategies for *T. parvispinus* in tropical and subtropical agriculture, with a focus on minimizing economic losses and preserving ecological balance.

Keywords: *Thrips parvispinus*, invasive pest, polyphagous, management, integrated approach

13.1 Introduction

Thrips constitute a significant group of sap-sucking pests that cause considerable economic damage, both through direct feeding and by acting as vectors of destructive plant viruses in a variety of horticultural crops (Mound and Teulon, 1995; Riley *et al.*, 2011). Outbreaks of thrips have been increasingly reported in various regions, often attributed to shifts in crop production practices, indiscriminate pesticide use and the effects of climate change (Reitz *et al.*, 2020).

The genus *Thrips* is one of the largest genera of the insect order Thysanoptera in the family Thripidae, with 295 species worldwide, of which 44 species are reported from India (Rachana and Varatharajan, 2017; Rachana *et al.*, 2018). *Thrips parvispinus* (Karny, 1922), a member of the *Thrips orientalis* group (Mound, 2005), belonging to the family Thripidae,

*Corresponding author: kadanavar@gmail.com

DOI: 10.1079/9781836991977.0013

has emerged as a serious pest in tropical and subtropical agroecosystems. It was described in 1922 from Southeast Asia. Initially considered a minor species, its pest status gained attention in the late 20th century when it began causing damage to horticultural crops in Indonesia, Thailand and Malaysia (Waterhouse, 1993; Murai *et al.*, 2010).

13.2 Distribution and Host Range

Thrips parvispinus was first reported outside its native range in Europe in the late 1990s, with greenhouse infestations in Greece (Mound and Collins, 2000) and subsequent detections in Spain (Lacasa *et al.*, 2019) and the Netherlands (Anonymous, 2019). Over the following decades, *T. parvispinus* rapidly expanded its geographic distribution, with established populations reported from Mauritius (Mound, 2010), Tanzania and Uganda (Moritz *et al.*, 2013), Hawaii (Sugano *et al.*, 2013), China, northern Australia, Hawaii and Ghana (Collins, 2000; Palanisamy *et al.*, 2023; EPPO, 2025). The increasing distribution of this species emphasizes its quarantine significance (Mound and Collins, 2000).

In India, *T. parvispinus* was first recorded on papaya (*Carica papaya*) in 2015 from Karnataka, where it was identified using both morphological traits and DNA barcoding methods (Tyagi *et al.*, 2015). Reports of *T. parvispinus* infestation in India have progressively expanded across various host plants and regions. Initial records include its presence on *Dahlia rosea* (Rachana *et al.*, 2018), *Brugmansia* sp. (Roselin *et al.*, 2021), cotton (*Gossypium* spp.) (Amutha and Rachana, 2023) and cowpea (*Vigna unguiculata*) (Nagdev *et al.*, 2022). These early detections were soon followed by severe outbreaks on chilli (*Capsicum annuum*), particularly in major chilli-producing states. The first significant infestation was reported from Karnataka (Basavaraj *et al.*, 2022), with subsequent reports emerging from Telangana and Andhra Pradesh (Hulagappa *et al.*, 2022; Veeranna *et al.*, 2022) and later from Gujarat (Patel *et al.*, 2022).

Thrips parvispinus is a highly polyphagous pest, infesting a broad range of fruits, vegetables and ornamental plants (Table 13.1). It was first recognized as a serious agricultural pest when it caused substantial damage to chilli pepper crops in Asia (Talekar, 1991). Since then, its host range has broadened significantly, now encompassing a variety of economically important fruit, vegetable and ornamental crops. These include coffee, papaya (*Carica papaya*), chilli and sweet pepper (*Capsicum* spp.), potato (*Solanum tuberosum*), tobacco (*Nicotiana tabacum*), green beans, *Vigna* sp., watermelon, cucumber, shallots, strawberries and a number of ornamental plants such as *Gardenia* sp., *Gerbera*, *Ficus benjamina*, *Schefflera* and *Dipladenia* (Moritz *et al.*, 2013; Hutasoit *et al.*, 2017; Lacasa *et al.*, 2019).

In Europe, the pest has been primarily found in greenhouses, where it infests ornamental species and citrus plants (Lacasa *et al.*, 2019). It was reported on *Gardenia* in Greece, papaya in Hawaii and vegetable crops including capsicum, eggplant (*Solanum melongena*) and potato in various countries (Murai *et al.*, 2010). In regions where the pest is well established, including Southeast Asia, the most severely affected crops include papaya, chilli and sweet peppers, eggplants, potatoes, beans, shallots, *Crotalaria*, cucurbits and coffee (Hutasoit *et al.*, 2017).

13.3 Factors Regulating the Invasion of *Thrips parvispinus*

The successful invasion and rapid spread of *Thrips parvispinus* in tropical and subtropical regions can be attributed to a combination of biological, ecological and anthropogenic factors. These include high fecundity, short generation time, cryptic feeding behaviour and ability to reproduce in the absence of males (parthenogenesis) (Mound, 2005; Tyagi *et al.*, 2015). Its small size and cryptic lifestyle make early detection difficult, especially in ornamental nurseries and polyhouses. This is compounded by the lack of awareness among farmers and plant quarantine officers in newly invaded regions.

The broad host range of *T. parvispinus* is another factor contributing to its invasion success. It feeds on and reproduces in a variety of economically important crops including chilli, papaya, pepper, beans, cucumber, tobacco,

Table 13.1. Host range of *Thrips parvispinus*. (Author's own table)

Host plant	Scientific name	Family	Reference
Pepper	*Piper nigrum*	Piperaceae	Johari *et al.* (2014); Hutasoit *et al.* (2017)
Anthurium	*Anthurium* sp.	Araceae	Johari *et al.* (2014)
Papaya	*Carica papaya*	Caricaceae	Tyagi *et al.* (2015)
Coffee	*Coffea* sp.	Rubiaceae	
Potato	*Solanum tuberosum*	Solanaceae	Hutasoit *et al.* (2017)
Tobacco	*Nicotiana tabacum*	Solanaceae	
Brinjal	*Solanum melongena*	Solanaceae	Hutasoit *et al.* (2017); Sridhar *et al.* (2021)
Chilli	*Capsicum annuum*	Solanaceae	Nagaraju *et al.* (2021); Sridhar *et al.* (2021); Basavaraj *et al.* (2022)
Dahlia	*Dahlia rosea*	Asteraceae	Rachana *et al.* (2018)
Chrysanths	*Chrysanthemum* sp.	Asteraceae	Sridhar *et al.* (2021)
Cowpea	*Vigna unguiculata*	Fabaceae	Hutasoit *et al.* (2017); Nagdev *et al.* (2022)
Mungbean	*Vigna radiata*	Fabaceae	Gothi and Panickar (2024)
Ridge gourd	*Luffa acutangula*	Cucurbitaceae	Fening *et al.* (2022)
Coriander	*Coriandrum sativum*	Apiaceae	Verghese *et al.* (2022)
Guava	*Psidium guajava*	Myrtaceae	Ranjith *et al.* (2022)
Onion	*Allium cepa*	Amaryllidaceae	Saini *et al.* (2023)
Cotton	*Gossypium* sp.	Malvaceae	Amutha and Rachana (2023)
Mulberry	*Morus* sp.	Moraceae	Kumar *et al.* (2024)

eggplant and several ornamental plants (Mound and Collins, 2000; Rachana *et al.*, 2018; Nagdev *et al.*, 2022). This polyphagy allows the insect to persist year-round by shifting between hosts depending on seasonal availability. Additionally, the pest is often misidentified due to morphological similarities with other common thrips species such as *Thrips palmi*.

Environmental adaptability, particularly to temperature and humidity, plays a significant role. The species thrives well in warm and humid environments, typical of tropical and subtropical zones, making large swathes of agricultural land vulnerable to colonization. Studies have shown that higher temperatures accelerate development and fecundity, with intrinsic rates of population increase rising from 0.18 at 20°C to 0.37 at 30°C (Murai *et al.*, 2010). Its ability to survive under greenhouse conditions also supports its persistence and dispersal in controlled environments.

13.4 Development and Biology of *Thrips parvispinus*

Thrips parvispinus completes its development through five distinct stages: egg, two nymphal instars, prepupa, pupa and adult (Mound and Collins, 2000; Hutasoit *et al.*, 2017). Under optimal conditions, especially in tropical and subtropical climates, the entire life cycle spans approximately 13–15 days, although variations occur depending on temperature, humidity, host quality and photoperiod (Lewis, 1973; Murai *et al.*, 2010; Hutasoit *et al.*, 2017). The durations for the egg stage, two nymphal instars, prepupal and pupal phases were recorded as 5.74, 2.16, 3.76, 1.83 and 4.17 days, respectively (Priyanka *et al.*, 2024). The pre-oviposition period lasted 1.11 days, with the complete life cycle completing in approximately 13.68 days (Maharijaya *et al.*, 2011; Hutasoit *et al.*, 2017).

The female inserts kidney-shaped eggs into the underside of leaves, flower buds, or young fruits using her saw-like ovipositor, typically in light-coloured tissues where eggs are more visible (Mound and Collins, 2000; Ahmed *et al.*, 2023). After 4–5 days, eggs hatch into larvae, which then undergo two moults over another 4–5 days, feeding voraciously on foliage and flowers (Ahmed *et al.*, 2023). These feeding stages play a critical role in rapid population build-up. Adults are small (1–1.2 mm), dark brown-to-black, and exhibit primarily sexual reproduction, although parthenogenesis may occur. Females typically live 8–9 days and lay approximately 15 eggs, while males live about 6 days (Hutasoit *et al.*, 2017; Ahmed *et al.*, 2023). However, temperature-dependent studies reveal that at higher temperatures (25–30°C), female fecundity increases dramatically to an average of 50–69 eggs, with generation times reduced to 18.8–24.8 days and intrinsic rates of increase reaching up to 0.37 fraction per day (Murai *et al.*, 2010).

13.5 Feeding Behaviour, Nature of Damage and Symptoms

The damage is primarily due to its rasping-sucking mode of feeding, which is carried out by both nymphs and adults. The pest targets tender plant tissues, including young leaves, floral parts and developing fruits, resulting in a variety of physiological and morphological abnormalities that impact plant growth and productivity. Early symptoms of infestation include deep punctures and fine abrasions on the underside of leaves, which eventually turn reddish-brown. The upper surface of affected leaves typically shows yellowing. Distortion of leaf lamina, necrotic patches and yellow streaks are also common, especially under heavy infestations (Sridhar *et al.*, 2021). Newly emerging leaves may appear blighted or dry out entirely, and flower petals often develop brownish streaks due to scraping by thrips (Fig. 13.1) (Mound and Collins, 2000). These symptoms are frequently accompanied by significant flower drop, reduced fruit set, and malformed fruit development, all of which directly reduce crop yields (Sireesha *et al.*, 2021).

In crops such as chilli, capsicum and papaya, *T. parvispinus* causes particular concern. The pest has been reported to infest the nectariferous regions of flowers, where adults of both male and female congregate and feed. Its preference for white, fragrant flowers is notable, as feeding in floral parts can lead to wilting and aborted fruit formation. Infestation in bell peppers, for instance, results in malformed, button-shaped fruits with rough, scarred surfaces (Maharijaya *et al.*, 2011). These symptoms not only affect the aesthetic and commercial value of produce but also expose plants to secondary infections. In papaya, thrips feeding creates entry points for saprophytic fungi such as *Cladosporium*, compounding the damage (Lim, 1989).

Feeding behaviour studies show that *T. parvispinus* larvae primarily remain on the underside of leaves, while adults prefer both the flowers and leaf undersides. Hutasoit *et al.* (2019) reported that *T. parvispinus* constituted 71% of thrips found on flowers and 56% on leaves of chilli and cayenne pepper plants. The observable damage may not fully reflect the pest's dietary preferences. While the physical symptoms suggest a primarily phytophagous feeding habit, studies have shown that many thrips species consume pollen, which improves their reproductive parameters. The inclusion of pollen in the diet has been shown to stimulate oviposition, reduce larval development time and increase female fecundity (Kirk, 1984, 1997; Zhi *et al.*, 2005; Riley *et al.*, 2007). Some thrips are also facultative predators, which broadens their ecological niche and contributes to their survival and invasive potential. In addition to vegetable crops, *T. parvispinus* also infests ornamental plants such as *Dahlia rosea*, chrysanthemums, gardenias, dipladenias and *Ficus* sp., where feeding results in severe aesthetic losses (Rachana *et al.*, 2018; Lacasa *et al.*, 2019).

13.6 Outbreak of *T. parvispinus* and its Economic Implications

In 2021, a significant outbreak of *Thrips parvispinus* caused devastating damage to chilli crops across three major Indian states, viz. Andhra Pradesh, Telangana and Karnataka with reported crop losses ranging from 70–100%

Fig. 13.1. Damage symptoms on chilli by *Thrips parvispinus*. (Author's own image)

in severely affected areas. The first evidence of thrips infestation on chilli flowers was recorded in January 2021, in Chilakaluripeta and Pratipadu mandals located in Guntur district, Andhra Pradesh. Following this, the pest rapidly spread throughout all chilli-producing regions in the state (Sireesha *et al.*, 2021). The pest was later found to have extended its reach into eastern Uttar Pradesh, with confirmed cases in Varanasi and Mirzapur districts (Sethy *et al.*, 2022). Its widespread presence also became evident in chilli-growing districts of Telangana, where infestations primarily occurred during the flowering stage, leading to significant impacts on fruit development and overall yield. A high level of infestation (8–12 thrips/ flower) was observed on chilli crops in Madhya Pradesh (Khargone, Khandwa, Satna, Narasinghpur, Dhar) and on capsicum hybrids in Haryana (Ambala, Panchkula and Panipat) grown under greenhouse (Timmanna *et al.*, 2023). Chatterjee *et al.* (2024) recorded infestation of *T. parvispinus* (1–7 thrips/flower) with damage ranging from 10–60% across different locations in Alipurduar district of West Bengal.

Apart from agricultural crops, the pest was found feeding on ornamental species such as Anthurium, Chrysanthemum, Dahlia, Dipladenia, Gardenia, and Ficus (NPPO, 2019). Surrounding vegetation and weed flora, including Parthenium, *Amaranthus* sp., *Axonopus* sp., *Ageratum* sp., *Alternanthera* sp. and *Thunbergia* sp., as well as neem and pongamia trees, were also found to be infested (Nagaraju *et al.*, 2021).

During its early stages of invasion, *T. parvispinus* entered a lag phase where its population remained relatively steady. However, over the subsequent four years, a dramatic increase in population was observed (Rachana *et al.*, 2022). This surge in population was attributed to its remarkable ability to adapt to diverse hosts and environmental conditions, along with a notable tendency to expand its geographic range across India. This adaptability has enabled *T. parvispinus* to replace the well-established chilli thrips species, *Scirtothrips dorsalis*, in many agricultural settings, particularly in Andhra Pradesh, Karnataka and Telangana (Sridhar *et al.*, 2021). Both species have been observed to coexist on chilli and capsicum crops, often alongside other thrips species such as *Thrips palmi*, *Thrips tabaci*, *Frankliniella schultzei* and *Megaleurothrips abdominalis* (Palanisamy *et al.*, 2023).

From an economic standpoint, *T. parvispinus* has emerged as a major threat to Indian agriculture, especially in chilli cultivation. The infestation causes flower drop, malformation and premature fruit shedding, which leads to drastic reductions in yield. The pest has emerged as a major threat to chilli cultivation in southern India. According to a joint survey by the Directorate of Plant Protection, Quarantine and Storage (DPPQ&S), and the State Agricultural Universities and Departments of Telangana and Andhra Pradesh, over 0.4 million hectare of chilli crops were infested, with yield losses ranging from 10–30% (DPPQ&S, 2021; Janyala, 2021). Particularly high levels of infestation of 10–20 thrips per flower were recorded in the districts of Warangal, Khammam and Guntur in Andhra Pradesh. Farmers in Andhra Pradesh reported financial losses as high as Rs. 0.4 million hectare per acre during the 2021 season alone due to the extensive damage caused by the pest (Anonymous, 2021a). In Indonesia, field studies highlighted that *T. parvispinus* causes yield losses between 23–60%, with some estimates noting a broader range of 10–46%, depending on crop and regional conditions (Johari *et al.*, 2014; Tasmin *et al.*, 2018). In India, similar severity has been documented, with chilli fields in Andhra Pradesh and Telangana recording infestation levels ranging from 40–80% (Anonymous, 2022). It also caused drastic yield losses in crops such as mungbean and mulberry, indicating its expanding impact across multiple cropping systems (Gothi and Panickar, 2024).

13.7 Management of *Thrips parvispinus*

13.7.1 Cultural methods

Early removal and destruction of heavily infested plants can significantly limit the pest's spread (Sridhar *et al.*, 2021). Planting should only be done using healthy and pest-free seedlings, ensuring the initial planting material does not introduce the pest into new areas. Another essential practice is routine monitoring and inspection through field surveys in chilli-growing regions to detect early infestations and track pest dynamics across new zones. Avoiding the excessive application of nitrogenous fertilizers and instead adhering to a balanced and recommended fertilizer regime has been shown to reduce pest pressure (Sireesha *et al.*, 2021).

To maintain plant health and support resistance, a comprehensive nutrient management strategy is advised. This includes applying organic fertilizers such as FYM at 10 tonnes/acre, neem cake at 200 kg/acre, vermicompost at 2 tonnes/acre and biofertilizers such as *Azospirillum* and phosphate-solubilizing bacteria at 2 kg/acre each, split over five applications during crop growth. Excessive nitrogen input can make plants more susceptible to thrips, while balanced fertilization promotes natural defence (Sireesha *et al.*, 2021). In terms of varietal resistance, several *Capsicum* accessions have been identified as resistant to *T. parvispinus*. These include *Capsicum annuum* AC 1979, *C. annuum* Bisbas, *C. annuum* CM 331, *C. annuum* Keystone Resistant Giant, *C. baccatum* no. 1553 and *C. baccatum* Aji Blanco Christal (Maharijaya *et al.*, 2011). Likewise, Mayang Ratih, a chrysanthemum genotype, was found to inherit resistance from one of its parents and is considered promising for resistance breeding (Musalamah *et al.*, 2021).

Strategic agronomic practices such as chilli cultivation using plastic mulch and without intercropping, paired with judicious pesticide use, have resulted in lower *T. parvispinus* populations during the vegetative phase (Haerul *et al.*, 2020). In an interesting ecological finding, coriander flowers were observed to attract more *T. parvispinus* individuals than nearby chilli plants, suggesting that coriander could serve effectively as a trap crop (Prasannakumar *et al.*, 2021). Reducing weed hosts and avoiding proximity to infested fields can also contribute to pest suppression.

13.7.2 Physical and mechanical methods

Studies conducted by Murai *et al.* (2010) revealed that *T. parvispinus* showed a stronger attraction to white traps compared with blue or yellow ones. However, in contrast, Sireesha *et al.* (2021) reported that blue and yellow sticky traps were more effective in attracting adult thrips in chilli fields, suggesting variability depending on field conditions or pest population dynamics. For mass trapping immediately after transplanting, it is recommended to install 25–35 blue sticky traps per acre, a method commonly adopted in chilli cultivation (Anonymous, 2021b). Meanwhile, for general monitoring and adult suppression, a deployment rate of 30 sticky traps per acre has proven effective in reducing thrips populations (Sireesha *et al.*, 2021). Exposure to a 60% carbon dioxide (CO_2) atmosphere at 30°C resulted in 100% mortality of several thrips species including *Frankliniella occidentalis* (Pergande), *Frankliniella intonsa* (Trybom), *Thrips tabaci* Lindeman, *Thrips palmi* Karny and *Thrips parvispinus* Karny (Seki and Murai, 2012).

13.7.3 Biological control

Biological control is increasingly being recognized as a key component of sustainable management for *Thrips parvispinus*, especially under integrated pest management (IPM) frameworks. Several natural enemies and microbial agents have demonstrated potential in both open and protected cultivation systems. Studies conducted in 2003 and 2005 in Indonesia focused on identifying natural enemies and setting threshold levels for thrips control in sweet pepper grown under tropical, protected conditions. The research identified two species of ladybird beetles, viz. *Menochilus sexmaculatus* and *Coccinella transversalis* as well as the entomopathogenic fungus *Lecanicillium lecanii* as effective biological control agents. Notably, the use of *M. sexmaculatus* and *L. lecanii* suppressed plant damage caused by thrips and led to sweet pepper yields comparable to those achieved through twice-weekly insecticide applications. Additionally, it also reduced insecticide spray frequency by about 90%, while keeping plant damage below 10% (Prabaningrum *et al.*, 2008).

In chilli crops, spraying biopesticides such as NBAIR-PFDWD (*Pseudomonas fluorescens*) and NBAIR-BATP (*Bacillus albus*) at 20 g/L on the flowers and fruits of chillies have shown effectiveness in controlling *T. parvispinus* populations (Anonymous, 2021b). The fungal agent *Lecanicillium lecanii* continues to be a reliable tool as a biocontrol agent, with field evidence supporting its role in suppressing thrips populations in multiple crops (Prabaningrum *et al.*, 2008).

13.7.4 Botanicals

In areas with heavy *Thrips parvispinus* infestation, the application of neem oil, pongamia oil, or even simple soap solution has been advised as an effective first-line defence (Sridhar *et al.*, 2021). As part of broader integrated pest management practices, neem cake can be applied to standing crops at a rate of 200 kg per acre to enhance soil health and suppress pest populations (Anonymous, 2021b). Several botanical extracts and oils have shown notable efficacy against *T. parvispinus* during the vegetative stage of crop growth. Field studies indicated that Indonesian mahogany (*Toona sureni*) at 3.0%, fish poison bean (*Tephrosia vogelii*) at 2.5 and 3.0%, and eucalyptus oil at 2.0% all delivered over 30% effectiveness, significantly lowering *T. parvispinus* populations up to 75 days after planting (DAP) and resulting in the highest yields of marketable chrysanthemum flowers (Rahardjo *et al.*, 2022).

13.7.5 Chemical control

Effective chemical control of *Thrips parvispinus* remains a key element in integrated pest management, especially when paired with cultural and biological methods. To delay the development of insecticide resistance, Sugano *et al.* (2013) recommended applying insecticides with different modes of action in rotation. Targeting the flowers and growing shoots of papaya, where the thrips predominantly feed and reside, is essential. Since young papaya leaves and fruits have waxy surfaces, the use of a surfactant such as Latron B-1956 is advised to enhance leaf wetting and spray coverage (Sugano *et al.*, 2013).

Several chemical options have been proposed on an ad hoc basis to combat *T. parvispinus* outbreaks. These include fipronil 80 WG at 40 g/acre, fipronil 40 + imidacloprid 40 at 40 g/acre, cyantraniliprole 10 at 240 ml/acre, acetamiprid 20 SP at 40 g/acre and spirotetramat 150 OD at 160 ml/acre, applied in rotation to reduce resistance buildup (Sireesha *et al.*, 2021). Anitha Kumari *et al.* (2021) recommended weekly sequential sprays using fipronil 80 WG at 0.2 g/L, cyantraniliprole at 1.25 ml/L, acetamiprid at 0.2 g/L or spinosad at 0.3 ml/L for effective suppression of the pest in chilli fields. Other promising chemical options include tolfenpyrad 15 EC at 1 ml/L and spinetoram 11.7 SC at 1 ml/L, which were observed to reduce thrips counts to 2.39 and 2.67 thrips per flower, respectively (Anonymous, 2022). However, excessive reliance on insecticides may inadvertently lead to thrips resurgence (Sireesha *et al.*, 2021).

13.8 Integrated Pest Management (IPM) Approaches for Thrips in Chilli Crop

Integrated Pest Management (IPM) strategies for managing thrips should begin with deep summer ploughing to eliminate resting pupae (Anonymous, 2022). It is advisable to initiate early planting and avoid staggered sowing, as well as consider cultivating resistant or early-maturing varieties, if available, to escape peak thrips activity. Regularly monitor thrips infestation not only in chilli crops but also on other host plants nearby. Severely infested plants should be collected and destroyed, and weeds such as *Parthenium hysterophorus*, *Cleome viscosa*, *Prosopis* sp., *Lantana camara*, *Calotropis* sp., *Tecoma* sp., *Abutilon* sp. and wild *Solanum* sp. present around field bunds should be uprooted, as they serve as alternate or off-season hosts for thrips. The plant debris from uprooted weeds and infested crops must be either buried or burnt to prevent further spread of the pest. Infested apical or top shoots should be nipped and destroyed.

Seeds should be treated with systemic insecticides approved under label claims to provide early-stage protection. Application of well-decomposed farmyard manure (FYM) or compost at 1 ton per acre, enriched with *Metarhizium anisopliae* or *Pseudomonas fluorescens* at 2 kg per ton, alongside the recommended 10–12 tons per acre of FYM, can enhance soil health and resistance. Incorporating 200 kg of neem cake and 500 kg of vermicompost per acre further strengthens the plant's defences. Balanced fertilization with adequate potash, along with nitrogen and phosphorus, helps induce resistance against thrips.

Mulching with silver-coloured polythene sheets of 25–30 micron thickness can suppress thrips pupation in the soil. To minimize pest movement, establishing border crops with 2–3 rows of tall-growing plants such as sorghum, maize, bajra or fodder grasses is beneficial. Intercropping chilli with maize or sorghum and cowpea in a 10:3:1 ratio creates a barrier and supports the multiplication of natural enemies, enhancing biological control. Regular inter-cultivation practices such as earthing up and soil raking help disrupt the soil-dwelling stages of thrips. Severely infested plants should be removed and destroyed through burying or burning. Blue sticky traps (25–30 per acre) set at canopy height are useful for both monitoring and mass trapping.

Using sprinkler irrigation instead of flood irrigation is recommended, as the water spray helps disrupt thrips development. Conservation of natural enemies is crucial and can be achieved by minimizing the use of chemical pesticides. Instead, botanical pesticides such as 5% neem seed kernel extract (NSKE), 3% neem oil at 2 ml/L, pongamia oil at 3 ml/L or *Vitex negundo* extract at 50–80 ml/L can be used. Microbial-based insecticides such as *Beauveria bassiana* at 4g or ml/L (spore load of 1×10^8 cfu/g or ml), *Pseudomonas fluorescens* – NBAIRPFDWD at 20 g/L or *Bacillus albus* – NBAIR-BATP at 20 g/L should be sprayed thoroughly on the plants. As a final measure, only need-based and judicious use of label claim insecticides should be considered (Table 13.2).

Table 13.2. CIB & RC approved registered insecticides for thrips in chilli (Anonymous, 2022).

Insecticides	Dose/ha in required water	Waiting period (in days)
Acephate 95% SG	790g in 500 L	7
Acetamiprid 20%	50–100g in 500–600 L	3
Carbofuran 3% CG	33.30 kg	–
Cyantraniliprole 10.26%	600g in 500 L	3
Dimethoate 30% EC	600 ml in 500–1000 L	–
Emamectin benzoate 5% SG	200g in 500 L	3
Emamectinbenzoate1.90% EC	375 ml in 500 L	14
Ethion 50% EC	1.50–2.00 L in 500–1000 L	5
Fenpropathrin 30% EC	250–340 ml in 750–1000 L	7
Fipronil 05% SC	800–1000g in 500 L	7
Fipronil 80% WG	50.00–62.50g in 500 L	5
Imidacloprid 70% WS	1.00–1.50 kg	–
Imidacloprid 30.50% m/m SC	125–150g in 500 L	5
Imidacloprid 17.80% SL	125–250 ml in 500–700 L	40
Lambda-cyhalothrin 4.90% CS	500ml in 500 L	5
Lambda-cyhalothrin 5% EC	300 ml in 400–600 L	5
Methomyl 40% SP	0.75–1.12 kg in 500–1000 L	5–6
Oxydemeton-methyl 25% EC	1 L in 500–1000 L	–
Spinosad 45% SC	160g in 500 L	
Thiacloprid 21.70% SC	225–300g in 500 L	5
Thiamethoxam 30% FS	Used as seed dresser	
Tolfenpyrad 15% EC	1 L in 500 L	7
Diafenthiuron 47% + Bifenthrin 9.40% w/w SC	625 ml in 500 L	7
Emamectin Benzoate 1.50% + Fipronil 3.50% SC	500–750g in 500 L	03
Emamectin benzoate 5% w/w + Lufenuron 40% w/w WG	60g in 500 L	03
Flubendiamide 19.92% + Thiacloprid 19.92 % w/w SC	200–250 ml in 500 L	05
Fipronil 7% + Hexythiazox 2% w/w SC	1 L in 500 L	07
Hexythiazox 3.5% + Diafenthiuron 42% WDG	650g in 500 L	07
Indoxacarb 14.5% + Acetamiprid 7.7% w/w SC	825–875 ml in 500 L	05
Profenofos 40% + Fenpyroximate 2.5% w/w EC	1 L in 500 L	07

13.9 Conclusion

Thrips parvispinus has rapidly evolved from a relatively obscure species into a formidable invasive pest causing widespread economic losses across a wide range of horticultural and ornamental crops, particularly in tropical and subtropical regions. Its polyphagous nature, high reproductive potential, adaptability to varied climatic conditions, and ability to thrive in both open fields and protected environments have contributed to its aggressive spread and pest status. In India, its impact has been especially severe in chilli cultivation, with infestations leading to substantial yield losses, reduced fruit quality and increased input costs for farmers. Despite the availability of multiple management strategies – including cultural practices, varietal resistance, botanicals, biological agents and selective chemical options – effective control of *T. parvispinus* remains challenging due to its rapid life cycle and resistance development. To manage this pest sustainably, an integrated pest management (IPM) approach is essential. This includes regular monitoring, use of resistant cultivars, conservation of natural enemies and the judicious application of insecticides with diverse modes of action. Greater awareness among farmers, combined with robust quarantine measures and coordinated research efforts, is vital to curb the spread of *T. parvispinus* and safeguard the productivity of affected cropping systems.

References

Ahmed, M.Z., Revynthi, A.M., McKenzie, C.L. and Osborne, L.S. (2023) *Thrips parvispinus* (Karny), an emerging invasive regulated pest in the United States. Available at: https://mrec.ifas.ufl.edu/lsolab/thrips/thrips-parvispinus/

Amutha, M. and Rachana, R.R. (2023) A new host record for the invasive thrips *Thrips parvispinus* Karny from India. *Indian Journal of Entomology* 1–3. DOI: 10.55446/IJE.2021.354.

Anitha Kumari, D., Bhasker, K. and Suresh, V. (2021) A new invasive chilli thrips (*Thrips parvispinus*) in Telangana State. *Insect Environment* 24(4), 520–522.

Anonymous (2019) *Thrips Parvispinus*, QS. Ent. *The Netherlands Plant Protection Organization* 1.

Anonymous (2021a) Thrips infestation threatens chilli production in 12,000 acres in Guntur district. *The New Indian Express*, 16 November, 7.

Anonymous (2021b) Pest Alert: Invasive thrips, *Thrips parvispinus* (Karny) threatening chilli cultivation in India. ICAR-NBAIR.

Anonymous (2022) Monitoring, diagnosis and management of South East Asian thrips, *Thrips parvispinus* in chilli. Technical booklet IPM, DPPQ&S, 1-26.

Basavaraj, K., Sreenivas, A.G., Prasad, P.B. and Rachana, R.R. (2022) First report of invasive thrips, *Thrips parvispinus* (Karny) (Thysanoptera: Thripidae) infesting chilli, *Capsicum annuum* L. in Kalaburagi, Karnataka, India. *Journal of Experimental Zoology* 25, 191–194. Available at: https://connectjournals.com/03895.2022.25.191

Chatterjee, M., Biswas, A., Chakraborty, D., Patra, B. and Sahoo, S.K. (2024) First report of invasive thrips, *Thrips parvispinus* (Karny)(Thysanoptera: Thripidae) in chilli from Alipurduar District of West Bengal, India. *Journal of Scientific Research and Reports* 30(5), 618–624. DOI: 10.9734/JSRR/2024/v30i51980.

Collins, L.A.M.D.W. (2000) A South East Asian pest species newly recorded from Europe: *Thrips parvispinus* (Thysanoptera: Thripidae), its confused identity and potential quarantine significance. *European Journal of Entomology* 97, 197–200.

DPPQ&S (Directorate of Plant Protection, Quarantine, and Storage) (2021) In: *Joint Survey Report on Chilli Flower Thrips Conducted in Telangana State*. Directorate of Plant Protection Quarantine and Storage, Faridabad, India.

EPPO (European and Mediterranean Plant Protection Organization) (2025) *EPPO Global Database: Thrips parvispinus* (THRIPV). Available at: https://gd.eppo.int

Fening, K.O., Amouzou, K., Hevi, W., Forchibe, E.E., Billah, M.K. *et al.* (2022) First report and population dynamics of the tobacco thrips, *Thrips parvispinus* (Karny) (Thysanoptera: Thripidae) on ridged gourd, *Luffa acutangula* (L.) Roxy in selected export fields in Southern Ghana. *Journal of Agriculture and Rural Development in the Tropics and Subtropics* 123(2), 235–245. DOI: 10.17170/kobra-202212057193.

Gothi, H.R. and Panickar, B.K. (2024) First report of *Thrips parvispinus* Karny infestation on mungbean in Gujarat, India. *Journal of Scientific Research and Reports* 30(6), 833–837. DOI: 10.9734/jsrr/2024/v30i62100.

Haerul, A.N., Nasruddin, A. and Gassa, A. (2020) The role of planting patterns to control thrips from red chili pepper plants. *International Journal of Recent Technology and Engineering* 8(5), 674–679. DOI: 10.35940/ijrte.D4419.018520.

Hulagappa, T., Baradevanal, G., Surpur, S., Raghavendra, D., Doddachowdappa, S. *et al*. (2022) Diagnosis and potential invasion risk of *Thrips parvispinus* under current and future climate change scenarios. *PeerJ* 10, e13868. DOI: 10.7717/peerj.13868.

Hutasoit, R.T., Triwidodo, H. and Anwar, R. (2017) Biology and demographic statistics of *Thrips parvispinus* Karny (Thysanoptera: Thripidae) in chili pepper (*Capsicum annuum* Linnaeus). *Indonesian Journal of Entomology* 14(3), 107–116. DOI: 10.5994/jei.14.3.107.

Hutasoit, R.T., Triwidodo, H. and Anwar, R. (2019) The abundance and diversity of thrips (Thysanoptera: Thripidae) on chili (*Capsicum annuum* L.) and cayenne (*Capsicum frutescens* L.) in Bogor. *Jurnal Hama Dan Penyakit Tumbuhan* 19(1), 33–41. DOI: 10.23960/j.hptt.11934-41.

Janyala, S. (2021) Insect hits 9 lakh acres of chilli crop in South India; experts point to use of pesticide. (accessed 20 June 2025).

Johari, A., Herlinda, S., Pujiastuti, Y., Irsan, C. and Sartiami, D. (2014) Morphological and genetic variation of *Thrips parvispinus* (Thysanoptera: Thripidae) in chili plantation (*Capsicum annuum* L.) in the lowland and highland of Jambi Province, Indonesia. *American Journal of Bioscience* 2, 17–21. DOI: 10.11648/j.ajbio.s.2014020601.14.

Karny H. (1922). Thysanoptera from Siam and Indo-China. *Journal of the Siam Society* 16(2), 91-153.

Kirk, W.D.J. (1984) Pollen-feeding in thrips (Insecta: Thysanoptera). *Journal of Zoology* (London) 204, 107–118. DOI: 10.1111/j.1469-7998.1984.tb02364.x.

Kirk, W.D.J. (1997) Feeding. In: Lewis, T. (ed.) *Thrips as Crop Pests*. CAB International, Wallingford, pp. 119–174.

Kumar, P.S., Rachana, R.R., Amarendra, B. and Sheela, N. (2024) Transient composition of the thrips species (Thysanoptera: Thripidae) infesting mulberry in southern India, first report of two, including the dominating invasive pest *Thrips parvispinus*. *Journal of Integrated Pest Management* 15(1), 32. DOI: 10.1093/jipm/pmae024.

Lacasa, A., Lorca, M., Martinez, M.C., Bielza, P. and Guirao, P. (2019) *Thrips parvispinus* (Karny 1922), un nuevo trips en cultivos de plantas ornamentales. *Phytoma España* 311, 62–69.

Lewis, T. (1973) *Thrips, Their Biology, Ecology and Economic Importance*. Academic Press Inc, London, pp. xv + 349.

Lim, W.H. (1989) Bunchy and malformed top of papaya cv. Eksotika caused by *Thrips parvispinus* and *Cladosporium oxysporum*. *MARDI Research Bulletin-Journal* 17(2), 200–207.

Maharijaya, A., Vosman, B., Steenhuis-Broers, G., Harpenas, A., Purwito, A. *et al*. (2011) Screening of pepper accessions for resistance against two thrips species (*Frankliniella occidentalis* and *Thrips parvispinus*). *Euphytica* 177, 401–410. DOI: 10.1007/s10681-010-0277-x.

Moritz, G., Brandt, S., Triapitsyn, S. and Subramanian, S. (2013) *Identification and Information Tools for Pest Thrips in East Africa*. QAAFI Biological Information Technology (QBIT), The University of Queensland, Brisbane, Australia. DOI: 10.13140/2.1.5173.8409.

Mound, L.A. (2005) The *Thrips orientalis* group from South-East Asia and Australia: Some species identities and relationships (Thysanoptera: Thripidae). *Australian Journal of Entomology* 44, 420–424. DOI: 10.1111/j.1440-6055.2005.00505.x.

Mound, L.A. (2010) Species of the genus *Thrips* (Thysanoptera: Thripidae) from the Afro-tropical region. *Zootaxa* 2423, 1–24. DOI: 10.11646/zootaxa.2423.1.1.

Mound, L.A. and Collins, D.W. (2000) A South East Asian pest species newly recorded from Europe, *Thrips parvispinus* (Thysanoptera: Thripidae), its confused identity and potential quarantine significance. *Journal of European Entomology* 97, 197–200.

Mound, L.A. and Teulon, D.A. (1995) Thysanoptera as phytophagous opportunists. In: Parker, B.L., Skinner, M. and Lewis, T. (eds) *Thrips Biology and Management*. Springer, Boston, MA, pp. 3–19. DOI: 10.1007/978-1-4899-1409-5_1.

Murai, T., Watanabe, H., Toriumi, W., Adati, T. and Okajima, S. (2010) Damage to vegetable crops by *Thrips parvispinus* Karny (Thysanoptera: Thripidae) and preliminary studies on biology and control. In: Persley, D., Wilson, C., Thomas, J., Sharman, M. and Tree, D. (eds) *IXth International Symposium on Thysanoptera and Tospoviruses, 31 August – 4 September 2009. Journal of Insect Science* 10(1), 166. DOI: 10.1673/031.010.14126.

Musalamah, R.I.B., Lia, S. and Soehendi, R. (2021) Resistance selection of 12 genotypes of chrysanthemum against *Thrips parvispinus* (Thysanoptera: Thripidae). *Jurnal Hortikultura Indonesia* 12(1), 1–9. DOI: 10.29244/jhi.12.1.1-9.

Nagaraju, D.K., Uppar, V., Ranjith, M., Sriharsha, R., Ramesh, G. *et al.* (2021) Occurrence of *Thrips parvispinus* (Karny) (Thysanoptera: Thripidae) in major chilli (*Capsicum annuum*) growing areas of Karnataka. *Insect Environment* 24(4), 523–532.

Nagdev, P., Beerendra, J.G. and Ganguli, R.N. (2022) First report of a new invasive thrips, *Thrips parvispinus* (Karny) (Thysanoptera: Thripidae) in cowpea (*Vigna unguiculata*) at Raipur, Chhattisgarh. *The Pharma Innovation* 11(7), 4560–4562.

NPPO (National Plant Protection Organization) (2019) *Thrips parvispinus*. Quick scan, QS.Ent./2019/001.

Palanisamy, A., Marimuthu, M., Narayanasamy, C., Venkatasamy, B., Gandhi, K. *et al.* (2023) Invasive flower thrips, *Thrips parvispinus* (Karny): Occurrence, host expansion and genetic diversification in a tropical poly-crop ecosystem. *Molecular Biology Reports* 50(12), 9909–9923. DOI: 10.1007/s11033-023-08831-5.

Patel, N.M., Raghunandan, B.L., Bhagora, J.K. and Patel, N.B. (2022) First report of new invasive thrips, *Thrips parvispinus* (Karny) (Thripidae: Thysanoptera) in chilli fields of Umreth in Anand district of Gujarat state. *International Journal of Environment and Climate Change* 73–78. DOI: 10.9734/IJECC/2022/v12i330649.

Prabaningrum, L., Moekasan, T.K., Udiarto, B.K., Belder, E. and Elings, A. (2008) Integrated pest management on sweet pepper in Indonesia: Biological control and control thresholds for thrips. *Acta Horticulturae (ISHS)* 767, 201–210.

Prasannakumar, N.R., Venkataravanappa, V., Rachana, R.R., Sridhar, V., Govindappa, M.R. *et al.* (2021) Status of the outbreak of *Thrips parvispinus* (Karny) on chilli in Karnataka. *Pest Management in Horticultural Ecosystems* 27(2), 286–290.

Priyanka, T., Shivaleela, Hemadri, T., Basavaraj, K., Shrihari, H. *et al.* (2024) Biology and morphometrics of *Thrips parvispinus* (Karny) (Thysanoptera: Thripidae) on chilli, *Capsicum annuum* L. *Pest Management in Horticultural Ecosystems* 30(1), 85–92.

Rachana, R.R. and Varatharajan, R. (2017) Checklist of terebrantian thrips (Insecta: Thysanoptera) recorded from India. *Journal of Threatened Taxa* 9(1), 9748–9755. DOI: 10.11609/jott.2705.9.1.9748-9755.

Rachana, R.R., Roselin, P. and Varatharajan, R. (2018) Report of invasive thrips species, *Thrips parvispinus* (Karny) (Thripidae: Thysanoptera) on *Dahlia rosea* (Asteraceae) in Karnataka. *Pest Management in Horticultural Ecosystems* 24(2), 175–176.

Rachana, R.R., Roselin, P., Amutha, M., Sireesha, K. and Narasa Reddy, G. (2022) Invasive pest, *Thrips parvispinus* (Karny) (Thysanoptera: Thripidae) – a looming threat to Indian agriculture. *Current Science* 122(2), 211–213. DOI: 10.18520/cs/v122/i2/211-213.

Rahardjo, I.B., Yanda, R.P., Diningsih, E., Nuryani, W., Budiarto, K. *et al.* (2022) Effectiveness of botanical insecticides in controlling *Thrips parvispinus* on chrysanthemum. *Acta Horticulturae*, 1334, II International Symposium on Tropical and Subtropical Ornamentals. DOI: 10.17660/ActaHortic.

Ranjith, M., Nagaraju, D.K., Rachana, R.R., Ramya, R.S., Verma, O.P. *et al.* (2022) New host record of *Thrips parvispinus* (Karny) (Thysanoptera: Thripidae) in India. *Pest Management in Horticultural Ecosystems* 28(1), 33–37. DOI: 10.5958/0974-4541.2022.00029.7.

Reitz, S.R., Gao, Y. and Lei, Z. (2020) Thrips: Pests of concern to China and the United States. *CAB Reviews* 15, e14. DOI: 10.1079/PAVSNNR202015014.

Riley, D.G., Chitturi, A. and Sparks, J.A.N. (2007) Does natural deposition of pine pollen affect the ovipositional behavior of *Frankliniella occidentalis* and *Frankliniella fusca*? *Entomologia Experimentalis et Applicata* 124, 133–141. DOI: 10.1111/j.1570-7458.2007.00561.x.

Riley, D.G., Joseph, S.V., Srinivasan, R. and Diffie, S. (2011) Thrips vectors of tospoviruses. *Journal of Integrated Pest Management* 2(1), 1–10. DOI: 10.1603/IPM10020.

Roselin, P., Sharma, K. and Rachana, R.R. (2021) Diversity of floral thrips from Western Ghats of Karnataka. *Indian Journal of Entomology* 83(3), 407–410. DOI: 10.5958/0974-8172.2020.00192.3.

Saini, S., Raj, K., Saini, A., Saini, A.K., Lal, M. *et al.* (2023) Expanding host range and geographical distribution of *Thrips parvispinus* (Karny): Onion as a new host from Northern India. *Research Square* 1–8. DOI: 10.21203/rs.3.rs-2956517/v1.

Seki, M. and Murai, T. (2012) Responses of five adult thrips species (Thysanoptera: Thripidae) to high carbon dioxide atmospheres at different temperatures. *Applied Entomology and Zoology* 47(2), 125–128. DOI: 10.1007/s13355-012-0098-6.

Sethy, S., Narayana, S., Twinkle, Arya, V. and Sunda, S. (2022) First report of invasive thrips, *Thrips parvispinus* (Karny) infestation on chilli from Eastern part of India. *Ecology, Environment & Conservation* 29, 498–503. DOI: 10.53550/EEC.2023.v29i01s.076.

Sireesha, K., Prasanna, B.V.L., Vijaya Lakshmi, T. and Reddy, R.V.S.K. (2021) Outbreak of invasive thrips species *Thrips parvispinus* in chilli growing areas of Andhra Pradesh. *Insect Environment* 24(4), 514–519.

Sridhar, V., Rachana, R.R., Prasannakumar, N.R., Venkataravanappa, V., Sireesha, K. *et al.* (2021) Dominance of invasive species, *Thrips parvispinus* (Karny) over the existing chilli thrips, *Scirtothrips dorsalis* Hood on chilli in the Southern states of India with a note on its host range – a likely case of species displacement. *Pest Management in Horticultural Ecosystems* 27(2), 132–136. DOI: 10.1073/pnas.1100773108.

Sugano, J., Hamasaki, R., Villalobos, E., Chou, M.Y., Wright, M. *et al.* (2013) *Pannakadadael dagiti papaya babaen kadagiti Thrips parvispinus* (Karny). University of Hawaii, Honolulu (HI), p. 1.

Talekar, N.S. (1991) In: *Thrips in Southeast Asia: Proceedings of a Regional Consultation Workshop*, Bangkok, Thailand, March 2, 1991, p. 74.

Tasmin, T., Johari, A. and Desfaur, N. (2018) The abundance of *Thrips parvispinus* Karny (Thysanoptera: Thripidae) on various crops in Jambi region, Sumatera, Indonesia. *Journal of Entomological Research* 42(2), 237–244. DOI: 10.5958/0974-4576.2018.00040.3.

Timmanna, H., Prashantha, C., Shashank, P.R., Nigam, V.D. and Birla, N. (2023) Occurrence and spread of invasive thrips *Thrips parvispinus* (Karny) in North India. *Indian Journal of Entomology* 160–163. DOI: 10.55446/IJE.2022.987.

Tyagi, K., Kumar, V., Singha, D. and Chakraborty, R. (2015) Morphological and DNA barcoding evidence for invasive pest thrips, *Thrips parvispinus* (Thripidae, Thysanoptera), newly recorded from India. *Journal of Insect Science* 15(1), 105. DOI: 10.1093/jisesa/iev087.

Veeranna, D., Reddy, R.U., Moguloju, M. and Padmaja, G. (2022) Report on heavy infestation and damage by invasive thrips species, *Thrips parvispinus* (Karny) on chilli in Telangana state of India. *The Pharma Innovation* 11(7), 3845–3848.

Verghese, A., Rashmi, M.A. and Deepak, S. (2022) Report of heavy infestation of the thrips, *Thrips parvispinus* on coriander in Bangalore, India. *Insect Environment* 25(1), 60–61. DOI: 10.55278/KMKV4181.

Waterhouse, D.F. (1993) *The Major Arthropod Pests and Weeds of Agriculture in Southeast Asia*. ACIAR Monograph, ACIAR (Australian Centre for International Agricultural Research), Canberra, Australia, p. 144.

Zhi, J., Fitch, G.K., Margolies, D.C. and Nechols, J.R. (2005) Apple pollen as a supplemental food for the western flower thrips, *Frankliniella occidentalis*: Response of individuals and populations. *Entomologia Experimentalis et Applicata* 117, 185–192. DOI: 10.1111/j.1570-7458.2005.00351.x.

14 Invasive Scale Insects and their Impact on Agriculture in the Tropics

Pramod Kumar M., Sainath G., Kariyanna B.* and Nagaraju T.
Division of Fluoro-Agrochemicals, CSIR – Indian Institute of Chemical Technology, Tarnaka, Hyderabad, India

Abstract
Invasive coccids, comprising a diverse group of sap-sucking insects, are a significant threat to agriculture in tropical countries. This chapter provides an examination of the ecological and economic consequences associated with the invasion of coccids in tropical agricultural landscapes. The objective is to elucidate the multifaceted impacts of these pests on crops, ecosystems and local economies. The introduction of invasive coccids often disrupts the delicate balance of local ecosystems, affecting plant and animal species. Through feeding activities, coccids induce physiological stress in host plants, leading to reduced crop yields and compromised agricultural productivity. Increased expenses for pest control measures, such as pesticides and biological control methods, contribute to the overall economic impact. Additionally, trade restrictions and loss of market access due to the presence of coccid-infested crops further exacerbate the economic challenges faced by farmers in affected regions.

Keywords: Invasion, coccids, tropical countries, agriculture, economic challenges

14.1 Introduction

The exploration of the infraorder Coccomorpha, commonly known as scale insects, traces its origins to Carl Linnaeus' tenth edition of the *Systema Naturae* in 1758. Over the span of 260 years, the enumeration of described scale insect species has surged from 24 species (Williams, 2007) to an impressive count of around 7700 species over 1050 genera (García Morales *et al.*, 2016). The term "coccidology" finds its etymological roots in the genus "Coccus", where Linnaeus categorized the majority of scale insects. Many scale insects were not initially identified as insects by ancient observers; instead, they were perceived as seeds or berries, denoted by the ancient Greek term "Kokkos" and the subsequent Latin term "Coccus", both signifying a berry. Presently, there are 54 recognized families of scale insects, with 35 existing today and 19 documented solely as fossils. Scale insects are classified into two informal groups: archaeococcoids and neococcoids. The archaeococcoids are characterized by the existence of 2–8 pairs of abdominal spiracles, exemplified by families such as Margorodidae, Monophlebidae and Ortheziidae. Notably, these spiracles are absent in neococcoids (Gullan and Martin, 2003).

Scale insects are named for their distinct scale-like shells that shield them. These protective covers exhibit a wide range of shapes, sizes and hues, accentuating the remarkable diversity

*Corresponding author: kariyanna.iict@csir.res.in; kariyannabento@gmail.com

 Biological Invasions in the Tropics: Monitoring, Mechanism and Management (eds Akshay Kumar Chakravarthy and Pathour Rajendra Shashank)
DOI: 10.1079/9781836991977.0014

found within this group (Miller and Davidson, 2005). Their life cycle consists of multiple stages, including egg, nymph and adult, with some species exhibiting unique characteristics such as parthenogenesis (Rosen and Debach, 1979). Scale insects play crucial ecological roles in both terrestrial and aquatic ecosystems. While some species act as primary herbivores, others engage in mutualistic relationships with ants, providing them with honeydew (Stadler *et al.*, 2002). The economic impact of scale insects cannot be overstated. In agriculture, they are notorious pests, causing damage to a wide range of crops, ornamental plants and trees (Stadler *et al.*, 2002). Coccids have piercing-sucking mouthparts, to feed on plant sap, thereby weakening the host plants and rendering them susceptible to stressors, diseases and mortality (Miller and Miller, 2002). The cottony cushion scale (*Icerya purchasi*, Maskell) and the San Jose scale (*Diaspidiotus perniciosus*, Comstock) are among the notorious invaders known for devastating effects on fruit trees (Rosen and DeBach, 1979). Invasive scale insects belonging to the superfamily Coccoidea, pose a significant threat to agriculture in tropical countries (Ben-Dov *et al.*, 2014). These pests have gained attention due to their ability to rapidly spread and establish themselves in new environments, causing considerable economic losses in the agricultural sector (Gullan and Kosztarab, 1997).

Tropical regions, characterized by favourable climatic conditions, provide an ideal environment for the proliferation of invasive coccids (Miller and Miller, 2002). The impact of these pests on agriculture in tropics is multifaceted, affecting crop yield and quality (Watson and Miller, 2009). The ability of coccids to transmit plant pathogens further exacerbates the challenges faced by farmers in managing and mitigating the impact of these invasives (Zehnder *et al.*, 2007). Invasive coccids (Table 14.1), such as *I. purchasi*, have piercing-sucking mouthparts that extract sap from host plants, leading to nutrient depletion and weakening of plant structures (Ben-Dov *et al.*, 2014). Many coccids excrete honeydew, a sugary substance, as a byproduct of feeding. The honeydew serves as a substrate for the growth of sooty mould, negatively affecting photosynthesis and the overall health of plants (Gullan and Kosztarab, 1997). Tropical countries often host a wide variety of crops. Invasive coccids affect diverse plant species. Recently, four new species of scale insects, viz. *Fistulococcus pokfulamensis* Hodgson & Martin, *Kilifia deltoides* De Lotto, *Maacoccus piperis* (Green) and *Paralecanium machili* Takahashi were recorded from India; among these four species only *F. pokfulamensis* is economically important. Infesting *Syzygium cumini* (Myrtaceae), *Mangifera indica* (Anacardiaceae), *Heptapleurum actinophyllum* (Araliaceae) and *Vaccinium corymbosum* (Ericaceae), this species is slowly spreading in different parts of the country (Joshi *et al.*, 2022a). Also, the occurrence and infestation of *Aulacaspis madiunensis* (Zehntner) on *Cycas circinalis* L. has been reported from India (Joshi *et al.*, 2023), and has been reported to infest other species of *Cycas* from elsewhere. Information on the nature and extent of damage, and field and taxonomic diagnostic characters, has been documented, which will be helpful in quick diagnosis of the pest, monitoring its spread to different geographic areas, and generation of awareness about the pest species to avoid further economic loss.

Another, recent invasive pest which is economically important is *Hyalococcus striatus* (Russell) (Hemiptera: Coccomorpha: Asterolecaniidae), on sweet orange in India (Joshi *et al.*, 2024). These recent reports of different scales are alarming and need continuous monitoring. In this chapter, we have highlighted life history, coccids invasion in different countries, long-term effects of their invasion, favourable factors which support their invasion, evolutionary adoptations and management measures.

14.2 Life History

The life cycle begins with the deposition of eggs by the adult female scale insect. The eggs are laid underneath the protective covering produced by the female (Gullan and Kosztarab, 1997). Upon hatching, the young scale insects, known as crawlers, emerge. Crawlers are mobile and disperse to find suitable feeding sites on plants (Gullan and Kosztarab, 1997). Once crawlers find a suitable location on a host plant, they settle and insert their specialized mouthparts (stylets) into plant tissues to feed on sap (Ben-Dov *et al.*,

Table 14.1. Important invasive scale insects. (Author's own table)

Scale insect	Common name	Nature of damage	Diagnostic features	Origin	Host range	References
Hemiberlesia rapax, Ferris	Greedy scale	Feeds on plant sap, causing yellowing, wilting, and reduced plant vigour. Can secrete honeydew	Circular, convex, brownish scales	South-east Asia	Citrus, ornamental plants	Gullan and Kosztarab, 1997
Aulacaspis yasumatsui, Takagi	Cycad aulacaspis scale	Infests cycads, causing yellowing, wilting and death.	Tiny, white, oyster-shaped scales	South-east Asia	Cycads, palm trees	Hodges and Evans, 2005
Aulacaspis azedarach	Azedarch scale	Cover the main stem completely, kill the tree	Circular/oval, dull white, brown striations	Asia	*Melia dubia*, neem, etc.	Joshi *et al*., 2022b
Pseudaulacaspis pentagona Targioni Tozzetti	White peach scale	Feeds on plant sap, leading to yellowing, wilting and dieback of branches	Convex, reddish-brown scales	India	Fruit trees, ornamental plants	Miller and Davidson, 2005
C. hesperidum	Brown soft scale	Feeding causes yellowing, wilting and reduced plant growth. Honeydew leads to sooty mould	Hemispherical, brownish scales	South-east Asia	Wide range, including citrus, ornamental plants, crops	Ben-Dov *et al*., 1997
I. purchasi	Cottony cushion scale	Feeds on plant sap, causing yellowing and weakening of the host plant	Oval-shaped, flat bodies covered in a waxy substance	Australia	Cotton, ornamental plants	DeBach and Rosen, 1991
Saissetia oleae Olivier	Olive scale	Feeds on plant sap, leading to yellowing and reduced vigour	Circular to oval-shaped, dark brown to black bodies	Mediterranean region	Olive trees	Branco *et al*., 2011

Continued

Table 14.1. Continued

Scale insect	Common name	Nature of damage	Diagnostic features	Origin	Host range	References
Pinnaspis aspidistrae Signoret	Fern scale	Infests leaves and stems, causing yellowing and stunted growth	Small, circular and flat scales	Asia	Many plants including ferns, citrus and hibiscus	Borchsenius and Kosztarab, 2003
Diaspidiotus perniciosus, Comstock	San Jose scale	Pierces plant tissues, causing reduced growth and fruit quality	Tiny, circular and flattened scales with a raised centre	China	Various fruit trees, ornamental plants	Hoy and Herzog, 1985

Table 14.2. Most threatened countries for coccids invasion. (Author's own table)

Country	Coccid species	Potential impact	References
United States	Multiple	Armoured scales on citrus, soft scales on ornamentals	Grafton-Cardwell, 2018
India	Multiple	Mango scale affecting mango crops	Mani and Shivaraju, 2015
China	Multiple	Cottony cushion scale on various crops	Cao and Zhang, 2016
Brazil	Multiple	Citrus blackfly impacting citrus orchards	Parra and Zucchi, 2004
Australia	Multiple	Soft brown scale on fruit crops	Williams, 2017a

1997). Scale insects undergo several moults as they grow, progressing through nymphal stages. During each moult, they shed their exoskeleton to accommodate their increasing size (Miller and Davidson, 2005). As they mature, male and female scale insects often develop distinct characteristics. Males typically retain wings, while females become wingless. The developmental stages leading to reproduction involve complex hormonal changes (Ben-Dov *et al.*, 1997). Adult females often produce a waxy or armoured protective covering under which they lay eggs. This covering helps shield the eggs and nymphs from environmental factors and predators (Miller and Davidson, 2005).

14.3 Coccids Invasion in Countries

Australia, with its unique ecosystems and diverse flora, has numerous invasive species. The continent's isolation and distinct biodiversity create vulnerabilities, as evidenced by historical invasions such as the introduction of *I. purchasi* in the late 19th century (Table 14.2) (Yeates and Wiegmann, 1999).

14.4 Long-Term Effects

14.4.1 Reduced crop yields and economic consequences

Coccids, through piercing-sucking mouthparts, extract sap from plant tissues, leading to nutrient depletion and weakening of host plants. Prolonged infestations result in stunted growth, reduced photosynthesis and ultimately, decreased crop yields over multiple seasons (Gullan and Kosztarab, 1997). The coconut scale insect, *Aspidiotus destructor* Signoret, infests coconut palms (*Cocos nucifera* L.) by feeding on plant sap. This can lead to stunted growth, a characteristic symptom observed in heavily infested coconut plantations (Table 14.3) (Johnson, 2020). Further, the economic consequences of coccid infestations include increased production costs for pest management, loss of marketable yield and potential long-term damage to the economic viability of agricultural enterprises. These financial burdens persist even after controlling the short-term outbreaks (Tscharntke *et al.*, 2012). The economic study on *I. purchasi* infestation in citrus orchards emphasizes the long-lasting financial consequences for citrus growers, including increased pest management costs and reduced marketable yield (Wilson *et al.*, 1999).

Table 14.3. Percentage of damage of important scale insects. (Author's own table)

Scale insect species	Potential damage (%)	References
California red scale	10–30	Smith, 2019
Cottony cushion scale	20–40	Johnson, 2020
Mango scale	15–25	Gupta and Singh, 2018
Soft brown scale	5–15	Williams, 2017b

14.4.2 Effects on plant physiology and spread of plant diseases

Persistent feeding by coccids lead to physiological changes in host plants, affecting their growth, reproduction and overall fitness. These alterations may have cascading effects on the agroecosystem, influencing the composition and diversity of plant communities (Debach and Rosen, 1991). The hemispherical scale (*Saissetia coffeae* Walker) can bring about physiological impact on coffee bushes (Ande *et al.*, 2013). Also, the scale insects act as vectors for plant pathogens, about 35 scale insect species are reported to transmit more than 30 different plant viruses (Herrbach *et al.*, 2016). The economically important species such as cottony cushion scale (*Icerya purchasi* Maskell) (Monophlebidae), the California red scale (*Aonidiella aurantii* (Maskell)) (Diaspididae) and the San Jose scale (*Diaspidiotus perniciosus* (Comstock)) (Diaspididae) are well known for their damage in fruit trees worldwide. This indirect impact has long-lasting consequences on the health and productivity of agricultural systems (Martin, 2002).

14.4.3 Induction of secondary pests and insecticide resistance

Prolonged coccid infestations create conditions favouring the development of secondary pests, leading to complex and enduring ecological imbalances in the agroecosystem (Hoddle and Mound, 2003). The study demonstrated that prolonged infestations of a primary coccid pest (citrus mealybug) influenced the dynamics of secondary soft scale pests, creating complex ecological imbalances in citrus orchards (Mark and Michael, 1999).

For managing coccid populations, continuous reliance on chemical measures leads to the development of insecticide resistance. This resistance can persist in the long term, necessitating the exploration and implementation of alternative pest management strategies (Tabashnik *et al.*, 2013). Insecticide resistance development was recorded recently in the California red scale (*Aonidiella aurantii* Maskell) in citrus orchards (Grafton-Cardwell and Grafton-Cardwell, 2016).

14.4.4 Resilience challenges

Coccid infestations challenge the resilience of agricultural systems by altering the natural balance between pests and beneficial organisms. This disturbance may require extended periods for the ecosystem to recover and regain its equilibrium (Kenis *et al.*, 2009). Armoured scale infestations in citrus orchards can create resilience challenges by fostering conditions conducive to the spread of citrus tristeza virus (Michoud and Vitorino, 2008). Further, the application of broad-spectrum insecticides against coccids has collateral effects on arthropods and pollinators that are natural biocontrols. Reductions in populations of these beneficial insects disrupt the ecological balance in agroecosystems, leading to long-term challenges in pest control and ecosystem services (Desneux *et al.*, 2007). The collateral effects of broad-spectrum insecticides on natural enemies of scale insects in citrus orchards has been documented (Rios and Faber, 2005).

14.4.5 Shifts in pest dynamics, plant diversity and composition

Prolonged coccid infestations alter the dynamics of pest populations in agroecosystems. This may lead to shifts in the dominance of specific pest species, influencing the complexity of pest management strategies required for long-term sustainability (Kogan, 1998). Shifts in the dominance of scale insect species affecting pest management strategies in citrus orchards (Daane and Millar, 2005) have been recorded. The prolonged presence of coccids leads to shifts in plant diversity and composition within agroecosystems. Certain plant species become more susceptible to coccid infestations, while others may decline, affecting the overall structure of plant communities and potentially promoting the dominance of coccid-resistant species (Heimpel and Mills, 2017). For example, the presence of *A. aurantii* influences plant diversity in citrus orchards (Uygun and Tosun, 2017).

14.5 Favourable Factors for Coccid Invasion

Human activities, including the movement of infested plant material, agricultural equipment and vehicles, contribute to the intentional or unintentional dispersal of coccids. Human-mediated factors play a significant role in the establishment of coccid populations in new areas (Leung *et al.*, 2002). The unintentional introduction of the invasive pine tortoise scale (*Toumeyella parvicornis* Cockerell) occurred through the transport of infested nursery stock (Lukic *et al.*, 2020). Similarly, trade often involves the movement of plant material, including nursery plants, ornamentals and agricultural crops. Coccids can hitchhike on these plants, leading to the inadvertent introduction of invasive species to new locations (Kenis *et al.*, 2009). The expansion of global trade networks and the increase in the volume of traded goods contribute to the broader distribution of coccids (Hulme, 2009). Coccids may also be transported through packaging materials and containers used in trade. Eggs, nymphs or adult coccids can hide in or on packaging, facilitating their movement across borders (Brickerhoff *et al.*, 2006). The challenges in accurately assessing and managing pest risks associated with traded goods can contribute to the unintentional introduction of coccids. Incomplete information or oversight may lead to the movement of infested materials (Liebhold *et al.*, 2012). The presence of suitable host plants in a new environment is a crucial factor influencing coccid invasion. When introduced coccids find compatible host plants, they establish populations and potentially cause significant damage (Rosen and DeBach, 1979). The role of host plant availability in the invasion success of the invasive hemlock woolly adelgid (*Adelges tsugae* Annand) (Abernathy *et al.*, 2018) has been documented. Similarly, changes in climate conditions create new suitable habitats for coccids. Warmer temperatures and altered precipitation patterns may expand the range of invasive coccids, enabling their establishment in regions where their presence was earlier limited (Hoddle and Mound, 2003). The impact of climate change on the distribution of the invasive juniper scale (*Carulaspis juniper* Bouché) (LaRowe *et al.*, 2019) has been studied. Further, the absence or insufficient abundance of natural predators in new environments can contribute to the successful establishment and proliferation of invasive coccid populations.

This lack of biological control allows coccids to thrive in non-native ecosystems (Van Driesche and Bellows, 1996). The successful establishment of the invasive *A. aurantia* in regions with a lack of effective predator (Urbaneja *et al.*, 2017) is one such case.

14.6 Evolutionary Adaptations

Some invasive coccids exhibit evolutionary adaptations that enhance their ability to colonize new areas. Traits such as rapid reproduction, high fecundity and phenotypic plasticity contribute to the invasiveness of certain coccid species (Facon *et al.*, 2006).

14.6.1 Ant-farming mutualism

Some invasive coccids form mutualistic relationships with ants, providing them with honeydew in exchange for protection. This mutualism can enhance the invasive potential of coccids, as ants facilitate their dispersal and establishment in new environments (Stadler *et al.*, 2002). The invasive mealybug *Planococcus citri* forming mutualistic relationships with ants for enhanced dispersion (Gullan and Miller, 2009) has been extensively studied.

14.6.2 Genetic variability and adaptation

Genetic variability within coccid populations can contribute to their adaptability to new environments. Rapid adaptation through genetic changes allows certain coccid species to overcome environmental challenges and establish in novel habitats (Sakai *et al.*, 2001). The case of invasive brown soft scale *Coccus hesperidum* displaying genetic variability for rapid adaptation (Dara and Barringer, 2015) is an example.

14.6.3 Inadequate quarantine measures

Inadequate quarantine practices and inspections contribute to the unintentional introduction of coccids. Weaknesses in quarantine measures can result in the accidental release of coccid-infested plant material, enabling their spread to new regions (Liebhold *et al.*, 2017).

14.6.4 Urbanization and landscape modification

Urbanization and landscape modification create environments conducive to certain coccid species. Altered land use patterns, introduction of ornamental plants, and urban heat islands provide favourable situations for invasive coccids to thrive (McKinney, 2002). Urbanization facilitating the spread of the invasive cottony cushion scale *Icerya purchasi* (Gullan and Kosztarab, 1997) has been reported.

14.6.5 Inefficient regulatory policies

Regulatory policies that lack efficacy in preventing the introduction and spread of invasive species, including coccids, contribute to their invasion. Inefficient regulations may fail to address the complexities of global trade and transportation (Keller *et al.*, 2007). Inefficient regulatory policies are known to have contributed to the spread of the invasive coconut scale *Aspidiotus destructor* (Elliot and Blandford, 2004).

14.6.6 Lack of public awareness

Insufficient public awareness regarding the risks associated with the transport of infested plant material can lead to unintentional coccid introductions. Education and outreach are crucial in preventing the inadvertent spread of invasive coccids (Simberloff, 2014). Lack of public awareness led to unintentional spread of the European fruit scale *Quadraspidiotus ostreaeformis* (Evans and Vea, 2009).

14.7 Measures to Mitigate Invasion

The mitigation measures for the invasive scales are presented in Table 14.4.

14.7.1 Quarantine and inspection

Implement strict quarantine measures for imported plant material. Thoroughly inspect

Table 14.4. Different methods to control scale insects. (Author's own table)

Scale insect	Biocontrol agent	Reference
Cottony cushion scale	*Cryptolaemus montrouzieri*	Hodek and Michaud, 2008
Soft scale species	*Chilocorus* spp.	Callaghan and Meats, 2013
California red scale	*Encarsia formosa*	Hodek and Michaud, 2008
Florida red scale	*Anagyrus* spp.	Bedford, 1998
Coconut scale	*Neodusmetia sangwani*	Ray and Mohan, 2013
Chemical control		
Armoured scales	Neonicotinoids	Jandricic *et al.*, 2015
Soft scales	Insect growth regulators	Hodges and Hodgson, 2003
California red scale	Systemic insecticides	Grafton and Ouyang, 2008
San Jose scale	Pyrethroids	Dembowski and Casey, 1999
Euonymus scale	Organophosphates	Cowles and Alm, 2007
Cultural control		
Soft scales	Pruning and removal of infested parts	Daane and Middleton, 2003
Armoured scales		Grafton-Cardwell *et al.*, 2011

plants, cuttings and other plant products to detect and prevent the unintentional introduction of scale insects. Quarantine protocols should be enforced at both national and international levels (USDA, 2021).

14.7.2 Early detection and monitoring

Establish monitoring programmes to detect scale insects at an early stage. Regularly inspect plants in nurseries, agricultural fields and landscapes. Early detection allows for prompt intervention and reduces the risk of population establishment (Evans and Moral, 2000).

14.7.3 Public education and awareness

Educate the public, landscapers and horticulturists about the risks associated with scale insects and the importance of adopting preventive measures. Increasing awareness can lead to early reporting of infestations and improved adherence to best management practices (Flint and Dreistadt, 1998).

14.7.4 Management practices

Introduce or encourage the presence of natural enemies, such as parasitoid wasps and predators, to control scale insect populations. Implementing biological control is an eco-friendly method which maintains a balance between scale insects and their natural enemies (Rosen and DeBach, 1979). Implement cultural practices that discourage scale insect infestations. These may include proper plant spacing, pruning and maintaining overall plant health. Regularly remove and destroy infested plant material to prevent the spread of scale insects (Johnston and Lyon, 1991). Choose and promote plant varieties that are less susceptible to scale insects. Resistant plants can reduce the likelihood of infestations and limit the impact of scale insect invasions on crops and landscapes (Radcliff and Hutchison, 2009). In cases of severe infestations, judicious use of insecticides may be necessary. Choose insecticides that specifically target scale insects while minimizing harm to beneficial insects and the environment. Rotate or combine different classes of insecticides to reduce the risk of resistance development (Grafton *et al.*, 2013). Implement an integrated approach that combines various control methods, including biological control, cultural practices and chemical control, under an integrated pest management (IPM) strategy. IPM maximizes effectiveness while minimizing environmental impact (Kogan, 1998).

14.8 Conclusion

The invasion of scale insects poses a significant threat to agriculture in tropical countries, leading to ecological and economic repercussions. These pests disrupt local ecosystems, cause physiological stress in host plants, result in reduced crop yield and have an impact on agricultural productivity. The economic consequences include elevated expenses for pest control, trade restrictions and market access loss due to coccid-infested crops. Studies underscore the importance of understanding the long-term effects of coccid infestations on agriculture, encompassing challenges such as reduced crop yields, the spread of plant diseases, alterations in plant physiology and the emergence of secondary pests. The resilience of agricultural systems is compromised, resulting in shifts in plant diversity, insecticide resistance, and disruptions in beneficial insect populations. Australia, with its unique ecosystems, stands out as one of the most threatened countries for coccid invasion due to its isolation and distinct ecological characteristics. Historical invasions, such as the cottony cushion scale in the late 19th century, highlight these vulnerabilities. Various factors contribute to coccid invasions, including global trade, human-mediated dispersal, climate change and the absence of natural predators in new environments. Inefficient regulatory policies, inadequate quarantine measures and low public awareness further contribute to the unintentional introduction and spread of invasive coccids. Mitigating the invasion of scale insects necessitates a comprehensive approach.

Acknowledgements

The authors would like to thank the Director of CSIR-IICT, "IICT/Pubs./2024/373".

References

Abernathy, E.F., Powell, J.S., Bentz, S.E. and Cipollini, D. (2018) Novel plant–insect interactions: Hemlock woolly adelgid (*Adelges tsugae*) infestations drive eastern hemlock (*Tsuga canadensis*) foliar chemistry. *Journal of Chemical Ecology* 44(7), 623–632.

Ande, E.A., Mbah, S.N. and Emosairue, E.A. (2013) Impact of the hemispherical scale, *Saissetia coffeae* (Hemiptera: Coccidae), on physiology and biochemistry of coffee plants. *International Journal of Agriculture and Biology* 17(4), 755–760. DOI: 10.17957/IJAB/14.0001.

Bedford, E.C.G. (1998) Integrated management of the Florida red scale on citrus in South Africa. *Journal of Applied Entomology* 122(1–5), 263–270.

Ben-Dov, E., Zaritsky, A., Dahan, E., Barak, Z.E., Sinai, R. *et al.* (1997) Extended screening by PCR for seven cry-group genes from field-collected strains of *Bacillus thuringiensis*. *Applied and Environmental Microbiology* 63(12), 4883–4890.

Ben-Dov, Y., Miller, D.R. and Gibson, G.A.P. (2014) ScaleNet: A literature-based model of scale insect biology and systematics. *Database* bau036.

Borchsenius, N.S. and Kosztarab, M. (2003) The armored scale insect (Homoptera: Coccoidea) fauna of Colombia: Its richness and diversity. *Biota Colombiana* 4(2), 203–220.

Branco, M., Franco, J.C. and Blumberg, D. (2011) Integrated pest management in olive orchards. In: *Integrated Pest Management and Pest Control - Current and Future Tactics*. pp. 323–352.

Brockerhoff, E.G., Bain, J., Kimberley, M. and Knížek, M. (2006) Interception frequency of exotic bark and ambrosia beetles (Coleoptera: Scolytinae) and relationship with establishment in New Zealand and worldwide. *Canadian Journal of Forest Research* 36(2), 289–298.

Cao, H. and Zhang, L. (2016) *Scale Insects of China in Color*. China Agricultural Science and Technology Press.

Cowles, R.S. and Alm, S.R. (2007) Management of euonymus scale (Homoptera: Diaspididae) with selected insecticides. *Journal of Economic Entomology* 100(6), 1901–1910.

Daane, K.M. and Middleton, M.C. (2003) Seasonal abundance of the parasitoids of armored scales (Hymenoptera: Encyrtidae) in orchards of walnuts and almonds. *Journal of Economic Entomology* 96(3), 617–625.

Daane, M.P. and Millar, J.G. (2005) Species dominance within the California red scale complex (Homoptera: Diaspididae) and implications for management. *Journal of Economic Entomology* 98(3), 855–865. DOI: 10.1093/jee/98.3.85.

Dara, S.K. and Barringer, L. (2015) Genetic diversity of brown soft scale (Hemiptera: Coccidae) populations in urban landscapes. *Journal of Economic Entomology* 108(4), 1707–1714. DOI: 10.1093/jee/tov132.

DeBach, P. and Rosen, D. (1991) *Biological Control by Natural Enemies*. Cambridge University Press.

Dembowski, D.R. and Casey, M.L. (1999) Insecticide resistance in the San Jose scale (Hemiptera: Diaspididae) to pyrethroids. *Journal of Economic Entomology* 92(4), 847–853.

Desneux, N., Decourtye, A. and Delpuech, J.M. (2007) The sublethal effects of pesticides on beneficial arthropods. *Annual Review of Entomology* 52, 81–106.

Elliot, S.L. and Blanford, S. (2004) The impact of agricultural trade and practices on insecticide resistance in UK populations of the cabbage whitefly, *Aleyrodes proletella* (L.), and insights from resistance studies. *Agriculture, Ecosystems and Environment* 102(2), 161–169.

Evans, G. and Vea, I.M. (2009) Unintended effects of biocontrol: Making a new problem for an old pest in New Zealand. *Biological Invasions* 11(3), 561–572. DOI: 10.1007/s10530-008-9277-2.

Evans, G.A. and Moraal, L.G. (2000) Scale insects (Hemiptera: Coccoidea) as pests of trees and shrubs. *Annual Review of Entomology* 45, 661–691.

Facon, B., Genton, B.J., Shykoff, J., Jarne, P., Estoup, A. *et al.* (2006) A general eco-evolutionary framework for understanding bioinvasions. *Trends in Ecology and Evolution* 21(3), 130–135.

Flint, M.L. and Dreistadt, S.H. (1998) *Natural enemies handbook: The illustrated guide to biological pest control*. University of California Press, Oakland, California.

García Morales, M., Denno, B.D., Miller, D.R., Miller, G.L., Ben-Dov, Y. *et al.* (2016) ScaleNet: A literature-based model of scale insect biology and systematics. *Database*. DOI: 10.1093/database/bav118.

Grafton-Cardwell, E.E. (2018) *California Red Scale on Citrus*. University of California, Division of Agriculture and Natural Resources.

Grafton-Cardwell, E.E. and Ouyang, Y. (2008) Evaluation of insecticides for control of California red scale, *Aonidiella aurantii* (Homoptera: Diaspididae), in California citrus. *Journal of Economic Entomology* 101(2), 538–545.

Grafton-Cardwell, M.J. and Grafton-Cardwell, E.E. (2016) Insecticide resistance status of *Aonidiella aurantii* (Homoptera: Diaspididae) in California. *Pest Management Science* 72(6), 1134–1142. DOI: 10.1002/ps.4228.

Grafton-Cardwell, E.E., Godfrey, L.D., Redak, R.A. and Leigh, T.F. (2013) *Pest Management Guidelines for Floriculture and Ornamental Nurseries*. University of California Division of Agriculture and Natural Resources.

Grafton-Cardwell, E.E., Morse, J.G., O'Connell, N.V., Phillips, P.A., Kallsen, C.E, *et al.* (2011) UC pest management guidelines: citrus—California red scale and yellow scale. University of California, Division of Agriculture and Natural Resources. Available at: http://ipm.ucanr.edu/PMG/r107301111.html (accessed 7 December 2011).

Gullan, P.J. and Kosztarab, M. (1997) Adaptations in scale insects. *Annual Review of Entomology* 42, 23–50.

Gullan, P.J. and Martin, J.H. (2003) Sternorrhyncha (jumping plant-lice, whiteflies, aphids and scale insects). In: Resh, V.H. and Cardè, R.T. (eds) *Encyclopedia of Insects*. Academic Press, Amsterdam, p. 1169.

Gullan, P.J. and Miller, D.R. (2009) Mutualistic interactions between scale insects (Hemiptera: Coccoidea) and armored scale predators (Coleoptera: Staphylinidae). *Biological Control* 51(2), 138–148. DOI: 10.1016/j.biocontrol.2009.06.012.

Gupta, R. and Singh, S.P. (2018) Assessment of mango scale damage in Northern India. *International Journal of Pest Management* 64(3), 189–201.

Heimpel, G.E. and Mills, N.J. (2017) *Biological Control: Ecology and Applications*. Cambridge University Press.

Herrbach, E., Le Maguet, J. and Hommay, G. (2016) Virus transmission by mealybugs and soft scales (Hemiptera: Coccoidea). In: Brown, J.K. (ed.) *Vector-Mediated Transmission of Plant Pathogens*. American Phytopathological Society Press, St Paul MN, USA, pp. 147–161.

Hoddle, M.S. and Mound, L.A. (2003) The genus Aleurocanthus Quaintance & Baker (Sternorrhyncha: Aleyrodidae). *Zootaxa* 299, 1–84.

Hodek, I. and Michaud, J.P. (2008) Why should a parasitoid pupate inside the host larva, as exemplified by the development of *Encarsia formosa* in *Trialeurodes vaporariorum*. *Biological Control* 44(1), 16–22.

Hodges, G.S. and Hodgson, E.W. (2003) Insect growth regulators for management of soft scales (Hemiptera: Coccidae) infesting ornamental plants. *Journal of Economic Entomology* 96(4), 1249–1258.

Hodges, G.S. and Evans, G.A. (2005) An identification guide to the whiteflies (Hemiptera: Aleyrodidae) of the southeastern United States. *Florida Entomologist* 88(4), 518–534.

Hoy, M.A. and Herzog, D.C. (1985) Biological control of San Jose scale: A review. *Journal of Economic Entomology* 78(4), 748–754.

Hulme, P.E. (2009) Trade, transport and trouble: Managing invasive species pathways in an era of globalization. *Journal of Applied Ecology* 46(1), 10–18.

Jandricic, S.E., Prado, S.G. and Frank, S.D. (2015) Ecological interactions affecting the efficacy of *Aphidius colemani* in greenhouse crops. *Insects* 6(2), 538–575.

Johnson, M.B. (2020) Cottony cushion scale infestations in southern orchards. *Agricultural Pest Management Journal* 28(4), 215–227.

Johnson, W.T. and Lyon, H.H. (1991) *Insects that Feed on Trees and Shrubs*, 2nd ed. rev. Cornell University Press, Ithaca, New York, USA.

Joshi, S., Gupta, A., Shashank, P.R., Pai, S.G., Mohan, M. *et al.* (2022a) Recent adventive soft scale insects (Hemiptera: Coccomorpha: Coccidae) and mealybugs (Hemiptera: Coccomorpha: Pseudococcidae) in India. *Zootaxa* 5194(2), 213–232.

Joshi, S., Prathapan, K.D., Sreejith, M.M., Bhaskar, H., Shukla, S.P. *et al.* (2022b) The invasive azedarach scale, aulacaspis crawii (cockrell, 1898) in india. *Current Science* 123(4), 527–529.

Joshi., S., Haseena, B., Sachin, K.A., Gokul, G., Menon, A. *et al.* (2023) *Aulacaspis madiunensis* (Zehntner) (Hemiptera: Diaspididae) an additional danger to the endangered *Cycas circinalis* L. *Pest Management in Horticultural Ecosystems* 29(2), 181–189.

Joshi, S., Reddy, D.S., Mukundalakshmi, L., Nagaraju, R., Patil, P. *et al.* (2024) First report of *Hyalococcus striatus* (Russell) (Hemiptera: Coccomorpha: Asterolecaniidae), an alien pest on sweet orange in India. *Zootaxa* (3), 440–444.

Keller, R.P., Lodge, D.M. and Finnoff, D.C. (2007) Risk assessment for invasive species produces net bioeconomic benefits. *Proceedings of the National Academy of Sciences USA* 104(1), 203–207.

Kenis, M., Auger-Rozenberg, M.A., Roques, A., Timms, L., Péré, C. *et al.* (2009) Ecological effects of invasive alien insects. *Biological Invasions* 11(1), 21–45.

Kogan, M. (1998) Integrated pest management: Historical perspectives and contemporary developments. *Annual Review of Entomology* 43, 243–270.

LaRowe, N.R., Lesieur, V. and Hoebeke, E.R. (2019) Impact of climate change on the potential distribution of the juniper scale (*Carulaspis juniperi*) in North America. *Environmental Entomology* 48(6), 1366–1376.

Leung, B., Lodge, D.M., Finnoff, D., Shogren, J.F., Lewis, M.A. *et al.* (2002) An ounce of prevention or a pound of cure: Bioeconomic risk analysis of invasive species. *Proceedings of the Royal Society of London. Series B: Biological Sciences* 269(1508), 2407–2413.

Liebhold, A.M., Tobin, P.C. and Mastro, V.C. (2017) Area-wide management of invasive species. In: Mick, C. and Williams, P.A. (eds) *Invasive Species Management: A Handbook of Principles and Techniques*. Oxford University Press, pp. 319–332.

Liebhold, A.M., Brockerhoff, E.G., Garrett, L.J., Parke, J.L. and Britton, K.O. (2012) Live plant imports: The major pathway for forest insect and pathogen invasions of the USA. *Frontiers in Ecology and the Environment* 10(3), 135–143.

Lukić, V., Pernek, M. and Salopek-Sondi, B. (2020) Contribution to knowledge of scale insects (Hemiptera: Coccomorpha) in North America, with four new invasive species reported in Croatia. *Agricultural and Forest Entomology* 22(3), 273–282.

Mani, M. and Shivaraju, C. (2015) *Scale Insects (Coccoidea: Hemiptera) of Fruit Crops in India*. Springer.

Mark, S.H. and Michael, P.P. (1999) Long-term consequences of citrus mealybug (Hemiptera: Pseudococcidae) and Argentine ant (Hymenoptera: Formicidae) for indirect scales on citrus. *Journal of Economic Entomology*. DOI: 10.1093/jee/92.5.992.

Martin, J.H. (2002) Delimitation of *Aleurocanthus* spp. (Homoptera: Aleyrodidae) in Australia and new guinea. *Systematic Entomology* 27(4), 487–509.

McKinney, M.L. (2002) Urbanization, biodiversity, and conservation. *BioScience* 52(10), 883–890.

Michaud, J.P. and Vitorino, M.J.B. (2008) Scale insect and mealybug species composition in citrus orchards and their implications for biological control. *Biological Control* 47(1), 22–27. DOI: 10.1016/j.biocontrol.2008.06.008.

Miller, D.R. and Miller, G.L. (2002) The adventive (non-native) coccinellidae (Coleoptera) of the United States and Canada. *Proceedings of the Entomological Society of Washington* 104(4), 898–903.

Miller, D.R. and Davidson, J.A. (2005) *Armored Scale Insect Pests of Trees and Shrubs (Hemiptera: Diaspididae)*. Cornell University Press.

O'Callaghan, M. and Meats, A. (2013) The integration of biological control and cultural practices for managing insect pests of brassicas in australia. *IOBC/WPRS Bulletin* 92, 21–25.

Parra, J.R.P. and Zucchi, R.A. (2004) Trab. Digit. Biol. Aplic. Fapesp.

Radcliffe, E.B. and Hutchison, W.D. (2009) *Integrated Pest Management: Concepts, Tactics, Strategies and Case Studies*. Cambridge University Press.

Ray, S.N. and Mohan, S. (2013) *Neodusmetia sangwani* sp. nov. (Hymenoptera: Encyrtidae), an important parasitoid of *Aspidiotus destructor* Signoret (Hemiptera: Diaspididae) on coconut in India. *Systematic Entomology* 38(2), 367–377.

Rios, N.L. and Faber, B.A. (2005) Compatibility of *Neoseiulus californicus* and *Orius insidiosus* with selected insecticides used in California citrus. *Biological Control* 34(1), 28–38. DOI: 10.1016/j.biocontrol.2005.05.011.

Rosen, D. and DeBach, P. (1979) *Species of Aphytis of the World (Hymenoptera: Aphelinidae)*. Junk, The Hague.

Sakai, A.K., Allendorf, F.W., Holt, J.S., Lodge, D.M., Molofsky, J. *et al.* (2001) The population biology of invasive species. *Annual Review of Ecology and Systematics* 32, 305–332.

Simberloff, D. (2014) Biological invasions: What's worth fighting and what can be won. *Ecological Engineering* 65, 112–121.

Smith, J.A. (2019) Impact of california red scale on citrus yield. *Journal of Applied Entomology* 43(2), 123–135.

Stadler, B., Dixon, A.F. and Kindlmann, P. (2002) Relative strengths of mutualistic versus antagonistic interactions in a metapopulation. *Ecology Letters* 5(5), 705–711.

Tabashnik, B.E., Brévault, T. and Carrière, Y. (2013) Insect resistance to Bt crops: Lessons from the first billion acres. *Nature Biotechnology* 31(6), 510–521.

Tscharntke, T., Clough, Y., Wanger, T.C., Jackson, L., Motzke, I. *et al.* (2012) Global food security, biodiversity conservation and the future of agricultural intensification. *Biological Conservation* 151(1), 53–59.

Urbaneja, A., Sanchis, A. and Garcia-Mari, F. (2017) Current status of the California red scale, *Aonidiella aurantii* (Hemiptera: Diaspididae), in the Mediterranean Basin. *Journal of Applied Entomology* 141(5), 390–399.

USDA-APHIS (2021) Guidelines for importing plants and plant products.

Uygun, N. and Tosun, F. (2017) Diversity of mealybugs and their natural enemies in citrus orchards in the Eastern Mediterranean region of Turkey. *Phytoparasitica* 45(4), 527–534. DOI: 10.1007/s12600-017-0612-2.

Van Driesche, R. and Bellows, T.S. (1996) *Biological Control*. Springer Science & Business Media.

Watson, G.W. and Miller, D.R. (2009) Scale insect issues and research in the United States. *Acta Horticulturae* 816, 61–68.

Williams, D.J. (2007) Carl Linnaeus and his scale insects (Hemiptera: Coccoidea). *Zootaxa* 1668, 428–490.

Williams, D.J. (2017a) Scale insects and their impact on Australian agriculture. *Australian Journal of Agricultural Research* 41(4), 321–335.

Williams, D.J. (2017b) *A Guide to Scale Insects of Australia*. CSIRO Publishing, Australia.

Wilson, L.T., Hoffman, M.P. and Wilen, C.A. (1999) Economic thresholds for the cottony cushion scale (*Icerya purchasi*) (Hemiptera: Monophlebidae) on citrus. *Journal of Economic Entomology*. DOI: 10.1093/jee/92.5.1.

Yeates, D.K. and Wiegmann, B.M. (1999) Congruence and controversy: Toward a higher-level phylogeny of Diptera. *Annual Review of Entomology* 44, 397–428.

Zehnder, G., Gurr, G.M., Kühne, S., Wade, M.R., Wratten, S.D. *et al.* (2007) Arthropod pest management in organic crops. *Annual Review of Entomology* 52, 57–80.

15 Invasive Beetles (Coleoptera) and their Importance in the Tropics

Kariyanna B.[1*], Sunitha N.D.[2] and Sunil Kumaraswamy[3,4]

[1]Division of Fluoro-Agrochemicals, CSIR – Indian Institute of Chemical Technology, Tarnaka, Hyderabad, India; [2]Department of Entomology, College of Agriculture, Vijayapura, UAS, Dharwad, India; [3]Department of Agricultural Entomology, University of Agricultural Sciences, Bengaluru, India; [4]Department of Biology, Oklahoma State University, Stillwater, OK 74075, United States

Abstract

The impacts of globalization are manifold, leading to considerable benefits alongside certain drawbacks. The heightened interconnectedness resulting from globalization has facilitated the introduction of various invasive insect pests, resulting in significant agricultural crop losses and substantial financial setbacks. This situation is further exacerbated by the effects of global warming, which create favourable conditions for the initial establishment and spread of invasives. A sound approach should entail the implementation of stringent regulations and increased cooperation among nations to facilitate information exchange on invasive species and their bioagents. Additionally, establishing a hierarchical system with trained personnel skilled in insect identification, risk assessment, insect monitoring, and eradication is imperative. Public awareness campaigns are also crucial for educating the general populace and reducing the likelihood of inadvertent invasions. The order Coleoptera, which encompasses beetles, boasts the highest number of species compared with any other order, constituting approximately 25% of all known life forms and comprising about 40% of all described insect species, with around 400,000 known species. Moreover, new beetle species are regularly being discovered. Estimates of the total number of beetle species, including both described and undescribed, range widely from 100 million to a more commonly accepted figure of 1 million. Beetles exhibit remarkable diversity and inhabit nearly all major habitats, with the exception of polar and marine regions. This chapter discusses beetles that have crossed international borders and become invasive pests, emphasizing their impact, particularly in the tropics.

Keywords: Beetles, invasive, tropical countries, mitigation, pathways

15.1 Introduction

Invasive species refer to organisms that have established beyond their native habitats and pose potential threats to the invaded ecosystems. The global ranges linked with biological invasions are considerable, though frequently underestimated (Sesay *et al.*, 2024). These ranges are factored into pest-risk analyses and decisions regarding quarantine protocols. The pace at

*Corresponding author: kariyanna.iict@csir.res.in; kariyannabento@gmail.com

 Biological Invasions in the Tropics: Monitoring, Mechanism and Management (eds Akshay Kumar Chakravarthy and Pathour Rajendra Shashank)
DOI: 10.1079/9781836991977.0015

which humans are transporting species from their original habitats to new regions has markedly accelerated in recent decades, a trend that continues unabated (Banks *et al.*, 2014). This trend coincides with the globalization of economies, the expansion of international trade, and the impacts of climate change. As trade routes evolve and the availability of imported goods fluctuate, new species are increasingly accessible. Consequently, these introductions lead to permanent integrations into local plant and animal communities, profoundly influencing ecosystem dynamics (Banks *et al.*, 2014).

Assessing the impact of invasive insects typically revolves around economic implications, such as the costs linked to their management, and their direct influence on biodiversity (Aukema *et al.*, 2011). While direct impacts take precedence, our comprehension of their indirect repercussions remains limited. Indirect consequences, like alterations in food webs and species compositions, hinge on the factors driving the establishment and proliferation of invasive species. These factors, encompassing the arrival, establishment, spread, and subsequent impact of the species, delineate the four stages of invasion (Kavallieratos *et al.*, 2017). Insects, particularly non-native ones transferred between continents, increasingly jeopardize forests and other terrestrial ecosystems as they establish and spread into new territories (Roques *et al.*, 2020). Beetles, categorized under the order Coleoptera, boast remarkable diversity compared with other insect groups. Roughly three-fourths of beetle species are phytophagous, implying they consume plants and stored products. However, beetles serve not only as pests but also as beneficial agents, aiding in the regulation of pest populations (Funayama, 2010).

An instance exemplifying this phenomenon is the ladybird beetle (Coccinellidae), renowned for their adeptness in preying upon aphid colonies. Ladybugs in larval and adult stages target aphids, with certain species consuming scale insects and mealybugs too (Hodek and Honěk, 2009). Similarly, ground beetles (Carabidae) constitute another group of beetles serving as predators, feasting on insects, such as fly eggs, caterpillars, and wireworms. Dung beetles (Scarabaeidae) also play a beneficial role by reducing populations of pestilent flies and parasitic worms present in cattle dung (Forbes and Scholtz, 2024). Nevertheless, despite the potential benefits of beetles, their ability to traverse borders poses a risk. If possessing traits akin to pests, they may become invasive, disturbing different ecosystems and inflicting significant economic consequences (De Wit *et al.*, 2022). In the Anthropocene epoch, a salient characteristic is the breach of geographic barriers by species through human-mediated introductions into novel regions. Some of these introduced species manage to establish self-sustaining populations, exerting adverse effects on native biota, leading to biodiversity loss and consequential ecological, economic, and sociological impacts (Blackburn *et al.*, 2011; Dawson *et al.*, 2017). This phenomenon of biological invasions is now observed globally and is recognized as a primary driver of biodiversity decline (Seebens *et al.*, 2016; Courchamp *et al.*, 2017). However, it particularly impacts developing countries characterized by high biodiversity levels (Frehse *et al.*, 2016). Despite increasing awareness regarding the ecological, economic, and sociological repercussions of human-mediated insect translocations, the rate of alien introductions and establishment continues to surge significantly (Ward and Masters, 2007; Seebens *et al.*, 2017).

Certain species exhibit a higher propensity for invasion compared with others. Analysis of biogeographic patterns in native and non-native beetle populations worldwide has unveiled disparities in representation (Liebhold *et al.*, 2021). Specifically, families such as Bostrichidae and Dermestidae are disproportionately over-represented in non-native beetle populations, while others such as Scarabaeidae, Buprestidae, and Carabidae are notably under-represented. This discrepancy suggests a phenomenon termed "invasion disharmony", highlighting imbalances in the global distribution of beetle families (Liebhold *et al.*, 2021).

15.2 Bio-Invasions

Alien species play a pivotal role in instigating global environmental shifts (Vitousek *et al.*, 1996). Not every species introduced to new environments becomes a successful invader. The "Tens rule" (Williamson, 1996) suggests that only one out of ten introduced species manages to establish itself, and from those, only one in ten turns into a pest (Williamson and Fitter, 1996; Ehler, 1998). Invasive species function either as "drivers" or "passengers" of environmental change (Didham *et al.*, 2005; MacDougall and Turkington, 2005). As drivers, they exert dominance over native diversity via their distinctive traits and mechanisms, while as passengers they proliferate in a region due to anthropogenic factors like habitat destruction. In both scenarios, they exert an influence on native biodiversity (Byers, 2002). Gray (2017) delineates various processes involved in biological invasions, commencing with the species' uptake in native range, followed by transportation and introduction. Upon introduction, an invasive species proceeds to colonize, establish itself, and disseminate (Colautti *et al.*, 2006; Hellmann *et al.*, 2008).

15.2.1 Pathways of invasion

The concept of invasiveness encompasses a series of stages, such as human-mediated arrival, establishment, naturalization, dispersal, and proliferation (Pokhrel *et al.*, 2020). For an alien species to thrive, it must successfully navigate obstacles as it progresses from its initial introduction to becoming established in a new environment. Initially, human assistance facilitates the introduction of species to new regions, enabling populations to overcome geographic barriers (Pyšek and Richardson, 2006). Subsequently, the success of establishment depends on factors such as the pressure exerted by the introduced species, its niche requirements, and interactions with the existing community. Once established, a species must overcome abiotic and biotic barriers to survive, reproduce consistently, and sustain its populations, a phase referred to as naturalization. If a naturalized species can generate sufficient offspring to sustain alien populations and expand into new territories, it is deemed invasive (Singh *et al.*, 2020).

Pathways denote the expected channels facilitating the movement of invasive species across diverse habitats. The categorization of pathways has a well-established history, with the most common ones being sea, land, and air routes (Hulme, 2009). Several cases of introduction have been recorded of species infiltrating India through these channels. For example, cashew was deliberately brought in during the 16th century by early Portuguese explorers, who recognized its significance, utilizing their maritime journeys. Today, owing to globalization, the frequency and impact of invasions have escalated considerably. Some species were intentionally transported into new ecosystems, while others were introduced inadvertently (Singh *et al.*, 2020). These pathways can be classified into three distinct types.

Understanding pathways of introduction of invasives introduced across different landscapes and time holds significant interest and importance. It yields crucial insights for preventive measures, monitoring protocols, and risk assessments concerning the diverse means of introduction (Clout and Williams, 2009; Essl *et al.*, 2015; Saul *et al.*, 2017). Seebens *et al.* (2017) have observed that 37% of documented cases of established alien species are relatively recent (1970–2014), suggesting an uptick in new species introductions over time. The invasion process commences with species introductions (Blackburn *et al.*, 2011), which may occur either deliberately or inadvertently (Essl *et al.*, 2015). Six introduction pathways were outlined by Essl *et al.* (2015), these being deliberate release into natural environments, escape from confinement such as zoos or botanical gardens, unintentional transportation via goods carrying contaminant species, transfer of organisms attached to transport vessels such as ships or boats, movement facilitated by transportation structures such as canals or tunnels, and secondary dispersal across borders subsequent to initial introduction via aforementioned pathways.

The effects of a pathway depend directly on the quantity of invasives it introduces (Wilson *et al.*, 2009). Pathways bringing in higher numbers and greater species diversity of invasives have greater potential to establish multiple alien species, unlike pathways with a smaller number of individuals and less species diversity (Lockwood *et al.*, 2009; Essl *et al.*, 2015; Pergl *et al.*, 2017). Taxa introduced through multiple pathways across habitats and regions are most likely to naturalize compared with those arriving via a single pathway (Pergl *et al.*, 2017). The naturalization process is significantly influenced by trade, considered a major contributing factor. With increasing globalization, the probability rises that numerous species will be introduced and settle beyond their native ranges (Dawson *et al.*, 2017). The presence of multiple alien species in native ecosystems is a concerning outcome of these heightened introductions and subsequent invasions, potentially leading to complexities in community interactions (Johnson *et al.*, 2009).

15.2.1.1 *Introduced as contaminants*

The timber trade is widely acknowledged as one of the most profitable industries, making significant contributions to foreign exchange earnings. However, this trade practice has its drawbacks, particularly in the introduction of alien species. When logs are stored in warehouses for extended periods without proper treatment, they become a favourable habitat for various insects and pathogens (O'Hanlon-Manners and Kotanen, 2004). Consequently, importing such logs also leads to the spread of foreign species. India has also experienced the importation of food products and seeds from other countries. In the past, when there were lax legal restrictions on food imports, the mixing of contaminants in imported food batches proved detrimental to the importing nation (Hulme, 2009). An example of invasive insects being passively transported without direct dissemination is the longhorn beetle (Cerambycidae) in Antarctica, introduced through infested timber logs from base camps (Osyczka *et al.*, 2012). The correlation between the biodiversity and distribution of longhorn beetles in Thailand, particularly in different regions and forest types at middle and low elevations, offers valuable quantitative insights for effective agricultural and forest plantation management, especially considering their potential as pests and invasive species (Yotkham *et al.*, 2021).

15.2.1.2 *Living industry pathway*

The notion of the living industry pertains to the trade of living organisms, encompassing plants utilized in landscaping or horticulture, agricultural produce, aquaculture, the pet industry, and the live seafood trade. When these organisms are transported internationally, there is a significant risk that they might unintentionally transport pests. For instance, the introduction of apple woolly aphids (*Eriosoma lanigerum*) into India is thought to have happened via importation of apple rootstocks from England (Thakur and Dogra, 1980).

15.2.1.3 *Transportation*

India's extensive cultural heritage draws in a significant influx of tourists each year, which in turn heightens the risk of potential invasions through various channels. These include the inadvertent introduction of insect pests or other organisms, the unintentional transfer of alien species between different locations, cargo shipments via air, land, and water, the exchange of food items, as well as the potential transportation of infested fruits and vegetables by tourists (Hall, 2015). Additionally, the nation's well-developed domestic transportation infrastructure, encompassing extensive railway, highway, flight, and harbour networks, further facilitates the unintentional spread of introduced insects across the country, often occurring alongside passengers during transit (Fahrner and Aukema, 2018).

15.2.2 Steps in invasion

A series of sequential stages occur during invasion by a species, namely introduction, establishment, propagation and assimilation (Singh *et al.*, 2020).

15.2.2.1 *Introduction*

To establish themselves in a new area, insects require transportation from their current environment. This form of insect movement, referred to as passive transport, relies on carriers. Presently, human-related activities serve as the primary carriers, resulting in the widespread and difficult-to-control passive transport of invasive species. Despite strict quarantine measures, detecting the incipient stages of insects, which may be inadvertently transported by tourists, poses significant challenges (Schwalbe, 1991). The introduction of single pregnant female insects into a new environment will have dire results if it is a highly reproductive species lacking its natural predators. For instance, timber beetles (Cerambycidae) have been found in Antarctica, believed to have been introduced through logs (Osyczka *et al.*, 2011). Some insects possess inherent migration capabilities, prompting them to traverse long distances when faced with unfavourable conditions (Lundmark, 2010).

Introductions involve the movement of organisms beyond their natural confinement due to human-mediated transportation, whether deliberate or unintentional (Falk-Petersen *et al.*, 2006; Richardson *et al.*, 2011). When introduced, insects can establish viable populations and subsequently spread from their initial introduction point, a phenomenon known as invasive spread. For instance, the ground beetle (*Merizodus soledadinus* Guérin-Méneville) invasive individual occupies an intertidal habitat compared with the species in its natural range which lives in forest litter in Tierra del Fuego and Patagonia (Lebouvier *et al.*, 2011; Hidalgo *et al.*, 2013). Despite this, there is a dearth of substantial data supporting the idea that invasive populations encounter new environmental conditions more frequently than native expanders. Intentional introductions often occur when biocontrol agents, such as the predatory harlequin ladybird (*Harmonia axyridis* Pallas), are released in various countries to manage hemipteran pests (Roques, 2010; Roques *et al.*, 2016).

15.2.2.2 Establishment

Short-term colonization is a commonplace phenomenon, yet an insect cannot be deemed invasive unless it effectively establishes in a new habitat, a feat achieved only when it surmounts environmental obstacles (Renault *et al.*, 2017). In contrast to local colonization, the establishment of alien species hinges on the quantity of individuals introduced into the fresh habitat. Human-induced disturbances in the recipient environment are thought to bolster the establishment of invasive species (Westphal *et al.*, 2007). These disruptions aid establishment either because indigenous species are ill-equipped to cope with the frequency, intensity, or disturbance by humans, or because invasive species often possess adaptations conducive to colonizing such environments. Insects are particularly adept at becoming invasive due to the increased resource availability and diminished resistance from native organisms (Fortuna *et al.*, 2022). Additionally, global warming serves as another catalyst for the rapid proliferation of invasive insects. This is because it alters the availability of resources and suitability of habitats, thus unsettling the ecological equilibrium for indigenous insects and providing an advantage for the alien species (Huang *et al.*, 2011).

Following the initial introduction, there exists a phenomenon known as carry-over effects, wherein past individual experiences or historical factors influence the current performance of organisms. These effects significantly impact the success of colonization and the subsequent proliferation of invasive species (Benard and McCauley, 2008; Piiroinen *et al.*, 2012; O'Connor *et al.*, 2014). Throughout invasions, factors such as poor habitat conditions, reduced carrying capacities, heightened levels of kin competition, and population growth due to the absence of natural enemies facilitate initial rapid spread under non-equilibrium conditions by encouraging increased dispersal (Strecker and Arnott, 2010). When populations arise from metapopulations characterized by spatial and temporal variability, they typically evolve heightened dispersal abilities, thereby facilitating range expansion in the introduced area. Similarly, stable populations established from a limited number of founders or from a highly isolated population may also develop elevated dispersal rates due to kin competition, consequently bolstering spread following introductions (Gandon, 1999; Leturque and Rousset, 2002).

15.2.2.3 Spread

Spreading denotes the process whereby a species initially settled in a specific area expands its distribution to other regions. This expansion is affected by diverse environmental factors, including weather patterns, microclimates, and habitat suitability. Within a community, spreading is further influenced by interactions like interspecific competition and predation (Svenning *et al.*, 2014). Individual insects' abilities to spread are shaped by behavioural traits such as aggressiveness or social tendencies, or morphological changes such as wing development, as seen in aphids (Sloggett and Weisser, 2002). Moreover, established insects can spread via human-mediated transportation, which conserves energy due to its swiftness. However, the success of this spread depends on how hospitable the new habitats are to the insect's presence (Banks *et al.*, 2014).

The impact of evolutionary history and native spatial structure significantly shapes the spread of species in new environments through genetic carry-over effects. For instance, the geographic expansion of native or alien insects favours the selection of individuals with enhanced abilities to cope with oxidative stress, thus profoundly influencing their reproductive success rate and subsequently, population dynamics (Yan and Sohal, 2000; Magwere *et al.*, 2006). These genetic carry-over effects, while less conspicuous than those resulting from adaptive evolution, lead to genetic divergence between populations at the periphery and core due to neutral or unfavourable mutations "riding the wave" and subsequent genetic drift in expanding populations, thus laying the genetic groundwork for adaptation to new environmental pressures (Excoffier and Ray, 2008).

The impact of birth habitat characteristics on the competitive interactions among dispersers, accounting for the density dependence of the colonized environment, is significant. Substandard habitats contribute to the competition between *Tribolium confusum* (confused flour beetle) and *T. castaneum* (red flour beetle). In environments of low quality, *T. confusum* tends to dominate, whereas *T. castaneum* exhibits higher competitiveness in areas with superior habitat quality (Van Allen and Rudolf, 2016). The energy reserves of dispersers have enduring effects on their subsequent dispersal. Insects with greater bodily reserves are capable of covering longer dispersal distances. For instance, the flight distances of the beetle, *Monochamus galloprovincialis* are influenced by the level of energy reserves allocated to the thorax (David *et al.*, 2013).

15.2.2.4 Successful establishment and influencing factors

Biological invasions exhibit considerable dynamism, with the majority attributed to human activities, a phenomenon known as "invasion debt" (Essl *et al.*, 2011; Ööpik *et al.*, 2013). Successful establishment (also known as ecesis) denotes the presence of a persistent and self-sustaining population of invasive species after their introduction into a new environment (Gray, 2017). The success of establishing an invasive species hinges on both the traits of the invading species and the conditions of the invaded environment. The intertwined concepts of invasiveness and invisibility are emphasized (Hui *et al.*, 2016).

Numerous explanations have been put forward to elucidate the success of invasive species. One such explanation is the "enemy release hypothesis", proposing that the escape of alien species from their native parasites aids in their invasion success (Darwin, 1859; Elton, 1958). Another notion is the "empty niche hypothesis", suggesting that invasive species prosper by exploiting resources in the introduced environments that are not utilized by native species (Elton, 1958; MacArthur, 1970; Levine and D'Antonio, 1999; Mack *et al.*, 2000). The "resource availability hypothesis" posits that invasion success hinges on the accessibility of resources, which can increase through resource provision or reduced competition from indigenous plants. The "species richness hypothesis" theorizes that communities with diverse species diversity are more resistant to invasion (Elton, 1958). The "limiting similarity hypothesis" suggests that invasive species encounter difficulties in establishing themselves in communities where native species possess comparable traits. The "novel weapons hypothesis" proposes that invasion success can be attributed to unique interactions between alien species and the native biota. The "disturbance hypothesis" contends that ecosystems subjected to disturbances are more vulnerable to invasion compared with less disturbed ecosystems (Hobbs, 1989). Additionally, there exists a phenomenon referred to as "invasional meltdown" or secondary invasion, whereby the successful establishment of one invasive species paves the way for subsequent invaders (Catford *et al.*, 2009; Jeschke *et al.*, 2012). In essence, these theories underscore the diverse independent or synergistic factors influencing the destiny of newly introduced and subsequently invading species.

15.3 Why Invasive Species are Problematic

The primary threat to agro-ecosystems likely stems from the profound impact of invasive pest species. When these pests invade new territories where they lack natural predators or effective suppressive methods, they quickly establish dominance and compete with native species (Paini *et al.*, 2016). This competition often leads to the displacement of indigenous species by non-native ones, ultimately risking significant alterations and potential collapse of entire ecosystems. Through resource competition and disruption of nutrient cycles, invasive pests profoundly alter the structure and species diversity of ecosystems, causing adverse effects on biodiversity, ecosystem health, economics, and facets of human well-being. Their occurrence often results in decreased agricultural productivity (Paini *et al.*, 2016). The impact of invasive species on ecosystems can yield both advantageous and detrimental outcomes, contingent upon the particular circumstances and human perceptions (Shackleton *et al.*, 2016; Kueffer and Kull, 2017). Hence, it is crucial to observe invasive species and their interactions within unfamiliar ecosystems. Although only a minority (5–20%) of introduced species pose issues, their effects on ecosystem organization and processes endure and hold significance (Vilà *et al.*, 2010; McGeoch *et al.*, 2016).

15.3.1 Characteristics of invasive insects

Forecasting the outcomes of introductions present considerable challenges due to the complex interactions between introduced alien species and native communities (Buckley, 2017). Key characteristics exhibited by invaders significantly influence the large-scale patterns of biological invasions (Iannone *et al.*, 2016). Traits linked to reproduction and dispersal are particularly critical, directly influencing an alien species' capacity to naturalize or become invasive. Invaders often possess pre-adaptations enabling them to thrive in specific climates or habitats, and they may also undergo local adaptations to coexist with native species. Furthermore, they are likely to have genetic modifications under selection pressures (Sakai *et al.*, 2001). The traits of invaders can exhibit temporal variations and differ across various introduction pathways (Essl *et al.*, 2015). Native beetle species themselves could potentially become invasive if introduced. For instance, bark beetles and borers exhibit four invasion strategies: inbred with inbreeding and flightless males; polyphagous with wood feeding and high fecundity; intermediate with mono- or oligophagy and short life cycles; and aggressive with monophagy and aggregation pheromones (Demidko *et al.*, 2021).

15.3.2 Major challenges for invasive insect science

The primary focus of invasion biology lies in practical applications, centring on three fundamental inquiries: (1) identifying the species most inclined to invade new territories, (2) pinpointing the locations where invasions are most likely to thrive, and (3) determining effective ways to utilize knowledge for managing biological invasions (Roura-Pascual *et al.*, 2024). As invasions get established via distinct phases of arrival, spread, and establishment, the strategies for management evolve from prediction and prevention to early detection and swift response, eventually transitioning to mitigation and comprehensive management (Roura-Pascual *et al.*, 2024). An analysis of the bioeconomic aspects typically confirms that efforts aimed at preventing or containing invasions prove more cost-effective than addressing damages or attempting ecosystem restoration post-invasion. Tackling issues necessitates multidisciplinary research and a concentration on objectives with broad public support and measurable success indicators. Proposed solutions, whether they entail novel sampling methods or control technologies, should be transparent, clearly elucidating their costs and benefits relative to existing practices (Venette and Hutchison, 2021).

15.4 Most Threatened Countries for Beetle Invasion and their Long-term Effects

Beetles pose significant challenges to perennial and annual crops, particularly in agroforestry and forested areas. They inflict considerable harm, internally and externally, by consuming wood materials and occasionally vegetative parts. One such polyphagous beetle, associated with a fungus known as ambrosia beetles

Table 15.1. Invasive beetles in the tropics and subtropics. (Author's own table)

Invasive beetle	Origin	Reported region	Hosts
Coffee berry borer, *Hypothenemus hampei* Ferrari (Curculionidae: Scolytinae; Coleoptera)	Africa	Northeast Africa, India and Sri Lanka	Arabica and robusta types of coffee
Asian long-horned beetle, *Anoplophora labripennis* (Motschulsky) (Cerambycidae: Coleoptera)	Eastern China and Korea	North America and parts of Eurasia	Hardwood trees
Tiger longicorn beetle, *Xylotrechus chinensis* (Chevrolat) (Cerambycidae: Coleoptera)	East Palaearctic region	Europe and USA	Mulberry trees
Citrus Long-horned Beetle, *Anoplophora chinensis* (Forster) (Cerambycidae: Coleoptera)	China, Japan, and Korea	USA and Canada	Natural vegetation
Gazella scarab, *Digitonthophagus gazella* (Fabricius) (Scarabaeidae: Coleoptera)	Afro-Asia	Australia, USA, Mexico and South America	Dung affects native dung beetle diversity
Red dung beetle, *Aphodius fimetarius* (L.) (Scarabaeidae: Coleoptera)	Europe and Asia	Australia, New Zealand, USA, and Canada	Dung affects native dung beetle diversity
Khapra beetle, *Trogoderma granarium* Everts (Dermestidae: Coleoptera)	Indian subcontinent	Asia, Middle East, Africa and Europe	Stored cereals and cereal products
Japanese beetle, *Popillia japonica* Newman (Scarabaeidae: Coleoptera)	Japan	USA and Canada	

(*Euwallacea fornicates* Eichhoff), has wreaked havoc on South Africa's forests and urban environments. This pest particularly targets maples, avocados, liquidambars, oaks, plane trees, and coral trees, causing severe damage (Bega, 2021). Another example is the introduction of *Rhynchophorus ferrugineus* (Olivier) to California in 2010, which led to the widespread death of the Canary Island date palm (*Phoenix canariensis* Chabaud). In India, invasive insects have caused staggering economic losses totalling approximately $127.3 billion over the past 60 years (Anonymous, 2022). The elusive habitats and concealed feeding habits of these alien beetles have caused significant concern among farmers and researchers. To address these challenges, it is crucial to identify and monitor key beetle species causing economic damage across tropical and subtropical regions (Table 15.1).

E. fornicatus species complex, comprising ambrosia beetles, has emerged as a significant concern for trees, showcasing a wide host range that encompasses crucial agricultural crops. Originally from Southeast Asia, the invasives were introduced into multiple regions, leading to substantial harm to numerous tree species. One study highlights several instances of outbreaks of *E. fornicatus* in Europe (Schuler *et al.*, 2023). The introduction of the ambrosia beetle *E. fornicatus* from Asia into South Africa in 2012 prompted an investigation into its dispersal pathways and population genetics. *E. fornicatus* individuals were sampled across South Africa, revealing a predominantly homogeneous population with one haplotype shared with California, Israel, and Vietnam, while a second haplotype present in two localities suggests multiple invasion events (Bierman *et al.*, 2022).

The Asian ambrosia beetle, *Xylosandrus crassiusculus* (Motschulsky), an invasive of Asian origin, spread globally, with first records in Queensland, Australia (2011) and New Zealand (2019); phylogenetic analysis revealed distinct populations in Australia, suggesting independent introductions (Tran *et al.*, 2022). *Xylosandrus compactus* (Eichhoff) and *X. crassiusculus*, originating from Asia, are invasive polyphagous ambrosia beetles found globally in circumtropical regions; MaxEnt future climate projections indicate potential expansion of *X. compactus* into currently unreported areas, particularly in Europe, Central America, Southeast Asia, and some Oceanian coastlines (Urvois *et al.*, 2021). *X. crassiusculus* exhibits two distinct genetic lineages (cluster 1 and cluster 2), with MaxEnt algorithm predicting cluster 2's suitability in South America, Europe, Central Asia, South Africa, and Oceania; future climate projections suggest a shift in suitability, decreasing in southern Europe and increasing in north-eastern Europe by 2071–2100 (Urvois *et al.*, 2024).

The success of Xyleborine ambrosia beetles in colonizing novel habitats globally is attributed to their cryptic lifestyle, inbreeding, fungal mutualism and broad host range. The invasive black timber bark beetle, *Xylosandrus germanus* (Blandford), originating from East Asia, particularly Honshu and/or Hokkaido, Japan, exhibits highly uniform genetic structure in non-native habitats such as North America and Europe, suggesting independent invasions of these continents despite a shared source lineage (Dzurenko *et al.*, 2021). In Romania, *Ips duplicatus* (Sahlberg) is established across most of the Norway spruce's range, *X. germanus* is spreading with high populations in certain areas, and *Neoclytus acuminatus* is confined to the warmest regions based on trap data with various attractants (Olenici *et al.*, 2022).

15.5 Mitigation of Beetle Invasion

The management approach for invasive insects involves addressing pest invasion across three distinct stages. Initially, before the pest is introduced, preventative actions are implemented to hinder their entry. These actions, such as pest risk analysis (PRA), quarantine, and monitoring, are widely recognized as the most effective means of managing invasive species (Venette and Hutchison, 2021). Secondly, if the species has been introduced but has not yet spread to neighbouring areas, post-quarantine procedures are enacted. These include rejecting shipments that harbour the pest and utilizing fumigation to eliminate the pests from the affected batch (Witcombe and Erskine, 1984). Lastly, curative measures employed against the established invasive insect. These measures encompass cultural, mechanical, biological, physical, and chemical methods of management (Ruzzier *et al.*, 2021).

15.5.1 Prediction and prevention

The primary recommended strategy for addressing the issue of alien invasive species revolves around anticipation and prevention. The core concept involves identifying potential species or pathways that present unacceptable risks and implementing measures to hinder their introduction into vulnerable areas. The success of this approach hinges on having a clear benchmark for effectiveness as arthropod species continue to increase steadily (Pysek *et al.*, 2020). However, the escalation in global transactions surpassed the rate of new species arrivals, indicating that biosecurity endeavours have achieved some efficacy. This does not provide a conclusive assessment of success (Venette and Morey, 2020). There is optimism regarding the potential integration of future systems to forecast threats and prevent introductions (Poland and Rassati, 2019).

The identification of insects during migration necessitates sophisticated instrumentation, possibly leveraging environmental DNA or the examination of volatile chemical signatures (Lucas, 2011; Larson *et al.*, 2020). Evaluating the threat posed by pests concerning spatial disinfection is pivotal for biosecurity endeavours. It serves the purpose of determining the suitability of importing a product, organizing surveillance initiatives, and formulating post-incursion strategies if by chance the species is infiltrating a susceptible area (Venette *et al.*, 2010). These evaluations predominantly concentrate on pinpointing the areas most susceptible to invasion and potential harm caused by pests, considering the requisite conditions for their establishment. Nonetheless, there exists an urgent demand for dependable, scalable, and cost-efficient methodologies to facilitate the development of such evaluations (Kim *et al.*, 2016). The growing recognition of the conceptual challenge involves the development of models capable of effectively adjusting to new spatial and temporal contexts, particularly within the realms of machine learning and statistical models (Morey and Venette, 2020). Although process-based models are appealing, their demanding data requirements make them difficult to use across numerous species of concern (Venette, 2017).

15.5.2 Early detection and rapid response

If prediction and prevention measures fail, the next feasible step is to promptly identify and either eliminate or manage emerging populations of high-risk species. Nevertheless, the current state of early detection of invasive insects in agricultural and forest ecosystems heavily relies on random discoveries, despite global advancements in detection methods (Venette, 2017). There is an urgent necessity for research to improve this approach by determining the optimal timing, locations, and methodologies for detecting new invasive species and confirming their identity upon detection. Pest risk maps provide valuable guidance by identifying areas vulnerable to invasion, assisting in selecting sites for early detection surveys. The boundaries of the endangered areas surpass available resources for comprehensive surveys. Therefore, further research is essential to accurately evaluate the likelihood of detecting low-density invasive arthropods using specific sampling techniques (Venette *et al.*, 2002). In the domain of plant pathogens, a "risk-based surveillance" method has been developed, focusing sampling efforts in areas where the most susceptible and valuable host plants are found (Mastin *et al.*, 2020).

Enhancing the affordability of identifying invasive species is a crucial objective that warrants further investigation, particularly concerning invasive insects. Although spatial optimization of sampling effort has been updated, these methodologies are computationally demanding (Yemshanov *et al.*, 2017, 2019).

Recent advancements in genomics, coupled with traditional taxonomic knowledge, offer a promising avenue for swiftly confirming species identities. Utilizing genomic methodologies also provides an added advantage by enabling the differentiation of various strains, biotypes, or haplotypes within a species. This capacity to distinguish specific haplotype information facilitates a more precise determination of the geographic origins of invasive populations. Numerous instances, such as the walnut twig beetle (*Pityophthorus juglandis* Blackman) (Rugman-Jones *et al.*, 2015), and the brown marmorated stink bug (*Halyomorpha halys* Stål) (Xu *et al.*, 2014), have illustrated how haplotype data accurately reflect the origins of pests. Through precise identification of a pest, targeted interventions can be promptly initiated to search for natural enemies, particularly species-specific parasitoids. These natural enemies can then undergo comprehensive assessments to ensure their safety and effectiveness before being deployed for biological control (Wyckhuys *et al.*, 2020).

Using the MaxEnt model, potential distribution of *Heteronychus arator, Oryctes boas*, and *Amphimallon majale* across continents was predicted; hence, species monitoring is imperative to mitigate financial losses ((Wang *et al.*, 2023b). Small hive beetles (*Aethina tumida* Murray) are originally parasites found in social bee colonies native to sub-Saharan Africa. However, they have transitioned into invasive species, spreading to various provinces in China, where they infest colonies of *Apis cerana* and *Apis mellifera* (Liu *et al.*, 2021). MaxEnt modelling predicts high risk of large hive beetle (*Oplostomus fuliginetus* Olivier) invasion into non-native regions, primarily southern Europe, due to temperature variations, aiding quarantine authorities in preventing its expansion beyond sub-Saharan Africa (Abou-Shaara *et al.*, 2021). Managing invasive pest species after their introduction is essential. Theoretical models of pest management options for the potential scenario of the Colorado potato beetle invading potato farms in the Philippines emphasize the importance of employing biological control agents and considering the use of chemical pesticides (Almarinez *et al.*, 2023).

15.5.3 Mitigation

Addressing invasive species requires both short-term and long-term strategies within integrated pest management (IPM) (UC IPM, 1996). Initially, emphasis is often placed on insecticidal control to safeguard growers' investments. Simultaneously, funding is allocated to research on biological control, pest-resistant varieties, cultural controls (Radcliffe *et al.*, 2009), physical barriers (Rogers *et al.*, 2016), and trap-based systems exploiting insect behaviour (Gregg *et al.*, 2018). Breeding for resistance against invasive pests is particularly crucial for trees and perennial crops. Additionally, emerging technologies such as transgenic insecticidal plants (Hutchison *et al.*, 2010) or genetic biocontrol agents, including gene-drives, hold promise to complement or potentially replace IPM programmes, pending regulatory approvals (Maselko *et al.*, 2017; Sudweeks *et al.*, 2019).

15.6 Invasive Beetles in the Tropics and Their Impact

Some important invasive beetles of tropics and subtropics are listed in Table 15.1.

15.6.1 *Hypothenemus hampei* Ferrari

The *H. hampei* is an immensely harmful insect that presents a significant peril to global coffee production. It inflicts substantial harm to both the quantity and quality of coffee, resulting in yearly losses surpassing US$500 million (Johnson *et al.*, 2020a). The female beetle burrows into the coffee fruit, commonly known as a berry, and penetrates the seed, or "bean", where it constructs tunnels for reproduction. Subsequently, the larvae feed on the seed's endosperm. This pest has spread worldwide, originating from the forests of Africa and extending to coffee-producing nations globally, with the exceptions of Australia, China, and Nepal (Johnson *et al.*, 2020a). *H. hampei* was first documented in Liberian coffee plantations in 1897 (Poland and Rassati, 2019), followed by sightings in the Democratic Republic of Congo in 1901 and Zaire in 1903 (Kim *et al.*, 2016). The precise origin and initial host of *H. hampei* in Africa remain uncertain; however, some scholars propose that it might have originated in Ethiopia alongside *Coffea arabica* (arabica) but wasn't detected in Ethiopian coffee farms until 1998 (Venette *et al.*, 2010; Lucas, 2011; Larson *et al.*, 2020). It is theorized that the original host of *H. hampei* was likely *Coffea canephora* (robusta), prevalent at lower altitudes in West and Central Africa (Venette, 2017; Morey and Venette, 2020). The hypothesis posits that *H. hampei* could have spread from infested robusta coffee berries in West Africa to arabica coffee in Saudi Arabia or Ethiopia, where it was initially introduced before the 15th century (Kim *et al.*, 2016).

Several studies employing genetic analysis techniques have investigated the transmission of *H. hampei* and pinpointed potential source regions for future outbreaks (Benavides *et al.*, 2005; Chapman *et al.*, 2015; Maselko *et al.*, 2017; Sudweeks *et al.*, 2019). Amplified fragment length polymorphism (AFLP) fingerprinting studies determined that *H. hampei* likely had three distinct introductions into the Americas, with West Africa emerging as the probable origin for introductions into America and Asia (Benavides *et al.*, 2005, 2006, 2007). A comprehensive analysis utilizing the mitochondrial cytochrome oxidase (COI) gene identified five distinct populations spanning Africa, Asia, Central and South America (Gauthier, 2010). It is believed that the pest entered in the infested beans to new regions during importation or unintentionally transported by plantation workers from infested farms (Gauthier, 2010; Chapman *et al.*, 2015).

The swift expansion of *H. hampei* in Ethiopia is directly correlated with rising temperatures (Jaramillo *et al.*, 2009). The Ethiopian highlands have witnessed a temperature increase of 0.39–0.65°C from 1979 to 2005 (Trenberth *et al.*, 2007). Previously, lower temperatures inhibited *H. hampei* from completing a full reproductive cycle per season. However, due to climate change, an estimated 0–2 reproductive cycles occur annually with a mere 1–2°C rise (Morey and Venette, 2020). Should temperatures rise by more than 2°C, the geographic distribution of the pest would shift in terms of altitude and latitude. *H. hampei* has already been detected at higher altitudes in East Africa, indicating the ongoing realization of these forecasted changes (Kyamanywa *et al.*, 2009; Jaramillo *et al.*, 2011). Farms situated at higher elevations may witness 5–10 reproductive cycles, while those at lower and mid-elevations could experience 11–16 cycles (Jaramillo *et al.*, 2011). The suitable area for cultivating arabica coffee might shrink by 39–59% by the end of the century (Moat *et al.*, 2017).

Shade trees planting in coffee plantations is recommended as a measure to counteract the adverse effects of increasing temperatures on coffee production. This approach offers numerous advantages, including erosion prevention, wind protection, improved nutrient circulation, temperature regulation, and supplementary income from growing other tree species (Jaramillo *et al.*, 2011; Vega *et al.*, 2015). However, the utilization of shade trees in coffee farms has sparked debate. According to some, trees might heighten infestations of *H. hampei* thus reducing coffee yields. Investigations have documented elevated levels of *H. hampei* infestation in shaded coffee environments (Bosselman *et al.*, 2009; Larsen and Philpott, 2010; Mariño *et al.*, 2016), whereas others have observed reduced infestation possibly due to alterations in environmental conditions and the presence of natural predators (Teodoro *et al.*, 2009; Mariño *et al.*, 2016). Conflicting findings also exist regarding the impact of shade on coffee yields, with research indicating higher, lower, or unaffected yields in shaded coffee settings (Soto-Pinto *et al.*, 2002; Jaramillo, 2013; Meylan *et al.*, 2017; Johnson *et al.*, 2020b). These discrepancies underscore the necessity for coffee-producing regions to evaluate various species and levels of shade trees to identify effective methods for addressing *H. hampei* impacts amid shifting climate patterns.

15.6.2 *Anoplophora glabripennis* Motschulsky

The Asian long-horned beetle, *A. glabripennis*, which is native to eastern China and Korea, poses a significant danger to hardwood trees across North America and parts of Eurasia. These beetles are identifiable by their large size, shiny black bodies adorned with irregular white or yellow-orange spots on wing covers, and lengthy antennae (Hu *et al.*, 2009). They have a life cycle spanning 1–2 years, remain active from April or May until October. During this time, beetles consume twigs and leaves while female beetles lay anywhere from 50 to 125 eggs, which typically hatch within 7–14 days. The resulting larvae burrow into the sapwood and heartwood, leaving behind waste resembling sawdust. Once they reach lengths of 30–50 mm, the larvae undergo pupation and subsequently emerge as adults through holes in the tree bark (Wang *et al.*, 2023a) *A. glabripennis* exhibits a preference for at least 15 tree genera, particularly poplar, maple, willow, and elm species. In its native habitat on the Korean peninsula, the Asian longhorned beetle naturally maintains low population densities without causing significant harm to trees (Hu *et al.*, 2009). However, in China, particularly in western regions, the beetle's population underwent a dramatic surge, resulting in extensive tree damage. This surge is attributed to the widespread cultivation of poplar trees as part of China's Three-North Shelterbelt Programme, initiated in the 1960s. This programme aimed to establish extensive tree shelterbelts to combat desertification, prevent soil erosion, and boost pulpwood production. The proliferation of host trees facilitated the transformation of *A. glabripennis* into a serious pest in China (Hu *et al.*, 2009).

The discovery of *A. glabripennis* in the USA and Canada dates back to 1992, but it was effectively eradicated. However, subsequent sightings have occurred in various locations including New Jersey, Illinois, Ohio, and Massachusetts; however, successful campaigns were launched in Illinois and New Jersey (Carter *et al.*, 2009). Additionally, the beetle has been detected in Japan, Austria, France, Germany, and Italy. Its primary mode of transport is through solid-wood packing materials, often containing larvae or pupae, such as pallets and crates. This invasive species inflicts substantial damage on trees, prompting affected countries to implement elimination strategies (Yasui *et al.*, 2024). These measures typically involve establishing an eradication zone where infected trees are removed, and unaffected ones are treated with insecticides. Areas beyond this zone undergo regular inspections for signs of infestation. The eradication process is deemed successful when no new infested trees are discovered after several years of monitoring (Wang *et al.*, 2023a). The USA granted $373 million towards eradication efforts, resulting in the removal of more than 110,000 trees and the application of pesticides. Citizens are urged to report sightings of infestations, refrain from transporting contaminated wood, and researchers are exploring biological control methods utilizing natural predators to manage the beetle population (Johnson *et al.*, 2024).

15.6.3 *Xylotrechus chinensis* Chevrolat

The tiger longicorn beetle (*Xylotrechus chinensis*), originating from the East Palaearctic region, was initially detected in Europe in Catalonia, Spain in 2018. However, there are indications that it had already established a presence in Catalonia as early as 2012 (Monteys and Tutusaus, 2018). While intercepted in Germany and the USA, it did not manage to establish itself successfully in those regions. The beetle's behaviour, with the damage it causes to mulberry trees, suggests potential further spread within Europe (Monteys and Tutusaus, 2018). In 2017, approximately 200 infested mulberry trees were discovered near the harbour of Heraklion, Crete, Greece, likely introduced through imported goods in 2014 or 2015. Subsequent cases of *X. chinensis* infestations were reported in July 2018 in the Valencia region of Spain. In France, the presence of *X. chinensis* was confirmed in October 2017, with indications of possible regional spread (Monteys and Tutusaus, 2018). The European and Mediterranean Plant Protection Organization (EPPO) included *X. chinensis* in its alert list in August 2018, emphasizing the importance of reporting any new outbreaks and implementing suitable control measures. As of now, *X. chinensis* has not been categorized as a quarantine pest in Europe, and mandatory control measures have not been enforced (Bragard *et al.*, 2021).

The larvae of *X. chinensis* primarily consume and burrow into mulberry trees, particularly those belonging to the *Morus* species. Experiments found that grapevines are not utilized as alternative hosts by *X. chinensis* (Bragard *et al.*, 2021). Currently, mulberry trees, especially *Morus alba*, are prevalent in numerous towns across the southern Mediterranean region, providing shade and enhancing aesthetics. Their widespread presence significantly contributes to the propagation and harm inflicted by *X. chinensis* (Monteys and Tutusaus, 2018). Researchers have warned about the detrimental effects on mulberry trees and the consequent economic impacts, particularly in Catalonia. The existence of infested trees not only results in their demise but also raises safety concerns in public parks and thoroughfares due to the heightened risk of falling branches (Sartoi Monteys *et al.*, 2021).

15.6.4 *Anoplophora chinensis* Forster

This specific pest, known by different names such as the citrus longhorn beetle, citrus root cerambycid, and mulberry white spotted longicorn, displayed capability to spread beyond its original habitat. It boasts a broad native range and thrives within it. Demonstrating high adaptability to diverse environments, it consumes a wide variety of food sources. Moreover, it tends to exhibit gregarious behaviour, increasing the likelihood of unintentional international transportation (Keena *et al.*, 2021). Due to confusion in taxonomy, the distribution of this pest, formally identified as *A. chinensis*, has been incorrectly associated with other *Anoplophora* species. Its primary habitats are China, Japan, and Korea, encompassing various regions within these countries (Ge *et al.*, 2014). Although interceptions have occurred in the USA and Canada, there is no evidence of established populations. Reports suggesting its establishment in the UK have been debunked; findings between 2005 and 2010 were traced back to imports from China (Van Der Gaag *et al.*, 2010). The pest was first noticed on natural vegetation in Europe in 2001, though it might have been present since 1997. Efforts to eradicate it from Italy have been ongoing since 2000, resulting in regions being free from infestation while others require containment measures to prevent further spread (Hérard *et al.*, 2006).

15.6.5 *Digitonthophagus gazella* Fabricius

The dung beetle species *D. gazella*, originally from Afro-Asia, has achieved global distribution. Initially introduced to Australia in 1968 for dung burial and pest control purposes, it gained popularity due to its effectiveness. Subsequently, it was transported to Hawaii in 1970 and later disseminated along the northern Atlantic coast of the USA, gradually expanding to other states (Noriega *et al.*, 2020). *D. gazella* has also established populations in South America, specifically in Mexico, Colombia, and Peru. Predictive models indicate its potential distribution spanning approximately 45° north to 45° south latitude (Noriega *et al.*, 2017). The species has exhibited varying rates of expansion across different regions. Its widespread presence is facilitated by the abundance of cattle dung and its ability to thrive in diverse climates. Recognized as one of the most pervasive invasive dung beetle species globally, its presence has contributed to a 40% decline in native dung beetle diversity in certain Brazilian regions over a span of 26 years (Filho *et al.*, 2017).

The widely distributed dung beetle species, *D. gazella*, is found in tropical, subtropical, and warm temperate regions. Its high reproductive capacity and effective dung burial behaviour contribute to its success. While typically found in open habitats, *D. gazella* has been observed in rainforests and deciduous forests. Its adaptability to different habitats can negatively impact native dung beetles and other ecological communities. In Colombia, the presence of *D. gazella* has led to changes in frog prey assemblages, with native frog species becoming significant predators of the beetle (Pokhrel *et al.*, 2020).

15.6.6 *Aphodius fimetarius* Linnaeus

The red dung beetle (*Aphodius fimetarius*), originally native to Europe and Asia, has expanded its habitat to include Australia, New Zealand, the USA, and Canada through unintentional introductions. It belongs to a group of closely related species, making it challenging to differentiate based solely on physical characteristics (Pokhrel *et al.*, 2020). Unlike other *Aphodius* species, *A. fimetarius* is present on every continent except Antarctica and is one of the largest within its genus, showcasing variations in colour and morphology. Despite minor morphological disparities such as the endophallus and pronotal punctation, it exhibits a high reproductive capacity, with females capable of laying over 100 eggs in their lifetime (Pokhrel *et al.*, 2020). *A. fimetarius* can be frequently found in significant numbers within a single dung pad, particularly in aged dung, and remains active throughout grazing seasons, with two generations annually (Pokhrel *et al.*, 2020).

This species exhibits a bimodal seasonal activity cycle, with adults appearing during both spring and autumn. The emergence of individuals that overwinter is influenced by factors such as habitat orientation and soil composition (Holter, 1982). Female members deposit eggs into the soil, where larvae subsequently develop within dung, eventually undergoing pupation either within the dung itself or in the soil. Eggs, larvae, prepupae, and adults are capable of surviving through the winter season. While the species endure a broad range of temperatures, elevated temperatures may induce heightened stress due to increased metabolic rates (Pokhrel *et al.*, 2020). *A. fimetarius* demonstrates a well-suited adaptation to cold environments by maintaining a low metabolic rate at lower temperatures and exhibiting resilience to cold conditions. Its considerable body size and preference for colder climates provide it with a competitive advantage over other dung-dwelling species within the Aphodine group. During cold and dry seasons, *A. fimetarius* effectively decomposes and utilizes large, aged dung pads, thereby avoiding competition with tunnelling and rolling dung beetles. Moreover, the species displays genetic diversity, suggesting its potential as an invasive dung beetle species (Pokhrel *et al.*, 2020).

15.6.7 *Trogoderma granarium* Everts

The khapra beetle, also known as the cabinet beetle, *T. granarium*, poses a significant quarantine threat and is widely acknowledged as a leading invasive species worldwide. Originating from the Indian subcontinent, it has proliferated across Asia, the Middle East, Africa, and Europe primarily through the transportation of durable goods (Yadav *et al.*, 2021). Flourishing in stored cereals and related products, it infests around 100 commodities. It is notorious for its "dirty feeding" habits, it damages more grain kernels than it consumes, leaving behind contamination in the form of body parts and setae. Severe infestations render the grain unappealing and unfit for commercial purposes, necessitating stringent quarantine measures (Athanassiou *et al.*, 2018). Detecting and controlling *T. granarium* present a formidable challenge due to its tendency to seek refuge and an extended diapause period lasting over 6 years. The diapausing larvae exhibit resilience against a range of control methods, including extreme temperatures and fumigants (Wilches *et al.*, 2018). Since the adults lack flight capabilities, the spread of the species across diverse ecological zones is primarily facilitated by the diapausing larvae. Moreover, even the non-diapausing larvae have demonstrated tolerance to commonly employed insecticides (Kavallieratos *et al.*, 2017; Al Hamadani *et al.*, 2025).

15.6.8 *Popillia japonica* Newman

Known as the Japanese beetle, *P. japonica* originally hails from Japan and was initially sighted in the USA in New Jersey in 1916. This invasive pest made its way to Halifax, Nova Scotia, Canada, by 1944 (Nardi *et al.*, 2024). Despite efforts to curb its spread through insecticide applications, populations managed to establish themselves in southern Ontario and Quebec by 1998. Subsequent detections occurred in British Columbia in 2017, and the beetles have since been identified in Portugal, Terceira Island, Italy, and Switzerland. Although Bermuda's climate was unsuitable for the beetles to thrive, they have found favourable conditions in various parts of Europe, Asia, the Pacific region, and potentially in Spain, France, Romania, Ireland, and the UK. Additionally, suitable habitats may also exist in South America, New Zealand, Australia, and Africa (Kistner-Thomas, 2019; Tuomola and Hannunen, 2024).

15.6.9 *Oryctes rhinoceros* Linnaeus

O. rhinoceros, commonly known as the coconut rhinoceros beetle (CRB), is a highly destructive pest that poses a serious threat to coconut and oil palm cultivation, particularly across Asia and the Pacific Islands. The beetle bores into the crowns of palm trees, damaging fronds and inhibiting new growth, often resulting in reduced yields or tree death. Historically, effective biological control measures, primarily involving the use of *Oryctes rhinoceros nudivirus* (OrNV) and entomopathogenic fungi from the *Metarhizium sorokiniae* species complex were successful in suppressing CRB populations (Paudel *et al.*, 2021). However, a new biotype known as CRB-G has recently emerged and spread rapidly across the Pacific, causing unprecedented levels of damage. This biotype exhibits a degree of resistance to previously effective biocontrol agents, undermining decades of pest management efforts (Paudel *et al.*, 2021). The initial introduction of CRB to the Pacific occurred accidentally in Samoa in 1909, from where it gradually spread to various countries and territories, severely affecting coconut production and, consequently, the livelihoods of local communities that depend on coconut-based economies. The resurgence and persistence of CRB have highlighted the urgent need to revise and strengthen IPM strategies. Experts have recommended the development of more sustainable, adaptive approaches to CRB control, including improved monitoring systems, deployment of more effective biocontrol strains, and greater international collaboration. In Samoa, where coconut is a culturally and economically vital crop, these efforts are particularly critical. As such, there is growing support for the Samoan government's initiatives aimed at coconut replantation and the restoration of affected plantations, ensuring the long-term viability of the coconut industry and rural livelihoods (Paudel *et al.*, 2022).

15.7 Conclusion

Biological invasions are a frequent phenomenon in nature, yet not every introduced species evolves into an invasive one. This discrepancy arises due to a variety of challenges they encounter, such as geographic, reproductive, and dispersal barriers. Species' ranges are in constant flux across different spatial scales, rendering the structuring of ecological communities a multifaceted process. Exotic species typically inhabit specific niches within degraded or unoccupied habitats, establishing self-sustaining populations. An optimal ecosystem facilitates interactions and resource sharing between local and invasive species without adverse impacts on the environment and local communities. The redistribution of exotic species heightens the invasion risk, as they may outcompete native species due to better adaptation to novel environments. Climate change, as an abiotic factor, can further modify native species distributions, potentially fostering range expansions and increased invasiveness. However, the impact of native species in such scenarios is generally less severe owing to their shared history and environmental context with coexisting species. Conversely, exotic species are often more invasive and pose greater threats due to lack of natural checks and balances. Non-indigenous invasive insects and diseases represent significant threats to ecosystems.

Acknowledgement

The authors would like to thank the Director of CSIR-IICT, "IICT/Pubs./2024/377".

References

Abou-Shaara, H., Alashaal, S.A., Hosni, E.M., Nasser, M.G., Ansari, M.J. *et al.* (2021) Modeling the invasion of the large hive beetle, *Oplostomus fuligineus*, into North Africa and South Europe under a changing climate. *Insects* 12(4), 275. DOI: 10.3390/insects12040275.

Al Hamadani, A.H.A., Al-fatlawi, M.K.M. and Hadi, M.H. (2025) Efficacy of some aquatic plant extracts on the Khapra beetle *Trogoderma granarium* Everts. *Indian Journal of Entomology* 87(1), 7–10. DOI: 10.55446/IJE.2024.1938.

Almarinez, B.J.M., Amalin, D.M., Aviso, K.B., Cabezas, H., Lao, A.R. *et al.* (2023) Network modeling for post-entry management of invasive pest species in the Philippines: The case of the colorado potato beetle, *Leptinotarsa decemlineata* (say, 1824) (Coleoptera: Chrysomelidae). *Insects* 14(9), 731. DOI: 10.3390/insects14090731.

Anonymous (2022) The cost of invasive species bears heavy on Indian economy, finds study. Mongabay Series. Mongabay. Available at: https://india.mongabay.com/2022/05/the-cost-of-invasive-species-bears-heavy-on-indian-economy-finds-study/

Athanassiou, C.G., Phillips, T.W. and Wakil, W. (2018) Biology and control of the Khapra beetle, *Trogoderma granarium*, a major quarantine threat to global food security. *Annual Review of Entomology* 64(1), 131–148. DOI: 10.1146/annurev-ento-011118-111804.

Aukema, J.E., Leung, B., Kovacs, K., Chivers, C., Britton, K.O. *et al.* (2011) Economic impacts of non-native forest insects in the Continental United States. *PLoS ONE* 6(9), e24587. DOI: 10.1371/journal.pone.0024587.

Banks, N.C., Paini, D.R., Bayliss, K.L. and Hodda, M. (2014) The role of global trade and transport network topology in the human-mediated dispersal of alien species. *Ecology Letters* 18(2), 188–199. DOI: 10.1111/ele.12397.

Bega, S. (2021) Tiny invasive beetle threatens millions of trees in South Africa. Environment. Available at: https://mg.co.za/environment/2021-06-06-tiny-invasive-beetle-threatens-millions-of-trees-in-south-africa

Benard, M.F. and McCauley, S.J. (2008) Integrating across life-history stages: Consequences of natal habitat effects on dispersal. *American Naturalist* 171(5), 553–567. DOI: 10.1086/587072.

Benavides, M.P., Stuart, J.J., Vega, F.E., Romero-Severson, J. and Bustillo, A.E. (2007) Biogeografía y aspectos genéticos de la broca del café *Hypothenemus hampei*. In: *Proceedings Manejo da Broca-do-Cafe*, Workshop Internacional, Londrina, Brasil, pp. 11–36.

Benavides, P., Vega, F.E., Romero-Severson, J., Bustillo, A.E. and Stuart, J.J. (2005) Biodiversity and biogeography of an important inbreed pest of coffee, coffee berry borer (Coleoptera: Curculionidae: Scolytinae). *Annals of the Entomological Society of America* 98, 359–366.

Benavides, P., Stuart, J.J., Vega, F.E., Romero-Severson, J., Bustillo, A.E. *et al.* (2006) Genetic variability of *Hypothenemus hampei* (Ferrari) in Colombia and development of molecular markers. In: *Proceedings, 21st International Scientific Colloquium on Coffee*, Association Scientifique Internationale du Café (ASIC), Montpellier, France, pp. 1301–1315.

Bierman, A., Roets, F. and Terblanche, J.S. (2022) Population structure of the invasive ambrosia beetle, *Euwallacea fornicatus*, indicates multiple introductions into South Africa. *Biological Invasions* 24(8), 2301–2312. DOI: 10.1007/s10530-022-02801-x.

Blackburn, T.M., Pyšek, P., Bacher, S., Carlton, J.T., Duncan, R.P. *et al.* (2011) A proposed unified framework for biological invasions. *Trends in Ecology and Evolution* 26(7), 333–339. DOI: 10.1016/j.tree.2011.03.023.

Bosselman, A.S., Dons, K., Oberthur, T., Smith, C., Raebild, A. *et al.* (2009) The influence of shade trees on coffee quality in smallholder coffee agroforestry systems in Southern Colombia. *Agriculture, Ecosystems and Environment* 129, 253–260.

Bragard, C., Baptista, P., Chatzivassiliou, E., Serio, F., Gonthier, P. *et al.* (2021) Pest categorisation of *Xylotrechus chinensis*. *EFSA Journal* 19(12), e07022. DOI: 10.2903/j.efsa.2021.7022.

Buckley, Y.M. (2017) Invasion ecology: Unpredictable arms race in a jam jar. *Nature Ecology and Evolution* 1(1), 0028.

Byers, J.E. (2002) Impact of non-indigenous species on natives enhanced by anthropogenic alteration of selection regimes. *Oikos* 97(3), 449–458.

Carter, M., Smith, M. and Harrison, R. (2009) Genetic analyses of the Asian longhorned beetle (Coleoptera, Cerambycidae, *Anoplophora glabripennis*) 12(5), 1165–1182. DOI: 10.1007/s10530-009-9538-9.

Catford, J.A., Jansson, R. and Nilsson, C. (2009) Reducing redundancy in invasion ecology by integrating hypotheses into a single theoretical framework. *Divers Distributions* 15(1), 22–40.

Chapman, E.G., Messing, R.H. and Harwood, J.D. (2015) Determining the origin of the coffee berry borer invasion of Hawaii. *Annals of the Entomological Society of America* 108(4), 585–592.

Clout, M.N. and Williams, P.A. (2009) *Invasive Species Management: A Handbook of Principles and Techniques*. Oxford University Press, Oxford.

Colautti, R.I., Bailey, S.A., Van Overdijk, C.D., Amundsen, K. and MacIsaac, H.J. (2006) Characterised and projected costs of nonindigenous species in Canada. *Biological Invasions* 8(1), 45–59.

Courchamp, F., Fournier, A., Bellard, C., Bertelsmeier, C., Bonnaud, E. *et al.* (2017) Invasion biology: Specific problems and possible solutions. *Trends in Ecology and Evolution* 32(1), 13–22.

Darwin, C. (1859) *On the Origin of Species by Means of Natural Selection or the Preservation of Favoured Races in the Struggle for Life*. John Murray London.

David, G., Giffard, B., Piou, D. and Jactel, H. (2013) Dispersal capacity of *Monochamus galloprovincialis*, the European vector of the pine wood nematode, on flight mills. *Journal of Applied Entomology* 138(8), 566–576. DOI: 10.1111/jen.12110.

Dawson, W., Moser, D., Van Kleunen, M., Kreft, H., Pergl, J. *et al.* (2017) Global hotspots and correlates of alien species richness across taxonomic groups. *Nature Ecology and Evolution* 1(7), 0186. DOI: 10.1038/s41559-017-0186.

Demidko, D.A., Demidko, N.N., Mikhaylov, PV. and Sultson, S.M. (2021) Biological strategies of invasive bark beetles and borers species. *Insects* 12(4). DOI: 10.3390/insects12040367.

De Wit, M.P., Crookes, D.J., Blignaut, J.N., De Beer, Z.W., Paap, T. *et al.* (2022) An assessment of the potential economic impacts of the invasive polyphagous shot hole borer (Coleoptera: Curculionidae) in South Africa. *Journal of Economic Entomology* 115(4), 1076–1086. DOI: 10.1093/jee/toac061.

Didham, R.K., Tylianakis, J.M., Hutchison, M.A., Ewers, R.M. and Gemmell, N.J. (2005) Are invasive species the drivers of ecological change? *Trends in Ecology and Evolution* 20(9), 470–474.

Dzurenko, M., Ranger, C.M., Hulcr, J., Galko, J. and Kaňuch, P. (2021) Origin of non-native *Xylosandrus germanus*, an invasive pest ambrosia beetle in Europe and North America. *Journal of Pest Science* 94(2), 553–562. DOI: 10.1007/s10340-020-01283-x.

Ehler, L. (1998) Conservation biological control. In: *Elsevier eBooks*. pp. 1–8. Available at: https://doi.org/10.1016/b978-012078147-8/50047-5

Elton, C.S. (1958) *The Ecology of Invasions by Animals and Plants*. Methuen, London.

Essl, F., Dullinger, S., Rabitsch, W., Hulme, P.E., Hülber, K. *et al.* (2011) Socioeconomic legacy yields an invasion debt. *Proceedings of the National Academy of Sciences USA* 108(1), 203–207.

Essl, F., Bacher, S., Blackburn, T.M., Booy, O., Brundu, G. *et al.* (2015) Crossing frontiers in tackling pathways of biological invasions. *BioScience* 65(8), 769–782.

Excoffier, L. and Ray, N. (2008) Surfing during population expansions promotes genetic revolutions and structuration. *Trends in Ecology and Evolution* 23(7), 347–351. DOI: 10.1016/j.tree.2008.04.004.

Fahrner, S. and Aukema, B.H. (2018) Correlates of spread rates for introduced insects. *Global Ecology and Biogeography* 27(6), 734–743. DOI: 10.1111/geb.12737.

Falk-Petersen, J., Bøhn, T. and Sandlund, O.T. (2006) On the numerous concepts in invasion biology. *Biological Invasions* 8(6), 1409–1424. DOI: 10.1007/s10530-005-0710-6.

Filho, W.M., Flechtmann, C.A.H., Godoy, W.a.C and Bjornstad, O.N. (2017) The impact of the introduced *Digitonthophagus gazella* on a native dung beetle community in Brazil during 26 years. *Biological Invasions* 20(4), 963–979. DOI: 10.1007/s10530-017-1603-1.

Forbes, A.B. and Scholtz, C.H. (2024) The impact of dung beetles on the free-living stages of ruminant parasites in faeces and their role as biological control agents in grazing livestock. *Veterinary Parasitology* 331, 110–267. DOI: 10.1016/j.vetpar.2024.110267.

Fortuna, T.M., Gall, P.L., Mezdour, S. and Calatayud, P. (2022) Impact of invasive insects on native insect communities. *Current Opinion in Insect Science* 51. DOI: 10.1016/j.cois.2022.100904.

Frehse, F. de. A., Braga, R.R., Nocera, G.A. and Vitule, J.R.S. (2016) Non-native species and invasion biology in a megadiverse country: Scientometric analysis and ecological interactions in Brazil. *Biological Invasions* 18(12), 3713–3725. DOI: 10.1007/s10530-016-1260-9.

Funayama, K. (2010) Influence of pest control pressure on occurrence of ground beetles (Coleoptera: Carabidae) in apple orchards. *Applied Entomology and Zoology* 46(1), 103–110. DOI: 10.1007/s13355-010-0015-9.

Gandon, S. (1999) Kin competition, the cost of inbreeding and the evolution of dispersal. *Journal of Theoretical Biology* 200(4), 345–364. DOI: 10.1006/jtbi.1999.0994.

Gauthier, N. (2010) Multiple cryptic genetic units in *Hypothenemus hampei* (Coleoptera: Scolytinae): evidence from microsatellite and mitochondrial DNA sequence data. *Biological Journal of the Linnean Society* 101, 113–129.

Ge, X., Zong, S., He, S., Liu, Y. and Kong, X. (2014) Areas of China predicted to have a suitable climate for *Anoplophora chinensis* under a climate-warming scenario. *Entomologia Experimentalis Et Applicata* 153(3), 256–265. DOI: 10.1111/eea.12247.

Gray, D.R. (2017) Climate change can reduce the risk of biological invasion by reducing propagule size. *Biol Invasions* 19(3), 913–923.

Gregg, P.C., Del Socorro, A.P. and Landolt, P.J. (2018) Advances in attract-and-kill for agricultural pests: Beyond pheromones. *Annual Review of Entomology* 63, 453–470. DOI: 10.1146/annurev-ento-031616-035040.

Hall, C.M. (2015) Tourism and biological exchange and invasions: A missing dimension in sustainable tourism? *Tourism Recreation Research* 40(1), 81–94. DOI: 10.1080/02508281.2015.1005943.

Hellmann, J.J., Byers, J.E., Bierwagen, B.G. and Dukes, J.S. (2008) Five potential consequences of climate change for invasive species. *Conservation Biology* 22(3), 534–543.

Hérard, F., Ciampitti, M., Maspero, M., Krehan, H., Benker, U. *et al.* (2006) *Anoplophora* species in Europe: Infestations and management processes1. *EPPO Bulletin* 36(3), 470–474. DOI: 10.1111/j.1365-2338.2006.

Hidalgo, K., Laparie, M., Bical, R., Larvor, V., Bouchereau, A. *et al.* (2013) Metabolic fingerprinting of the responses to salinity in the invasive ground beetle *Merizodus soledadinus* at the Kerguelen Islands. *Journal of Insect Physiology* 59(1), 91–100. DOI: 10.1016/j.jinsphys.2012.10.017.

Hobbs, R.J. (1989) The nature and effects of disturbance relative to invasions. In: Drake, J.A., Mooney, H.A. and Castri, F. (eds) *Biological Invasions: A Global Perspective*. John Wiley and Sons, New York, pp. 389–405.

Hodek, I. and Honěk, A. (2009) Scale insects, mealybugs, whiteflies and psyllids (Hemiptera, Sternorrhyncha) as prey of ladybirds. *Biological Control* 51(2), 232–243. DOI: 10.1016/j.biocontrol.2009.05.018.

Holter, P. (1982) Resource utilization and local coexistence in a guild of Scarabaeid Dung Beetles (*Aphodius* spp.). *Oikos* 39(2), 213. DOI: 10.2307/3544488.

Huang, D., Haack, R.A. and Zhang, R. (2011) Does global warming increase establishment rates of invasive alien species? A centurial time series analysis. *PLoS ONE* 6(9). DOI: 10.1371/journal.pone.0024733.

Hui, C., Richardson, D.M., Landi, P., Minoarivelo, H.O., Garnas, J. *et al.* (2016) Defining invasiveness and invasibility in ecological networks. *Biological Invasions* 18(4), 971–983.

Hu, J., Angeli, S., Schuetz, S., Luo, Y. and Hajek, A.E. (2009) Ecology and management of exotic and endemic Asian longhorned beetle *Anoplophora glabripennis*. *Agricultural and Forest Entomology* 11(4), 359–375. DOI: 10.1111/j.1461-9563.2009.00443.

Hulme, P.E. (2009) Trade, transport and trouble: Managing invasive species pathways in an era of globalization. *Journal of Applied Ecology* 46(1), 10–18. DOI: 10.1111/j.1365-2664.2008.01600.

Hutchison, W.D., Burkness, E.C., Mitchell, P.D., Moon, R.D., Leslie, T.W. *et al.* (2010) Areawide suppression of European corn borer with Bt maize reaps savings to non-Bt maize growers. *Science* 330, 222–225. DOI: 10.1126/science.1190242.

Iannone, B.V., Potter, K.M., Guo, Q., Liebhold, A.M., Pijanowski, B.C. *et al.* (2016) Biological invasion hotspots: A trait-based perspective reveals new sub-continental patterns. *Ecography* 39(10), 961–969.

Jaramillo, J. (2013) Coffee insects in a changing world. In: *Proceedings 24th International Scientific Colloquium on Coffee*, Association Scientifique Internationale du Café (ASIC), San José, CA, USA, pp. 645–652.

Jaramillo, J., Chabi-Olaye, A., Kamonjo, C., Jaramillo, A., Vega, F.E. *et al.* (2009) Thermal tolerance of the coffee berry borer *Hypothenemus hampei*: Predictions of climate change impact on a tropical insect pest. *PLoS ONE* 4, 64–87.

Jaramillo, J., Muchugu, E., Vega, F.E., Davis, A., Borgemeister, C. *et al.* (2011) Some like it hot: Theinfluence and implications of climate change on coffee berry borer (*Hypothenemus hampei*) and coffee production in East Africa. *PLoS ONE* 6, 524–528.

Jeschke, J., Aparicio, L.G., Haider, S., Heger, T., Lortie, C. *et al.* (2012) Support for major hypotheses in invasion biology is uneven and declining. *NeoBiota* 14, 1–20.

Johnson, P.T., Olden, J.D., Solomon, C.T. and Zanden, M.J. (2009) Interactions among invaders: Community and ecosystem effects of multiple invasive species in an experimental aquatic system. *Oecologia* 159(1), 161–189.

Johnson, M.A., Ruiz-Diaz, C.P., Manoukis, N.C. and Rodrigues, J.C.V. (2020a) Coffee berry borer (*Hypothenemus hampei*), a global pest of coffee: Perspectives from historical and recent invasions, and future priorities. *Insects* 11(12), 882. DOI: 10.3390/insects11120882.

Johnson, M.A., Fortna, S. and Manoukis, N.C. (2020b) Evaluation of exclusion netting for coffee berry borer (*Hypothenemus hampei*). *Management. Insects* 11, 364.

Johnson, C.L., Coyle, D.R., Duan, J.J., Lee, S., Lee, S. *et al.* (2024) A review of non-microbial biological control strategies against the Asian longhorned beetle (Coleoptera: Cerambycidae). *Environmental Entomology*. DOI: 10.1093/ee/nvae116.

Kavallieratos, N.G., Athanassiou, C.G., Guedes, R.N.C., Drempela, J.D. and Boukouvala, M.C. (2017) Invader competition with local competitors: Displacement or coexistence among the invasive Khapra beetle, *Trogoderma granarium* Everts (Coleoptera: Dermestidae), and two other major stored-grain beetles? *Frontiers of Plant Science* 8, 18–37. DOI: 10.3389/fpls.2017.01837.

Keena, M.A., Moore, P.M. and Bradford, G. (2021) Effects of temperature on *Anoplophora chinensis* (Coleoptera: Cerambycidae) adult survival, reproduction, and egg hatch. *Forests* 12(4), 432. DOI: 10.3390/f12040432.

Kim, B.S., Park, C.G., Moon, Y.M., Sung, B.K., Ren, Y. *et al.* (2016) Quarantine treatments of imported nursery plants and exported cut flowers by phosphine gas (PH3) as methyl bromide alternative. *Journal of Economic Entomology* 109, 2334–2340. DOI: 10.1093/jee/tow200.

Kistner-Thomas, E.J. (2019) The potential global distribution and voltinism of the Japanese beetle (Coleoptera: Scarabaeidae) under current and future climates. *Journal of Insect Science* 19(2). DOI: 10.1093/jisesa/iez023.

Kueffer, C. and Kull, C.A. (2017) Non-native species and the aesthetics of nature. In: Vilà, M. and Hulme, P. (eds) *Impact of Biological Invasions on Ecosystem Services*. Invading Nature – Springer series in invasion ecology, Springer, pp. 311–324.

Kyamanywa, S., Kucel, P., Uringi, N., Kovach, J., Roberts, A. *et al.* (2009) Arabica coffee pest profiles in the Mt. Elgon area of Uganda. Report on Arabica IPM-CRSP Project in Uganda. Blacksburg, VA, USA: Virginia Tech.

Larsen, A. and Philpott, S.M. (2010) Twig-nesting ants: The hidden predators of the coffee berry borer in Chiapas, Mexico. *Biotropica* 42, 342–347.

Larson, E.R., Graham, B.M., Achury, R., Coon, J.J., Daniels, M.K. *et al.* (2020) From eDNA to citizen science: Emerging tools for the early detection of invasive species. *Frontiers in Ecology and Environment* 18, 194–202. DOI: 10.1002/fee.2162.

Lebouvier, M., Laparie, M., Hullé, M., Marais, A., Cozic, Y. *et al.* (2011) The significance of the sub-Antarctic Kerguelen Islands for the assessment of the vulnerability of native communities to climate change, alien insect invasions and plant viruses. *Biological Invasions* 13(5), 1195–1208. DOI: 10.1007/s10530-011-9946-5.

Leturque, H. and Rousset, F. (2002) Dispersal, kin competition, and the ideal free distribution in a spatially heterogeneous population. *Theoretical Population Biology* 62(2), 169–180. DOI: 10.1006/tpbi.2002.1600.

Levine, J.M. and D'Antonio, C.M. (1999) Elton revisited: A review of evidence linking diversity and invasibility. *Oikos* 87, 15–26.

Liebhold, A.M., Turner, R.M., Blake, R.E., Bertelsmeier, C., Brockerhoff, E.G. *et al.* (2021) Invasion disharmony in the global biogeography of native and non-native beetle species. *Diversity and Distributions* 27(11), 2050–2062. DOI: 10.1111/ddi.13381.

Liu, Y., Han, W., Gao, J., Su, S., Beaurepaire, A. *et al.* (2021) Out of Africa: Novel source of small hive beetles infesting Eastern and Western honey bee colonies in China. *Journal of Apicultural Research* 60(1), 108–110. DOI: 10.1080/00218839.2020.1816686.

Lockwood, J.L., Cassey, P. and Blackburn, T.M. (2009) The more you introduce the more you get: The role of colonization pressure and propagule pressure in invasion ecology. *Diversity and Distributions* 15(5), 904–910.

Lucas, J.A. (2011) Advances in plant disease and pest management. *Journal of Agricultural Science* 149, 91–114. DOI: 10.1017/S0021859610000997.

Lundmark, C. (2010) Long-distance insect migration. *BioScience* 60(5), 400. DOI: 10.1525/bio.2010.60.5.15.

MacArthur, R. (1970) Species packing and competitive equilibrium for many species. *Theoretical Population Biology* 1(1), 1–11.

MacDougall, A.S. and Turkington, R. (2005) Are invasive species the drivers or passengers of change in degraded ecosystems? *Ecology* 86(1), 42–55.

Mack, R.N., Simberloff, D., Lonsdale, W.M., Evans, H., Clout, M. *et al.* (2000) Biotic invasions: Causes epidemiology global consequences and control. *Ecological Applications* 10(3), 689–710.

Magwere, T., West, M., Riyahi, K., Murphy, M.P., Smith, R.A.J. *et al.* (2006) The effects of exogenous antioxidants on lifespan and oxidative stress resistance in *Drosophila melanogaster*. *Mechanisms of Ageing and Development* 127(4), 356–370. DOI: 10.1016/j.mad.2005.12.009.

Mariño, Y.A., Pérez, M.E., Gallardo, F., Trifilio, M., Cruz, M. *et al.* (2016) Sun vs. shade affects infestation, totalpopulation and sex ratio of the coffee berry borer (*Hypothenemus hampei*) in Puerto Rico. *Agriculture, Ecosystems & Environment* 222, 258–266.

Maselko, M., Heinsch, S.C., Chacon, J.M., Harcombe, W.R. and Smanski, M.J. (2017) Engineering species-like barriers to sexual reproduction. *National Communication* 8, 7. DOI: 10.1038/s41467-017-01007-3.

Mastin, A.J., Gottwald, T.R., Bosch, F., Cunniffe, N.J. and Parnell, S. (2020) Optimising risk-based surveillance for early detection of invasive plant pathogens. *PLoS Biology* 18. DOI: 10.1371/journal.pbio.3000863.

McGeoch, M.A., Genovesi, P., Bellingham, P.J., Costello, M.J., McGrannachan, C. *et al.* (2016) Prioritizing species pathways and sites to achieve conservation targets for biological invasion. *Biological Invasions* 18(2), 299–314.

Meylan, L., Gary, C., Allinne, C., Ortiz, J., Jackson, L. *et al.* (2017) Evaluating the effect of shade trees on provision of ecosystem services in intensively managed coffee plantations. *Agriculture, Ecosystems & Environment* 245, 32–42.

Moat, J., Williams, J., Baena, S., Wilkinson, T., Gole, T.D. *et al.* (2017) Resilience potential of the Ethiopian coffee sector under climate change. *Nature Plants* 3, 1–14.

Monteys, V.S.I. and Tutusaus, G.T.I. (2018) A new alien invasive longhorn beetle, *Xylotrechus chinensis* (Cerambycidae), is infesting mulberries in Catalonia (Spain). *Insects* 9(2), 52. DOI: 10.3390/insects9020052.

Morey, A.C. and Venette, R.C. (2020) Minimizing risk and maximizing spatial transferability: Challenges in constructing a useful model of potential suitability for an invasive insect. *Annals of the Entomological Society of America* 113. DOI: 10.1093/aesa/saz049.

Nardi, F., Boschi, S., Funari, R., Cucini, C., Cardaioli, E. *et al.* (2024) The direction, timing and demography of *Popillia japonica* (Coleoptera) invasion reconstructed using complete mitochondrial genomes. *Scientific Reports* 14(1). DOI: 10.1038/s41598-024-57667.

Noriega, J.A., Delgado, O., Blanco, J.I., Gámez, J. and Clavijo, J. (2017) Introduction, establishment, and invasion of *Digitonthophagus gazella* (Fabricius, 1787) (Coleoptera: Scarabaeinae) in the Savannas of Venezuela and Colombia. *Natural Resources* 8(5), 370–381. DOI: 10.4236/nr.2017.85023.

Noriega, J.A., Floate, K.D., Génier, F., Reid, C.A., Kohlmann, B. *et al.* (2020) Global distribution patterns provide evidence of niche shift by the introduced African dung beetle *Digitonthophagus gazella*. *Entomologia Experimentalis et Applicata* 168(10), 766–782. DOI: 10.1111/eea.12961.

O'Connor, C.M., Norris, D.R., Crossin, G.T. and Cooke, S.J. (2014) Biological carryover effects: Linking common concepts and mechanisms in ecology and evolution. *Ecosphere* 5(3), 28. DOI: 10.1890/es13-00388.1.

O'Hanlon-Manners, D.L. and Kotanen, P.M. (2004) Logs as refuges from fungal pathogens for seeds of Eastern hemlock (*Tsuga canadensis*). *Ecology* 85(1), 284–289. DOI: 10.1890/03-3030.

Olenici, N., Duduman, M.L., Popa, I., Isaia, G. and Paraschiv, M. (2022) Geographical distribution of three forest invasive beetle species in Romania. *Insects* 13(7), 7. DOI: 10.3390/insects13070621.

Ööpik, M., Bunce, R.G.B. and Tischler, M. (2013) Horticultural markets promote alien species invasions: An Estonian case study of herbaceous perennials. *NeoBiota* 17, 19–37.

Osyczka, P., Mleczko, P., Karasiński, D. and Chlebicki, A. (2011) Timber transported to Antarctica: A potential and undesirable carrier for alien fungi and insects. *Biological Invasions* 14(1), 15–20. DOI: 10.1007/s10530-011-9991-0.

Osyczka, P., Mleczko, P., Karasiński, D. and Chlebicki, A. (2012) Timber transported to Antarctica: A potential and undesirable carrier for alien fungi and insects. *Biological Invasions* 14(1), 15–20.

Paini, D.R., Sheppard, A.W., Cook, D.C., De Barro, P.J., Worner, S.P. *et al.* (2016) Global threat to agriculture from invasive species. *Proceedings of the National Academy of Sciences USA* 113(27), 7575–7579. DOI: 10.1073/pnas.1602205113.

Paudel, S., Mansfield, S., Villamizar, L.F., Jackson, T.A. and Marshall, S.D.G. (2021) Can biological control overcome the threat from newly invasive coconut rhinoceros beetle populations (Coleoptera: Scarabaeidae)? A review. *Annals of the Entomological Society of America* 114(2), 247–256. DOI: 10.1093/aesa/saaa057.

Paudel, S., Marshall, S.D.G., Richards, N.K., Hazelman, G., Tanielu, P. *et al.* (2022) Coconut Rhinoceros Beetle in Samoa: Review of a century-old invasion and prospects for control in a changing future. *Insects* 13(5), 5. DOI: 10.3390/insects13050487.

Pergl, J., Pyšek, P., Bacher, S., Essl, F., Genovesi, P. *et al.* (2017) Troubling travellers: Are ecologically harmful alien species associated with particular introduction pathways? *NeoBiota* 32, 1–20.

Piiroinen, S., Lyytinen, A. and Lindström, L. (2012) Stress for invasion success? Temperature stress of preceding generations modifies the response to insecticide stress in an invasive pest insect. *Evolutionary Applications* 6(2), 313–323. DOI: 10.1111/eva.12001.

Pokhrel, M.R., Cairns, S.C. and Andrew, N.R. (2020) Dung beetle species introductions: When an ecosystem service provider transforms into an invasive species. *PeerJ* 8. DOI: 10.7717/peerj.9872.

Poland, T.M. and Rassati, D. (2019) Improved biosecurity surveillance of non-native forest insects: A review of current methods. *Journal of Pest Science* 92, 37–49. DOI: 10.1007/s10340-018-1004-y.

Pyšek, P. and Richardson, D.M. (2006) The biogeography of naturalization in alien plants. *Journal of Biogeography* 33(12), 2040–2050. DOI: 10.1111/j.1365-2699.2006.01578.

Pysek, P., Hulme, P.E., Simberloff, D., Bacher, S., Blackburn, T.M. *et al.* (2020) Scientists' warning on invasive alien species. *Biological Reviews* 95, 1511–1534. DOI: 10.1111/brv.12627.

Radcliffe, E.B., Hutchison, W.D. and Cancelado, R. (2009) *Integrated Pest Management: Concepts, Tactics, Strategies and Case Studies*. Cambridge University Press, Cambridge. DOI: 10.1017/CBO9780511626463.

Renault, D., Laparie, M., McCauley, S.J. and Bonte, D. (2017) Environmental adaptations, ecological filtering, and dispersal central to insect invasions. *Annual Review of Entomology* 63(1), 345–368. DOI: 10.1146/annurev-ento-020117-043315.

Richardson, D.M., Pysek, P. and Carlton, J.T. (2011) A compendium of essential concepts and terminology in invasion ecology. In: Richardson, D.M. (ed.) *Fifty Years of Invasion Ecology: The Legacy of Charles Elton*. Wiley-Blackwell, Oxford, pp. 409–420.

Rogers, M.A., Burkness, E.C. and Hutchison, W.D. (2016) Evaluation of high tunnels for management of *Drosophila suzukii* in fall-bearing red raspberries: Potential for reducing insecticide use. *Journal of Pest Science* 89, 815–821. DOI: 10.1007/s10340-016-0731-1.

Roques, A. (2010) Introductory notes to factsheets. Chapter 14. *BioRisk* 4, 855–1021. DOI: 10.3897/biorisk.4.69.

Roques, A., Auger-Rozenberg, M.A., Blackburn, T.M., Garnas, J., Pyšek, P. *et al.* (2016) Temporal and interspecific variation in rates of spread for insect species invading Europe during the last 200 years. *Biological Invasions* 18(4), 907–920. DOI: 10.1007/s10530-016-1080-y.

Roques, A., Shi, J., Auger-Rozenberg, M.A., Ren, L., Augustin, S. *et al.* (2020) Are invasive patterns of non-native insects related to woody plants differing between Europe and China? *Frontiers in Forests and Global Change* 2, 91. DOI: 10.3389/ffgc.2019.

Roura-Pascual, N., Saul, W., Pérez-Granados, C., Rutting, L., Peterson, G.D. *et al.* (2024) A scenario-guided strategy for the future management of biological invasions. *Frontiers in Ecology and the Environment* 22(4). DOI: 10.1002/fee.2725.

Rugman-Jones, P.F., Seybold, S.J., Graves, A.D. and Stouthamer, R. (2015) Phylogeography of the walnut twig beetle, *Pityophthorus juglandis*, the vector of thousand cankers disease in North American walnut trees. *PLoS ONE* 10, 118–264. DOI: 10.1371/journal.pone.0118264.

Ruzzier, E., Galli, A. and Bani, L. (2021) Monitoring exotic beetles with inexpensive attractants: A case study. *Insects* 12(5), 5. DOI: 10.3390/insects12050462.

Sakai, A.K., Allendorf, F.W., Holt, J.S., Lodge, D.M., Molofsky, J. *et al.* (2001) The population biology of invasive species. *Annual Review of Ecology, Evolution, and Systematics* 32(1), 305–332.

Sartoi Monteys, V., Costa Ribes, A. and Savin, I. (2021) The invasive longhorn beetle *Xylotrechus chinensis*, pest of mulberries, in Europe: Study on its local spread and efficacy of abamectin control. *PLoS ONE* 16(1), 245–527. DOI: 10.1371/journal.pone.024552.

Saul, W.C., Roy, H.E., Booy, O., Carnevali, L., Chen, H.J. *et al.* (2017) Assessing patterns in introduction pathways of alien species by linking major invasion data bases. *Journal of Applied Ecology* 54(2), 657–669.

Schuler, H., Witkowski, R., Van de Vossenberg, B., Hoppe, B., Mittelbach, M. *et al.* (2023) Recent invasion and eradication of two members of the *Euwallacea fornicatus* species complex (Coleoptera: Curculionidae: Scolytinae) from tropical greenhouses in Europe. *Biological Invasions* 25(2), 299–307. DOI: 10.1007/s10530-022-02929-w.

Schwalbe, C.P. (1991) Challenges in managing introduced pests. In: *SPIE Books*. DOI: 10.4182/nvhz5051.12.495.

Seebens, H., Essl, F. and Blasius, B. (2016) The intermediate distance hypothesis of biological invasions. *Ecology Letters* 20(2), 158–165. DOI: 10.1111/ele.12715.

Seebens, H., Blackburn, T.M., Dyer, E.E., Genovesi, P., Hulme, P. *et al.* (2017) No saturation in the accumulation of alien species worldwide. *National Communication* 8, 14435.

Sesay, R.E.V., Sesay, F., Azizi, M.I. and Rahmani, B. (2024) Invasive species and biodiversity: Mechanisms, impacts, and strategic management for ecological preservation. *Asian Journal of Environment and Ecology* 23(9), 82–95. DOI: 10.9734/ajee/2024/v23i9600.

Shackleton, C.M., Ruwanza, S., Sanni, G.S., Bennett, S., Lacy, P. *et al.* (2016) Unpacking pandora's box: Understanding and categorising ecosystem disservices for environmental management and human wellbeing. *Ecosystems* 19(4), 587–600.

Singh, S., Hind Sharma, J., Udikeri, A. and Ansari, H. (2020) Invasive insects in India. *IntechOpen*. DOI: 10.5772/intechopen.91986.

Sloggett, J.J. and Weisser, W.W. (2002) Parasitoids induce production of the dispersal morph of the pea aphid, *Acyrthosiphon pisum*. *Oikos* 98(2), 323–333. DOI: 10.1034/j.1600-0706.2002.980213.

Soto-Pinto, L., Perfecto, I. and Caballero-Nieto, J. (2002) Shade over coffee: Its effects on berry borer, leaf rust andvspontaneous herbs in Chiapas. *Mexico Agroforestry Systems* 55, 37–45.

Strecker, A.L. and Arnott, S.E. (2010) Complex interactions between regional dispersal of native taxa and an invasive species. *Ecology* 91(4), 1035–1047. DOI: 10.1890/08-0065.1.

Sudweeks, J., Hollingsworth, B., Blondel, D.V., Campbell, K.J. and Dhole, S. (2019) Locally fixed alleles: A method to localize gene drive to island populations. *Scientific Reports* 9, 10. DOI: 10.1038/s41598-019-51994-0.

Svenning, J., Gravel, D., Holt, R.D., Schurr, F.M., Thuiller, W. *et al.* (2014) The influence of inter-specific interactions on species range expansion rates. *Ecography* 37(12), 1198–1209. DOI: 10.1111/j.1600-0587.2013.

Teodoro, A., Klein, A., Reist, P. and Tscharntke, R.A. (2009) Agroforestry management affects coffee pests contingent onseason and developmental stage. *Agricultural and Forest Entomology* 11, 295–300.

Thakur, J.R. and Dogra, G.S. (1980) Woolly apple aphid, *Eriosoma lanigerum*. *Research in India. Tropical Pest Management* 26(1), 8–12. DOI: 10.1080/09670878009414276.

Tran, H.X., Doland Nichols, J., Li, D., Le, N.H. and Lawson, S.A. (2022) Seasonal flight and genetic distinction among *Xylosandrus crassiusculus* populations invasive in Australia. *Australian Forestry* 85(4), 224–231. DOI: 10.1080/00049158.2022.2151722.

Trenberth, K.E., Jones, P.D., Ambenje, P., Bojariu, R., Easterling, D. *et al.* (2007) Observations: Surface and atmospheric climate change. In: *Climate Change the Physical Science Basis Contribution of WG 1 to the Fourth Assessment Report of the Intergovernmental Panel on Climate Change*.

Tuomola, J. and Hannunen, S. (2024) Assessment of the suitability of the Finnish climate for *Popillia japonica*. *Food Risk Assess Europe* 2(2). DOI: 10.2903/fr.efsa.

UC IPM (1996) What is integrated pest management (IPM)? University of California, College of Agriculture and Natural Resources, Davis, CA. Available at: https://www2.ipm.ucanr.edu/What-is-IPM/ (accessed 13 December 2020).

Urvois, T., Auger-Rozenberg, M.A., Roques, A., Rossi, J.P. and Kerdelhue, C. (2021) Climate change impact on the potential geographical distribution of two invading *Xylosandrus ambrosia* beetles. *Scientific Reports* 11(1), 1339. DOI: 10.1038/s41598-020-80157-9.

Urvois, T., Auger-Rozenberg, M.A., Roques, A., Kerdelhué, C. and Rossi, J.P. (2024) Intraspecific niche models for the invasive ambrosia beetle *Xylosandrus crassiusculus* suggest contrasted responses to climate change. *Oecologia*. DOI: 10.1007/s00442-024-05528-9.

Van Allen, B.G. and Rudolf, V.H.W. (2016) Carryover effects drive competitive dominance in spatially structured environments. *Proceedings of the National Academy of Sciences USA* 113(25), 6939–6944. DOI: 10.1073/pnas.1520536113.

Van Der Gaag, D.J., Sinatra, G., Roversi, P.F., Loomans, A., Hérard, F. *et al.* (2010) Evaluation of eradication measures against *Anoplophora chinensis* in early-stage infestations in Europe. *EPPO Bulletin* 40(2), 176–187. DOI: 10.1111/j.1365-2338.2010.

Vega, F.E., Infante, F. and Johnson, A.J. (2015) The genus *Hypothenemus*, with emphasis on *H. hampei*, the coffee berry borer. In: *Bark Beetles: Biology and Ecology of Native and Invasive Species*. Academic Press, San Diego, CA, USA, pp. 427–494.

Venette, R.C. (2017) Climate analyses to assess risks from invasive forest insects: Simple matching to advanced models. *Current Forestry Reports* 3, 255–268. DOI: 10.1007/s40725-017-0061-4.

Venette, R.C. and Hutchison, W.D. (2021) Invasive insect species: Global challenges, strategies and opportunities. *Frontiers in Insect Science* 1. DOI: 10.3389/finsc.2021.650520.

Venette, R.C., Moon, R.D. and Hutchison, W.D. (2002) Strategies and statistics of sampling for rare individuals. *Annual Review of Entomology* 143–174. DOI: 10.1146/annurev.ento.47.091201.145147.

Venette, R.C., Kriticos, D.J., Magarey, R.D., Koch, F.H., Baker, R.H.A. *et al.* (2010) Pest risk maps for invasive alien species: A roadmap for improvement. *Bioscience* 60, 349–362. DOI: 10.1525/bio.2010.60.5.5.

Venette, R.C. and Morey, A.C. (2020) Advances in understanding the ecology of invasive crop insect pests and their impact on IPM. In: Kogan, M. and Heinrichs, E.A. (eds) *Integrated Management of Insect Pests: Current and Future Developments*. Burleigh Dodds Science Publishing, Cambridge, pp. 161–190. DOI: 10.19103/AS.2019.0047.06.

Vilà, M., Basnou, C., Pyšek, P., Josefsson, M., Genovesi, P. *et al.* (2010) How well do we understand the impacts of alien species on ecosystem services? A pan-European cross-taxa assessment. *Frontiers in Ecology and the Environment* 8(3), 135–144.

Vitousek, P.M., Antonio, C.M., Loope, L.L. and Westbrooks, R. (1996) Biological invasions as global environmental change. *American Scientist* 84(5), 468.

Wang, L., Li, C., Luo, Y., Wang, G., Dou, Z. *et al.* (2023a) Current and future control of the wood-boring pest *Anoplophora glabripennis*. *Insect Science* 30(6), 1534–1551. DOI: 10.1111/1744-7917.13187.

Wang, S., Lu, Y., Han, M., Li, L., He, P. *et al.* (2023b) Using maxent model to predict the potential distribution of three potentially invasive scarab beetles in China. *Insects* 14(3), 3. DOI: 10.3390/insects14030239.

Ward, N.L. and Masters, G.J. (2007) Linking climate change and species invasion: An illustration using insect herbivores. *Global Change Biology* 13(8), 1605–1615. DOI: 10.1111/j.1365-2486.2007.

Westphal, M.I., Browne, M., MacKinnon, K. and Noble, I. (2007) The link between international trade and the global distribution of invasive alien species. *Biological Invasions* 10(4), 391–398. DOI: 10.1007/s10530-007-9138-5.

Wilches, D.M., Laird, R.A., Floate, K.D. and Fields, P.G. (2018) Control of *Trogoderma granarium* (Coleoptera: Dermestidae) using high temperatures. *Journal of Economic Entomology* 112(2), 963–968. DOI: 10.1093/jee/toy379.

Williamson, M. (1996) *Biological Invasions*. Chapman and Hall, London.

Williamson, M. and Fitter, A. (1996) The varying success of invaders. *Ecology* 77(6), 1661–1666.

Wilson, J.R., Dormontt, E.E., Prentis, P.J., Lowe, A.J. and Richardson, D.M. (2009) Something in the way you move: Dispersal pathways affect invasion success. *Trends in Ecology & Evolution* 24(3), 136–144.

Witcombe, J.R. and Erskine, W. (1984) Genetic resources and their exploitation Chickpeas, Faba beans and Lentils. In: *Springer eBooks*. DOI: 10.1007/978-94-009-6131-9.

Wyckhuys, K.A.G., Lu, Y., Zhou, W., Cock, M.J.W., Naranjo, S.E. *et al.* (2020) Ecological pest control fortifies agricultural growth in Asiapacific economies. *Nature Ecology & Evolution* 4, 1522. DOI: 10.1038/s41559-020.

Xu, J.W., Fonseca, D.M., Hamilton, G.C., Hoelmer, K.A. and Nielsen, A.L. (2014) Tracing the origin of US brown marmorated stink bugs, *Halyomorpha halys*. *Biological Invasions* 16, 153–166. DOI: 10.1007/s10530-013-0510-3.

Yadav, S.K., Bhowmik, S., Yadav, P.C. and Sharma, K.C. (2021) Identification and control of *Trogoderma granarium* (Coleoptera: Dermestidae), a potential threat to stored products and international trade. *International Journal of Tropical Insect Science* 42(2), 999–1017. DOI: 10.1007/s42690-021-00635-z.

Yan, L.J. and Sohal, R.S. (2000) Prevention of flight activity prolongs the life span of the housefly, *Musca domestica*, and attenuates the age-associated oxidative damamge to specific mitochondrial proteins. *Free Radical Biology and Medicine* 29(11), 1143–1150. DOI: 10.1016/s0891-5849(00)00423-8.

Yasui, H., Fujiwara-Tsujii, N., Kugimiya, S., Shibuya, K., Mishiro, K. *et al.* (2024) *Anoplophora glabripennis*, an invasive longhorned beetle, has the potential to damage fruit trees in Japan. *Scientific Reports* 14(1). DOI: 10.1038/s41598-024-63548-0.

Yemshanov, D., Haight, R.G., Koch, F.H., Lu, B., Venette, R. *et al.* (2017) Robust surveillance and control of invasive species using a scenario optimization approach. *Ecological Economics* 133, 86–98. DOI: 10.1016/j.ecolecon.2016.11.018.

Yemshanov, D., Haight, R.G., Koch, F.H., Venette, R.C., Swystun, T. *et al.* (2019) Optimizing surveillance strategies for early detection of invasive alien species. *Ecological Economics* 162, 87–99. DOI: 10.1016/j.ecolecon.2019.04.030.

Yotkham, S., Suttiprapan, P., Likhitrakarn, N., Sulin, C. and Srisuka, W. (2021) Biodiversity and spatiotemporal variation of longhorn beetles (Coleoptera: Cerambycidae) in tropical forest of Thailand. *Insects* 12(1), 1. DOI: 10.3390/insects12010045.

16 Invasion of Fall Armyworm, *Spodoptera frugiperda* (J. E. Smith) (Lepidoptera: Noctuidae) in the Tropics

Nandhini D.[1], Mogili Ramaiah[1], Kishore Chandra Sahoo[2], Aarthi Helen P.[1], Ramya N.[3], Keerthi M.C.[4], Chaitanya B.[5], Eurekha S.[5] and Pathour Rajendra Shashank[1*]

[1]Division of Entomology, ICAR – Indian Agricultural Research Institute, New Delhi, India; [2]ICAR – Indian Agricultural Research Institute, Assam, India; [3]ICAR – Indian Institute of Agricultural Biotechnology, Ranchi, Jharkhand, India; [4]Division of Entomology and Nematology, ICAR – Indian Institute of Horticultural Research, Bangalore, India; [5]University of Agricultural Sciences, Raichur, Karnataka, India

Abstract

The fall armyworm, *Spodoptera frugiperda* (J. E. Smith) (Lepidoptera: Noctuidae), is a notorious invasive pest of maize and other grasses. Though native to the Americas, the fall armyworm invasion of African and Asian countries in recent years has made it one of the most damaging pests to agricultural and horticultural crops. Despite numerous scientific studies conducted to control this invasive pest, the damage imposed on crops is increasing rapidly and alarmingly in many countries. This chapter provides valuable insights into the study of *S. frugiperda* with special emphasis on distribution, host range, taxonomy, morphology, the extent of the crop damage it inflicts, exploration of its genetic diversity, and specially designed strategies for managing its impact in tropical regions.

Keywords: Distribution, genetic diversity, management, biocontrol, IPM, novel management, invasive pest, fall armyworm

16.1 Introduction

Spodoptera frugiperda (J.E. Smith), commonly referred to as the fall armyworm (FAW), belongs to the Noctuidae family within the order Lepidoptera. Originally native to tropical and subtropical regions of the Americas, where it has long been a major agricultural problem, this invasive species rapidly was distributed all over the globe, causing extensive damage to major crops, especially maize and sorghum (Nagoshi and Meagher, 2008). This pest acquired the name "fall armyworm" due to its exceptionally destructive impact on various crops in Florida during the late autumn of 1896 and the early spring of 1897 (Quaintance, 1897). The scientific name *"frugiperda"* means fruit destroyer in Latin, highlighting its tendency to damage crops during the reproductive or fruiting stage (Sarkowi and Mokhtar, 2021). It has emerged

*Corresponding author: spathour@gmail.com

 Biological Invasions in the Tropics: Monitoring, Mechanism and Management (eds Akshay Kumar Chakravarthy and Pathour Rajendra Shashank)
DOI: 10.1079/9781836991977.0016

as a significant agricultural pest in tropical regions, seriously threatening food security and livelihoods. Until 2016, this pest was limited to the Americas, but in early 2016, it was identified in the central and western regions of Africa (Goergen *et al.*, 2016). In the subsequent years, the pest rapidly invaded Asia, Australia and Europe, resulting in massive yield losses, especially of maize, and is now a global threat to agriculture (Kenis *et al.*, 2023).

Although maize is the most preferred host plant, it is a dreaded polyphagous pest documented to feed on 353 host plants belonging to 76 plant families. Plants, including non-crop and weed species, belonging to the family Poaceae are the preferred hosts, followed by Asteraceae and Fabaceae (Montezano *et al.*, 2018). An estimate of crops worth over US$13 billion per annum throughout sub-Saharan Africa is at risk of damage by this pest (Harrison *et al.*, 2019). The FAW occurs in two races: corn strain "C" which infests maize, sorghum and cotton and rice strain "R", which prefers rice and turf grass (Nagoshi and Meagher, 2016; Swamy *et al.*, 2018).

It is a complex pest with irregular lengths of life cycle across seasons (about a month in summer, two months in spring and autumn, and three months during winter). The population rapidly multiplies under warm, humid conditions with moderate rainfall with an optimal temperature range between 11°C and 30°C, adding to the complexity of its distribution mechanism (Nagoshi *et al.*, 2012; Day *et al.*, 2017; Kumela *et al.*, 2019). The adult moth can fly for a distance of about 1000 km in a lifetime and is reported to be able to cover nearly 1600 km distance from the southern state of Mississippi in the USA to south Canada in 30 hours (Rose *et al.*, 1975; Prasanna *et al.*, 2018). Such evolutionary strategies acquired by the pest make it able to inhabit different geographic locations across the globe and it is thus considered an invasive pest of agriculture.

The high migration ability, prolific reproduction rate, polyphagous feeding habit, rapid evolution of insecticidal resistance and high physiological and behavioural adaptability make the implementation of control strategies challenging (Paredes-Sánchez *et al.*, 2021). This chapter provides an up-to-date comprehensive examination of FAW in tropical countries, encompassing its geographic distribution and spread, taxonomy and morphology, nature of damage, biology and strategies for its control and management.

16.2 Distribution, Spread and Economic Loss

The fall armyworm, native to tropical and subtropical Americas, was a sporadic pest of cultivated crops in Georgia since 1797. In 1845, it was found causing damage to grass, corn, sugarcane and rice in Florida. Later, in 1870, it was found damaging oat, wheat and grass fields in Missouri and Illinois. Until 1899, sporadic infestations occurred every few years. Extensive outbreaks of this pest in different crops, such as rice, cowpea and corn, were observed throughout South Carolina, North Carolina, Virginia, West Virginia, Indiana, Illinois, Missouri, Kansas, and other western and northern states. However, in 1912, the most severe outbreak occurred, covering the entire USA east of the Rocky Mountains, causing extensive havoc to crops including corn, cotton, millets, and even the grass on lawns in cities, in parts of many southern states (Walton and Luginbill, 1916).

The first report of the invasion of FAW outside the Americas came in 2016 from western and central Africa (Goergen *et al.*, 2016) and later in Ghana in 2017 (Cock *et al.*, 2017). The pest rapidly spread across other African countries, covering all of sub-Saharan Africa, and also reached the Indian Ocean islands, including the island countries of Cape Verde, Madagascar and the Seychelles (FAO, 2018). In 2018, the pest invaded Asia, being reported from Yemen (EPPO, 2019) and Karnataka, India (Ganiger *et al.*, 2018; Sharanabasappa *et al.*, 2018a). Subsequently, the pest was reported in other South Asian countries including Bangladesh (Alam *et al.*, 2018), Thailand (Guo *et al.*, 2018), Sri Lanka (Perera *et al.*, 2019), Myanmar (Yee *et al.*, 2019), Nepal (Bajracharya *et al.*, 2019) and China (Sun *et al.*, 2021).

Later, by June 2019, FAW's first invasion was reported in Kagoshima of Japan and by November 2019, in Sudan, Egypt (IPPC, 2019a) as well as in many Asian countries, including Cambodia (Hadi, 2019), Laos (Ryu *et al.*, 2019), Malaysia

(IPPC, 2019b) and the Philippines (Navasero *et al.*, 2019). In 2020, FAW invaded Jeju Island in Korea (Lee *et al.*, 2020), Indonesia (Sartiami *et al.*, 2020) and Vietnam (Hang *et al.*, 2020). Later, it was recorded in Australia (EPPO, 2020a), Mauritania (EPPO, 2020b), Timor Leste (EPPO, 2020c), and the United Arab Emirates between February and May 2020 and from Jordan (IPPC, 2020), Papua New Guinea (Bourke and Sar, 2020) and Syria (Heinoun *et al.*, 2021).

During January 2021, it was detected in New Caledonia (IPPC, 2021) and by April in the Canary Islands of Spain (EPPO, 2021), Europe. Recently, it was confirmed in Türkiye (Pehlivan and Atakan, 2022) and Cyprus (EPPO, 2023). Within the last few years since its report of the first invasion in Africa in 2016, the pest has spread to more than 105 countries, and most of the invaded countries are tropical (CABI, 2019). The pest can shift its potential areas and even different host plants with the changing climatic conditions. Zacarias (2020) predicted the potential of the spread of FAW and its associated impacts on major host plants globally. According to the simulations, there is a significant climatic potential for the species to expand, with predicted rises of between 12–44% potentially hitting border regions between the USA and Canada, sub-Saharan Africa and central Europe.

The seasonal shifts in the environment drive significant seasonal activities of this pest, and numerical simulations of its dispersal patterns and distribution across space and time help coordinate comprehensive pest control strategies. The Hybrid Single-Particle Lagrangian Integrated Trajectory (HYSPLIT) model simulated the migratory flight patterns of FAW moths, originating from different winter-breeding source areas, adding valuable insights into the population dynamics of FAW at a continental scale and will be valuable for predicting and understanding the annual fluctuations in insect migration routes (Westbrook *et al.*, 2016). In Zacatecas, Mexico, the number of FAW generations varies based on climate maps depicting different environmental conditions. Currently, three to four generations are common and the projections indicate that by 2060, most of the states could experience four to five FAW generations per year. Certain localized areas in Mexico already see up to six FAW generations, which may even reach up to seven (Ramírez-Cabral *et al.*, 2020).

Using remote sensing and meteorological data integration, a mechanistic model known as FAW–DDM (Fall Armyworm Dynamic Distribution Model) was developed in China, which combines the weather-influenced flight patterns of FAW with host plant growth stages and environmental suitability to predict FAW distribution dynamics (Huang *et al.*, 2022). Timilsena *et al.* (2022) re-parameterized the existing CLIMEX model to assess the invasion threat of FAW. They depicted that, under the current circumstances, FAW can establish itself in nearly all eastern and central African countries and a significant portion of western Africa. Climatic restraints like heat and dry conditions may constrain the expansion of FAW to northern and southern Africa. Future forecasts indicate that the invasive range of the FAW will move southward and northward towards the equator. However, a sizable portion of eastern and central Africa is anticipated to have the ideal climate for FAW persistence. These regions act as "FAW hotspots", from which it may migrate to the north and south during favourable seasons and eventually threaten the world economy. FAW has immense potential to change its hosts and biogeography under changing climatic conditions. Although the pest is currently more severe in the tropics, it can further spread to temperate regions of the globe with increasing temperatures in the future. The advanced modelling tools help predict such shifts in the pestiferous behaviour of the FAW, including its spread and distribution.

Spodoptera frugiperda, a migratory pest, has been identified as a crucial threat in its native and invaded range, inducing substantial reported losses in crop yield, thereby with significant economic impacts. The quantifiable yield loss has been reported from different parts of the world. In the USA, the annual crop yield losses due to FAW are estimated to be around US$300 million, rising to US$500 million or more during outbreak periods (Mitchell, 1979). In 2009, Brazil spent US$600 million to control FAW (Ferreira Filho *et al.*, 2010). Surveys in Ghana and Zambia showed that about 98% of farmers reported the pest in maize crops, whereas damage to Napier grass, sorghum or millet was only 2–4%. The average loss in maize of yield was about 26.6% and 35% in Ghana and Zambia, respectively.

16.3 Biology and Morphology

Being a holometabolous insect, it undergoes four distinct developmental stages: egg, larva, pupa and adult (Fig. 16.1). The duration of each stage varies significantly depending on the different host plants they feed on and the rearing conditions (Dumas *et al.*, 2015; He *et al.*, 2021; Muniappan, 2024).

The eggs are typically laid in clusters, with a preference for the lower surface of the leaf lamina or the inner side of the whorl. However, occasionally, they may be found on the upper surface of the leaf during the early night hours (Dew, 1913;

Fig. 16.1. (A, B) Egg masses, (C) Neonates of FAW, (D, E) Later instar larva, (F) Prepupa, (G) Pupa, (H) Male pupa, (I) Female pupa. (Author's own images)

Luginbill, 1928; Bhavani *et al.*, 2019). Each cluster of eggs contains oblate-spheroidal eggs. Eggs are densely covered with pale yellowish scales from the female moth. The number of eggs in each cluster varies from 44 to 593, and the cluster is arranged in 2–3 layers. Eggs are green, but as time progresses they change to a cream colour and eventually turn black just before hatching (Sparks, 1979). After 2–4 days of incubation, the neonates or first instar larvae hatch, depending on the prevailing environmental conditions (Luginbill, 1928).

The fall armyworm larvae can be easily distinguished from other noctuid larvae by their external appearance. They have an elongated body, with a dark head and a pale green to dark brown body featuring pale dorsal longitudinal stripes. These larvae have three pairs of thoracic legs and five pairs of abdominal prolegs, each equipped with crochets arranged semi-circularly. One prominent characteristic of this species is the inverted Y-shaped suture on the frons. Additionally, four noticeable black spots or pinaculae on the pale-colour dorsal surface. Black spots are located at various positions on the thoracic and abdominal segments. These black spots and setae create a distinct pattern on the larva's body: a line on the second and third thoracic segments, a trapezoid shape on the first seven abdominal segments, and a larger square on the eighth abdominal segment. The distinctive pinkish lines are found above the spiracles of the posterior abdominal segments (Bajracharya *et al.*, 2019; Bhavani *et al.*, 2019; Navasero *et al.*, 2019; Rajisha *et al.*, 2021; Higo *et al.*, 2022).

The fall armyworm larva experiences 5–6 moults, which can vary depending on the host plants it feeds on (Murúa and Virla, 2004). Right after hatching, the neonates consume the egg-shells, carefully avoiding the scales. This provides them with a nutrient-rich first meal, and afterward, they conserve energy by staying immobile for 4–10 hours and disperse later in various directions by ballooning, searching for the most tender plant parts (Dew, 1913; Luginbill, 1928). During the first and second instar stages, the larvae cause elongated papery windows on the leaves by scraping the leaf epidermis. As they progress to the third instar, they feed inside the whorl, creating a series of holes and depositing faecal matter in the unfurling leaves. As the larvae develop, the rate of feeding, hole size, and the volume of faecal matter increase. By the time they reach the sixth instar, the larvae cause significant defoliation, leaving behind a large quantity of faecal matter in the plant whorl. When older larvae infest plants at an early whorl stage, it leads to the complete defoliation and death of the host plant (Suby *et al.*, 2020).

The duration of each larval instar varies from 2 to 4 days, depending on the host plants. The larval phase lasts 14–19 days, when the larvae are reared on maize (Sharanabasappa *et al.*, 2018b; Soujanya *et al.*, 2022). After the third moulting, the larva displays cannibalistic behaviour and feeds on its conspecifics when found in the group. The cannibalism rate can also be influenced by the type of food available (Slansky and Scriber, 1985; Raffa, 1987). Upon completing the last larval instar, the larval body shrinks and curls to enter the prepupal stage, which lasts for about a day.

The obtect pupae have a reddish-brown colour and an oval shape. Pupation mainly occurs in the soil, and sometimes in the leaf folds, with the pupal period lasting around 7–8 days. However, under certain atmospheric conditions, it can extend up to 45 days ((Luginbill, 1928; Soujanya *et al.*, 2022). A shorter distance between these two indicates a male pupa, whereas a longer distance signifies a female pupa (Sharanabasappa *et al.*, 2018b; Navasero and Navasero, 2020).

Adult moths display sexual dimorphism and are active during the night. The male moths have greyish-brown forewings with a triangular white patch at the apical region and a circular spot in the middle of the wings (Fig. 16.2A). In contrast, the forewings of females are uniformly greyish-brown, with a slight mottling of grey and brown (Fig. 16.2B). Both male and female moths have silver-white hind wings with a narrow dark border. Adult females have different durations of reproductive stages: the pre-oviposition period ranges from 3 to 4 days, the oviposition period spans 2–3 days, and the post-oviposition period takes 4–5 days (Sharanabasappa *et al.*, 2018b; Bajracharya *et al.*, 2019; Navasero and Navasero, 2020).

Fig. 16.2. (A) Adult male, (B) Adult female, (C) Larval feeding, (D–F) Papery window symptoms caused by the larvae, (G, H) Severe defoliation caused by the larvae. (Author's own images)

16.4 Host Range

FAW is highly polyphagous and has been reported to feed on 353 plant species belonging to 76 families (Dumas *et al.*, 2015; Montezano *et al.*, 2018). Though it is known for its polyphagous feeding behaviour, it prefers certain crops and can adapt and utilize various plants as food sources. Two strains, the corn/maize strain and rice strain are reported based on their preference for different host plants.

It primarily targets cereal and forage crops such as maize, rice, sorghum, wheat, etc. and causes significant damage to these crops by affecting the quality and yield (Keerthi *et al.*, 2021). However, sugarcane, millets including finger millet, barnyard millet (Poaceae), okra, hollyhock, cotton, hibiscus (Malvaceae), groundnut, pigeon pea, soyabean, chickpea, cowpea, lucerne, clover, common bean, pea, velvet bean (Fabaceae), chilli, pepper, tomato, potato, tobacco (Solanaceae), onion, garlic, (Liliaceae), amaranthus (Amaranthaceae), chrysanthemum, daisy, sunflower (Asteraceae), garden croton, castor, rubber (Euphorbiaceae), beetroot, sugarbeet, spinach (Chenopodiaceae), grape (Vitaceae), ginger (Zingiberaceae), bananas (Musaceae) and pineapple (Bromeliaceae) (CABI, 2019; Chikkarugi *et al.*, 2023; Moorthy *et al.*, 2023). Furthermore, FAW infests various grasses, including turf grasses commonly used for lawns and sports fields (Sisay *et al.*, 2022; Tura *et al.*, 2025).

Some plants may exhibit natural resistance or tolerance mechanisms that limit the impact of FAW infestation. Understanding the host range of FAW is crucial for monitoring and managing its spread and impact. Effective pest management strategies often involve a combination of cultural practices, biological control, and judicious use of pesticides, emphasizing integrated pest management (IPM) principles. Farmers and researchers can implement targeted measures to mitigate FAW damage and minimize economic losses in agricultural systems by considering the host range and specific vulnerabilities of crops. Life table studies of FAW feeding on different hosts are also necessary to understand critical parameters such as larval growth, survival, reproduction and intrinsic rate of increase (Ashok *et al.*, 2020). The potential for an outbreak of FAW is high for the other least preferred crops (He *et al.*, 2021).

16.5 Nature of Damage

The neonates initially scrape the chlorophyll and later, they move on to other leaves for feeding (Fig. 16.2C–H). As they progress, they enter the leaf whorls of maize plants and unopened whorl leaves, resulting in shot holes and windowpane-like damage on the inner whorls. The larvae leave moist, sawdust-like frass matter plugged around the inner lamina and upper leaves, where they conceal themselves while continuing to feed. In later stages, the larvae feed voraciously, leading to the skeletonization of the leaves and a distinct ragged appearance of the entire shoot, resembling damage caused by grasshoppers (Deole and Paul, 2018; Sisodiya *et al.*, 2018). The larvae burrow into the tassels and ears of maize, causing considerable damage to the reproductive structures (Abrahams *et al.*, 2017; Capinera, 2017). The severity of damage caused by *S. frugiperda* can be assessed on a scale of 0–9 given by Davis *et al.*, 1992, based on leaf damage and the presence of frass in the whorl.

The 7-day and 14-day rating scales were originally developed for studying maize resistance and have been specifically designed for mid-whorl stage maize that has been intentionally infested with FAW neonates (Davis *et al.*, 1992).

16.6 Genetic Diversity of Fall Armyworm

Genetic diversity refers to the variability in genetic characteristics among individuals within a population and it plays a crucial role in the adaptation and resilience of species to environmental changes, as well as their ability to respond to selection pressures and survival threats. Genetic diversity in FAW provides insights into biology and behaviour and hence, researchers investigated the genetic diversity of FAW to understand host plant preferences and insecticide resistance potential.

Two morphologically identical strains or populations of FAW exist that differ in host preference and exhibit mating incompatibility due to the changes in pheromone composition and mating behaviour (Pashley and Martin, 1987; Clark *et al.*, 2007; Lima and McNeil, 2009). Further corn strain “C” is highly associated with the *triosephosphate isomerise* (Tpi) gene located on the Z chromosome, and the rice strain “R” associated with the cytochrome oxidase subunit I (COI) gene (Nagoshi *et al.*, 2016; Acharya *et al.*, 2021). Examination of genetic makeup of FAW populations across four geographic regions, America, Africa, India and Asia-II, suggested that India and Asia-II exhibit greater genetic similarity to the African population in contrast to the American. This indicates a significant variation between native American and invasive populations (Nayyar *et al.*, 2021). In India, in addition to the ancestral rice and corn strain haplotype, there were records of inter-strain hybrid haplotypes of rice and corn strains, possibly due to inter-strain matings (Acharya *et al.*, 2021).

The molecular genetic diversity studies indicated that the FAW population in India belongs to the "R" strain based on variations in the COI gene (Swamy *et al.*, 2018). In Africa, the molecular markers were linked to mitochondrial rice strain COI and nuclear corn strain Tpi genes located on the Z chromosome did not effectively determine the genetic background of the strains with precision (Nagoshi *et al.*, 2019). However, the common ancestral haplotypes recorded in India, Africa and Asia suggest a common founder population for the invaded populations.

Researchers have developed microsatellites to determine population structure and genetic diversity across Africa and to overcome the limited genetic diversity and unreliability of the COI and Tpi markers. The higher levels of genetic diversity of the microsatellites revealed that the African population is predominantly a panmictic, or randomly interbreeding, population (Withers *et al.*, 2021). The variations between the two populations revealed through the sequenced genomes of the "C" and "R" strains of FAW were 438 Mb and 371 Mb, respectively (Kakumani *et al.*, 2014; Gouin *et al.*, 2017). The "C" strain contains 21,700 protein-coding genes, while the "R" strain is predicted to have 26,329 genes. The expansions of the genes are associated with chemo sensation and detoxification, primarily resulting from tandem duplications, enabling polyphagy. In Yunnan Province of China, the chromosome level of FAW genomes of the male moths was 542.42 Mb, while the female genome measured 530.77 Mb, both with a total of 22,201 genes. This was due to the expansion of cytochrome P450 (CYP) and glutathione S-transferase genes, often associated with pesticide detoxification and tolerance (Liu *et al.*, 2019). Similarly, the CYP gene family plays a role in intensified detoxification. The expansion of the CYP gene is associated with the highly polyphagous feeding behaviours of FAW and helps deactivate or break down foreign substances (xenobiotics) into harmless compounds (Cheng *et al.*, 2017; Gui *et al.*, 2022). C-strain larvae exhibit higher tolerance to pesticides such as methyl parathion, cypermethrin and *Bacillus thuringiensis* (Bt) ∂-endotoxins than the R-strain (Adamczyk *et al.*, 1997).

The genetic diversity of fall armyworm encompasses variations in genetic characteristics within and between populations. Investigating this diversity provides valuable insights into the pest's biology, behaviour, evolutionary history, and potential management strategies. By studying the genetic traits associated with pesticide resistance scientists developed approaches to combat the pest. Additionally, knowledge of FAW genetic diversity can aid in developing crop varieties with enhanced resistance or tolerance to the pest.

16.7 Behaviour, Ecology and Dispersal of Fall Armyworm (*Spodoptera frugiperda*)

The fall armyworm (FAW) (*Spodoptera frugiperda*) is a highly adaptable and destructive pest, with a broad geographic distribution and capacity to severely damage a wide range of economically important crops. This adaptability is due to its behaviour, ecology, dispersal abilities, and the presence of distinct strains. Understanding FAW's behaviour, ecology and dispersal patterns is critical for developing effective control strategies and minimizing its impact on agriculture worldwide.

16.7.1 Behavioural differences between strains

As mentioned earlier, one of the complexities of fall armyworm biology lies in the presence of two genetically distinct strains, commonly referred to as the rice-strain (R-strain) and corn-strain (C-strain) (Quisenberry, 1991; Nagoshi and Meagher, 2004). Despite being morphologically identical, these strains exhibit clear differences in their host plant preferences, behaviour and some life history traits. Understanding these differences is crucial for developing effective pest management strategies.

16.7.1.1 *Host plant specialization*

The R-strain of FAW prefers grasses such as rice (*Oryza sativa*) and Bermuda grass (*Cynodon dactylon*), as well as other members of the Graminaceae family. The C-strain specializes in crops like corn (*Zea mays*) and cotton (*Gossypium hirsutum*), making it a significant threat to key staple crops globally (Nagoshi and Meagher, 2004). Genetic markers and allozyme variants have helped distinguish between these two strains, providing insight into their population dynamics and enabling more tailored pest control measures. Studies have shown that carbon isotope ratios can be used to distinguish the two strains, especially in cotton-growing environments where the C-strain dominates (Nagoshi *et al.*, 2007). The C-strain develops in substantial numbers compared with the R-strain. However, populations often consist of a mix of both strains, with mating between them occurring at varying levels of success, and the presence of both strains in mixed populations complicates pest management efforts (Pashley, 1986).

16.7.1.2 *Mating behaviour*

Intra-strain mating occurs more frequently than inter-strain mating, suggesting some level of reproductive isolation. R-strain females will mate with C-strain males, which can result in mixed populations, while C-strain females rarely mate with R-strain males, since they are reproductively incompatible (Whitford *et al.*, 1988; Quisenberry, 1991). This partial reproductive isolation could lead to different strain compositions in agricultural regions, impacting pest control efforts (Nagoshi *et al.*, 2007).

16.7.1.3 *Implications for pest management*

The two strains vary in their resistance to insecticides and susceptibility to *Bacillus thuringiensis* (Bt) crops. For instance, the C-strain tends to show higher resistance to Bt toxins than the R-strain. Differences in development rates on specific crops also impact control strategies (Botha *et al.*, 2019). The presence of mixed populations can affect how the pest responds to chemical treatments, insecticide resistance and Bt crops, requiring different strategies depending on the local strain composition. Variation in larval development, feeding behaviour, and use of food resources between the two strains means that integrated pest management (IPM) strategies must be tailored to the strain present in a given area (Nagoshi and Meagher, 2004; Botha *et al.*, 2019).

16.7.2 Ecological adaptability

FAW is a highly adaptable pest, thriving in diverse ecosystems, ranging from dry savannas to tropical rainforests. Its adaptability allows it to establish in a wide range of climates and habitats, contributing to its status as one of the most economically damaging pests worldwide. Some key factors that contribute to its ecological success include:

Polyphagy: FAW has a broad host range, feeding on over 353 plant species, including key crops like maize, rice, sorghum and cotton. This ability to feed on a variety of hosts means that even if one food source becomes scarce or unsuitable, FAW can switch to other available crops (Sharanabasappa *et al.*, 2018).

Resistance Development: FAW populations have shown the ability to develop resistance to insecticides and Bt toxins, which poses significant challenges for control programmes. Managing FAW effectively requires the use of integrated pest management (IPM) strategies that incorporate cultural practices, biological controls and judicious use of chemicals (Gui *et al.*, 2020).

Life Cycle Flexibility: The life cycle of FAW varies depending on environmental conditions. In warm, summer conditions, the pest can complete its life cycle in about 30 days, while in cooler conditions, such as in spring and autumn, the life cycle can take up to 60 days. In winter, development is slower, with the life cycle taking 80–90 days. This flexibility allows FAW populations to grow and spread even in less-than-optimal conditions (Sharanabasappa *et al.*, 2023).

Non-Diapause Species: Unlike some other moth species, FAW does not enter diapause (a period of suspended development) during unfavourable conditions. This trait makes the pest more resilient and capable of continuous reproduction throughout the year in regions with mild winters.

16.7.3 Dispersal and migration patterns

FAW's ability to migrate long distances contributes to its status as a global pest. The species is known for its remarkable flight capacity, which enables it to spread over vast geographic areas in relatively short periods of time. FAW is a highly migratory species, capable of travelling over long distances in search of favourable breeding and feeding grounds. Adult moths are strong fliers, primarily active during the night and their migration patterns are largely influenced by wind currents. Studies have documented tethered flight durations of 16–30 hours for FAW males, showing the species' ability to sustain long flights (van Handel, 1974). FAW moths generally disperse from seasonally dry areas to more humid habitats in search of better food sources and suitable conditions for laying eggs. In favourable wind conditions, moths can travel up to 500 km before oviposition.

One famous study recorded FAW moths travelling 1,600 km from Mississippi to southern Canada in 30 hours (Rose *et al.*, 1975; Young, 1979), driven by a weather front. This demonstrates the species' ability to migrate great distances rapidly, which has implications for its spread across continents. FAW flies above the boundary layer, a part of the atmosphere where wind speeds are faster and more consistent, aiding its dispersal (Rose *et al.*, 1975). The direction and extent of FAW migration largely depend on prevailing winds, and moths often migrate in response to weather systems.

16.7.4 Global spread and invasion

The ability of FAW to migrate long distances has facilitated its spread across the world, causing extensive agricultural damage in new regions. After its introduction into West Africa in 2016, FAW spread rapidly across the continent, moving into southern, eastern and western regions within just 18 months (Zhang *et al.*, 2021). The most probable means of introduction into new regions, such as Egypt, is unaided dispersal through flight, although human-assisted spread through trade and transport (e.g. stowaway in aircraft) is also likely (Li *et al.*, 2023). These factors make FAW a highly mobile pest that is difficult to manage once established in a region.

16.7.4.1 Methods of spread

Unaided Dispersal: FAW's ability to fly long distances unaided is considered the most probable cause of its spread within and between regions. For example, the moth likely entered Egypt via unaided flight from neighbouring countries like Sudan (Li *et al.*, 2023).

Stowaway Dispersal: Another potential method of spread is human-assisted dispersal, where FAW eggs or larvae are transported via international trade, aeroplanes or cargo shipments.

Trade: Global trade and movement of agricultural products between countries may have also contributed to FAW's rapid invasion of new continents, as larvae or eggs could have been introduced through contaminated shipments (Heinrichs *et al.*, 2018).

Risk of Invasion into New Regions: Once FAW arrives in a new region, it can quickly establish itself due to its high reproductive rate, wide host range and long-distance dispersal capabilities. For example, it was feared that FAW would spread into Egypt from neighbouring Sudan due to its proximity and the pest's migration habits. Risk assessments were conducted to estimate the rate of movement and guide mitigation efforts (Day *et al.*, 2017).

FAW's global spread is a major concern for agriculture worldwide. Its invasion of Africa and subsequent spread to Asia and other parts of the world underscores the pest's ability to colonize new regions quickly and cause significant economic damage. The fall armyworm is a major agricultural pest with remarkable adaptability, high dispersal potential and a broad host range. Its ability to migrate over long distances, coupled with its rapid population growth and capacity for resistance development, makes it a global threat to food security. Integrated pest management approaches, which combine biological control, cultural practices and judicious use of chemical controls, are essential for sustainable management of this invasive pest (Wang *et al.*, 2024).

16.8 Overwintering and Diapause

The relationship between temperature and the development rate significantly influences the pest's biology, distribution and abundance (Tobin *et al.*, 2003). Similarly, the environmental conditions and migration are related, as discussed above. Thus, it becomes crucial to understand the impact of temperature on the development and migration of FAW under changing climate conditions. Such understanding is vital for conducting risk analyses, forecasting, and implementing effective management strategies to minimize pest infestation levels (Calvo and Molina, 2005). Unlike other migratory insects, FAW does not exhibit diapause but migrates to regions with favourable environmental conditions. FAW survives the winter in the southernmost regions where plants are consistently available, and the temperature is above 9.9°C (Luginbill, 1928). The winter planting of host crops also influences the extent of effective overwintering areas, as these crops serve as the primary food source and habitat for the FAW during the overwintering period (Jiang *et al.*, 2019).

The outbreaks of FAW in North America start with overwintering populations of South Florida and Texas that migrate northward during spring and early summer. The southern Texas populations migrate northeast, into the Mississippi River basin and Oklahoma, whereas the southern Florida populations migrate through northern Florida to Georgia, then south to South Carolina and reach middle Tennessee in July (Luginbill, 1928; Sparks, 1979). However, it has been observed that temperatures below 13°C at the overwintering sites may not support larval and pupal survival of FAW (Perkins, 1979); a minimum temperature of 10°C is required for their survival (Sparks, 1979). The adult emergence rate during overwintering positively correlates with the average soil temperature (Wood *et al.*, 1979). The total developmental duration is approximately 4 weeks during the summer (June to August), and 3 months during winter (December to February) (Wu *et al.*, 2019).

The FAW outbreak largely depends on the prevailing environmental conditions in the winter months. If host plants are available to sustain continuous generations with no significant fluctuations in temperature, FAW may be capable of overwintering in latitudes between 28°N and 29°N. Studying the possible overwintering environments for FAW that support their survival is essential for understanding the insect population's scope and formulating early management strategies. The cold hardiness directly determines geographic distribution by adaptation to winter temperatures of different regions (Zhang *et al.*, 2021). Research on the potential overwintering distribution of fall armyworms in Yunnan, with an investigation into the influence of environmental factors on distribution, serves as a basis for the prevention and management. The insights gained from such studies help in decision-making processes, potentially resulting in significant reductions in the use of pesticides (Huang *et al.*, 2020). In addition, a combination of pheromone traps and Polymerase Chain Reaction-Restriction Fragment Length Polymorphism (PCR-RFLP) analysis can provide insights into the population ecology and strain isolation mechanisms of FAW populations across habitats, including the overwintering areas located in southern Florida (Meaghe and Gallo-Meaghe, 2003).

Cultivation, irrigation and the elimination of straw in regions where FAW overwinters all have a substantial impact on reducing FAW populations. Employing field monitoring techniques also enables the estimation of adult FAW peak emergence in the following spring when weather conditions are favourable (Zhang *et al.*, 2021).

16.9 Integrated Pest Management

It is vital to comprehend the fall armyworm's management options and strategies due to its swift spread across the globe (Overton *et al.*, 2021).

16.9.1 Monitoring, Sampling and Pest Forecasting

FAW can be monitored using regular field inspection, pheromones and light traps. Regional monitoring can signal early pest presence. Although pheromone trap catches offer an early indication of moth activity within a region, monitoring and inspecting plants affected by pests provide valuable insights into the extent of field infestations (Kenis *et al*., 2023). This information is pivotal in determining the necessity of implementing control measures (Westbrook *et al*., 2016). Barfield *et al.* (1980) noted the knowledge-intensive nature of developing such a system due to the erratic nature of pest occurrences. The FAO introduced the FAW Monitoring and Early Warning System (FAMEWS) in 2020, gathering data and offering insights into FAW population changes. FAMEWS is a global platform accessible online, and it creates maps using the data collected through a mobile app during field scouting and pheromone trap checks (FAO, 2021). It provides real-time situational overviews, including maps and analytics of FAW infestations at global and regional levels.

16.9.2 Trapping

Detecting fall armyworm infestation before it causes economic damage is the key to management. During the first 40 days after planting, it is necessary to scout crop fields regularly every 3–4 days. For instance, on maize, if 5% of seedlings are cut or 20% of whorls of small plants (during the first 30 days) are infested with FAW (Assefa and Ayalew, 2019) it is recommended to apply an effective control measure to prevent further damage. The moths of FAW are attracted to light sources (Haftay and Fissiha, 2020). Therefore, the use of night-time light traps can be one of the monitoring mechanisms for FAW. Monitoring methods for FAW mainly involve use of specific pheromone traps, involving (Z)−7-dodecenyl acetate (Z)−7–12: Ac, (Z)−9-dodecenyl acetate (Z)−9–12: Ac, (Z)−9-tetradecenyl acetate (Z)−9–14: Ac, and (Z)−11- hexadecenyl acetate (Cruz *et al*., 2012). Installation rate of five pheromone traps per acre on or before germination for monitoring purposes can be followed (Tumlinson *et al*., 1986).

16.9.3 Cultural methods

Push-and-pull strategy is one of the most effective cultural controls. For example, maize is intercropped with silver-leaf or green-leaf desmodium that repel *S. frugiperda* and Napier, Sudan or Molasses grasses that attract *S. frugiperda*. When compared to those regions where maize was the only crop planted, push-and-pull areas showed around 83% reduction in the number of larvae per plant and 87% reduction in plant damage per plot (Midega *et al*., 2018). Use of clean planting material (seeds), early planting or growing early maturing cultivars, handpicking of larvae and the sprinkling of ash into the maize whorls, proper fertilizer management, stubble burning, regularly weeding and ploughing the soil deeply to expose larvae and pupae to the upper surface of the soil effectively decrease the pest population (Assefa, 2018).

16.9.4 Genetically modified crops

Transgenic plants expressing insecticidal proteins, particularly *Bacillus thuringiensis* Berliner (Bt) proteins, have contributed significantly to the global IPM programmes. To combat the yield losses caused by *S. frugiperda*, genetically modified crops such as Bt maize, Bt cotton and Bt soybean have been commercialized in Brazil (Machado *et al*., 2020). In Brazil and the USA, Bt maize is regarded as one of the most popular and successful genetic strategies to control *S. frugiperda*. When compared with the use of a single gene, transgenes with multiple modes of action, such as the Cry + Vip genes, may be more effective and provide long-lasting control. Cry1A, Cry1Ab and Cry1F against *S. frugiperda* are three crystal protein genes incorporated from Bt (Horikoshi *et al*., 2016).

16.9.5 Biological control

Biological control appears as a potential alternative to the chemical management of fall armyworm. Many natural enemies have been discovered related to this pest. Parasitoids and predators that attack fall armyworm are listed in Tables 16.1 and 16.2.

Entomopathogens infect and cause diseases in insects include viruses, bacteria, fungi nematodes, protozoans, and others that are well known against FAW (Table 16.3).

16.9.6 Action threshold levels

Precise determination of economic threshold levels for FAW control remains elusive (Overton *et al.*, 2021), and existing guidelines serve as the foundation for decision making (McGrath *et al.*, 2018). Considerable variability is observed in the damage caused by a specific degree of FAW infestation and in plants' responses to such damage (Cruz and Turpin, 1983; Hruska and Gladstone, 1988; Baudron *et al.*, 2019; Van der Berg *et al.*, 2021a). Field studies conducted in Zimbabwe indicated that predicting yield loss solely from isolated assessments of infestation and leaf damage is unreliable (Baudron *et al.*, 2019). To address the considerable uncertainty surrounding the interplay between infestation levels, plant impairment and yield loss, McGrath *et al.* (2018) and Van der Berg *et al.* (2021a) proposed the adoption of more cautious thresholds, especially when considering the circumstances of smallholder farmers.

16.10 Chemical Control

In the 1970s, organophosphates and carbamates were commonly employed, and pyrethroids succeeded in American countries. Around 2000, there was a substantial increase in the usage of spinosyns, avermectins, oxadiazines, insect growth regulators and diamides in the Americas (Yu, 1991, 1992). This intensive usage was particularly prominent in countries such as Brazil, the USA and Mexico (Carvalho *et al.*, 2013). However, as a consequence of resistance that developed due to the extensive use of chemicals, including organophosphates, carbamates, pyrethroids, and benzoylureas, farmers began to adopt Bt maize as an alternative approach (Burtet *et al.*, 2017; Kenis *et al.*, 2023).

In Africa, the recommendations initially followed those of the Americas, and most of the available pesticides were purchased and distributed. Pyrethroids and organophosphates were the most abundantly used chemicals for controlling fall armyworms followed by avermectins due to their easy availability and affordable prices. African countries also use chemicals belonging to other classes such as carbamates, benzoylureas, oxadiazine, spinosyn, neonicotinoids, combination chemicals, etc. (Otim *et al.*, 2021).

In most Asian countries, synthetic pesticides are considered to be a major hazard to sustainable agriculture. Hence, people adopted biological control measures to enhance plant and human health. However, under severe conditions, chemical pesticides give quick control of invasive insect pests. In the case of FAW, cypermethrin, lambda-cyhalothrin, bifenthrin, beta-cycflatherine, deltamethrin, chlorantraniliprole and flubendiamide were effective. In India, Central Insecticides Board and Registration Committee (CIB&RC) recommended insecticides are spinetoram, chlorantraniliprole and emamectin benzoate. In contrast toxicities of flubendiamide, indoxacarb, lambda-cyhalothrin and novaluron were at par according to the leaf-dip bioassay. The seed treatment with chlorantraniliprole + lambda-cyhalothrin and cyantraniliprole + thiamethoxam was effective and reduced the need for foliar sprays against fall armyworms.

16.11 New Control Techniques

In recent years, there has been a notable upsurge in the development of innovative methods encompassing RNA interference (RNAi), CRISPR/Cas9 genome editing and nanopesticides to address challenges posed by pests. Among these cutting-edge techniques, a highly promising advancement involves the fusion of RNAi with a nanocarrier. These fusions have yielded a novel, stable and secure strategy with substantial potential to transform the field of pest management (Ma *et al.*, 2020; Yan *et al.*, 2020). Scientists honed in on the prospects of CRISPR/Cas9 technology in pest control programmes. Wu *et al.* (2018) delved into the realm of possibilities by investigating the modification of the abdominal-A (Sfabd-A)

Table 16.1. List of parasitoids recorded on *Spodoptera frugiperda*. (Author's own table)

S.No.	Scientific name	Family: Order	References
Egg parasitoids			
1	*Telenomus remus*	Hymenoptera: Platygastridae	Shylesha *et al*., 2018; Keerthi *et al*., 2023
2	*Trichogramma pretiosum*	Hymenoptera: Trichogrammatidae	Shylesha *et al*., 2018; Keerthi *et al*., 2023
3	*Trichogramma chilonus*	Hymenoptera: Trichogrammatidae	Keerthi *et al*., 2023
4	*Trichogrammatoidea armigera*	Hymenoptera: Trichogrammatidae	Karimoune *et al*., 2018
Larval parasitoids			
5	*Aleiodes* sp.	Hymenoptera: Braconidae	Keerthi *et al*., 2023
6	*Cotesia icipe*	Hymenoptera: Braconidae	Agboyi *et al*., 2020
7	*Cotesia ruficrus*	Hymenoptera: Braconidae	Keerthi *et al*., 2023
8	*Bracon hebeter*	Hymenoptera: Braconidae	Agboyi *et al*., 2020
9	*Glyptapanteles creatonoti*	Hymenoptera: Braconidae	Shylesha *et al*., 2018
10	*Coccygidium melleum*	Hymenoptera: Braconidae	Sharanabasappa *et al*., 2019
11	*Coccygidium* sp.	Hymenoptera: Braconidae	Sagar *et al*., 2022
12	*Campoletis chlorideae*	Hymenoptera: Ichneumonidae	Sharanabasappa *et al*., 2019
13	*Eriborus* sp.	Hymenoptera: Ichneumonidae	Sharanabasappa *et al*., 2019
14	*Odontepyris* sp.	Hymenoptera: Bethylidae	Sharanabasappa *et al*., 2019
15	*Exorista sorbillans*	Diptera: Tachinidae	Sharanabasappa *et al*., 2019
16	*Winthemia trinitatis*	Diptera: Tachinidae	Bortolotto *et al*., 2014
17	*Drino quadrizonula*	Diptera: Tachinidae	Bortolotto *et al*., 2014
18	*Archytas marmoratus*	Diptera: Tachinidae	Bortolotto *et al*., 2014
19	*Lespesia archippivora*	Diptera: Tachinidae	Bortolotto *et al*., 2014
20	*Temelucha sp*	Hymenoptera: Ichneumonidae	Sagar *et al*., 2022
Egg larval parasitoids			
21	*Chelonus insularis*	Hymenoptera: Braconidae	Koffi *et al*., 2023
22	*Chelonus nr. blackburni*	Hymenoptera: Braconidae	Sagar *et al*., 2022
23	*Chelonus formosanus*	Hymenoptera: Braconidae	Sagar *et al*., 2022
24	*Chelonus bifoveolatus*	Hymenoptera: Braconidae	Koffi *et al*., 2023
Larval pupal parasitoids			
25	*Lespesia archippivora*	Diptera: Tachinidae	Sisay *et al*., 2018
26	*Archytas marmoratus*	Diptera: Tachinidae	Sisay *et al*., 2018

Table 16.2. List of predators against *Spodoptera frugiperda*. (Author's own table)

S. No.	Scientific name	Order: Family	Reference
1	*Doru luteipes*	Forficulidae: Dermaptera	Abbas *et al*., 2022
2	*Euborellia annulipes*	Anisolabididae: Dermaptera	Barros *et al*., 2018
3	*Eocanthecona furcellata*	Hemiptera: Pentatomidae	Keerthi *et al*., 2020
4	*Harmonia octomaculata*	Coleoptera: Coccinellidae	Sharanabasappa *et al*., 2019
5	*Coccinella transversalis*	Coleoptera: Coccinellidae	Sharanabasappa *et al*., 2019
6	*Calosoma calidum*	Coleoptera: Carabidae	Knutson, 2008
7	*Podisus maculiventris*	Pentatomidae: Hemiptera	Kneeland *et al*., 2012
8	*Orius similis*	Anthocoridae: Hemiptera	Zeng *et al*., 2021
9	*Pristhesancus plagipennis*	Reduviidae: Hemiptera	Grundy, 2007
10	*Rhynocoris marginatus*	Reduviidae: Hemiptera	Pradeep *et al*., 2022

Table 16.3. List of entomopathogens against FAW. (Author's own table)

Entomopathogenic virus			
S. No.	Name	Effective dosage	Reference
1	SfGV	1.5–3.0 × 10^{12} occlusion bodies/ha	Hussain *et al*., 2021
	SpfrNPV	1.5–3.0 × 10^{12} occlusion bodies/ha	Singh *et al*., 2023
2	SfMNPV	1.5–3.0 × 10^{12} occlusion bodies/ha	Hussain *et al*., 2021
Entomopathogenic bacteria			
3	Bt *aizawai*	Spray 2 g/litre or 400 g/acre	Guo *et al*., 2020
4	Bt *thuringiensis*	Spray 2 g/litre or 400 g/acre	Guo *et al*., 2020
5	Bt *kurstaki*	Spray 2 g/litre or 400 g/acre	Guo *et al*., 2020
Entomopathogenic fungi			
6	*Beauveria bassiana*	1 × 10^8 cfu/g or 5 g/litre as whorl application	Paredes-Sánchez *et al*., 2021
7	*Nomuraea rileyi*	1 × 10^8 cfu/g or 5 g/litre as whorl application	Paredes-Sánchez *et al*., 2021
8	*Metarhizium anisopliae*	1 × 10^8 cfu/g or 5 g/litre as whorl application	Paredes-Sánchez *et al*., 2021
Entomopathogenic nematode			
9	*Steinernema carpocapsae*	4000 crore infective juveniles/acre	Abbas *et al*., 2022
10	*Heterorhabditis bacteriophora*	5000 crore infective juveniles/acre	Abbas *et al*., 2022

gene using the CRISPR/Cas9 system, paving the way for innovative control strategies. Similarly, Jin *et al.* (2021) took a step forward by creating a SfABCC2 knockout strain of FAW through the CRISPR/Cas9 system and this groundbreaking development furnished additional functional insights into the role of this gene in susceptibility and resistance to Cry1F. Furthermore, the exploration of the Odorant Receptor Co-Receptor (ORCO) gene in FAW through CRISPR techniques has emerged as a significant area of study because of its precision in targeting specific olfactory genes; the potential applications of this research are vast. This may lead to the formulation of fresh, alternative strategies employing insecticides and microbial sprays for FAW control (Ayra-Pardo and Borras-Hidalgo, 2019). A sensor-based pheromone trap, developed as part of a project supported by the UK Foreign, Commonwealth and Development Office's (FCDO) Science and Innovation Network in India, enables remote operation worldwide to combat the FAW population in India (CABI, 2023).

16.12 Conclusion

FAW has been advancing into new territories at an alarming rate, posing a significant threat to agriculture and global food security. The Food and Agriculture Organization sounded the alarm for the potential impact of this recent pest outbreak in Asia on the world's food and nutritional security. Consequently, we must prioritize the adoption of effective management strategies to address the FAW issue. Even though FAW has a long history of causing substantial damage and there are extensive research efforts worldwide, several critical aspects of its ecology and management still demand a deeper understanding. This is particularly important to minimize the devastating consequences of its invasion in affected regions.

Acknowledgements

The authors gratefully acknowledge Prof V.V. Ramamurthy former Professor, ICAR–IARI, New Delhi for his guidance and Dr M.K. Dhillon, Head, Division of Entomology, ICAR–IARI, New Delhi for providing the necessary facilities. Dr Shashank acknowledges support from the project titled "Development of pest and disease management strategies for major crops (CRSCIARISIL20140332 65)".

References

Abbas, A., Ullah, F., Hafeez, M., Han, X., Dara, M.Z.N. *et al.* (2022) Biological control of fall armyworm, *Spodoptera frugiperda*. *Agronomy* 12(11), 2704.

Abrahams, P., Bateman, M., Beale, T., Clottey, V., Cock, M. *et al.* (2017) Fall armyworm: Impacts and implications for Africa; Evidence note (2); Report to DFID. Available at: https://www.cabi.org/ISC/FullTextPDF/2018/20187200428 (accessed 20 November 2019).

Acharya, R., Akintola, A.A., Malekera, M.J., Kamulegeya, P., Nyakunga, K.B. *et al.* (2021) Genetic relationship of fall armyworm (*Spodoptera frugiperda*) populations that invaded Africa and Asia. *Insects* 12(5), 439.

Adamczyk Jr, J.J., Holloway, J.W., Leonard, B.R. and Graves, J.B. (1997) Susceptibility of fall armyworm collected from different plant hosts to selected insecticides and transgenic BT cotton. *Journal of Cotton Science* 1, 21–28.

Agboyi, L.K., Goergen, G., Besehm, P., Mensah, S.A., Clottey, V.A. *et al.* (2020) Parasitoid complex of fall armyworm, *Spodoptera frugiperda*, in Ghana and Benin. *Insects* 11, 68.

Alam, S.N., Sarker, D., Pradhan, M.Z., Rashid, M.H., Sarkar, M.A. *et al.* (2018) First report of occurrence of fall armyworm *Spodoptera frugiperda* in Bangladesh. *Bangladesh Journal of Entomology* 28(1), 97–101.

Ashok, K., Kennedy, J.S., Geethalakshmi, V., Jeyakumar, P., Sathiah, N. *et al.* (2020) Lifetable study of fall army worm *Spodoptera frugiperda* (J.E. Smith) on maize. *Indian Journal of Entomology* 82(3), 574–579.

Assefa, F. (2018) Status of fall armyworm (*Spodoptera frugiperda*), biology and control measures on maize crop in Ethiopia: A Review. *International Journal of Entomological Research* 6(2), 75–85.

Assefa, F. and Ayalew, D. (2019) Status and control measures of fall armyworm (*Spodoptera frugiperda*) infestations in maize fields in Ethiopia: A review. *Cogent Food and Agriculture* 5, 1641902. DOI: 10.1080/23311932.2019.1641902.

Ayra-Pardo, C. and Borras-Hidalgo, O. (2019) Fall armyworm (FAW; Lepidoptera: Noctuidae): Moth oviposition and crop protection. In: Picimbon, J.F. (ed.) *Olfactory Concepts of Insect Control – Alternative to Insecticides*. Springer, Cham, pp. 93–116.

Bajracharya, A.S., Bhat, B., Sharma, P., Shashank, P.R., Meshram, N.M. *et al.* (2019) First record of fall army worm *Spodoptera frugiperda* (J.E. Smith) from Nepal. *Indian Journal of Entomology* 81(4), 635–639.

Barfield, C.S., Stimac, J.L. and Keller, M.A. (1980) Fall armyworm symposium: State-of-the-art for predicting damaging infestations of fall armyworm. *Florida Entomologist* 63(4), 364–375.

Barros, E.M., Da Silva-Torres, C.S.A., Torres, J.B. and Rolim, G.G. (2018) Short-term toxicity of insecticides residues to key predators and parasitoids for pest management in cotton. *Phytoparasitica* 46, 391–404.

Baudron, F., Zaman-Allah, M.A., Chaipa, I., Chari, N. and Chinwada, P. (2019) Understanding the factors influencing fall armyworm (*Spodoptera frugiperda* J.E. Smith) damage in African smallholder maize fields and quantifying its impact on yield. A case study in Eastern Zimbabwe. *Crop Protection* 120, 141–150. DOI: 10.1016/j.cropro.2019.01.028.

Bhavani, B., Chandra Sekhar, V., Kishore Varma, P., Bharatha Lakshmi, M., Jamuna, P. *et al.* (2019) Morphological and molecular identification of an invasive insect pest, fall armyworm, *Spodoptera frugiperda* occurring on sugarcane in Andhra Pradesh. *India. Journal of Entomology and Zoology Studies* 7(4), 12–18.

Bortolotto, O.C., Menezes, A.D.O., Hoshino, A.T., Carvalho, M.G., Pomari-Fernandes, A. *et al.* (2014) Sugar solution treatment to attract natural enemies and its impact on fall armyworm *Spodoptera frugiperda* in maize fields. *Interciencia* 39, 416.

Botha, A.M., Kunert, K.J., Malinga, J. and Foyer, C.H. (2019) Defining biotechnological solutions for insect control in Sub-Saharan Africa. *Food and Energy Security* 9, e191. DOI: 10.1002/fes3.191.

Bourke, M. and Sar, S. (2020) Fall armyworm in Papua New Guinea: How big is the risk? Australian National University, Canberra.

Burtet, L.M., Bernardi, O., Melo, A.A., Pes, M.P., Strahl, T.T. *et al.* (2017) Managing fall armyworm, *Spodoptera frugiperda* (Lepidoptera: Noctuidae), with Bt maize and insecticides in southern Brazil. *Pest Management Science* 73(12), 2569–2577.

CABI (2019) Community-based fall armyworm (spodoptera frugiperda) monitoring, early warning and management. *Training of Trainers Manual* 21.

CABI (2023) CABI science report 2022. Centre for Agriculture and Biosciences International, Wallingford, UK, p. 54.

Calvo, D. and Molina, J.M. (2005) Fecundity–longevity trade-off in the fall armyworm *Spodoptera frugiperda*. *Entomologia Experimentalis et Applicata* 116, 205–211.

Capinera, J.L. (2017) Fall armyworm, *Spodoptera frugiperda* (J.E. Smith) (Insecta: Lepidoptera: Noctuidae). EENY098, IFAS Extension, University of Florida, 1–6. Available at: https://edis.ifas.ufl.edu/in255

Carvalho, R.A., Omoto, C., Field, L.M., Williamson, M.S. and Bass, C. (2013) Investigating the molecular mechanisms of organophosphate and pyrethroid resistance in the fall armyworm *Spodoptera frugiperda*. *PLoS One* 8(4), e62268.

Cheng, T., Wu, J., Wu, Y., Chilukuri, R.V., Huang, L. *et al.* (2017) Genomic adaptation to polyphagy and insecticides in a major East Asian noctuid pest. *Nature Ecology & Evolution* 1, 1747–1756.

Chikkarugi, N.M., Vijaykumar, L., Yeshwanth, H.M., Raveendra, H.R. and Shivanna, B. (2023) Occurrence of ear head caterpillar species complex on Kenyan finger millet (*Eleusine coracana* L. Gaertn.) genotypes. *Journal of Environmental Biology* 44(2), 207–213.

Clark, P.L., Molina-Ochoa, J., Martinelli, S., Skoda, S.R., Isenhour, D.J. *et al.* (2007) Population variation of the fall armyworm, *Spodoptera frugiperda*, in the Western Hemisphere. *Journal of Insect Science* 7, 1–10.

Cock, M.J.W., Beseh, P.K., Buddie, A.G., Cafá, G. and Crozier, J. (2017) Molecular methods to detect *Spodoptera frugiperda* in Ghana, and implications for monitoring the spread of invasive species in developing countries. *Scientific Reports* 7, 4103.

Cruz, I. and Turpin, F.T. (1983) Yield impact of larval infestations of the fall armyworm (Lepidoptera: Noctuidae) to midwhorl growth stage of corn. *Journal of Economic Entomology* 76(5), 1052–1054.

Cruz, I., Figueiredo, M.D.L.C., Silva, R.B.D., Silva, I.F.D., Paula, C.D. *et al.* (2012) Using sex pheromone traps in the decision-making process for pesticide application against fall armyworm (*Spodoptera frugiperda* [Smith] [Lepidoptera: Noctuidae]) larvae in maize. *International Journal of Pest Management* 58(1), 83–90.

Davis, F.M., Ng, S.S. and Williams, W.P. (1992) Visual rating scales for screening whorl-stage corn for resistance to fall armyworm. *Mississippi Agricultural and Forestry Experiment Station Technical Bulletin* 186, 1–9.

Day, R., Abrahams, P., Bateman, M., Beale, T., Clottey, V. *et al.* (2017) Fall armyworm: Impacts and implications for Africa. *Outlooks on Pest Management* 28(5), 196–201.

Deole, S. and Paul, N. (2018) First report of fall army worm, *Spodoptera frugiperda* (JE Smith), their nature of damage and biology on maize crop at Raipur, Chhattisgarh. *Journal of Entomology and Zoology Studies* 6(6), 219–221.

Dew, J.A. (1913) Fall army worm. *Journal of Economic Entomology* 6(4), 361–366.

Dumas, P., Legeai, F., Lemaitre, C., Scaon, E., Orsucci, M. *et al.* (2015) *Spodoptera frugiperda* (Lepidoptera: Noctuidae) host-plant variants: Two host strainsor two distinct species? *Genetica* 143, 305–316.

EPPO (2019) *Spodoptera frugiperda* continues to spread in Asia. In EPPO Reporting Service. Available at: https://gd.eppo.int/reporting/article-6483 (accessed 21 May 2024).

EPPO (2020a) First detection of *Spodoptera frugiperda* (fall armyworm) in Torres Strait (AUS-96/1). Available at: https://www.ippc.int/en/countries/australia/pestreports/2020/02/first-detection-of-spodoptera-frugiperda-fall-armyworm-in-torres-strait/ (accessed 25 May 2024).

EPPO (2020b) *Spodoptera frugiperda* in Southern Mauritania (2020/143). Available at: https://gd.eppo.int/taxon/LAPHFR/distribution/MR (accessed 25 May 2024).

EPPO (2020c) *Spodoptera frugiperda* in East Timor (2020/142). Available at: https://gd.eppo.int/taxon/LAPHFR/distribution/TP (accessed 25 May 2024).

EPPO (2021) *Spodoptera frugiperda* in Spain (2021/053). Available at: https://gd.eppo.int/taxon/LAPHFR/distribution/ES_ci (accessed 25 May 2024).

EPPO (2023) *Spodoptera frugiperda* in Cyprus (2023/034). Available at: https://gd.eppo.int/taxon/LAPHFR/distribution/ES_ci (accessed 25 May 2024).

FAO (2018) *Briefing note on fall armyworm in Africa: Status and strategy for FAO*. Food and Agriculture Organization of the United Nations, Rome, Italy.

FAO (2021) *Global action for fall armyworm control: Third progress report (July–December 2020)*. Food and Agriculture Organization of the United Nations, Rome, Italy.

Ferreira Filho, J.B.D.S., Alves, L.R.A., Gottardo, L.C.B. and Georgino, M. (2010) Dimensionamento do custo econômico representado por Spodoptera Frugiperdana cultura do milho no Brasil. *Tecnologias, Desenvolvimento e Integração Social; Anais*.

Ganiger, P.C., Yeshwanth, H.M., Muralimohan, K., Vinay, N., Kumar, A.R. *et al.* (2018) Occurrence of the new invasive pest, fall armyworm, *Spodoptera frugiperda* (J.E. Smith) (Lepidoptera: Noctuidae), in the maize fields of Karnataka, India. *Current Science* (4), 621–623.

Goergen, G., Kumar, P.L., Sankung, S.B., Togola, A. and Tamò, M. (2016) First report of outbreaks of the fall armyworm *Spodoptera frugiperda* (J.E. Smith) (Lepidoptera, Noctuidae), a new alien invasive pest in West and Central Africa. *PLoS ONE* 11(10), e0165632.

Gouin, A., Bretaudeau, A., Nam, K., Gimenez, S., Aury, J.M. *et al.* (2017) Two genomes of highly polyphagous lepidopteran pests (*Spodoptera frugiperda*, Noctuidae) with different host-plant ranges. *Scientific Reports* 7(1), 11816.

Grundy, P. (2007) Utilizing the assassin bug, *Pristhesancus plagipennis* (Hemiptera: Reduviidae), as a biological control agent within an integrated pest management programme for *Helicoverpa* spp. (Lepidoptera: Noctuidae) and *Creontiades* spp. (Hemiptera: Miridae) in cotton. *Bulletin of Entomological Research* 97, 281–290.

Gui, F., Lan, T., Zhao, Y., Guo, W., Dong, Y. *et al.* (2022) Genomic and transcriptomic analysis unveils population evolution and development of pesticide resistance in fall armyworm *Spodoptera frugiperda*. *Protein and Cell* 13(7), 513–531.

Gui, F.R., Lan, T., Zhao, Y., Guo, W., Dong, Y. *et al*. (2020) Chromosome-level genome assembly of the fall armyworm *Spodoptera frugiperda*. *Molecular Ecology Resources* 20, 1–13.

Guo, J., Zhao, J., He, K., Zhang, F. and Wang, Z. (2018) Potential invasion of the crop-devastating insect pest fall armyworm *Spodoptera frugiperda* to China. *Plant Protection* 44(6), 1–10.

Guo, J., Wu, S., Zhang, F., Huang, C., He, K. *et al*. (2020) Prospects for microbial control of the fall armyworm *Spodoptera frugiperda*: A review. *BioControl* 65, 647–662.

Hadi, B. (2019) Fall armyworm in cambodia: Surveillance, detection and response. Abstract of a paper presented at the XIX international plant protection congress IPPC 2019 (Hyderabad, India, 2019-11-10/14). In: *Abstract Book*. p. 5.

Haftay, G.G. and Fissiha, G.G. (2020) Effect of integrating night-time light traps and push-pull method on monitoring and deterring adult fall armyworm (*Spodoptera frugiperda*). *International Journal of Entomology Research* 5(1), 28–32.

Hang, D.T., Liem, N.V., Lam, P.V., Wyckhuys, K.A., Thuy, N.T. *et al*. (2020) First record of fall armyworm *Spodoptera frugiperda* (J.E. Smith) (Lepidoptera: Noctuidae) on maize in Viet Nam. *Zootaxa* 4772(2), 396–400.

Harrison, R.D., Thierfelder, C., Baudron, F., Chinwada, P., Midega, C. *et al*. (2019) Agro-ecological options for fall armyworm (*Spodoptera frugiperda* J.E. Smith) management: Providing low-cost, smallholder friendly solutions to an invasive pest. *Journal of Environmental Management* 243, 318–330.

He, L.M., Wang, T.L., Chen, Y.C., Ge, S.S., Wyckhuys, K.A. *et al*. (2021) Larval diet affects development and reproduction of East Asian strain of the fall armyworm, *Spodoptera frugiperda*. *Journal of Integrative Agriculture* 20(3), 736–744.

Heinoun, K., Muhammad, E., Abdullah Smadi, H., Annahhas, D. and Abou Kubaa, R. (2021) First record of fall armyworm (*Spodoptera frugiperda*) in Syria. *EPPO Bulletin* 51(1), 213–215.

Heinrichs, E.A., Sidhu, J., Muniappan, R., Fayad, A., Adiga, A. *et al*. (2018) Pest risk assessment of the fall armyworm, *Spodoptera frugiperda* in Egypt. In: *Feed the Future: The U.S. Government'S Global Hunger and Food Security Initiative*.

Higo, Y., Sasaki, M. and Amano, T. (2022) Morphological characteristics to identify fall armyworm, *Spodoptera frugiperda* (Lepidoptera: Noctuidae) from common polyphagous noctuid pests for all instar larvae in Japan. *Applied Entomology and Zoology* 57(3), 263–274.

Horikoshi, R.J., Bernardi, D., Bernardi, O., Malaquias, J.B., Okuma, D.M. *et al*. (2016) Effective dominance of resistance of *Spodoptera frugiperda* to Bt maize and cotton varieties: Implications for resistance management. *Scientific Reports* 6(1), 34864.

Hruska, A.J. and Gladstone, S.M. (1988) Effect of period and level of infestation of the fall armyworm, *Spodoptera frugiperda*, on irrigated maize yield. *The Florida Entomologist* 71(3), 249–254. DOI: 10.2307/3495428.

Huang, Y., Dong, Y., Huang, W., Ren, B., Deng, Q. *et al*. (2020) Overwintering distribution of fall armyworm (*Spodoptera frugiperda*) in Yunnan, China, and influencing environmental factors. *Insects* 11(11), 805.

Huang, Y., Lv, H., Dong, Y., Huang, W., Hu, G. *et al*. (2022) Mapping the spatio-temporal distribution of fall armyworm in China by coupling multi-factors. *Remote Sensing* 14(17), 4415.

Hussain, A.G., Wennmann, J.T., Goergen, G., Bryon, A. and Ros, V.I. (2021) Viruses of the fall armyworm *Spodoptera frugiperda*: A review with prospects for biological control. *Viruses* 13(11), 2220.

IPPC (2019a) Report of first detection of *Spodoptera frugiperda* – Fall armyworm (FAW) in Egypt (EGY-01/1). Available at: https://www.ippc.int/en/countries/egypt/pestreports/2019/06/report-of-first-detection-of-spodoptera-frugiperda-fall-armyworm-faw-in-egypt/ (accessed 22 May 2024).

IPPC (2019b) Fall Armyworm (*Spodoptera frugiperda*) in Malaysia (MYS-02/4). Available at: https://www.ippc.int/en/countries/malaysia/pestreports/2019/11/fall-armyworm-spodoptera-frugiperda-control/ (accessed 22 May 2024).

IPPC (2020) Report of first detection of *Spodoptera frugiperda* - Fall Armyworm (FAW) in Jordan (JOR-04/1). Available at: https://www.ippc.int/fr/countries/jordan/pestreports/2020/09/report-of-first-detection-of-spodoptera-frugiperda-fall-armyworm-faw-in-jordan-3/ (accessed 25 May 2024).

IPPC (2021) New Caledonia (NCL-04/1 and NCL-04/2) Detection of *Spodoptera frugiperda* (fall armyworm) in New Caledonia. Available at: https://www.ippc.int/fr/countries/new-caledonia/pestreports/2021/01/-0/ (accessed 25 May 2024).

Jiang, Y., Liu, J., Xie, M., Li, Y., Zhang, M. *et al*. (2019) Observation on law of diffusion damage of *Spodoptera frugiperda* in China in 2019. *Plant Protection* 45, 10–19.

Jin, M.H., Tao, J.H., Li, Q., Cheng, Y., Sun, X.X. *et al*. (2021) Genome editing of the SfABCC2 gene confers resistance to Cry1F toxin from *Bacillus thuringiensis* in *Spodoptera frugiperda*. *Journal of Integrative Agriculture* 20(3), 815–820.

Kakumani, P.K., Malhotra, P., Sunil, K., Mukherjee, R. and Bhatnagar, K. (2014) A draft genome assembly of the armyworm, *Spodoptera frugiperda*. *Genomics* 104(2), 134–143.

Karimoune, L., Ba, M.N., Baoua, I.B. and Muniappan, R. (2018) The parasitoid *Trichogrammatoidea armigera* Nagaraja (Hymenoptera: Trichogrammatidae) is a potential candidate for biological control of the millet head miner *Heliocheilus albipunctella* (de Joannis) (Lepidoptera: Noctuidae) in the Sahel. *Biological Control* 127, 9–16.

Keerthi, M.C., Sravika, A., Mahesha, H.S., Gupta, A. and Bhargavi, H.A. (2020) Performance of the native predatory bug, *Eocanthecona furcellata* (Wolff) (Hemiptera: Pentatomidae), on the fall armyworm, *Spodoptera frugiperda* (J.E. smith) (Lepidoptera: Noctuidae), and its limitation under field condition. *Egyptian Journal of Biological Pest Control* (1), 1–4.

Keerthi, M.C., Mahesha, H.S., Kulkarni, N.S., Shivakumar, B.G., Shivakumara, K.T. *et al*. (2021) The threat of fall armyworm invasion and its host expansion on fodder crops. *Forage Research* 47(3), 372–375.

Keerthi, M.C., Mahesha, H.S., Manjunatha, N., Gupta, A., Saini, R.P. *et al*. (2023) Biology and oviposition preference of fall armyworm, *Spodoptera frugiperda* (J.E. Smith) (Lepidoptera: Noctuidae) on fodder crops and its natural enemies from Central India. *International Journal of Pest Management* 69(3), 215–224.

Kenis, M., Benelli, G., Biondi, A., Calatayud, P.A., Day, R. *et al*. (2023) Invasiveness, biology, ecology, and management of the fall armyworm, *Spodoptera frugiperda*. *Entomologia Generalis* 43(2), 187–241.

Kneeland, K., Coudron, T.A., Lindroth, E., Stanley, D. and Foster, J.E. (2012) Genetic variation in field and laboratory populations of the spined soldier bug. *Podisus Maculiventris*. *Entomologia Experimentalis Et Applicata* 143(2), 120–126. DOI: 10.1111/j.1570-7458.2012.01239.x.

Knutson, A. (2008) The fall armyworm—pest of pasture, hayfields and small grains, agrilife extension; texas a&m research and extension center: college station, TX, USA.

Koffi, D., Agboka, K., Adjevi, M.K.A., Adom, M., Tounou, A.K. *et al*. (2023) Natural control agents of the fall armyworm, *Spodoptera frugiperda*, in Togo: Moderating insecticide applications for natural control of the pest. *Journal of Pest Science* 96(4), 1405–1416.

Kumela, T., Simiyu, J., Sisay, B., Likhayo, P., Mendesil, E. *et al*. (2019) Farmers' knowledge, perceptions and management practices of the new invasive pest, fall armyworm, *Spodoptera frugiperda*, in Ethiopia and Kenya. *International Journal of Pest Management* 65(1), 1–9.

Lee, G.S., Seo, B.Y., Lee, J., Kim, H., Song, J.H. *et al*. (2020) First report of the fall armyworm, *Spodoptera frugiperda* (Smith, 1797) (Lepidoptera: Noctuidae), a new migratory pest in Korea. *Korean Journal of Applied Entomology* 59, 73–78.

Lima, E.R. and McNeil, J.N. (2009) Female sex pheromones in the fall armyworm *Spodoptera frugiperda*. *Journal of Chemical Ecology* 35, 137–145.

Liu, H., Lan, T., Fang, D., Gui, F., Wang, H. *et al*. (2019) Chromosome-level draft genome of the fall armyworm *Spodoptera frugiperda*. *Molecular Ecology Resources* 19, 1–12.

Li, Y.P., Yao, S.Y., Feng, D., Haack, R.A., Yang, Y. *et al*. (2023) Dispersal behavior characters of *Spodoptera frugiperda* larvae. *Insects* 14(6), 488.

Luginbill, P. (1928) *The fall army worm* (No. 34). US Department of Agriculture, Washington, D.C. DOI: 10.22004/ag.econ.156281.

Machado, E.P., Rodrigues Junior, G.L.S., Führ, F.M., Zago, S.L., Marques, L.H. *et al*. (2020) Cross-crop resistance of *Spodoptera frugiperda* selected on Bt maize to genetically modified soybean expressing Cry1Ac and Cry1F proteins in Brazil. *Scientific Reports* 10, 10080.

Ma, Z.Z., Zhou, H., Wei, Y.L., Yan, S. and Shen, J. (2020) A novel plasmid–*Escherichia coli* system produces large batch dsRNAs for insect gene silencing. *Pest Management Science* 76(7), 2505–2512.

McGrath, D., Huesing, J.E., Beiriger, R., Nuessly, G., Tepa-Yotto, T.G. *et al*. (2018) Monitoring, surveillance, and scouting for fall armyworm. *Fall Armyworm in Africa: A Guide for Integrated Pest Management* 11–28.

Meagher Jr, R.L. and Gallo-Meagher, M. (2003) Identifying host strains of fall armyworm (Lepidoptera: Noctuidae) in Florida using mitochondrial markers. *Florida Entomologist* 86(4), 450–455.

Midega, C.A., Pittchar, J.O., Pickett, J.A., Hailu, G.W. and Khan, Z.R. (2018) A climate-adapted push-pull system effectively controls fall armyworm, *Spodoptera frugiperda* (JE Smith), in maize in East Africa. *Crop Protection* 105, 10–15.

Mitchell, E.R. (1979) Fall armyworm symposium: Preface. *Florida Entomologist* 81–81.

Montezano, D.G., Sosa-Gómez, D.R., Specht, A., Roque-Specht, V.F., Sousa-Silva, J.C. *et al.* (2018) Host plants of *Spodoptera frugiperda* (Lepidoptera: Noctuidae) in the Americas. *African Entomology* 26(2), 286–300.

Moorthy, A.V., Shanthi, M. and Rani, S. (2023) Synthesis of N alkyl chitosan and their impact on biology and physiology of fall armyworm, *Spodoptera frugiperda*. Available at: https://www.researchsquare.com/article/rs-2875480/v1

Muniappan, R. (2024) Management of fall army worm *Spodoptera frugiperda* (JE Smith) in South Asia-current status and future strategies. *Indian Journal of Entomology* e22782, 295–299.

Murúa, G. and Virla, E.G. (2004) Contribution to the biological knowledge of *Euplectrus platyhypenae* (Hymenoptera: Eulophidae), a parasitoid of *Spodoptera frugiperda* (Lepidoptera: Noctuidae) in Argentina. *Folia Entomológica Mexicana* 43(2), 171–180.

Nagoshi, R.N. and Meagher, R.L. (2016) Using intron sequence comparisons in the triose-phosphate isomerase gene to study the divergence of the fall armyworm host strains. *Insect Molecular Biology* 25(3), 324–337.

Nagoshi, R.N. and Meagher, R.L. (2008) Review of fall armyworm, *Spodoptera frugiperda* (Lepidoptera: Noctuidae), genetic complexity and migration. *Florida Entomologist* 91(4), 546–554.

Nagoshi, R.N. and Meagher, R.L. (2004) Seasonal distribution of fall armyworm (Lepidoptera: Noctuidae) host strains in agricultural and turf grass habitats. *Environmental Entomology* 33, 881–889.

Nagoshi, R.N., Adamczyk, J.J., Jr., Meagher, R.L., Gore, J. and Jackson, R. (2007) Using stable isotope analysis to examine fall armyworm (Lepidoptera: Noctuidae) host strains in a cotton habitat. *Journal of Economic Entomology* 100, 1569–1576.

Nagoshi, R.N., Goergen, G., Tounou, K.A., Agboka, K., Koffi, D. *et al.* (2016) Genetic studies of fall armyworm indicate strain-specific migration in africa. *PLoS ONE* 11(4), e0153492.

Nagoshi, R.N., Meagher, R.L. and Hay-Roe, M. (2012) Inferring the annual migration patterns of fall armyworm (*Spodoptera frugiperda*) in the United States from mitochondrial haplotypes. *Ecology and Evolution* 2(7), 1458–1467.

Nagoshi, R.N., Htain, N.N., Boughton, D., Zhang, L., Xiao, Y. *et al.* (2019) Southeastern Asia fall armyworm populations are closely related to populations in Africa and India. *PLoS ONE* 14(9), e0221422.

Navasero, M.M. and Navasero, M.V. (2020) Life cycle, morphometry and natural enemies of fall armyworm, *Spodoptera frugiperda* (JE Smith) (Lepidoptera: Noctuidae) on *Zea mays* L. in the Philippines. *Journal of the International Society for Southeast Asian Agricultural Sciences* 26(2), 17–29.

Navasero, M.V., Navasero, M.M., Burgonio, G.A.S., Ardez, K.P., Ebuenga, M.D. *et al.* (2019) Detection of the fall armyworm, *Spodoptera frugiperda* (J.E. Smith) (Lepidoptera: Noctuidae), using larval morphological characters and observations on its current local distribution in the Philippines. *Philippine Entomologist* 33(2), 171–184.

Nayyar, N., Gracy, R.G., Ashika, T.R., Mohan, G., Swathi, R.S. *et al.* (2021) Population structure and genetic diversity of invasive fall armyworm after 2 years of introduction in India. *Scientific Reports* 11(1), 7760.

Otim, M.H., Fiaboe, K.K.M., Akello, J., Mudde, B., Obonyo, M.A.T. *et al.* (2021) Managing a transboundary pest: The fall armyworm on maize in Africa. *Moths and Caterpillars* 1–16. DOI: 10.5772/intechopen.96637.

Overton, K., Maino, J.L., Day, R., Umina, P.A., Bett, B. *et al.* (2021) Global crop impacts, yield losses and action thresholds for fall armyworm (*Spodoptera frugiperda*): A review. *Crop Protection* 145, 105641.

Paredes-Sánchez, F.A., Rivera, G., Bocanegra-García, V., Martínez-Padrón, H.Y., Berrones-Morales, M. *et al.* (2021) Advances in control strategies against *Spodoptera frugiperda*. A review. *Molecules* 26(18), 5587.

Pashley, D.P. (1986) Host-associated genetic differentiation in fall armyworm (Lepidoptera: Noctuidae): A sibling species complex. *Annals of the Entomological Society of America* 79, 898–904.

Pashley, D.P. and Martin, J.A. (1987) Reproductive incompatibility between host strains of the fall armyworm (Lepidoptera: Noctuidae). *Annals of the Entomological Society of America* 80(6), 731–733.

Pehlivan, S. and Atakan, E. (2022) First record of the fall armyworm, *Spodoptera frugiperda* (J.E. Smith, 1797) (Lepidoptera: Noctuidae) in Türkiye. *Çukurova Tarım ve Gıda Bilimleri Dergisi* 139–145.

Perera, N., Magamage, M., Kumara, A., Galahitigama, H., Dissanayake, K. *et al.* (2019) Fall armyworm (FAW) epidemic in Sri Lanka: Ratnapura district perspectives. *International Journal of Entomological Research* 7(1), 9–18.

Perkins, W.D. (1979) Laboratory rearing of the fall armyworm. *The Florida Entomologist* 62, 87–91.

Pradeep, P., Deshmukh, S.S., Kalleshwaraswamy, C.M. and Rajan, S.J. (2022) Biology and predation potential of the hemipteran predator, *Rhynocoris marginatus* (Fab. 1794) on the fall armyworm, *Spodoptera frugiperda* (J.E. Smith) (Lepidoptera: Noctuidae). *Egyptian Journal of Biological Pest Control* (1), 1–6.

Prasanna, B.M., Huesing, J.E., Eddy, R. and Peschke, V.M. (2018) Fall armyworm in Africa: A guide for integrated pest management. Mexico. CIMMYT. USAID.

Quaintance, A.L. (1897) Brief notes on the more injurious insects of the year. *Annual Report of the Florida Agricultural Experiment Station* 57–59.

Quisenberry, S.S. (1991) Fall armyworm (Lepidoptera: Noctuidae) host strain reproductive compatibility. *Florida Entomologist* 74, 194–199.

Raffa, K.F. (1987) Effect of host plant on cannibalism rates by fall armyworm (Lepidoptera: Noctuidae) larvae. *Environmental Entomology* 16(3), 672–675.

Rajisha, P.S., Muthukrishnan, N., Nelson, S.J., Jerlin, R., Marimuthu, P. *et al.* (2021) Biology and nutritional indices of the fall armyworm *Spodoptera frugiperda* (J.E. Smith) on maize. *Indian Journal of Entomology* 83, 1–5.

Ramírez-Cabral, N., Medina-García, G. and Kumar, L. (2020) Increase of the number of broods of fall armyworm (*Spodoptera frugiperda*) as an indicator of global warming. *Revista Chapingo Serie Zonas Áridas* 19(1), 1–16.

Rose, A.H., Silversides, R.H. and Lindquist, O.H. (1975) Migration flight by an aphid, *Rhopalosiphum maidis* (Hemiptera: Aphididae), and a noctuid, *Spodoptera frugiperda* (Lepidoptera: Noctuidae). *The Canadian Entomologist* 107(6), 567–576.

Ryu, M., Lee, S.J., Lee, H.S., Kim, J.R., Choi, E.J. *et al.* (2019) First report of fall armyworm, *Spodoptera frugiperda* and its molecular characteristics. In: *Spring International Conference of KSAE* (Vol. 66).

Sagar, D., Suroshe, S.S., Keerthi, M.C., Poorani, J., Gupta, A. *et al.* (2022) Native parasitoid complex of the invasive fall armyworm, *Spodoptera frugiperda* (J.E. Smith) from Northern India. *International Journal of Tropical Insect Science* 42(3), 2773–2778.

Sarkowi, F.N. and Mokhtar, A.S. (2021) The fall armyworm (FAW) *Spodoptera frugiperda*: A review on biology, life history, invasion, dispersion and control. *Outlooks on Pest Management* 32(1), 27–32.

Sartiami, D., Dadang, 0.0, Harahap, I.S., Kusumah, Y.M. and Anwar, R. (2020) First record of fall armyworm (*Spodoptera frugiperda*) in Indonesia and its occurrence in three provinces. In: *IOP Conference Series: Earth and Environmental Science*, Vol. 468. IOP Publishing, p. 012021.

Sharanabasappa, D., Kalleshwaraswamy, C.M., Asokan, R., Swamy, H.M.M., Maruthi, M.S. *et al.* (2018) First report of the fall armyworm *Spodoptera frugiperda* (J.E. Smith) on maize in India. *Pest Management in Horticultural Ecosystems* 24(1), 23–29.

Sharanabasappa, D., Kalleshwaraswamy, C.M., Poorani, J., Maruthi, M.S. and Pavithra, H.B. (2023) Spread and management of fall armyworm *Spodoptera frugiperda* in india. *Indian Journal of Entomology* 85(1), 1–12.

Sharanabasappa, K., Kalleshwaraswamy, C.M., Asokan, R., Mahadeva Swamy, H.M., Maruthi, M.S. *et al.* (2018a) First report of the fall armyworm, *Spodoptera frugiperda* (J.E. Smith) (Lepidoptera: Noctuidae), an alien invasive pest on maize in India. *Pest Management in Horticultural Ecosystems* (1), 23–29.

Sharanabasappa, K., Kalleshwaraswamy, C.M., Maruthi, M.S. and Pavithra, H.B. (2018b) Biology of invasive fall armyworm *Spodoptera frugiperda* (J.E. Smith) (Lepidoptera: Noctuidae) on maize. *Indian Journal of Entomology* 80(3), 540–543.

Sharanabasappa, K., Kalleshwaraswamy, C.M., Poorani, J., Maruthi, M.S., Pavithra, H.B. *et al.* (2019) Natural enemies of *Spodoptera frugiperda* (J.E. Smith) (Lepidoptera: Noctuidae), a recent invasive pest on maize in South India. *The Florida Entomologist* 102(3), 619–623.

Shylesha, A.N., Jalali, S.K., Gupta, A., Varshney, R., Venkatesan, T. *et al.* (2018) Studies on new invasive pest *Spodoptera frugiperda* (J.E. Smith) (Lepidoptera: Noctuidae) and its natural enemies. *Journal of Biological Control* 32(3), 1–7.

Singh, S., Raghuraman, M., Keerthi, M.C., Das, A., Kar, S.K. *et al.* (2023) Occurrence, distribution, damage potential, and farmers' perception on fall armyworm, *Spodoptera frugiperda* (J.E. Smith): Evidence from the Eastern Himalayan Region. *Sustainability* 15(7), 5681.

Sisay, B., Simiyu, J., Malusi, P., Likhayo, P., Mendesil, E. *et al.* (2018) First report of the fall armyworm, *Spodoptera frugiperda* (Lepidoptera: Noctuidae), natural enemies from Africa. *Journal of Applied Entomology* 142, 800–804.

Sisay, B., Sevgan, S., Weldon, C.W., Krüger, K., Torto, B. *et al.* (2022) Responses of the fall armyworm (*Spodoptera frugiperda*) to different host plants: Implications for its management strategy. *Pest Management Science* 79(2), 845–856.

Sisodiya, D.B., Raghunandan, B.L., Bhatt, N.A., Verma, H.S., Shewale, C.P. *et al.* (2018) The fall armyworm, *Spodoptera frugiperda* (J.E. Smith) (Lepidoptera: Noctuidae); First report of new invasive pest in maize fields of Gujarat, India. *Journal of Entomology and Zoology Studies* 6(5), 2089–2091.

Slansky, F., Jr. and Scriber, J.M. (1985) Food consumption and utilization. In: Kerkut, G.A. and Gilbert, L.I. (eds) *Comprehensive Insect Physiology, Biochemistry and Pharmacology Pergamon Press*. Oxford, pp. 87–163.

Soujanya, P.L., Sekhar, J.C., Yathish, K.R., Karjagi, C.G., Rao, K.S. *et al.* (2022) Leaf damage-based phenotyping technique and its validation against fall armyworm, *Spodoptera frugiperda* (J.E. Smith), in maize. *Frontiers in Plant Science* 13.

Sparks, A.N. (1979) A review of the biology of the fall armyworm. *The Florida Entomologist* 62, 82–87.

Suby, S.B., Soujanya, P.L., Yadava, P., Patil, J., Subaharan, K. *et al.* (2020) Invasion of fall armyworm (*Spodoptera frugiperda*) in India: Nature, distribution, management and potential impact. *Current Science* 119(1), 44–51.

Sun, X.X., Hu, C.X., Jia, H.R., Wu, Q.L., Shen, X.J. *et al.* (2021) Case study on the first immigration of fall armyworm, *Spodoptera frugiperda* invading into China. *Journal of Integrative Agriculture* 20(3), 664–672.

Swamy, H.M., Asokan, R., Kalleshwaraswamy, C.M., Prasad, Y.G., Maruthi, M.S. *et al.* (2018) Prevalence of "R" strain and molecular diversity of fall armyworm *Spodoptera frugiperda* (J.E. Smith) (Lepidoptera: Noctuidae) in India. *Indian Journal of Entomology* 80(3), 544–553.

Timilsena, B.P., Niassy, S., Kimathi, E., Abdel-Rahman, E.M. and Seidl-Adams, I. (2022) Potential distribution of fall armyworm in Africa and beyond, considering climate change and irrigation patterns. *Scientific Reports* 12(1), 539.

Tobin, P.C., Nagarkatti, S. and Saunders, M.C. (2003) Phenology of grape berry moth (Lepidoptera: Tortricidae) in cultivated grape at selected geographic locations. *Environmental Entomology* 32, 340–346.

Tumlinson, J.H., Mitchell, E.R., Teal, P.E.A., Heath, R.R. and Mengelkoch, L.J. (1986) Sex pheromone of fall armyworm, *Spodoptera frugiperda* (J.E. Smith): Identification of components critical to attraction in the field. *Journal of Chemical Ecology* 12, 1909–1926.

Tura, C.L., Aceres, L.V. and Joshi, R.C. (2025) Life history of fall armyworm, *Spodoptera frugiperda* (J.E. Smith) on hedgehog sedge, *Cyperus compressus* L. in Philippines. *Hexapoda* 32(1), 19–25. DOI: 10.55446/hexa.2025.488.

Van der Berg, J., Britz, C. and Plessis, H. (2021a) Maize yield response to chemical control of *Spodoptera frugiperda* at different plant growth stages in South Africa. *Agriculture* 11(9), 826.

van Handel, E. (1974) Lipid utilization during sustained flight of moths. *Journal of Insect Physiology* 20(12), 2329–2332.

Walton, W.R. and Luginbill, P. (1916) The fall army worm, or 'grass worm' and its control. *Farmers Bulletin* 752, U.S. Department of Agriculture.

Wang, Y., Li, Z., Zhang, S., Liu, H. and Wu, K. (2024) Global invasion pathways and adaptive evolution of fall armyworm. *Annual Review of Entomology* 69, 1–19.

Westbrook, J.K., Nagoshi, R.N., Meagher, R.L., Fleischer, S.J. and Jairam, S. (2016) Modeling seasonal migration of fall armyworm moths. *International Journal of Biometeorology* 60, 255–267.

Whitford, F., Quisenberry, S.S., Moellenbeck, D.J. and Davis, F.M. (1988) Nutritional basis for resistance to fall armyworm in corn. *Journal of Economic Entomology* 81, 947–951.

Withers, A.J., de Boer, J., Chipabika, G., Zhang, L., Smith, J.A. *et al.* (2021) Microsatellites reveal that genetic mixing commonly occurs between invasive fall armyworm populations in Africa. *Scientific Reports* 11(1), 20757.

Wood, J.R., Poe, S.L. and Leppla, N.C. (1979) Winter survival of fall armyworm pupae in Florida. *Environmental Entomology* 8(2), 249–252.

Wu, K., Shirk, P.D., Taylor, C.E., Furlong, R.B., Shirk, B.D. *et al.* (2018) CRISPR/Cas9 mediated knockout of the abdominal-A homeotic gene in fall armyworm moth (*Spodoptera frugiperda*). *PLoS ONE* 13, e0208647.

Wu, Q.L., Jiang, Y.Y., Hu, G. and Wu, K.M. (2019) Analysis on spring and summer migration routes of fall armyworm (*Spodoptera frugiperda*) from tropical and southern subtropical zones of China. *Plant Protection* 45, 1–9.

Yan, S., Qian, J., Cai, C., Ma, Z., Li, J. *et al.* (2020) Spray method application of transdermal dsRNA delivery system for efficient gene silencing and pest control on soybean aphid *Aphis glycines*. *Journal of Pest Science* 93, 449–459.

Yee, K.N., Aye, M.M., Htain, N.N., Oo, A.K., Kyi, P.P. *et al.* (2019) First detection report of the fall armyworm *Spodoptera frugiperda* (Lepidoptera: Noctuidae) on maize in Myanmar. IPPC Official Pest Report.

Young, J.R. (1979) *Fall armyworm: Control with insecticides*. Florida Entomologist, pp. 130–133.

Yu, S.J. (1991) Insecticide resistance in the fall armyworm, *Spodoptera frugiperda* (J.E. Smith). *Pesticide Biochemistry and Physiology* 39, 84–91.

Yu, S.J. (1992) Detection and biochemical characterization of insecticide resistance in fall armyworm (Lepidoptera: Noctuidae). *Journal of Economic Entomology* 85(3), 675–682. DOI: 10.1093/jee/85.3.675.

Zacarias, D.A. (2020) Global bioclimatic suitability for the fall armyworm, *Spodoptera frugiperda* (Lepidoptera: Noctuidae), and potential co-occurrence with major host crops under climate change scenarios. *Climatic Change* 161(4), 555–566.

Zeng, G., Zhi, J., Zhang, C., Zhang, T., Zhou, L. *et al.* (2021) *Orius similis* (Hemiptera: Anthocoridae): A promising candidate predator of *Spodoptera frugiperda* (Lepidoptera: Noctuidae). *Journal of Economic Entomology* 114(2), 582–589.

Zhang, D.D., Zhao, S.Y., Wu, Q.L., Li, Y.Y. and Wu, K.M. (2021) Cold hardiness of the invasive fall armyworm, *Spodoptera frugiperda* in China. *Journal of Integrative Agriculture* 20(3), 764–771.

17 Invasion and Management of Aphid Pests in the Tropics

Velusamy Naveen[1], Bhimashankar Gotyal[2*], Subrata Goswami[3], Palanisamy Dhanapriya[4], Velmurugan Anithaa[2], Anbuchezhiyan Sneka[4], Elango Santhoshinii[5], Poiyamozhi Harikrishnan[6], Gundappa B.[2], Shivakumara K.T.[2] and Yandigeri M.S.[2]

[1]Department of Entomology, University of Agricultural Sciences, Dharwad, Karnataka, India; [2]Division of Germplasm Conservation and Utilisation, Indian Council of Agricultural Research – National Bureau of Agricultural Insect Resources, Bengaluru, Karnataka, India; [3]Department of Entomology and Agricultural Zoology, Institute of Agricultural Sciences, Banaras Hindu University, Varanasi, Uttar Pradesh, India; [4]Department of Entomology, Faculty of Agriculture, Annamalai University, Chidambaram, Cuddalore, India; [5]Department of Plant Pathology, University of Agricultural Sciences, Dharwad, Karnataka, India; [6]Department of Biotechnology, University of Agricultural Sciences, Dharwad, Karnataka, India

Abstract

This chapter reviews the biology, ecology, and management strategies of aphid pests in tropical agroecosystems. Emphasis is placed on the invasive potential of aphids, their role as virus vectors, and the ecological adaptations facilitating year-round persistence in tropical climates. Key IPM strategies including cultural control, host plant resistance, and the use of biologicals and botanicals are critically assessed. Advances in molecular diagnostics, virome analysis, and digital surveillance tools are also discussed. The chapter highlights case studies from India and other tropical regions, offering practical insights into sustainable aphid management under climate-affected scenarios.

Keywords: Aphid diversity, virus transmission, integrated pest management, tropical agroecosystems, molecular diagnostics

17.1 Introduction

Aphids are among the most economically important insect pests in tropical agroecosystems due to their sap-feeding habits and role as vectors of over 165 plant viruses. Their polyphagy, rapid reproduction, and adaptability to environmental fluctuations enable persistent infestations across diverse crops such as cotton, legumes, okra, chilli, banana, and sorghum (Singh, 2023). Key species such as *Aphis gossypii*, *Myzus persicae*, and *Rhopalosiphum maidis* are particularly destructive. Tropical climates, with high humidity and consistent warmth, favour aphid reproduction and

*Corresponding author: gotyalento@gmail.com

DOI: 10.1079/9781836991977.0017

dispersal. Parthenogenesis and high fecundity contribute to rapid population growth, while winged morphs and wind dispersal support regional spread (Dixon, 1998). Global trade further accelerates their introduction into new regions. Climate change is altering aphid phenology, overwintering patterns, and distribution, exacerbating their pest status (Hullé *et al.*, 2020). Aphids exhibit specific ecological adaptations in the tropics, including endosymbiotic relationships with *Buchnera aphidicola*, which enhance thermal tolerance and nutrient acquisition (Douglas, 1998). Interactions with natural enemies – ladybird beetles, lacewings, parasitoids like *Aphidius colemani*, and fungi like *Beauveria bassiana* – also influence their dynamics. Damage includes chlorosis, stunted growth, and yield loss, compounded by honeydew-induced sooty mould. More critically, aphids transmit viruses such as PRSV, PVY, and BCMV through both persistent and non-persistent modes, fuelling disease outbreaks (Ziebell and Carr, 2010).

Management in the tropics relies on early detection via sticky and suction traps, direct sampling, and molecular diagnostics. Decision-support tools and forecasting models enhance precision management (Hullé *et al.*, 2020). IPM strategies remain central, incorporating crop rotation, intercropping, trap crops, biological control, and botanical insecticides (Singh, 2023). Judicious use of chemicals is necessary to avoid resistance and preserve beneficial arthropods. Innovative approaches such as RNAi, CRISPR, nanotechnology, and semiochemical-based tactics are advancing aphid suppression (Bhattacharya, 2019). AI-enabled diagnostics and mobile advisory tools support real-time decision making. Case studies from India, Brazil, and Kenya highlight the efficacy of integrated approaches and farmer participation (Emaru *et al.*, 2024).

Sustainable aphid management in the tropics depends on understanding aphid biology, ecology, and virus transmission under tropical conditions, while promoting innovations and empowering stakeholders through effective extension frameworks. This chapter consolidates the current understanding of aphid diversity, biology, virus transmission dynamics, monitoring tools, and IPM strategies tailored for tropical contexts. It also presents emerging research innovations to guide the future of aphid management under tropical agricultural scenarios.

17.2 Aphid Diversity and Distribution in the Tropics

17.2.1 Key aphid species in tropical ecosystems

Globally, approximately 5,000 aphid species across 510 genera have been recorded, with around 250 causing significant agricultural and horticultural damage (Singh and Singh, 2021). Damage includes nutrient depletion, gall formation, and promotion of sooty mould via honeydew excretion. Aphids also transmit numerous plant viruses, compounding their impact. Aphid diversity is highest in temperate zones, owing to seasonal life cycles and host synchrony. The tropics' limited seasonality, sporadic host availability, and chemically defended plants restrict aphid specialization.

17.2.2 Host range and specificity

Aphid host specificity ranges from monophagy to broad polyphagy (Table 17.1). Most tropical species remain monophagous, influenced by ecological and physiological constraints. Phylogenetic analyses show strong associations with angiosperms, fewer with gymnosperms (Peccoud *et al.*, 2010). Host selection is driven by specialized olfactory and chemosensory proteins. For example, *Myzus persicae* detects (E)-β-farnesene via OBPs MperOBP3/7/9 (Wang *et al.*, 2021). Aphids also detoxify plant allelochemicals using enzymes like glutathione S-transferases and cytochrome P450s. Some plants, such as *Arabidopsis* pad3-1 mutants, can even enhance aphid fecundity through altered secondary metabolism (Kuśnierczyk *et al.*, 2008). These adaptations, shaped by evolutionary pressures, enable aphids to exploit specific hosts, evade plant defences, and sustain population growth across generations.

17.2.3 Geographic and climatic influences on aphid distribution

Aphids are more widely distributed in temperate regions than in the tropics owing to their

Table 17.1. List of key aphid species, their host specificity and global distribution. (Author's own table)

Key aphid species	Regions found	Host specificity	References
Aphis craccivora	India, Honduras, Sri Lanka, Philippines, Thailand, Southern USA, Tropical Africa, Latin America, Australia, New Zealand, Pacific Islands	Oligophagous (legumes and some Brassica)	Mille *et al.*, 2020; Singh and Singh (2020); Singh and Agrawal (2022)
Myzus persicae	India, Pakistan, Australia, Sri Lanka, Southeast Asia, Africa, South America	Highly polyphagous (over 40 families)	
Brevicoryne brassicae	India, Pakistan, East Africa, Sri Lanka, Southeast Asia, Africa, South America	Specialist on Brassicaceae	
Aphis gossypii	India, Pakistan, Honduras, Australia, Sri Lanka, Southeast Asia, Africa, South America	Polyphagous (cotton, cucurbits, citrus, etc.)	
Lipaphis erysimi	India, Pakistan, Sri Lanka, Southeast Asia, Africa, South America	Specialist on Brassicaceae	
Sitobion avenae	India, Australia, Sri Lanka, Southeast Asia, Africa, South America	Mainly cereals (wheat, barley)	
Aphis nerii	Pakistan, East Africa, Sri Lanka, Southeast Asia, Africa, South America	Specialist (oleander, milkweed)	
Uroleucon sonchi	East Africa, Sri Lanka, Southeast Asia, Africa, South America	On Sonchus (Asteraceae)	
Rhopalosiphum maidis	India, East Africa, Pakistan, Sri Lanka, Southeast Asia, Africa, South America	Cereals (mainly maize)	
Toxoptera aurantii	India, Africa, South America, central America, Eastern Asia, Australia, the Mediterranean region, Netherland	Citrus, coffee	Piron *et al.* (2019)
Pentalonia nigronervosa	India, Africa, South America, central America, Eastern Asia, Australia	Banana, ginger, cardamom, *Heliconia* spp., *Caladium* spp., *Alpinia* spp., and *Dieffenbachia* spp.	Robson *et al.*, 2014
Diuraphis noxia	Australia, East Africa, North Africa, Middle East, Central Asia	Mainly cereals (wheat, barley, oats)	EPPO (European and Mediterranean Plant Protection Organization) (2024)

biology, reproductive adaptation, and host availability which should be synchronized with the seasons. Within tropical regions, landscape of the field plays a major role in abundance and composition of species. Aphid abundance is higher and non-colonizing species that act as plant virus vectors are frequently preferred in landscapes where intensive agriculture predominates. Aphid community dynamics alter through time as a result of intra-annual fluctuation, which is further influenced by seasonal variations in temperature, precipitation, and crop development (Mahas *et al.*, 2023). The important reasons for restricted distribution of aphids to tropical regions is host alternation for life cycle completion which cannot be sustained in tropical regions (Yamamoto *et al.*, 2020).

17.3 Mechanisms of Invasion and Spread

Globalization has accelerated aphid invasions through contaminant and hitchhiking pathways involving traded plants, luggage, and cargo (Gippet *et al.*, 2019). These introductions disrupt agroecosystems, compromise pest management, and reduce crop yields (Yazdani *et al.*, 2018), particularly in cereals. Aphids' unique reproductive modes of parthenogenesis, zygogenesis, paedogenesis, and polymorphism (Singh and Singh, 2021) enable rapid population growth. Holocyclic species alternate sexual and asexual reproduction, whereas anholocyclic ones reproduce solely through parthenogenesis. Invasion success is often decoupled from genetic diversity, relying instead on phenotypic plasticity (Nibouche *et al.*, 2014) and developmental polyphenism (Brisson and Davis, 2016). Wind-assisted migration of alates facilitates long-range dispersal and new colony establishment (Parry, 2013), with alates showing enhanced resistance to starvation and environmental stress (Hazell *et al.*, 2005). Abiotic factors such as temperature, precipitation, photoperiod and biotic cues such as crowding and predators influence morph production and invasion dynamics (Purandare *et al.*, 2014). Temperature is a key regulator: low temperatures often induce alates (Chen *et al.*, 2019), while high temperatures may suppress or stimulate alate formation depending on species. Aphids' thermal sensitivity positions them as climate-responsive invaders, warranting deeper research to forecast their spread under warming scenarios (Sentis *et al.*, 2017).

17.4 Biology and Ecology of Aphids in Tropical Climates

17.4.1 Life cycle under tropical conditions

In tropical climates, aphids reproduce parthenogenetically year-round due to the absence of overwintering constraints. Continuous generations are sustained by viviparous females, with species such as *Aphis gossypii* completing a life cycle in under a week at 25–30°C, leading to rapid population growth (Van Emden and Harrington, 2007). Without diapause-inducing photoperiods, dispersive morphs (alates) are triggered by crowding and declining host quality, facilitating swift colonization. Year-round host availability further supports uninterrupted aphid development and virus transmission.

17.4.2 Adaptations to high temperature and humidity

Tropical aphids experience strong selection from high temperature and humidity. They adapt via cuticular hydrocarbon changes to reduce desiccation and behaviourally by settling on leaf undersides to avoid heat (Hagen *et al.*, 1999). Reproductive plasticity in response to microclimates is common. Facultative symbionts like *Serratia symbiotica* enhance thermotolerance by mitigating heat-related protein damage (Russell and Moran, 2006), supporting survival in fluctuating tropical conditions.

17.4.3 Aphid–Host–Symbiont Relationships

Aphids depend on microbial symbionts to meet nutritional needs and resist stress. *Buchnera aphidicola* provides essential amino acids, while secondary symbionts such as *Hamiltonella defensa*, *Regiella insecticola*, and *Serratia symbiotica* confer parasitoid resistance, broaden host range, and enhance heat tolerance (Henry *et al.*, 2013). In tropical regions, diverse symbiont profiles are common, likely due to high biotic and abiotic stress (Frago *et al.*, 2012).

17.4.4 Interaction with natural enemies and trophic networks

Aphids are central to tropical food webs, serving as prey for predators like ladybird beetles, hoverflies, and lacewings and as hosts to parasitoids such as *Aphidius* and *Lysiphlebus* spp. (Van Emden and Harrington, 2007). Ant–aphid mutualisms, driven by honeydew rewards, offer aphids protection but can hinder biological control. Additionally, aphids are key virus vectors, transmitting both persistent and non-persistent plant viruses (Ng and Falk, 2006), positioning them as pivotal players in tropical agroecosystems and IPM strategies.

17.5 Crop Vulnerability and Plant Resistance

17.5.1 Host range

Aphids exhibit varying host specificities – monophagy, oligophagy, or polyphagy – shaped by ecological pressures. Despite high plant diversity, tropical regions paradoxically support lower aphid species richness due to factors such as host apparency and environmental variability

(Bista, 2020). Many tropical aphids are associated with cultivated plants, including both specialists and generalists. The banana aphid *Pentalonia nigronervosa* is a *Musa* specialist in tropical plantations (Abrol, 2015), while polyphagous species such as *Aphis gossypii* and *Myzus persicae* thrive on diverse crops (Singh and Singh, 2021). *A. gossypii* displays cotton- and cucurbit-specialized biotypes (Liu *et al.*, 2017) and infests over 459 species in India alone (Edde, 2022). *A. craccivora* affects >400 plant species, especially pulses (Singh and Singh, 2017). *M. persicae*, a major potato pest, colonizes hosts in 40 families, including Rosaceae, Solanaceae, and Brassicaceae (Verma *et al.*, 2018). Other tropical aphids include *Rhopalosiphum maidis* on Poaceae (Zhang *et al.*, 2024), *Lipaphis erysimi* on Brassicaceae (Capinera, 2001), and species like *Aphis spiraecola*, *A. glycines*, *R. padi*, *Aulacorthum solani*, *Hyperomyzus lactucae*, *Sitobion miscanthi*, and *Brachycaudus amygdalinus*.

17.5.2 Damage

Aphids cause both direct (sap feeding, leaf curling, stunting, yield loss) and indirect damage (honeydew, virus transmission) (Fig. 17.1). Of 620 known plant viruses, 206 are arthropod-transmitted, with aphids responsible for over 165 (Singh and Singh, 2017). *Aphis craccivora* causes puckering, pod shrivelling, wilting, and up to 80% yield loss in pulses; its honeydew supports sooty mould growth, impairing photosynthesis. *A. gossypii* infests cotton, leading to leaf curling, boll drop, sticky cotton, and poor fibre development, while transmitting viruses such as bunchy top, anthocyanosis, and blue disease. *Rhopalosiphum maidis* damages maize by stunting, tassel and silk contamination with honeydew, and vectoring viruses such as MDMV, PVY, and maize chlorotic dwarf virus (Edde, 2022). *Myzus persicae* induces premature senescence and spreads over 150 viruses in Solanaceae crops (Verma *et al.*, 2018). *Lipaphis erysimi* causes leaf cupping and yellowing in brassicas and vectors viruses such as cauliflower mosaic and turnip mosaic (Capinera, 2001).

17.5.3 Role of host plant resistance in aphid suppression

Host plant resistance (HPR) suppresses aphids through structural, biochemical, and genetic defences, including phytohormone signalling (salicylic acid (SA) and jasmonic acid (JA)), rapid

Fig. 17.1. Damage of crops by aphids in the tropics. (A) *Aphis odinae*, (B) *Aphis spiraecola*, (C) *Brevicoryne brasicae*, (D) *Lipaphis erysimi*, (E) *Toxoptera aurantia*, (F) *Rhopalosiphum maidis*. Copyright © 2013, ICAR – National Bureau of Agricultural Insect Resources.

recognition of aphid molecules by R proteins, hypersensitive responses, and phloem occlusion, reducing aphid feeding and fitness (Jaouannet *et al.*, 2014). HPR integrates well into IPM, offering a sustainable alternative to insecticides, lowering costs, and preventing pesticide resistance while supporting natural enemies. Examples include *Solanum berthaultii* glandular trichomes producing farnesene to repel aphids, the Mi-1.2 gene conferring resistance to potato aphid and other pests, Rag1 in soybean resisting *Aphis glycines*, and the Vat gene in melon resisting *Aphis gossypii* and blocking virus transmission (Kumar, 2019). Aphids feeding on resistant plants show longer probing, reduced phloem ingestion, diminished reproduction, and increased vulnerability to predators.

17.5.4 Breeding for resistance in tropical crop varieties

Breeding for aphid resistance in tropical crops combines traditional and molecular methods to reduce yield losses and chemical use (Abang *et al.*, 2021). Understanding the resistance mechanisms antibiosis, antixenosis, and tolerance is vital, with wild relatives and landraces providing valuable genetic diversity for durable resistance (MacWilliams, 2020). Conventional breeding techniques like pedigree selection and backcrossing, coupled with field screening under natural aphid pressure, remain important (Reddy *et al.*, 2017). Molecular tools such as marker-assisted selection (MAS) using SSRs and SNPs, genomic selection, and transcriptomics enable early and precise identification of resistance genes, accelerating breeding (Arora, 2017). For example, cowpea near-isogenic lines with QTLs QAC-vu7.1 and QAC-vu1.1 confer antibiosis and antixenosis (MacWilliams *et al.*, 2023). Despite advances, traditional breeding remains essential, complemented by modern technologies to speed improvement.

17.6 Aphid-Transmitted Viruses in the Tropics

Aphids transmit plant viruses via distinct modes: non-circulative (non-persistent and semi-persistent) and circulative (non-propagative and propagative). Non-persistent viruses, like Papaya ringspot and Turnip mosaic virus, are acquired and transmitted quickly during brief feeding and retained only on the stylet (Ng and Falk, 2006). Semi-persistent viruses, such as Cauliflower mosaic virus, persist in the foregut for hours without replication. Circulative, non-propagative viruses like Potato leafroll virus circulate through the aphid's body but do not replicate, enabling lifelong transmission. Rare propagative viruses, including Alfalfa dwarf virus, replicate inside the vector (Bejerman *et al.*, 2011). These modes influence virus spread and control strategies.

17.6.1 Role of aphids in disease epidemics

Aphids are major vectors, transmitting over 200 plant viruses due to high reproduction, polyphagy, and winged dispersal (Van Emden and Harrington, 2007). Species such as *Myzus persicae* and *Aphis gossypii* transmit viruses via both persistent and non-persistent modes, infecting diverse hosts. Non-persistent transmission allows rapid spread during brief probes (Fereres and Moreno, 2009), while persistent viruses such as Potato leafroll virus cause sustained epidemics. Early aphid infestations are often asymptomatic, complicating control, thus making aphid management vital in integrated virus control. Aphids' adaptability and transmission efficiency drive agricultural virus outbreaks.

17.6.2 The aphid virome

Viruses have been linked to plant diseases since 1898. The "virome" refers to the collective viral genomes within an organism or environment and is part of the broader microbiome (Zárate *et al.*, 2017). Plant viruses significantly impact agriculture, causing half of emerging plant diseases. Traditional detection methods such as ELISA and PCR target known viruses but have limitations (Fig. 17.2). High-throughput sequencing (HTS) and next-generation sequencing (NGS) have revolutionized virus discovery, revealing new viruses, viroids, and host-virus interactions (Pecman *et al.*, 2017). Virome

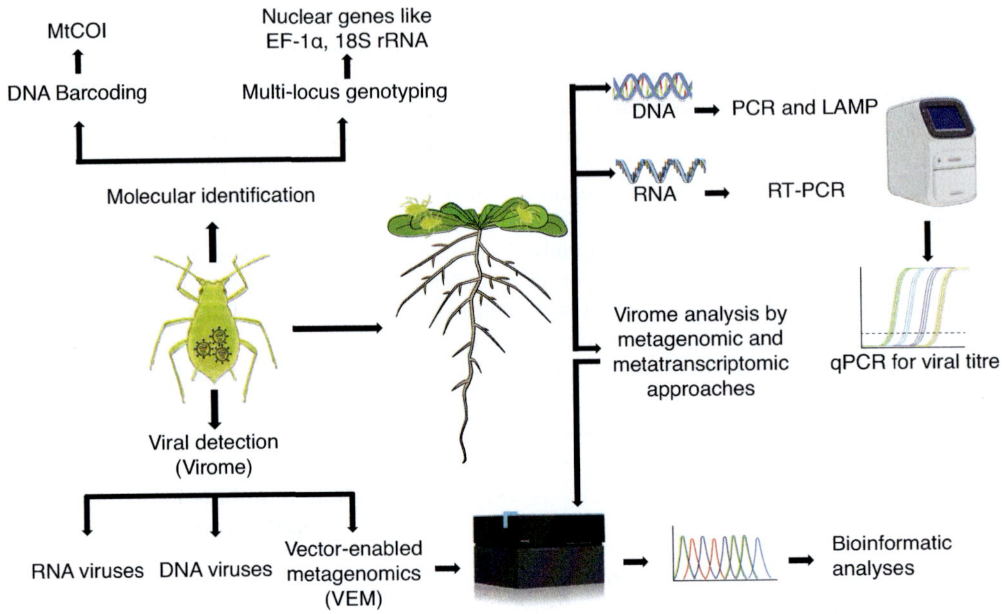

Fig. 17.2. Workflow depicting insect vector identification and virome analysis using molecular tools, PCR-based detection, and bioinformatics.

studies in aphids through transcriptome deep sequencing have identified known virus families (nege/kita, flavi, tombus) and novel RNA viruses, leading to new virus groups such as Centivirus and Aphiglyvirus, suggesting plants may act as reservoirs (Kondo *et al.*, 2020). Research on *Myzus persicae* detected densovirus, flavivirus-like virus, and polerovirus, also showing RNA interference (RNAi) varies with diet, indicating environmental modulation of antiviral defence (Sukhikh *et al.*, 2024). Feng *et al.* (2025) analysed wheat aphids' RNA virome and endogenous viral elements (EVEs), discovering 43 RNA viruses (12 new insect-specific viruses), and 90 EVEs with transcriptional activity, highlighting their ecological and evolutionary roles. Their study also noted geographic variation and potential plant virus transmission by aphids, including Broad bean wilt virus 2 and Maize chlorotic mottle virus.

17.7 Monitoring and early detection

Effective aphid management requires timely detection and reliable monitoring, especially in tropical climates favouring year-round aphid survival. Monitoring integrates visual, mechanical, and biological methods for real-time or predictive data on aphid populations. Yellow sticky traps attract flying aphids, are inexpensive, and effectively signal immigration onset. Suction traps (e.g. 12.2 m Rothamsted traps) sample aerial aphids over large areas, enabling insights into dispersal and species diversity. Although costly, they provide long-term data useful for forecasting. Manual sampling (leaf sampling, beat sheets, quadrats) remains essential for local detection and economic threshold assessment despite its labour demands (Kumar *et al.*, 2014). Forecasting models use temperature, humidity, photoperiod, and host phenology – degree-day models are common in the tropics to predict aphid generations. Decision support systems (DSS) integrate trap catches, weather, and cropping data to guide management; examples include Aphid Forecast and tropical DSS for soybean and cotton aphids (Singh and Singh 2020). Mobile and GIS-based apps are increasingly integrated into IPM systems in tropical regions.

17.7.1 Molecular tools for aphid identification

Aphid species often show limited morphological variation, especially among immature instars, complicating accurate identification essential for managing aphid-borne viruses. Traditional morphological methods face challenges due to phenotypic plasticity, homoplasy across developmental stages, and cryptic species complexes. Consequently, molecular tools have become indispensable for precise taxonomic resolution. DNA barcoding, which targets a conserved region of the genome, is widely used for species discrimination. In insects, the mitochondrial cytochrome c oxidase subunit I (COI) gene (~658 bp) serves as the standard barcode (Hebert *et al.*, 2003). This method is particularly effective for aphids (Hemiptera: Aphididae), a group comprising over 5,000 described species, many of which are key agricultural pests and virus vectors in tropical and temperate regions. Morphological similarity between winged and wingless forms, host-induced phenotypic plasticity, cryptic species complexes (e.g. *Aphis gossypii* complex), and parthenogenetic reproduction generating clonal diversity undermine morphology-based identification (Foottit *et al.*, 2008).

17.7.2 Applications of molecular characterization in aphid taxonomy and pest management

DNA barcoding is a crucial tool for distinguishing morphologically similar and cryptic aphid species by targeting the mitochondrial COI gene. It reliably separates important pests such as *Aphis gossypii* and *Aphis craccivora*, *Myzus persicae* and *Myzus antirrhinii* (Foottit *et al.*, 2008), and rose aphids *Macrosiphum euphorbiae* and *Macrosiphum rosae* (Lee *et al.*, 2011). Barcoding also uncovers cryptic diversity within species such as *Sitobion avenae* (Wang and Qiao, 2009) and reveals geographic haplotypes in *Toxoptera citricida* linked to virus transmission. Molecular identification supports early detection and monitoring of invasive aphids such as *Myzus ascalonicus* and *Rhopalosiphum maidis* (Rugman-Jones *et al.*, 2010), and helps pinpoint key virus vectors in mixed populations. However, single-locus barcoding faces challenges from mitochondrial introgression and incomplete lineage sorting, limiting resolution among recently diverged species (Simon *et al.*, 2010). Multi-locus genotyping using microsatellites and nuclear genes (EF-1α, ITS, 18S rRNA) improves species complex resolution and reveals population structure, clonal diversity, and gene flow. Advanced diagnostics such as qPCR with High-Resolution Melting analysis enable rapid species differentiation based on minor sequence variations. Loop-Mediated Isothermal Amplification (LAMP) offers a quick, field-friendly method for detecting aphids directly from plant tissue without complex equipment, with demonstrated success in *Myzus persicae* and *Aphis glycines*. Together, these molecular tools enhance aphid taxonomy and pest management, aiding integrated pest management and virus vector control in tropical agroecosystems.

17.8 Integrated Pest Management (IPM) Strategies

Persistent aphid pressure demands effective and eco-friendly control strategies. IPM provides a sustainable, multi-faceted framework combining cultural, biological, botanical, and chemical methods to maintain aphid populations below economic thresholds while preserving ecosystem health.

17.8.1 Cultural control

Cultural practices reduce aphid survival and reproduction by disrupting their life cycles. Key strategies include crop rotation with non-host plants to interrupt food continuity (Hooks and Fereres, 2006), intercropping to create physical and chemical barriers, and trap cropping to concentrate aphids on more attractive or tolerant plants for targeted control (Way and van Emden, 2000). Adjusting planting times helps avoid peak aphid infestations by aligning crop phenology with lower pest abundance (Dent and Binks, 2020).

17.8.2 Biological control

Biological control leverages natural enemies, i.e. predators, parasitoids, and pathogens, to suppress aphids in tropical agroecosystems, where biodiversity and climatic conditions influence their effectiveness.

17.8.2.1 Predators and parasitoids

Ladybird beetles (Coccinellidae), such as *Coccinella septempunctata* and *Harmonia axyridis*, are voracious aphid predators throughout their life stages (Emden and Harrington, 2017). Lacewing larvae (*Chrysoperla carnea*) and syrphid fly larvae also significantly reduce aphid populations due to their mobility and feeding efficiency (Ghaith *et al.*, 2023). Parasitoid wasps of the genus *Aphidius* (Braconidae) attack aphids by oviposition inside hosts, leading to mummification and death. Species such as *A. colemani*, *A. ervi*, and *A. matricariae* are widely used in commercial and field settings (Straub *et al.*, 2008; Emden and Harrington, 2017). Augmentative biological control involves releasing mass-reared natural enemies: inundative releases apply large numbers to suppress pests immediately, while inoculative releases aim for population establishment supported by ecological engineering (Emden and Harrington, 2017; Ghaith *et al.*, 2023).

17.8.2.2 Entomopathogenic fungi

Entomopathogenic fungi such as *Beauveria bassiana*, *Metarhizium anisopliae*, and *Lecanicillium lecanii* are effective biorational alternatives to chemical aphicides. They infect aphids by adhering to the cuticle, germinating under humid conditions, penetrating via enzymatic and mechanical means, and proliferating internally to cause death (Straub *et al.*, 2008). Tropical climates with high humidity and warmth favour fungal efficacy. Studies report successful control of *Aphis gossypii*, *Myzus persicae*, and *Brevicoryne brassicae* in both greenhouse and field conditions (Khulbe and Batra, 2024). Formulations with wetting agents, oils, or UV protectants improve fungal persistence under tropical sunlight. *B. bassiana* and *M. anisopliae* at 2 kg/ha were effective in field trials against *Lipaphis erysimi*, delivering substantial aphid suppression with favourable cost–benefit outcomes (Tomar *et al.*, 2023).

17.8.3 Botanical and biorational pesticides

Botanical pesticides, particularly neem-based products containing azadirachtin, are vital to sustainable IPM due to their selective toxicity and biodegradability. Azadirachtin disrupts aphid development by inhibiting ecdysone 20-monooxygenase and suppressing juvenile hormone, reducing feeding, growth, and reproduction. Neem oil treatments consistently reduce aphid survival and fecundity, with low resistance risk and minimal non-target impact (Shannag *et al.*, 2014). Horticultural oils and insecticidal soaps kill aphids via suffocation and membrane disruption and degrade quickly to minimize harm to beneficial insects (Cranshaw, 2004; Emden and Harrington, 2017). Other effective botanicals include oxymatrine from *Sophora* spp. and essential oils from mint and papaya (Mwanauta *et al.*, 2021). Biorationals such as insect growth regulators (e.g. buprofezin) and microbial agents such as *Beauveria bassiana* provide targeted aphid control with low mammalian toxicity (Wend *et al.*, 2024).

17.8.4 Chemical control

Insecticides such as neonicotinoids (imidacloprid) and butenolides (flupyradifurone) are effective against sap-feeding aphids such as *Myzus persicae* and *Aphis craccivora*, acting on their nervous systems to cause paralysis and death. Thiamethoxam 25 WG at 100 g/ha proved most effective, significantly reducing mustard aphid density (Kaur *et al.*, 2023). However, resistance due to overuse threatens their long-term efficacy (Bass *et al.*, 2014). To delay resistance, chemical control must be integrated with IPM, including rotating insecticides with different modes of action (e.g. neonicotinoids and IGRs) and combining non-chemical tactics like biological control, cultural practices, and resistant varieties (Denholm *et al.*, 2002). Monitoring aphid populations and applying chemicals only above economic thresholds further reduces resistance risk. While temperate regions have well-established aphid IPM guidelines (Flint, 2018), tropical areas face ecological complexities such as year-round cropping and diverse agroecosystems that complicate aphid management and research (Luo *et al.*, 2022).

17.9 Emerging Technologies and Approaches for Aphid Management

17.9.1 Biotechnological approaches for aphid management

17.9.1.1 RNAi

RNA interference (RNAi) is a biological process that can be triggered by endogenously expressed or exogenously applied double-stranded RNAs (dsRNAs). RNA interference (RNAi) has emerged as a targeted and eco-friendly strategy for aphid control, mainly through host-induced (HIGS) and spray-induced gene silencing (SIGS) (Bhattacharya, 2019). Silencing genes related to aphid development, reproduction, and immunity – such as *MpDhc64C, CHS1, Ace1* – using transgenic plants, synthetic dsRNA, or virus-induced gene silencing (VIGS) significantly reduced survival and fecundity in species like *M. persicae, S. avenae,* and *S. miscanthi* (Zhao *et al.*, 2018; Faisal *et al.*, 2019; Dong *et al.*, 2022). Spray-based delivery (SIGS) and microinjection have also shown success in silencing critical genes such as *GNBP1, CP19,* boosting aphid mortality and susceptibility to entomopathogens (Ye *et al.*, 2022). Nanocarriers further enhance RNAi efficacy by enabling stable delivery of dsRNA/siRNA via transdermal or oral routes, successfully targeting genes like *flightin, Sm9723,* etc. while minimizing off-target effects and conserving natural enemies (Zhang *et al.*, 2022). RNAi biopesticides silence vital pest genes via dsRNA, offering species-specific control while sparing humans and beneficial organisms. However, resistance evolution is a major concern, necessitating integration with IPM, judicious application, farmer education, and adherence to IRAC's resistance management protocols (Gao *et al.*, 2024). Table 17.2 summarizes some of the recent examples of successful use of RNAi for aphid management in various crops.

17.9.1.2 CRISPR/Cas9-based Approaches

While CRISPR/Cas9 has been widely applied in model insects such as *Drosophila, Aedes, Tribolium,* and *Bombyx,* its use in Hemiptera, particularly aphids, has been limited due to their complex life cycle involving cyclical parthenogenesis and obligatory diapause (Javed *et al.*, 2025). *Acyrthosiphon pisum* served as a model for one of the first successful aphid genome-editing protocols, overcoming challenges of seasonal egg production and diapause. Le Trionnaire *et al.* (2019) developed a 7- month CRISPR/Cas9 protocol involving embryo microinjection, diapause incubation, and clonal propagation from G0 founders, achieving mutations in the stylin-01 gene with 70–80% efficiency, though stable inheritance remains unconfirmed. A recent breakthrough by Takenaka *et al.* (2023) introduced cell membrane-penetrating peptides (CMPPs), particularly penetrating (PEN), for delivering recombinant proteins into aphid tissues via haemolymph injection. Fluorescent marker assays confirmed successful systemic delivery, including to bacteriocytes and embryos, offering a promising alternative to direct embryonic injection for genome editing in aphids. Despite its potential for precisely editing genome in aphids, CRISPR/Cas9 faces challenges in delivery, ethics, and unintended effects. Overcoming these requires improved delivery methods, robust ethical guidelines, and comprehensive risk assessments. Ultimately, responsible and transparent innovation involving all stakeholders is key to its sustainable application in aphid control.

17.9.2 Nanotechnology for aphid management

Nanotechnology is being increasingly applied in sustainable pest management, with bioengineered nanoparticles (BNPs) emerging as effective, green alternatives to conventional pesticides for aphid control. These particles disrupt insect physiology by binding to phosphorus and sulfur in proteins and DNA, leading to enzyme denaturation, membrane damage, and cell death (Bhattacharyya *et al.*, 2016a). AgNPs synthesized from

Table 17.2. Application of RNAi for aphid management

Type of gene silencing	Species of aphid	Plant species	Mode of delivery	Target genes	Main effects	References
HIGS	*M. persicae*	Tomato	Transgenic tomato (Agrobacterium-mediated transformation)	Acetylcholine esterase 1 (*Ace1*)	Silenced *Ace1* gene and reduced fecundity	Faisal *et al.*, 2019
	M. persicae	Tobacco	Transgenic tobacco (Plastid-mediated RNA interference)	*MpDhc64C*	Reduced weight, impaired fecundity, and increased insect mortality	Dong *et al.*, 2022
	S. avenae	Wheat	Transgenic wheat (Particle bombardment)	Chitin synthase1 (*CHS1*)	Hampered molting	Zhao *et al.*, 2018
SIGS	*A. pisum*	Pea	Feeding	*Stylin-01*	Hampered Cauliflower mosaic virus transmission	Webster *et al.*, 2018
	A. citricidus	Citrus	Feeding and citrus stem dipping	*AcGNBP1*	Reduced immune-related phenoloxidase activity	Ye *et al.*, 2022
	A. glycines	Soybean	Nanocarrier-based dsRNA delivery system	*ATPD, TREH, CHS1, ATPE*	Increased mortality due to target gene silencing	Yan *et al.*, 2020
	A. pisum	Pea	Injection and nanocarrier delivery	*Flightin*	Malformed wings, deformed dorsal longitudinal muscle shapes, and wider and looser dorsoventral flight muscles were observed	Chang *et al.*, 2022
	S. miscanthi	Wheat	Topical application and nanocarrier-mediated transdermal dsRNA delivery	*Sm9723*	Reduced feeding, fecundity, and survival	Zhang *et al.*, 2022

entomopathogenic fungi like *Metarhizium anisopliae, Isaria fumosorosea, and Aspergillus foetidus* showed high aphicidal activity (63–93%) by boosting the fungal pathogenicity (Nawaz *et al.*, 2022). Over 90% mortality was achieved in *Aphis nerii, Rhopalosiphum padi,* and *Sitobion avenae*, within 2 days using rice husk-derived AgNPs (70–80 nm, 200–600 ppm) (Amjad *et al.*, 2022). Similar outcomes were reported for *Lipaphis erysimi* using *B. bassiana*-derived AgNPs (Kamil *et al.*, 2017), and for *Macrosiphum rosae* with *Solanum lycopersicum*-based AgNPs (Bhattacharyya *et al.*, 2016b). The silica nanoparticles induce oxidative stress in aphids, impacting their overall physiological balance. SiO_2-based NPs synthesized from TEOS, also caused significant mortality in *S. avenae*, with >64% killed in 24 hours at 3 mM and complete mortality at 300–425 mg/L after 15 days, while enhancing plant volatile emissions that attract natural enemies (Thabet *et al.*, 2021; Qi *et al.*, 2024;). Larger SiO_2-coated BNPs (~9 μm) also achieved 100% mortality in *A. fabae* within 24 hours (Agrafioti *et al.*, 2020). Natural polymers like chitin and chitosan have also been used to produce BNPs. Nano chitosan and cationic nanochitin whiskers interfere with the growth and development leading to >80% mortality in *S. avenae* within 7 days at 600–6000 mg/L. FeS-based NPs, in combination with *Artemis herba-alba* extracts, showed >80% efficacy against *Myzus persicae* at 251–302 ppm within 72 hours (Asoufi *et al.*, 2018). These BNPs exhibit strong aphid-targeting potential, are biodegradable, and can be safely integrated into IPM strategies. However, challenges remain in optimizing formulations, improving field application, understanding ecological impacts, and scaling cost-effectiveness. Future research should focus on new materials and biological synthesis methods for broader agricultural adoption.

17.9.3 Artificial intelligence (AI) for aphid management

AI-driven predictive models enhance insect monitoring and management by forecasting climate-driven changes in pest behavior, enabling smarter decisions and reduced insecticide use (Batz *et al.*, 2023). Advanced AI algorithms have achieved precise aphid colony detection using datasets like AphColDatto (Brassica with *Myzus persicae*). Faster R-CNN showed highest accuracy, while SSD and EfficientNet provided faster speeds (Amrani *et al.*, 2023). The lightweight SSV2-YOLO model, trained on sugarcane aphid images, achieved 97.5% accuracy and 400 FPS, outperforming YOLOv7-tiny and others, making it ideal for mobile field use. A fusion super-pixel method reached 90% accuracy under natural light, surpassing benchmark models (Xu *et al.*, 2022). Recent deep learning approaches surpass traditional models like SSD MobileNet, demonstrating strong potential for reliable pest monitoring (Batz *et al.*, 2023).

17.10 Conclusion

Aphids remain a significant and evolving threat to tropical agriculture due to rapid reproduction, broad host range, and efficient virus transmission. Their prevalence in crops such as cotton, legumes, banana, and vegetables is driven by favourable climates and global trade. This chapter reviewed aphid diversity, biology, ecology, and their role in virus epidemics, highlighting their impact on crop yields. Integrated Pest Management (IPM) remains the cornerstone of control, combining cultural, biological, molecular, and biotechnological methods adapted to tropical conditions. Sustainable options include natural enemies, botanical pesticides, and habitat management, while emerging tools like molecular diagnostics, RNAi, nanotechnology, and AI offer precision solutions. Successful regional programs in India, Brazil, and Kenya illustrate the effectiveness of locally tailored, participatory approaches. However, sustained management requires ongoing research, farmer education, resistant cultivars, and institutional support. With climate change affecting aphid dynamics, proactive, resilient, and system-level strategies are essential. Ultimately, a holistic, data-driven, farmer-focused approach supported by regional collaboration is vital for safeguarding food security and managing aphid threats in tropical agroecosystems.

Acknowledgements

The authors gratefully acknowledge the use of the following openly licensed icons in the preparation of figures for this manuscript: *peach_aphid* and *PCR_cycle* icons by DBCLS (https://togotv.dbcls.jp/en/pics.html), licensed under CC-BY 4.0 Unported; *Arabidopsis_plant* icon by Frédéric Bouché, available via Plant Illustrations on Figshare, also licensed under CC-BY 4.0 Unported; and the *RNA* icon by Servier, accessed from Servier Medical Art, licensed under CC-BY 3.0 Unported. We thank these creators for making their artwork freely available, thereby enhancing the visual quality of scientific communication. All these images were taken from the website, Bioicon. The authors also acknowledge the creators for the use of aphid damage pictures by ICAR - NBAIR, available via https://databases.nbair.res.in/insectpests/imageindex.php?keyword=A, copy right under Copyright © 2013, ICAR-National Bureau of Agricultural Insect Resources.

References

Abang, A.F., Srinivasan, R., Hanna, R., Fotso, A.K., Kekeunou, S. *et al.* (2021) Productivity and resistance of okra (*Abelmoschus* spp.) to the cotton aphid *Aphis gossypii* Glover (Hemiptera: Aphididae) under tropical conditions. *International Journal of Tropical Insect Science* 41, 197–208.

Abrol, D.P. (2015) Tropical fruits. In: *Pollination Biology*, Vol. 1. Cham: Springer, pp. 161–198.

Agrafioti, P., Faliagka, S., Lampiri, E., Orth, M., Pätzel, M. *et al.* (2020) Evaluation of silica-coated insect proof nets for the control of *Aphis fabae*, *Sitophilus oryzae*, and *Tribolium confusum*. *Nanomaterials* 10(9), 1658. DOI: 10.3390/nano10091658.

Amjad, T., Afsheen, S. and Iqbal, T. (2022) Nanocidal effect of rice husk-based silver nanoparticles on antioxidant enzymes of aphid. *Biological Trace Element Research* 200, 4855–4864. DOI: 10.1007/s12011-021-03067-5.

Amrani, A., Sohel, F., Diepeveen, D., Murray, D. and Jones, M.G. (2023) Deep learning-based detection of aphid colonies on plants from a reconstructed brassica image dataset. *Computers and Electronics in Agriculture* 205, 107587. DOI: 10.1016/j.compag.2022.107587.

Arora, R. (2017) *Breeding Insect Resistant Crops for Sustainable Agriculture*. Springer Nature Singapore Pte Ltd.

Asoufi, H., Al-Antary, T. and Awwad, A. (2018) Biosynthesis and characterization of iron sulfide (FeS) nanoparticles and evaluation of their aphicidal activity on the green peach aphid *Myzus persicae* (Homoptera: Aphididae). *Fresenius Environmental Bulletin* 27, 7767–7775.

Bass, C., Puinean, A.M., Zimmer, C.T., Denholm, I., Field, L.M. *et al.* (2014) The evolution of insecticide resistance in the peach-potato aphid, *Myzus persicae*. *Insect Biochemistry and Molecular Biology* 51, 41–51.

Batz, P., Will, T., Thiel, S., Ziesche, T.M. and Joachim, C. (2023) From identification to forecasting: The potential of image recognition and artificial intelligence for aphid pest monitoring. *Frontiers in Plant Science* 14, 1150748. DOI: 10.3389/fpls.2023.1150748.

Bejerman, N., Nome, C., Breuil, S. and Lenardon, S. (2011) Alfalfa dwarf virus: A novel cytorhabdovirus infecting alfalfa. *Archives of Virology* 156(8), 1291–1295.

Bhattacharya, S. (2019) Brassica-aphid interaction: Challenges and prospects of genetic engineering for integrated aphid management. *Physiological and Molecular Plant Pathology* 108, 101442. DOI: 10.1016/j.pmpp.2019.101442.

Bhattacharyya, A., Duraisamy, P., Govindarajan, M., Buhroo, A.A. and Prasad, R. (2016a) Nano-biofungicides: Emerging trend in insect pest control. In: Prasad, R. (ed.) *Advances and Applications through Fungal Nanobiotechnology*. Springer International Publishing, Cham, pp. 307–319. DOI: 10.1007/978-3-319-42990-8_15.

Bhattacharyya, A., Prasad, R., Buhroo, A.A., Duraisamy, P., Yousuf, I. *et al.* (2016b) One-pot fabrication and characterization of silver nanoparticles using *Solanum lycopersicum*: An eco-friendly and potent control tool against rose aphid, *Macrosiphum rosae*. *Journal of Nanoscience* 4679410. DOI: 10.1155/2016/4679410.

Bista, M. (2020) Diversity of aphid pests on the crops in Kanchanpur district, Nepal. *Journal of Natural History Museum* 33, 52–60.

Brisson, J.A. and Davis, G.K. (2016) The right tools for the job: Regulating polyphenic morph development in insects. *Current Opinion in Insect Science* 13, 1–6. DOI: 10.1016/j.cois.2015.09.011.

Capinera, J.L. (2001) *Green peach aphid, Myzus persicae (Sulzer) (Insecta: Hemiptera: Aphididae)*. University of Florida Cooperative Extension Service, Institute of Food and Agricultural Sciences, EDIS, Panama City, FL.

Chang, M., Cheng, H., Cai, Z., Qian, Y., Zhang, K. *et al.* (2022) *miR-92a-1-p5* modulated expression of the *flightin* gene regulates flight muscle formation and wing extension in the pea aphid, *Acyrthosiphon pisum* (Hemiptera: Aphidoidea). *Journal of Insect Science* 22, 14.

Chen, Y., Verheggen, F.J., Sun, D., Wang, Z., Francis, F. *et al.* (2019) Differential wing polyphenism adaptation across life stages under extreme high temperatures in corn leaf aphid. *Scientific Reports* 9(1), 8744.

Cranshaw, W. (2004) *Garden Insects of North America: The Ultimate Guide to Backyard Bugs*. Princeton University Press, Princeton, NJ.

Denholm, I., Devine, G.J. and Williamson, M.S. (2002) Insecticide resistance on the move. *Science* 297, 2222–2223.

Dent, D. and Binks, R.H. (2020) *Insect Pest Management*, 3rd edn. CABI, Wallingford.
Dixon, A.F.G. (1998) *Aphid Ecology: An Optimization Approach*, 2nd ed. Springer, Dordrecht.
Dong, Y., Wu, M., Zhang, Q., Fu, J., Loiacono, F.V. *et al.* (2022) Control of a sap-sucking insect pest by plastid-mediated RNA interference. *Molecular Plant* 15, 1176–1191.
Douglas, A.E. (1998) Nutritional interactions in insect–microbial symbioses: Aphids and their symbiotic bacteria *Buchnera*. *Annual Review of Entomology* 43, 17–37.
Edde, P.A. (2022) Arthropod pests of cotton (*Gossypium hirsutum* L.). In: *Field Crop Arthropod Pests of Economic Importance*. Academic Press, pp. 208–274.
Emaru, A., Nyaanga, J.G. and Saidi, M. (2024) Integrating *Metarhizium anisopliae* with border cropping. *Heliyon* 10(13), e33037.
Emden, H.V. and Harrington, R. (eds) (2017) *Aphids as Crop Pests*, 2nd edn. CABI, Wallingford.
EPPO (European and Mediterranean Plant Protection Organization) (2024) *Diuraphis noxia* distribution. Available at: https://gd.eppo.int/taxon/DIURNO (accessed 20 May 2025).
Faisal, M., Abdel-Salam, E.M., Alatar, A.A., Saquib, Q., Alwathnani, H.A. *et al.* (2019) Genetic transformation and siRNA-mediated gene silencing for aphid resistance in tomato. *Agronomy* 9, 893. DOI: 10.3390/agronomy9120893.
Feng, K.H., Qi, Y.H., Ye, Z.X., Li, T., Jiao, G.Y. *et al.* (2025) Diversity and evolution analysis of RNA viruses in three wheat aphid species. *BMC Genomics* 26(1), 1–15.
Fereres, A. and Moreno, A. (2009) Behavioural aspects influencing plant virus transmission by homopteran insects. *Virus Research* 141(6), 158–168. DOI: 10.1111/j.1755-0998.2008.02297.
Flint, M.L. (2018) *Pests of the Garden and Small Farm: A Grower's Guide to Using Less Pesticide*, 3rd edn, Vol. 248. CA: University of California, Agriculture and Natural Resources.
Foottit, R.G., Maw, H.V., Von Dohlen, C.D. and Hebert, P.D.N. (2008) Species identification of aphids through DNA barcodes. *Molecular Ecology Resources* 8(6), 1189–1201. DOI: 10.1111/j.1755-0998.2008.02297.
Frago, E., Dicke, M. and Godfray, H.C.J. (2012) Insect symbionts as hidden players in insect–plant interactions. *Trends in Ecology and Evolution* 27(12), 705–711. DOI: 10.1016/j.tree.2012.08.013.
Gao, Y., Alyokhin, A., Zhang, R., Smagghe, G., Palli, S.R. *et al.* (2024) Proactive resistance management for sustaining the efficacy of RNA interference for pest control. *Journal of Economic Entomology* 117(4), 1306–1308. DOI: 10.1093/jee/toae099.
Ghaith, M., Farag, A.A. and Youssef, A.A. (2023) Effect of different methods and natural enemies on the control of aphids of some herbage grass field crops in Nepal. *Zagazig Journal of Agricultural Research* 50, 55–65.
Gippet, J.M., Liebhold, A.M., Fenn-Moltu, G. and Bertelsmeier, C. (2019) Human-mediated dispersal in insects. *Current Opinion in Insect Science* 35, 96–102. DOI: 10.1016/B978-012257305-7/50063-1.
Hagen, K.S., Mills, N.J., Gordh, G. and McMurtry, J.A. (1999) Terrestrial arthropod predators of insect and mite pests. In: *Handbook of Biological Control: Principles and Applications of Biological Control*. pp. 383–503. DOI: 10.1016/B978-012257305-7/50063-1.
Hazell, S.P., Gwynn, D.M., Ceccarelli, S. and Fellowes, M.D.E. (2005) Competition and dispersal in the pea aphid: Clonal variation and correlations across traits. *Ecological Entomology* 30(3), 293–298. DOI: 10.1111/j.0307-6946.2005.00703.x.
Hebert, P.D., Cywinska, A., Ball, S.L. and deWaard, J.R. (2003) Biological identifications through DNA barcodes. *Proceedings of the Royal Society B* 270(1512), 313–321.
Henry, L.M., Peccoud, J., Simon, J.C., Hadfield, J.D., Maiden, M.J.C. *et al.* (2013) Horizontally transmitted symbionts and host colonization of ecological niches. *Current Biology* 23(17), 1713–1717. DOI: 10.1016/j.cub.2013.07.029.
Hooks, C.R. and Fereres, A. (2006) Protecting crops from non-persistently aphid-transmitted viruses: A review on the use of barrier plants as a management tool. *Virus Research* 120, 1–16.
Hullé, M., Coeur d'Acier, A., Bankhead-Dronnet, S. and Harrington, R. (2020) Aphids in the face of global changes. *Annual Review of Entomology* 65, 519–540.
Jaouannet, M., Rodriguez, P.A., Thorpe, P., Lenoir, C.J., MacLeod, R. *et al.* (2014) Plant immunity in plant–aphid interactions. *Frontiers in Plant Science* 5, 663.
Javed, K., Smagghe, G., Hussain, B., Javed, H., Shixian, Z. *et al.* (2025) Exploring innovative strategies to control aphids: Meta-analysis and a critical view on what we have and what the future is. *Journal of Pest Science* 97, 1–37.

Kamil, D., Prameeladevi, T., Ganesh, S., Prabhakaran, N., Nareshkumar, R. *et al.* (2017) Green synthesis of silver nanoparticles by entomopathogenic fungus *Beauveria bassiana* and their bioefficacy against mustard aphid (*Lipaphis erysimi* Kalt.). *Indian Journal of Experimental Biology* 55, 555–561.

Kaur, H., Khadda, B.S. and Singh, P. (2023) Efficacy of some botanical pesticides against mustard aphid *Lipaphis erysimi* (Kalt.) in *Brassica napus*. *Indian Journal of Entomology* 86(4), 1355–1358. DOI: 10.55446/IJE.2023.1047.

Khulbe, A. and Batra, P. (2024) Insect-pests and diseases in greenhouse cultivation and their biological control. In: *Protected Cultivation*. Oakville, Canada: Apple Academic Press, pp. 217–254.

Kondo, H., Fujita, M., Hisano, H., Hyodo, K., Andika, I.B. *et al.* (2020) Virome analysis of aphid populations that infest the barley field: The discovery of two novel groups of Nege/Kita-like viruses and other novel RNA viruses. *Frontiers in Microbiology* 11, 509.

Kumar, S., Tripathi, P. and Singh, S.K. (2014) Role of monitoring tools and action thresholds in pest management. *Journal of Plant Protection Research* 54(3), 234–241.

Kumar, S. (2019) Aphid–plant interactions: Implications for pest management. In: *Plant Communities and Their Environment*. London: IntechOpen.

Kuśnierczyk, A., Winge, P.E.R., Jørstad, T.S., Troczyńska, J., Rossiter, J.T. *et al.* (2008) Towards global understanding of plant defence against aphids–timing and dynamics of early *Arabidopsis* defence responses to cabbage aphid (*Brevicoryne brassicae*) attack. *Plant, Cell and Environment* 31(8), 1097–1115.

Lee, W., Kim, H., Lim, J., Choi, H. R., Kim, Y. *et al.* (2011) A DNA barcode for *Aphididae* (Insecta: Hemiptera: Sternorrhyncha). *Molecular Ecology Resources*, 11(1), 32–38.

Le Trionnaire, G., Tanguy, S., Hudaverdian, S., Gléonnec, F., Richard, G. *et al.* (2019) An integrated protocol for targeted mutagenesis with CRISPR-Cas9 system in the pea aphid. *Insect Biochemistry and Molecular Biology* 110, 34–44. DOI: 10.1016/j.ibmb.2019.04.016.

Liu, X.D., Xu, T.T. and Lei, H.X. (2017) Refuges and host shift pathways of host-specialized aphids *Aphis gossypii*. *Scientific Reports*, 7(1), 2008.

Luo, K., Zhao, H., Wang, X. and Kang, Z. (2022) Prevalent pest-management strategies for grain aphids: Opportunities and challenges. *Frontiers in Plant Science*, 12, 790919.

MacWilliams, J.R. (2020) *Identification and Characterization of Cowpea Aphid Salivary Proteins and the Cowpea Resistance Mechanism to the Cowpea Aphid*, PhD dissertation, University of California, Riverside.

MacWilliams, J.R., Nabity, P.D., Mauck, K.E. and Kaloshian, I. (2023) Transcriptome analysis of aphid-resistant and susceptible near-isogenic lines reveals candidate resistance genes in cowpea (*Vigna unguiculata*). *BMC Plant Biology*, 23(1), 22.

Mahas, J.W., Mahas, J.B., Ray, C., Kesheimer, A., Steury, T.D. *et al.* (2023) The spatiotemporal distribution, abundance, and seasonal dynamics of cotton-infesting aphids in the southern U.S. *Insects* 14(7), 639. DOI: 10.3390/insects14070639.

Mille, C., Jourdan, H., Cazères, S., Maw, E. and Foottit, R. (2020) New data on the aphid (Hemiptera, Aphididae) fauna of New Caledonia: Some new biosecurity threats in a biodiversity hotspot. *ZooKeys* 943, 53–89. DOI: 10.3897/zookeys.943.47785.

Mwanauta, R.W., Ndakidemi, P.A. and Venkataramana, P. (2021) A review on papaya mealybug identification and management through plant essential oils. *Environmental Entomology*, 50, 1016–1027.

Nawaz, A., Razzaq, F., Razzaq, A., Gogi, M.D., Fernández-Grandon, G.M. *et al.* (2022) Compatibility and synergistic interactions of fungi, *Metarhizium anisopliae*, and insecticide combinations against the cotton aphid, *Aphis gossypii* Glover (Hemiptera: Aphididae). *Scientific Reports* 12, 4843. DOI: 10.1038/s41598-022-08841-6.

Ng, J.C.K. and Falk, B.W. (2006) Virus–vector interactions mediating nonpersistent and semipersistent transmission of plant viruses. *Annual Review of Phytopathology* 44, 183–212. DOI: 10.1146/annurev.phyto.44.070505.143325.

Nibouche, S., Fartek, B., Mississipi, S., Delatte, H., Reynaud, B. *et al.* (2014) Low genetic diversity in *Melanaphis sacchari* aphid populations at the worldwide scale. *PLoS ONE* 9(8), e106067.

Parry, H.R. (2013) Cereal aphid movement: General principles and simulation modelling. *Movement Ecology*, 1(1), 1–15.

Peccoud, J., Simon, J.C., von Dohlen, C., Cœurd'acier, A., Plantegenest, M. *et al.* (2010) Evolutionary history of aphid–plant associations and their role in aphid diversification. *Comptes Rendus Biologies*, 333(6), 474–487.

Pecman, A., Kutnjak, D., Gutiérrez-Aguirre, I., Adams, I., Fox, A. *et al*. (2017) Next generation sequencing for detection and discovery of plant viruses and viroids: comparison of two approaches. *Frontiers in Microbiology*, 8, 1998.

Piron, P.G.M., Haas, M.C.D. and Sonnemans, M.A.H.M. (2019) The presence of *Aphis* (toxoptera) *aurantia* (Homoptera: Aphididae) in the Netherlands. *Entomologische Berichten* 79(5), 162–164.

Purandare, S.R., Bickel, R.D., Jaquiery, J., Rispe, C. and Brisson, J.A. (2014) Accelerated evolution of morph-biased genes in pea aphids. *Molecular Biology and Evolution* 31(8), 2073–2083. DOI: 10.1093/molbev/msu149.

Qi, X., Xue, X., Wang, Z., Li, S., Zhang, Z. *et al*. (2024) Silicon application enhances wheat defence against *Sitobion avenae* F. by regulating plant physiological-biochemical responses. *Basic and Applied Ecology* 74, 13–23. DOI: 10.1016/j.baae.2023.11.003.

Reddy, S.P., Bhagwat, V.R., Prasad, G.S. and Tonapi, V.A. (2017) Breeding for insect resistance in sorghum and millets. In: *Breeding Insect Resistant Crops for Sustainable Agriculture*. Springer Nature Singapore Pte Ltd, Singapore, pp. 231–264.

Robson, J.D., Wright, M.G. and Almeida, R.P.P. (2014) Biology of *Pentalonia nigronervosa* (Hemiptera: Aphididae) on banana using different rearing methods. *Environmental Entomology* 36(1), 46–52.

Rugman-Jones, P. F., Hoddle, M. S. and Stouthamer, R. (2010) Nuclear-mitochondrial barcoding exposes cryptic species in an invasive aphid complex. *Biological Invasions*, 12(12), 3665–3677.

Russell, J.A. and Moran, N.A. (2006) Costs and benefits of symbiont infection in aphids: Variation among symbionts and across temperatures. *Proceedings of the Royal Society B: Biological Sciences* 273(1586), 603–610. DOI: 10.1098/rspb.2005.3348.

Sentis, A., Hemptinne, J.L. and Brodeur, J. (2017) Non-additive effects of simulated heat waves and predators on prey phenotype and transgenerational phenotypic plasticity. *Global Change Biology* 23(11), 4598–4608. DOI: 10.1111/gcb.13674.

Shannag, H.S., Capinera, J.L. and Freihat, N.M. (2014) Efficacy of neem-based biopesticides against the green-peach aphid, *Myzus persicae* (Hemiptera: Aphididae). *Journal of Applied Entomology*, 138, 78–86.

Simon, J.C. *et al* (2010) Ecological genetics and evolution of sexual and cyclically parthenogenetic aphids. *Annual Review of Ecology, Evolution, and Systematics* 41, 529–553.

Singh, H. (2023) Insights into biointensive management of aphids: A broad perspective. *Biological Forum – An International Journal*, 15(3), 474–482.

Singh, G. and Singh, R. (2017) Distribution and economic importance of *Aphis* (Aphis) *craccivora* Koch, 1854 (Aphidini: Aphidinae: Aphididae: Hemiptera) and its food plants in India. *International Journal of Recent Advances in Multidisciplinary Research* 4(2), 2274–2286.

Singh, R. and Singh, G. (2020) Species diversity of Indian aphids (Hemiptera: Aphididae). *International Journal of Biological Innovations*, 1(1), 41–52.

Singh, R. and Singh, G. (2021) Aphids. In: *Polyphagous Pests of Crops*. Springer, Singapore, pp. 105–182. DOI: 10.1007/978-981-15-8075-8_3.

Singh, R. and Agrawal, R. (2022) An updated checklist of aphids infesting brassica plants in India. *Journal of Applied Entomologist*, 2(2),14–19.

Straub, C.S., Finke, D.L. and Snyder, W.E. (2008) Are conservation of natural-enemy biodiversity and biological control compatible goals? *Biological Control*, 45, 225–237.

Sukhikh, N., Golyaev, V., Laboureau, N., Clavijo, G., Rustenholz, C. *et al*. (2024) Deep sequencing analysis of Virome components, viral gene expression and antiviral RNAi responses in *Myzus persicae* Aphids. *International Journal of Molecular Sciences* 25(23), 13199.

Takenaka, A., Konno, H. and Kikuta, S. (2023) *In vivo* direct cell-penetrating peptide-mediated protein transduction system in *Acyrthosiphon pisum*. *BMC Research Notes* 16, 1–6. DOI: 10.1186/s13104-023-06514-9.

Thabet, A.F., Boraei, H.A., Galal, O.A., El-Samahy, M.F., Mousa, K.M. *et al*. (2021) Silica nanoparticles as pesticide against insects of different feeding types and their non-target attraction of predators. *Scientific Reports* 11(1), 14484. DOI: 10.1038/s41598-021-93518-9.

Tomar, M., Yadav, H.S. and Kumar, A. (2023) Efficacy of some insecticides and biopesticides against mustard aphid *Lipaphis erysimi* kalt. *Indian Journal of Entomology*, 86(1), 221–223. DOI: 10.55446/IJE.2023.845.

Van Emden, H.F. and Harrington, R. (2007) *Aphids as Crop Pests*. Wallingford: CABI. DOI: 10.1079/9780851998190.0000.

Verma, S.C., Sharma, P.L., Chandel, R.S. and Negi, S. (2018) Spatial distribution of green peach aphid, *Myzus persicae* Sulzer and its parasitoid, *Aphelinus asychis* Walker in bell pepper under polyhouse conditions. *Journal of Entomology and Zoology Studies*, 6(3), 776–780.

Wang, J.J. and Qiao, G.X. (2009) DNA barcoding of genus *Sitobion* (Hemiptera: Aphididae): Cryptic species and barcode divergence. *Molecular Ecology Resources*, 9(Suppl. 1), 157–166.

Wang, Q., Liu, J.T., Zhang, Y.J., Chen, J.L., Li, X.C. *et al.* (2021) Coordinative mediation of the response to alarm pheromones by three odorant binding proteins in the green peach aphid *Myzus persicae*. *Insect Biochemistry and Molecular Biology* 130, 103528. DOI: 10.1016/j.ibmb.2021.103528.

Way, M.J. and van Emden, H.F. (2000) Integrated pest management in practice: Pathways towards successful application. *Crop Protection* 19, 81–103.

Webster, C.G., Pichon, E., Van Munster, M., Monsion, B., Deshoux, M. *et al.* (2018) Identification of plant virus receptor candidates in the stylets of their aphid vectors. *Journal of Virology*, 92, e00432-18.

Wend, K., Zorrilla, L., Freimoser, F.M. and Gallet, A. (2024) Microbial pesticides – challenges and future perspectives for testing and safety assessment with respect to human health. *Environmental Health*, 23, 49.

Xu, W., Thomasson, J.A., Su, Q., Ji, C., Shi, Y. *et al.* (2022) A segmentation algorithm incorporating super pixel block and holistically nested edge for sugarcane aphids images under natural light conditions. *Biosystems Engineering* 216, 241–255. DOI: 10.1016/j.biosystemseng.2022.02.011.

Yamamoto, T., Hattori, M. and Itino, T. (2020) Seasonal migration in the aphid genus *Stomaphis* (Hemiptera: Aphididae): Discovery of host alternation between woody plants in subfamily Lachninae. *Journal of Insect Science* 20, 5–10. DOI: 10.1093/jisesa/ieaa103.

Yan, S., Qian, J., Cai, C., Ma, Z., Li, J. *et al.* (2020) Spray method application of transdermal dsRNA delivery system for efficient gene silencing and pest control on soybean aphid *Aphis glycines*. *Journal of Pest Science*, 93, 449–459.

Yazdani, M., Baker, G., DeGraaf, H., Henry, K., Hill, K. *et al.* (2018) First detection of Russian wheat aphid *Diuraphis noxia* Kurdjumov (Hemiptera: Aphididae) from Australia: A major threat to cereal production. *Austral Entomology* 57, 410–417. DOI: 10.1111/aen.12292.

Ye, C., Wang, Z.W., Sheng, Y.L., Wang, Z.G., Smagghe, G. *et al.* (2022) GNBP1 as a potential RNAi target to enhance the virulence of *Beauveria bassiana* for aphid control. *Journal of Pest Science* 95, 1–14. DOI: 10.1007/s10340-022-01524-y.

Zárate, S., Taboada, B., Yocupicio-Monroy, M. and Arias, C.F. (2017) Human virome. *Archives of Medical Research*, 48(8), 701–716.

Zhang, M., Kang, C., Xing, K. and Zhao, F. (2024) Complete mitochondrial genome of *Rhopalosiphummaidis* (Hemiptera: Aphididae) and its phylogenetic implications. *Mitochondrial DNA Part B: Resources*, 9(1), 51–55.

Zhang, Y., Liu, X., Fu, Y., Crespo-Herrera, L., Liu, H. *et al.* (2022) Salivary effector *Sm9723* of grain aphid *Sitobion miscanthi* suppresses plant defense and is essential for aphid survival on wheat. *International Journal of Molecular Sciences*, 23, 6909.

Zhao, Y., Sui, X., Xu, L., Liu, G., Lu, L. *et al.* (2018) Plant-mediated RNAi of grain aphid *CHS1* gene confers common wheat resistance against aphids. *Pest Management Science*, 74, 2754–2760.

Ziebell, H. and Carr, J.P. (2010) Cross-protection: A century of mystery. *Advances in Virus Research*, 76, 211–264.

18 Invasion and Management of Fruit Flies (Diptera: Tephritidae) in the Tropics

Godavari, Preeti Sharma, Santhosh Naik, Tulasi B., Samreen, Suman Barman, Mukesh Kumar Dhillon, Suresh M. Nebapure and Pathour Rajendra Shashank*

Division of Entomology, ICAR – Indian Agriculture Research Institute, New Delhi, India

Abstract

Fruit flies (Diptera: Tephritidae) are major invasive pests threatening global agriculture, especially in tropical regions. Species such as *Bactrocera dorsalis*, *Ceratitis capitata*, *Zeugodacus cucurbitae*, and *Anastrepha grandis* spread rapidly due to trade, climate change, and habitat disruption. These polyphagous pests reproduce quickly, disperse widely, and infest diverse host plants, leading to severe yield losses, reduced fruit quality, and trade restrictions. Smallholder farmers are disproportionately affected, facing economic hardship and market exclusion. Ecologically, invasions harm native biodiversity and worsen environmental degradation. Sustainable management requires integrated strategies including biological control, Sterile Insect Technique (SIT), Male Annihilation Technique (MAT), biopesticides, and cultural methods. Advances in genetics, microbiome studies, and AI-based surveillance improve early detection and targeted control. Success stories from India, Africa, and Latin America highlight the value of area-wide management and global cooperation. Strengthening biosecurity and empowering local communities are vital for resilient, long-term fruit fly management in tropical horticulture.

Keywords: Invasive fruit flies, Tephritidae, diversity, invasive pathways, pest management, biosecurity, tropical agriculture

18.1 Introduction

In recent decades, insect pest invasions have increased sharply across the globe, posing significant biosecurity threats to both natural ecosystems and agricultural systems. The expansion of transboundary trade, combined with the effects of climate change, has greatly amplified the scale of these invasions, thereby facilitating the spread and establishment of invasive species. Notably, human-mediated movement alone has led to the introduction of over 6,700 insect species worldwide, and this number is expected to rise substantially in the coming years (Early *et al.*, 2016; Biondi *et al.*, 2018; Qin *et al.*, 2019; Roy *et al.*, 2024; Tortorici *et al.*, 2025). In severe cases, climate-driven invasions can profoundly alter both natural and agricultural ecosystems, often resulting in invasive alien species outcompeting and displacing native ones (Walther *et al.*, 2009). The invasion process of phytophagous (plant-feeding) insects typically unfolds in a series of stages beginning with introduction, followed by colonization, establishment, and ultimately spread, largely influenced by environmental conditions such as temperature, precipitation, and the availability of

*Corresponding author: spathour@gmail.com

 Biological Invasions in the Tropics: Monitoring, Mechanism and Management (eds Akshay Kumar Chakravarthy and Pathour Rajendra Shashank)
DOI: 10.1079/9781836991977.0018

suitable host plants in new habitats (Blackburn *et al.*, 2011; Stokstad, 2018). Once propagules are introduced into a new environment, the resulting increase in invasive alien species can disrupt local biodiversity and alter food web dynamics (Zeng *et al.*, 2019). These invasive pests often cause direct agricultural losses through reduced crop yields and elevated management expenses, primarily due to increased reliance on synthetic pesticides. However, heavy pesticide use can lead to further complications, including the development of insecticide resistance and the breakdown of ecological relationships essential for natural pest control and pollination. Additionally, infestations by invasive species may result in trade restrictions and market access barriers due to quarantine regulations (Ekesi and Billah, 2007; Early *et al.*, 2016; Biondi *et al.*, 2018; Karsten *et al.*, 2018).

Fruit flies (Diptera: Tephritidae) are the most destructive among non-native agricultural pests, comprising approximately 5,000 species globally. These insects infest a wide range of fruits and vegetables, causing particularly severe impacts in tropical regions (Norrbom *et al.*, 1999). The tropical environment, characterized by consistently warm temperatures, high humidity, and the year-round availability of host plants, creates ideal conditions for their proliferation. In such settings, fruit flies have become increasingly invasive due to their adaptability, high reproductive potential, and polyphagous feeding habits, which enable them to exploit diverse ecological niches and infest a wide variety of crops.

18.1.1 Biology of fruit flies

Fruit flies are holometabolous insects that undergo complete metamorphosis through egg, larva, pupa, and adult stages. Females typically deposit eggs just beneath the skin of ripening fruits or vegetables, and the larvae pass through three feeding instars before forming a puparium, within which pupation occurs usually in soil, though some species pupate inside the fruit (Christenson and Foote, 1960). Dacine fruit flies follow a similar life cycle, with adults maturing shortly after emergence; males can mate repeatedly, while females become unreceptive post-mating and require protein-rich food for egg development, in addition to carbohydrates and water for survival (Tzanakakis *et al.*, 1968; Bateman, 1972; Fay and Meats, 1983). Most polyphagous, oligophagous, and even many monophagous species are multivoltine, though some, like *Dacus opiliae*, are univoltine due to host seasonality (Fitt, 1981a, b). While dacines do not undergo true diapause, adults may enter facultative reproductive dormancy during unfavourable conditions, aggregating in refuges and reverting to an immature state (Fitt, 1981a; Hancock, 1985).

Egg size and structure vary among species but are not consistently correlated with body size or ovariole number (Fitt, 1983). In cucurbit feeders such as *Zeugodacus cucurbitae* and *Z. cucumis*, eggs are relatively large and hatch faster than those of other species (Vargas *et al.*, 1984). Dacine larvae are acephalic cyclorrhaphan maggots with a recessed head, three thoracic, and eight abdominal segments; identification relies on changes in mouth hooks and spiracles across instars (Anderson, 1963). Egg hatching duration ranges from 1 day in summer to 7–20 days in winter (Xie and Zhang, 2005), while larval development lasts 9–23 days. Scarce or deteriorating food significantly increases mortality in early instars, whereas third instars may pupate early, develop slowly, or emerge smaller (Zhang *et al.*, 1991). Pupal development in tephritid fruit flies accelerates with rising temperatures, ranging from 5 to 33 days depending on thermal conditions (Yuan *et al.*, 2005), with prepupae transitioning to pupae within 1–2 days (He WanZhong *et al.*, 2002). Soil moisture influences pupation depth and survival (Xie and Zhang, 2005). In *Bactrocera minax*, low temperatures can induce diapause yet also support pupation and emergence under field conditions, though diapause termination appears independent of photoperiod (He *et al.*, 2023). Adult emergence occurs throughout the day, peaking between 6:00 and 10:00 am, after which adults rest in shaded areas for 2–3 h until their wings expand and feeding begins (Zhang *et al.*, 1998; Xie and Zhang, 2005). Sexual maturation follows nutritional intake, with mating occurring 8–15 days post-eclosion, mainly during

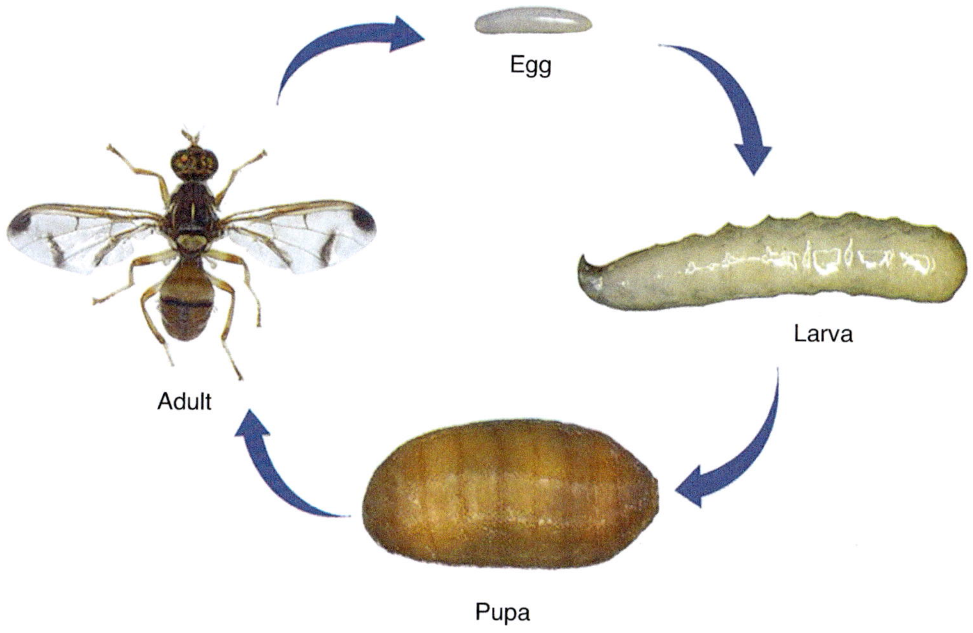

Fig. 18.1. Different life stages of a fruit fly (*Zeugodacus cucurbitae*). (Authors' own image)

early morning hours (Zhang *et al.*, 1991; Ma *et al.*, 2013). In *Bactrocera dorsalis*, high flight capacity coincides with reproductive maturity without the typical "oogenesis-flight syndrome", enabling long-distance dispersal and contributing to their widespread distribution, persistence after eradication, and invasion of new areas (Chen *et al.*, 2007; Shi *et al.*, 2010; Yuan RuiLing *et al.*, 2014) (Fig. 18.1).

18.1.2 Importance of fruit flies in tropical agriculture

The family Tephritidae includes some of the world's most destructive fruit and vegetable pests, especially species from the genera *Anastrepha*, *Bactrocera*, *Zeugodacus*, *Ceratitis*, and *Rhagoletis*, native to diverse biogeographic regions (White and Elsonharris, 1992; Norrbom *et al.*, 1999; Courtney *et al.*, 2009). These invasive fruit flies possess traits that enhance invasiveness, such as internal larval development in fruits via egg-laying with a specialized ovipositor, followed by soil pupation and adult emergence within 1–2 weeks. Tropical species can complete 8–12 generations annually, enabling rapid population growth, especially where natural enemies are absent, causing ecological disruption and biodiversity loss (Ekesi and Billah, 2007). Their survival, development, and population dynamics vary widely, further aiding their success (Carey *et al.*, 1988; Vargas *et al.*, 2000; Malacrida *et al.*, 2007; Papadopoulos, 2014). In tropical countries, fruit flies threaten food security and incomes, particularly for smallholder farmers, through yield losses, increased management costs, and quarantine-driven export restrictions. These restrictions lead to trade bans, shipment rejections, and costly phytosanitary treatments such as vapour heat or irradiation, undermining exports and foreign exchange (Lux *et al.*, 2003). Smallholders often rely heavily on chemical pesticides, raising concerns over resistance, food safety, and health (Clarke *et al.*, 2005). Invasive species from *Bactrocera*, *Ceratitis*, and *Anastrepha* have profoundly challenged pest management by outcompeting natives and overwhelming natural enemies, causing ecological imbalance (Clarke *et al.*, 2005). Their presence disrupts trade and regulatory systems due to quarantine classifications, imposing economic burdens from crop losses, control costs, and limited market access (Siebert and Cooper, 1995; Mumford, 2002; Papadopoulos, 2014). These invasions have historically spread through international trade and passenger luggage (Kiritani and Yamamura, 2003; Liebhold *et al.*, 2006; Karsten *et al.*, 2015), prompting biosecurity measures such as import bans, inspections, and phytosanitary treatments (Li *et al.*, 2013; Phillips, 2013). Effective management also requires continuous surveillance and rapid eradication of incursions (Quilici and Donner, 2012; Suckling *et al.*, 2016). Given the high economic stakes and limited resources, enhancing risk assessment and predictive tools is urgently needed to improve biosecurity outcomes (Qin *et al.*, 2015). Their importance in tropical agriculture arises from a combination of biological, ecological, and economic factors.

18.1.2.1 Polyphagy and continuous infestation

Due to their polyphagous nature, many fruit fly species attack a wide range of host plants. For instance, *B. dorsalis*, one of the most notorious species, infests over 300 plant species, including mango (*Mangifera indica*), guava (*Psidium guajava*), citrus (*Citrus* spp.), papaya (*Carica papaya*), and tomato (*Solanum lycopersicum*) (White and Elsonharris, 1992; Dhillon *et al.*, 2005). This broad host range allows for year-round population persistence in tropical regions, where cultivated and wild hosts are continuously available, complicating control efforts (Clarke *et al.*, 2005).

18.1.2.2 Reproductive capacity and crop damage

Fruit flies reproduce rapidly under favourable conditions, completing 8–12 generations per year (White and Elsonharris, 1992). Females deposit eggs under the skin of ripening fruits, where larvae feed internally, causing fruit drop, reduced marketability, and diminished fruit quality. Larval damage also promotes secondary infections by fungi and bacteria, compounding postharvest losses (Ekesi *et al.*, 2006).

18.1.2.3 Economic impacts on smallholder farmers

The economic consequences are particularly harsh for smallholder farmers. Pest control requires intensive labour and high costs Trapping, fruit bagging, sanitation, and chemical sprays are often unaffordable or impractical. In severe infestations, entire crops may be lost, threatening food security and livelihoods (Ekesi and Billah, 2007).

18.1.2.4 Barriers to trade

Many fruit fly species are recognized as quarantine pests by importing countries. Their detection in exported goods can lead to trade bans, expensive treatments (fumigation, irradiation, cold disinfestation), and loss of market access (Lux *et al.*, 2003). For countries dependent on agricultural exports, these pests significantly impact national economies.

18.1.2.5 Ecological disruption

Invasive species such as *B. dorsalis* can displace native fruit flies and disrupt ecological balances. This affects interactions with natural enemies, reduces biodiversity, and undermines natural pest regulation, often increasing the need for chemical controls (White and Elsonharris, 1992; Ekesi and Billah, 2007).

18.1.3 Economic and ecological consequences of infestations

The impact of fruit fly infestations in tropical regions extends far beyond direct crop damage. These pests cause wide-ranging economic losses and ecological disruptions, some of which are long-lasting and difficult to reverse.

18.1.3.1 Ecological consequences

Invasive Tephritid species often outcompete native fruit flies, altering insect community structure and reducing biodiversity (Clarke *et al.*, 2005; Drew *et al.*, 2005). Their ability to exploit diverse hosts and reproduce rapidly gives them a competitive edge in ecological niches once occupied by native herbivores or pollinators.

This displacement weakens food web stability, as native natural enemies are often ineffective against these invaders (White and Elson-Harris, 1992). Moreover, infestations trigger increased pesticide use, leading to unintended environmental impacts. Non-selective insecticides can harm pollinators, natural enemies, and soil microbes. Runoff may contaminate aquatic systems, further disturbing ecosystems (Ekesi and Billah, 2007; Vayssières *et al.*, 2009).

18.1.3.2 Economic consequences

Uncontrolled infestations of fruit flies can lead to devastating direct yield losses, with damage reaching up to 80–100% during peak fruiting seasons (Dhillon *et al.*, 2005). This significant threat forces farmers to increase their production expenditures, as they must invest heavily in various control measures such as trapping, sanitation, baiting, and protective coverings for fruits (Ekesi and Billah, 2007). Beyond the field, the economic impact continues into the postharvest phase. Fruit flies are often categorized as quarantine pests, which results in stringent international trade restrictions, including shipment rejections, export bans, and mandatory phytosanitary treatments (Clarke and Measham, 2022). These trade barriers severely limit market access for affected growers. Additionally, fruit fly outbreaks cause market volatility; infested produce often saturates local markets, driving prices down. Conversely, during severe infestations, reduced supply can lead to price spikes, which in turn affects affordability and access for consumers (Mwatawala *et al.*, 2009).

18.2 Diversity and Biogeography of Fruit Flies in the Tropics

The family Tephritidae comprises over 5000 species worldwide, with about 1400 developing in fleshy fruits and inhabiting all major biogeographic regions except extreme deserts and polar zones lacking suitable hosts (Savaris *et al.*, 2016). Within Tephritidae, the tribe Dacini includes about 880 described species, 629 in the genus *Bactrocera* and 248 in *Dacus*. Most *Bactrocera* species, except around ten, are native from India through Southeast Asia to the Pacific Islands, while *Dacus* species are primarily African (Drew, 2004). Species richness declines westward toward India and eastward and southward from Southeast Asia and Papua New Guinea. About 80% of *Dacini* species occur across Peninsular Malaysia, Thailand, and Vietnam, marking this continental region, along with Indonesia, Philippines, Borneo, and Papua New Guinea, as a major diversification hotspot (Drew, 2004).

Several fruit fly species, including *Z. cucurbitae*, *Z. tau*, *B. dorsalis*, *B. zonata*, *Ceratitis capitata*, and *Anastrepha grandis*, are notorious for their invasiveness and economic impact worldwide. *B. dorsalis*, first recorded in Taiwan in 1912 (Hardy, 1973; Drew and Hancock, 1994; Aketarawong *et al.*, 2014), has since spread across Asia-Pacific (Aketarawong *et al.*, 2007), east, west, and southern Africa after its detection in Kenya in 2003 (Mwatawala *et al.*, 2006; Magagula *et al.*, 2015; Manrakhan *et al.*, 2015), and Oceania and the Americas, including Hawaii (1945), Guam (1947), Florida (1960), California (1960), and Suriname (1975) (Pemberton, 1946; White and Elsonharris, 1992; Aketarawong *et al.*, 2014). In China, it has invaded most southern provinces and is expanding northward (Liang *et al.*, 2003; Chen *et al.*, 2011; Li *et al.*, 2012), threatening fruit industries by raising production costs and quarantine restrictions (Aketarawong *et al.*, 2014). Its rapid spread is driven by high adaptability, dispersal ability, global trade, and travel (Wan *et al.*, 2012; Kriticos *et al.*, 2013), while it also disrupts ecosystems by displacing native fruit flies (Duyck *et al.*, 2004; Clarke *et al.*, 2005; Aketarawong *et al.*, 2007). *B. zonata*, native to South and Southeast Asia, has spread to over 20 countries including India, Bangladesh, Bhutan, Laos, Myanmar, Nepal, Pakistan, Sri Lanka, Thailand, and Vietnam, expanded into the Arabian Peninsula in the 1980s (White, 2006), and after its 1997 detection in Egypt, further spread to Sudan, Libya, and Israel (Ekesi *et al.*, 2016). Due to its broad host range, high fecundity, and adaptability, *B. zonata* remains a major quarantine pest (Zingore *et al.*, 2020), with unconfirmed reports in China's Yunnan Province.

The Mediterranean fruit fly, *C. capitata*, is one of the most destructive and widespread pests, infesting over 200 host species worldwide (Liquido *et al.*, 1991). Native to Africa

(White and Elsonharris, 1992), it spread to the Mediterranean in the early 1800s and is now globally established (Headrick and Goeden, 1996). Another significant pest, *A. grandis* (South American cucurbit fly), primarily attacks melons, watermelons, zucchini, squash, pumpkins, and cucumbers (Silva and Malavasi, 1993), and occurs throughout South America and Panama, from tropical forests in Colombia, Venezuela, Ecuador, Peru, Guyana, Suriname, and Bolivia to temperate zones in Paraguay, Uruguay, Argentina, and Chile; in Brazil, it inhabits the Midwest, Southeast, and South regions, including the pampa and cerrado biomes (Arias *et al.*, 2014). *Z. tau* is widely distributed across Asia, including China, India, Korea, Japan, Vietnam, Myanmar, Thailand, Laos, Bhutan, Brunei, the Philippines, Cambodia, Nepal, Singapore, Bangladesh, Malaysia, Sri Lanka, and Indonesia (Prabhakar, 2011; Jaleel *et al.*, 2018), with significant crop damage reported in southern China (Fujian, Guangdong, Hainan) (Li *et al.*, 2020) and across several Indian states such as Delhi, Kerala, Tamil Nadu, and Maharashtra (Nair *et al.*, 2017). Globally, the highest numbers of established non-native Tephritidae species occur in the USA and Mauritius, followed by Reunion and the Hawaiian Islands with seven species each; Australia, France, and several Middle Eastern countries report six species each, and at least one tephritid invasion has been documented in every biogeographic region except Antarctica (Trombik *et al.*, 2023).

18.3 Invasion Pathways

Phytophagous insects from the Tephritidae family, are known for several successful invasions, mostly caused by human activity. These flies are found in nearly all fruit-growing regions around the world (White and Elsonharris, 1992). The five major genera in this family, *Ceratitis*, *Bactrocera*, *Anastrepha*, *Zeugodacus*, and *Rhagoletis*, include several important pest species (White and Elsonharris, 1992). Below are the invasive pathways that lead to species spread across the tropical countries.

18.3.1 Globalization and international trade

Globalization has increased the movement of people, goods, and agricultural products across continents. This has unintentionally opened up new ways for invasive pests like fruit flies to spread. Eggs and larvae often travel inside infested fruits and vegetables, whether through commercial shipments or personal luggage. The growth of international trade has played a major role in spreading the Oriental fruit fly, *B. dorsalis* to Africa and the Americas. Similarly, the Mediterranean fruit fly, *C. capitata* has been found in many countries because it can survive in different climates and feeds on a wide range of host plants. Malacrida *et al.* (2007) described how *C. capitata* has spread along global trade routes. To control this, the FAO and IPPC have created phytosanitary guidelines such as ISPM-15 and ISPM-26 to regulate pest movement. Trade regions most at risk include Southeast Asia, sub-Saharan Africa, and Latin America, where inspections are limited and trade in fresh produce is high.

18.3.2 Climatic suitability and climate change

Climate change and biological invasions are closely linked global challenges, with invasive species causing major ecological disruption and economic losses in new environments (Cook *et al.*, 2007; Hulme, 2009). Climate change facilitates their spread by making previously unsuitable areas more hospitable (Early *et al.*, 2016; Hulme, 2017), especially for insects whose survival depends on local temperatures (Cornelissen *et al.*, 2019). Understanding how climate change affects pest insect distributions is critical, as these species severely impact agriculture and readily invade new regions (Santana *et al.*, 2019). Climate change alters species distributions and population dynamics, but invasive species must overcome biological constraints like resource competition and predation to establish successfully (White, 2006; Broennimann *et al.*, 2007; David *et al.*, 2017). For instance, *C. capitata* has expanded northward and to higher altitudes in California, highlighting climate-driven range shifts and the need for adaptive pest management. Similarly, Australia's Queensland fruit fly (*B. tryoni*), the country's most damaging horticultural pest, has extended its range southward alongside fruit cultivation, with its distribution limited by rainfall and temperature. Northern populations may hybridize with related species, and suitable climates in regions such as Perth have allowed outbreaks since the late 1980s.

18.3.3 Anthropogenic activities and landscape change

Anthropogenic activities and landscape changes significantly contribute to the invasion and spread of fruit flies in tropical regions by altering habitats, facilitating dispersal, and increasing resource availability. These human-induced factors create new ecological opportunities that invasive fruit flies exploit, leading to rapid establishment and population growth. Many human activities lead to invasion of fruit fly species. These include agricultural expansion and monoculture: the intensification of fruit cultivation, especially in monocultures, provides abundant and predictable food sources for fruit flies. In tropical Asia, large-scale cultivation of mango, guava, and cucurbits has supported the proliferation of *B. dorsalis* and *Z. cucurbitae* (Drew *et al.*, 2005; Vayssières *et al.*, 2008). Continuous cropping and overlapping fruiting periods reduce natural bottlenecks that might otherwise control pest populations.

Urban areas often have a mix of ornamental and fruiting plants, backyard gardens, and waste fruit, which can serve as refuges for invasive species. Urban heat islands also create microclimates that allow tropical fruit flies to persist in otherwise unsuitable areas (Grimm *et al.*, 2008). *B. dorsalis* has been observed thriving in peri-urban regions of India and Southeast Asia due to such favourable conditions (Prabhakar, 2012). Clearing of natural forests disrupts ecological balances, reduces predator populations, and creates edge habitats rich in host plants. Such disturbed landscapes often favour invasive species over native ones (Didham *et al.*, 2007). In Papua New Guinea and Indonesia, forest fragmentation has coincided with increases in *Bactrocera* species abundance in adjacent agricultural lands (Clarke *et al.*, 2001). Human movement of fruits and planting materials facilitates accidental introductions of invasive fruit flies. *B. zonata*, for example, spread from South Asia to the Arabian Peninsula and North Africa through infested fruit trade (White, 2006; Ekesi *et al.*, 2016). Research in Eastern Central Tanzania observed that the abundance and diversity of both indigenous and invasive fruit fly species varied along an altitudinal transect. The study found that anthropogenic factors, such as land use changes and agricultural practices, influenced the distribution patterns of fruit flies, with certain species becoming more prevalent in altered landscapes (Geurts *et al.*, 2012).

18.3.4 Weak quarantine and surveillance systems

Fruit flies are major quarantine pests worldwide, and importing countries enforce strict measures to prevent the introduction of exotic species or strains, especially through fruits such as mangoes. However, inadequate quarantine and surveillance often allow undetected introductions and delayed responses, facilitating invasions. For instance, Australia's surveillance system, established in the 1990s, relies on outdated data and methods and requires updates with new lure types, trap designs, optimal grid spacing, suitable pesticides, and species distribution models integrated with climate forecasts, focusing on key entry points (Hoskins *et al.*, 2023a). Moreover, the lack of a specific emergency response plan delays action after post-border detections, worsening infestations. Tropical countries face additional challenges in implementing International Standards for Phytosanitary Measures (ISPMs) due to limited resources and institutional capacity, creating biosecurity gaps exploited by invasive fruit flies (Meyerson and Reaser, 2002). Early intervention depends on public and stakeholder participation in surveillance and reporting, yet many tropical regions lack outreach to farmers and traders, who are often the first to detect new pests (Hulme, 2006).

18.4 History of Key Invasive Species

Tephritidae, a family of invasive fruit flies, is a remarkable illustration of how small insects can cause significant ecological and economic upheavals worldwide. Over the past century, these flies, which have their origins in tropical and subtropical regions, have silently invaded fields and orchards, decimating fruit and vegetable crops. Some of the most invasive and economically detrimental species among them are *B. dorsalis*, *C. capitata*, *Z. cucurbitae*, *B. zonata*, and *B. carambolae*. The Mediterranean fruit fly (*C. capitata*), also referred to as the medfly, was discovered in Madeira, Portugal, and the Azores, Spain, in the 1830s, marking the first documented occurrence of an invasive tephritid.

Genetically distinct populations of this species, believed to have originated in East Africa and spread via Mediterranean trade routes to the Americas (de Deus *et al.*, 2024), marked the beginning of Tephritidae invasions. Only 20 further invasions occurred in the following century, doubling by the 1930s, with at least 189 incursions reported worldwide by 2018. While *C. capitata* facilitated the global spread of tephritids, other aggressive species soon followed. The melon fly (*Z. cucurbitae*) became a major cucurbit pest in Asia and Africa due to rapid reproduction and preference for warm, humid climates (Clarke *et al.*, 2005). The peach fruit fly (*B. zonata*), native to South Asia, now threatens citrus and peach orchards across the Middle East, North Africa, and parts of Europe (European Food Safety Authority (EFSA), 2024). Southeast Asia's *B. carambolae* has recently invaded South America, particularly Suriname, likely via fruit imports from Indonesia (Essl *et al.*, 2019). The invasion burden increased with *B. latifrons*, a Solanaceae specialist, found in Hawaii and East Africa. Among the most studied and widespread invaders is *B. dorsalis*, first described in Taiwan in 1912 (Papadopoulos *et al.*, 2024), now established in over 75 countries across the Americas, Asia, Africa, and Oceania, infesting more than 250 host plants including mango, guava, banana, papaya, citrus, and stone fruits (Ullah *et al.*, 2023).

Bactrocera dorsalis has invaded 42 African nations, accounting for approximately 56% of the continent's fruit fly invasions (European Food Safety Authority (EFSA) *et al.*, 2024). These species exert substantial economic impact: in India, infestations of mango, guava, and citrus fruit flies are estimated to cause annual losses of US$356 million (Sharma *et al.*, 2015). In Thailand, *B. dorsalis* reduces mango yields by up to 15.5% (Russell *et al.*, 2017; Zhai *et al.*, 2018), while *Ceratitis capitata* in the Mediterranean imposes heavy costs related to export restrictions, chemical controls, and surveillance (Malacrida *et al.*, 2007). Beyond direct fruit damage, these flies pose major phytosanitary risks that jeopardize international trade, often triggering export bans and strict quarantine measures. The European Food Safety Authority (European Food Safety Authority, 2023) identifies southern Europe including Spain, Italy, and Greece as highly vulnerable to *B. dorsalis* establishment due to favourable climate and abundant hosts.

Knowledge regarding these species' introduction and spread has been aided by genetic research, which has shown intricate invasion paths that frequently require several introductions. The significance of historical trade and human migration in promoting invasions is highlighted by the fact that populations of *C. capitata* in Latin America, for example, most likely originate from various European origins (Hoskins *et al.*, 2023b). Together, tephritid fruit flies pose a serious risk to tropical and subtropical agriculture today (Table 18.1).

18.5 Impact Assessment

18.5.1 Crop damage and yield losses

The destructive feeding behaviours of tephritid fruit flies, such as *B. dorsalis*, *C. capitata*, and *Z. cucurbitae*, are well-known for causing crop damage and yield losses. Fruit drop, rotting, and premature ripening result from the larvae feeding on the pulp of the host fruit after the females place their eggs inside it. Fruit quality is impacted in addition to the marketable yield being decreased by this damage (He *et al.*, 2023). In Africa, *B. dorsalis* infestations have resulted to mango yield losses ranging from

Table 18.1. Record of key species of invasive fruit flies and their hosts in the tropics. (Authors' own table)

Species	Regions	Native range	Host range	References
Bactrocera correcta	East South Southwest	India and South-east Asia	*Mangifera indica* (+++) *Annona squamosa* (++) *Musa nana* (++) *Psidium guajava* (+++) *Syzygium samarangense* (++) *Averrhoa carambola* (+++) *Ziziphus mauritiana* (++) *Pseudocydonia sinensis* (++) *Citrus sinensis* (++) *Momordica charantia* (++) *Diospyros morrisiana* (++) *Ziziphus jujuba* (++++)	Yan *et al*., 2016
Bactrocera dorsalis	Central East South Southwest	South-east China	*Actinidia fulvicoma* (+) *Mangifera indica* (+ / ++++) *Desmos chinensis* (+) *Diospyros kaki* (+) *Diospyros morrisiana* (++) *Diospyros tuherii* (+) *Phyllanthus emblica* (+) *Melastoma dodecandrum* (+) *Broussonetia kaempferi* (+) *Broussonetia papyrifera* (+) *Ficus hirta* (+) *Ficus sagittata* (+) *Musa nana* (nd) *Myrica rubra* (+++) *Acmena acuminatissima* (+) *Cleistocalyx operculatus* (++) *Psidium guajava* (+++ / ++++) *Rhodomyrtus tomentosa* (+++) *Syzygium jambos* (+++) *Syzygium leineei* (+) *Syzygium samarangense* (++++) *Averrhoa carambola* (+++) *Punica granatum* (+++) *Ziziphus jujuba* (++++) *Ziziphus* spp. (nd) *Carallia brachiata* (++) *Amygdalus davidiana* (++) *Duchesnea indica* (+) *Eriobotrya fragrans* (+) *Eriobotrya japonica* (++ / ++++) *Malus pumila* (+) *Prunus mume* (+) *Prunus persica* (++ / ++++) *Prunus phaeosticta* (+) *Prunus salicina* (+) *Pseudocydonia sinensis* (+) *Pyrus calleryana* (+) *Pyrus pyrifolia* (+) *Rubus leucanthus* (+) *Rubus reflexus* (+) *Rubus rosifolius* (+) *Rubus sumatranus* (+) *Citrus limon* (+) *Citrus maxima* (++) *Citrus reticulata* (+++) *Clausena lansium* (++) *Fortunella hindsii* (++) *Manilkara zapota* (++) *Cayratia japonica* (++) *Vitis quinquangularis* (+) *Vitis vinifera* (+)	Yan *et al*., 2016

Continued

Table 18.1. Continued

Species	Regions	Native range	Host range	References
Bactrocera latifrons (only captured by bait traps)	East South Southwest	South-east Asia	*Capsicum annuum* (+) *Solanum melongen* (+)	Yan *et al.*, 2016
Bactrocera minax	Northwest East Central South Southwest	China	*Citrus maxima* (+++/++++) *Citrus medica* (+++/++++) *Citrus poonensis* (+/++++) *Citrus sinensis* (+/++/+++/++++) *Citrus tangerina* (+/+++) *Citrus unshiu* (+/++/+++/++++)	Andongma *et al.*, 2019
Bactrocera tsuneonis	East Central Southwest	East Asia	*Citrus sinensis* (+) *Citrus tangerina* (+) Citrus unshiu (+)	Al Baki *et al.*, 2017
Zeugodacus scutellatus (only captured by bait traps)	North Northwest East Central South Southwest	East Asia	Cucurbitaceae flowers (nd)	Guo *et al.*, 2019
Carpomyia vesuviana	Northwest	India	*Ziziphus* spp. (++)	He *et al.*, 2011
Zeugodacus cucurbitae	East Central South Southwest	India	*Benincasa hispida* (+) *Citrullus lanatus* (++) *Cucumis sativus* (++++) *Cucurbita moschata* (nd) *Cucurbita pepo* (nd) *Luffa aegyptiaca* (++++) *Momordica charantia* (++) *Sechium edule* (++)	McQuate *et al.*, 2017 Deng *et al.*, 2021
Zeugodacus tau	East Central South Southwest	Asia	*Benincasa hispida* (nd) *Citrullus lanatus* (++) *Cucumis sativus* (++++) *Cucurbita moschata* (+++) *Cucurbita pepo* (nd) *Luffa aegyptiaca* (++++) *Momordica charantia* (++) *Sechium edule* (++)	Guo *et al.*, 2019

Notes: "+" represents the degree of damage (<10%: +; 10–30%: ++; 30–50%: +++; >50%: ++++); "nd" indicates there is no record of the degree of harm although there is a host.

0.34 to 6.5 t/ha, depending on the cultivar and season, with losses exceeding 70% by mid-June in some places (EFSA *et al.*, 2020). Fruit flies are thought to cause an average of 5.65 t/ha of mango output losses in Ghana, which translates to losses of about $3,428.97/ha. Similarly, in Thailand, *B. dorsalis* infestations in mango farms have caused average annual yield losses of 15.5%. *Z. cucurbitae* significantly affects cucurbit crops, causing up to 26% of fruits to be damaged or dropped prematurely and 14% damaged during harvest (Cugala *et al.*, 2020). In India, fruit fly infestations have led to reductions in marketable yields of mango by 25–30%, with estimated economic losses in guava, sapota, citrus fruits, and mango amounting to US$356 million (Jaffar *et al.*, 2023).

18.5.2 Market access and trade restrictions

International trade is severely impacted by the existence of invasive fruit flies. Strict quarantine laws are enforced by many nations to stop the spread of these pests, which restricts trade and costs exporting countries money. Less than 10% of Ghana's output is exported, and more than

60% of it is wasted every year as a result of fruit fly infestations. Ghana lost more than $30 million in 2015 when it was prohibited from exporting specific vegetables to the European Union (Opoku *et al.*, 2025). Exports of Spanish clementines in the USA fell from 77,000 tons in 2001 to 45,000 tons in 2002, resulting in losses for the fruit exporting sector estimated at €80–107 million. The establishment of invasive fruit fly species in new regions often necessitates costly eradication campaigns. For instance, a single eradication campaign against *B. dorsalis* in Florida in 2015 cost several million US dollars. The USDA has secured $129.2 million in emergency funds to combat exotic fruit fly outbreaks, highlighting the significant financial burden these pests impose (Papadopoulos *et al.*, 2024).

18.5.3 Ecological disruption of native fauna

By outcompeting native species for resources and upsetting ecological balances, invading fruit fly species disturb local ecosystems. For instance, *B. dorsalis* has diminished biodiversity in various African countries by displacing local fruit fly species. Because of their aggressive behaviour and high rates of reproduction, invasive fruit flies can swiftly take over new settings (Li *et al.*, 2013). They can take advantage of a variety of host plants according to their polyphagous eating habits, which intensifies their effects on native species and ecosystems. Further ecological imbalances may result from the widespread use of chemical pesticides to eradicate these pests, which may have unexpected effects on beneficial insects and pollinators, among other non-target creatures (Serrão *et al.*, 2022).

18.5.4 Socio-economic implications for farmers

Beyond crop losses and trade restrictions, invasive fruit flies have a substantial negative economic impact on farmers' livelihoods, particularly for smallholders in developing nations. Fruit fly infestations in Ghana have resulted in employment losses in the horticulture industry, lower earnings, and greater poverty (Lengai *et al.*, 2022). Due to lower yields, farmers frequently accrue debt and are unable to pay employees and creditors, which has an impact on the economic stability of the larger society. Small-scale farmers frequently cannot afford the large expenses involved in putting into practice efficient pest management techniques, such as the construction of postharvest treatment facilities. Due to a lack of resources, they are unable to comply with international phytosanitary requirements, which restricts their access to markets and revenue streams (Papadopoulos *et al.*, 2024).

18.6 Management Strategies

Effective management of fruit flies requires a comprehensive and integrative approach due to their high fecundity, wide host range, and strong dispersal ability (Muriithi *et al.*, 2016). A single method cannot provide sustainable control; therefore, it is necessary to combine and use multiple strategies in an area-wide Integrated Pest Management (IPM) programme (IAEA/FAO (International Atomic Energy Agency and Food and Agriculture Organization), 2013).

18.6.1 Detection, Monitoring, and Surveillance

Effective detection, monitoring, and surveillance are the basic components of fruit fly management, allowing for timely interventions and population suppression. A key strategy involves the use of species-specific trapping systems and attractants. Parapheromone-based male lures, such as Methyl Eugenol (ME) and Cue-Lure (CL) are widely utilized for mass trapping and surveillance. ME is highly attractive to *B. dorsalis*, while CL targets species like *B. cucurbitae* and *B. zonata* (IAEA (International Atomic Energy Agency), 2003). These lures, often incorporated into Jackson or Steiner traps, help in estimating population densities and distribution trends. Along with male-targeted pheromone lures, food-based attractants also aid in monitoring both sexes of fruit flies, particularly females. Traditional proteinaceous baits, hydrolysed torula yeast combined with borax, have been extensively used due to their effectiveness in attracting a broad spectrum of tephritid species (Shelly *et al.*, 2023). However, these baits require frequent replacement due to their short-lived attractiveness and susceptibility to microbial degradation (Vazquez *et al.*, 2022). Despite the advancements in synthetic lures, torula yeast-borax mixtures have been reported to capture higher numbers of certain species, such as *B. dorsalis* and *Z. cucurbitae*, compared with synthetic counterparts (Shelly *et al.*, 2022). Moreover, the integration of insecticidal agents with food-based lures in bait facilitates attract-and-kill strategies, reducing impact on non-target organisms (Lasa and Williams, 2022). The selection of lure type, formulation, and deployment strategy must be optimized based on the target fruit fly species and local agro-ecological conditions to maximize capture efficiency and support effective management practices (Enkerlin *et al.*, 2025).

18.6.2 Spatial and temporal monitoring strategies

Effective fruit fly management necessitates advanced spatial and temporal monitoring strategies to ensure timely detection and intervention which includes integration of digital and remote sensing technologies. Geographic Information Systems (GIS) and remote sensing technologies have been integrated to revolutionize the spatial tracking and predictive modelling of fruit fly outbreaks. By analysing environmental variables and land use patterns, these tools facilitate the identification of potential hotspots and the assessment of pest spread dynamics (Aziz *et al.*, 2025). Incorporation of artificial intelligence (AI) and machine learning algorithms into these systems, have enhanced their predictive capabilities in recent years (Diller *et al.*, 2023). A study demonstrated the use of computer vision and machine learning techniques, such as Random Forest and Support Vector algorithms, to detect and classify the olive fruit fly with high accuracy, indicating the potential of these technologies in real-time pest monitoring (Molina-Rotger *et al.*, 2023). Furthermore, the fusion of deep learning methods with GIS has shown promise in improving the accuracy and efficiency of spatial data analysis, which is vital for monitoring dynamic changes in pest populations (Afroosheh and Askari, 2024). Hence, the combination of molecular diagnostics and advanced digital tools offers a comprehensive approach to the spatial and temporal monitoring of fruit flies. These strategies not only enhance the precision of pest detection but also enable proactive management interventions, thereby mitigating potential agricultural losses.

18.6.3 Regulatory and quarantine measures

Quarantine protocols and risk mitigation procedures were enforced to control the spread of fruit fly species (Liquido and Griffin, 2015). Global trade compliance is regulated through standards established under the International Plant Protection Convention (IPPC) and guided by the International Standards for Phytosanitary Measures (ISPMs) (Afriansyah and Ardiastuti, 2017). Irradiation is the most important tactic used in quarantine. Grapefruit, *Citrus paradisi* Macfayden, infested by Mexican fruit fly, *Anastrepha ludens* showed effective control when irradiated with gamma rays (Hallman and Thomas, 2010). The other methods which aid in control during quarantine include modified atmosphere, fumigation, vapour-heat and cold treatments (Dias *et al.*, 2018). Thus, implementing strict quarantine protocols, adhering to international phytosanitary standards, and applying appropriate treatment methods minimizes the risk of pest invasion and damage.

18.6.4 Biological control

Biological control is a cornerstone of sustainable fruit fly management, employing natural enemies such as parasitoids, entomopathogenic fungi (EPF), and nematodes to suppress pest populations. The introduction of exotic parasitoids has been effective in reducing fruit fly populations in various regions. Notably, *Fopius arisanus* and *Diachasmimorpha longicaudata* have been successfully established in areas such as Hawaii and French Polynesia, leading to significant declines in *B. dorsalis* populations. These parasitoids exhibit high parasitism rates, with *F. arisanus* targeting the egg stage and *D. longicaudata* attacking larvae, thereby disrupting the pest's life cycle effectively (Vargas *et al.*, 2013). In addition to this, entomopathogenic fungi such as *Beauveria bassiana* and *Metarhizium anisopliae* have shown efficacy against various fruit fly species when integrated with other control methods. *Anastrepha fraterculus* (Diptera: Tephritidae) was the most important fruit pest in South America and recent studies demonstrated that *Beauveria bassiana* and *Metarhizium anisopliae* caused high larval mortality (93.3 and 96.7%, respectively) and moderate pupal mortality, with lethal concentration (LC_{50}) values of 22.56 g/L for *B. bassiana* and 23.45 g/L for *M. anisopliae*. Additionally, infected adults showed significantly reduced longevity, indicating strong potential for these fungi in fruit fly biocontrol programmes (Chaneiko *et al.*, 2019). Laboratory and field studies have demonstrated the efficacy of *Steinernema carpocapsae* against various fruit fly species. For instance, in experiments targeting *B. zonata*, *S. carpocapsae* strains achieved significant larval mortality rates, outperforming other nematode species (Sallam *et al.*, 2024). Similarly, research focusing on the Mediterranean fruit fly, *C. capitata*, revealed that *S. carpocapsae* exhibited higher pathogenicity compared with *Heterorhabditis bacteriophora*, with lower LC_{50} values, indicating greater virulence (Elqdhy *et al.*, 2024). The integration of EPNs with entomopathogenic fungi has been explored to enhance biocontrol efficacy. Combined applications of *S. carpocapsae* with fungi such as *B. bassiana* have shown synergistic effects, leading to higher mortality rates in fruit fly populations than when either agent is used alone (Wakil *et al.*, 2022). But, success of such combinations can vary based on environmental conditions and the specific strains used. *S. carpocapsae* presents a promising avenue for the biological control of fruit flies, only when integrated with other biocontrol agents such as EPF. Their combined use can offer a more robust and sustainable approach to managing fruit fly infestations in various agricultural programmes.

18.6.5 Chemical control

Fruit flies can be controlled with chemical insecticides along with bait sprays. Protein bait sprays, such as GF-120 and an insecticide such as spinosad can be combined to target adult fruit flies, especially protein-deprived females (Prokopy *et al.*, 2003). Methyl eugenol and malathion-treated wooden blocks can attract male *Bactrocera* (Reddy *et al.*, 2020). Monitoring the use of insecticide and rotation of insecticides with different mode of action is essential to avoid resistance build-up (Rajput *et al.*, 2020). Different insecticides are used against various species of fruit flies. For example, spinosad and alphacypermethrin for *C. capitata* (Burns *et al.*, 2001; Baronio *et al.*, 2023), dipterex for *B. dorsalis*, *B. zonata*, and *Carpomyia incomplete* (Khan *et al.*, 2005), decamethrin and fenvalerate for *Carpomyia vesuviana* (Vadivelu, 2014), alpha methrin for *B. tryoni* (Reynolds *et al.*, 2017), and melathion+methyleugenol for *B. dorsalis* (Sumathi *et al.*, 2019). However, use of insecticides alone can result in residue and resistance problems, but its efficiency can be increased when combined with other methods that has less impact on non-target organisms.

18.6.6 Sterile Insect Technique (SIT)

The Sterile Insect Technique (SIT) is an autocidal method of insect pest control in which sterilized males are released in large numbers to induce sterility in wild female populations through mating, thereby resulting in the gradual suppression and eradication of pest population (Bourtzis and Vreysen, 2021; Pla *et al.*, 2021). SIT is most effective when implemented as part of an area-wide integrated pest management (AW-IPM) programme, ensuring comprehensive coverage and sustained suppression of the target species (Dyck *et al.*, 2021). Combining SIT with sanitation practices, bait sprays and biological control agents such as parasitoids has shown synergistic effects. For instance, the concurrent application of SIT and the release of the parasitoid, *D. longicaudata* resulted in approximately 98% suppression of *A. ludens* populations in commercial mango orchards (Cancino *et al.*, 2023). A study was conducted in Thailand for the control of Oriental fruit fly (*B. dorsalis*) using the SIT method. The programme achieved long-term suppression of the pest population through integration of SIT with methyl eugenol traps and cultural control methods (Vreysen *et al.*, 2007). In the late 1970s, a programme was initiated in

Moscamed for the control of Mediterranean fruit fly, *C. capitata* and it became a landmark for the SIT initiative. Through the mass release of sterile males, the programme achieved eradication in parts of Mexico (Enkerlin, 2021). Although SIT has various advantages, there are certain challenges that must be overcome to obtain success. It requires costly facilities for mass rearing, sterilization, and large-scale operations, often unaffordable for smallholder-dominated regions (Parker and Mehta, 2007). Effective SIT depends on precise releasing time and methods. Ensuring sterility while maintaining male competitiveness is totally demanding (Rull *et al.*, 2012). Environmental factors such as temperature, humidity, and unpredictable weather can reduce sterile male performance and interfere with releases and trap-based monitoring. It requires coordinated, area-wide management across entire pest populations, which can be difficult to achieve (Pereira *et al.*, 2021).

18.6.7 Successful integrated pest management strategies in tropical regions

In India, a study conducted by Singh *et al.* (2013) showed that continuous use of the Male Annihilation Technique (MAT) with methyl eugenol-malathion traps in mango orchards reduced *B. dorsalis* populations by over 80% in 2007. Combining MAT with sanitation, soil drenching, and bait sprays provided up to 94.5% protection against fruit fly infestation, demonstrating the effectiveness of integrated approaches in mango pest management. A study conducted with mango farmers in Meru County, Kenya, evaluated the impact of Integrated Pest Management (IPM) strategies – including parasitoids, *M. anisopliae*-based biopesticides, orchard sanitation, food bait spot sprays, and MAT – on pesticide expenditure, yield losses, and profits. Results showed a 48% average increase in mango net income and a 19% reduction in fruit fly-related yield losses among IPM users (Muriithi *et al.*, 2016). Zida *et al.* (2023) conducted a study in Burkina Faso's South-Sudanian zone to evaluate four IPM approaches combining orchard sanitation with bait components and protein baits to control *B. dorsalis* and *Ceratitis cosyra* over two mango seasons (2018–2019). Among the treatments, sanitation combined with Timaye and GF-120 protein bait (STG) was the most effective, significantly reducing fruit fly damage. Continued research, farmer education, and adaptive management are key to enhancing the effectiveness and sustainability of IPM programmes.

18.7 Advances and Future Directions

18.7.1 Uses of genomics and microbiome research

Genomics helps in elucidating the key mechanisms occurring in an organism. The use of whole genome sequencing of selected *Bacillus thuringiensis* strains (NBAIR Bt151, NBAIR Bt152, NBAIR Bt153, NBAIR Bt154, and NBAIR BtPl) with reference strain 4Q1 against melon fruit fly, *Z. cucurbitae*, revealed that the presence of *cry70Aa1* along with other *cry* genes in NBAIR BtPl may have a significant impact on toxicity to *Z. cucurbitae* (Aarthi *et al.*, 2024). Additionally, the NBAIR BtPl strain has gene clusters that encode secondary metabolites, which in turn encode the primary bioactive compounds. For example, a gene that shares 40% of its DNA with fengycin, a biocontrol agent that can stop the spread of a variety of fungal diseases in plants (Touré *et al.*, 2004). Furthermore, bacillibactin clusters are present in NBAIR BtPl. This siderophore chelates iron and reduces its bioavailability, potentially serving as a biocontrol agent against fungi in plants (Dimopoulou *et al.*, 2021).

On the other hand, gut microbes play an important role in insect morphogenesis, nutrition, development of resistance against parasitoids, and detoxification of toxic compounds (Gujjar *et al.*, 2017; Raza *et al.*, 2020). To analyse the gender specificity of gut bacterial communities of two major fruit fly species of India, viz. *B. dorsalis* (Hendel) and *B. cucurbitae* (Coquillett) (Diptera: Tephritidae), the molecular identification has been done. It showed that the bacterial species *Enterobacter cloacae*, *Enterobacter asburiae*, and *Citrobacter freundii* were primarily present in *B. dorsalis* females, whereas *Providencia rettgerii*, *Klebsiella oxytoca*, *Enterococcus faecalis*, and *Pseudomonas aeruginosa* were found to be present in *B. dorsalis* males. *Bacillus cereus*, *B. licheniformis*, and *B. subtilis* were the aerobic endospore formers that colonized the majority of *B. cucurbitae* males, whereas *Morganella morganii* and *Bacillus pumilis* made up the majority of the culturable diversity from females (Gujjar *et al.*, 2017).

Furthermore, the different developmental stages associated with microbiota in *B. dorsalis* were characterized using 16S rRNA gene (V3–V4 region) metabarcoding on the Illumina HiSeq platform. This revealed that *Enterobacter* and *Pseudomonas* were the most prevalent bacterial genera throughout the larval and adult stages, respectively. Out of the 2645 operational taxonomic units (OTUs) that were found, 151 OTUs (core microbiota) were ubiquitous across all the phases of *B. zonata* development. *Enterobacter*, *Klebsiella*, and *Pantoea* were the most prevalent genera in the core microbiota, and PICURSt analysis predicted that the microbiota linked to *B. zonata* may have roles in signalling, cellular functions, membrane transport, reproduction and repair, and the metabolism of carbohydrates and amino acids (Naaz *et al.*, 2020). Hence, microbial symbionts can also be used as biopesticides, biosources of semiochemicals for lure and kill strategies, and vectors for para-transgenesis that can be incorporated into fruit fly control initiatives that are both precise and sustainable management strategies (Raza *et al.*, 2020).

18.7.2 Gene editing and RNAi for fruit fly control

Fruit flies have traditionally been managed through various cultural practices, pesticidal spraying, bagging, postharvest techniques, and methyl eugenol baiting techniques; however, it is difficult to tackle this serious pest, hence it is of great concern (Verghese *et al.*, 2002). Therefore, the present scenario warrants the need to go for a novel pest management such as genome editing, predominantly mediated by CRISPR/Cas9, widely used as it allows precise modifications to the DNA of organisms (Zhao *et al.*, 2019) and RNA interference (RNAi). It has wide applications in gene function research, disease modelling, drug development, and agricultural research, etc. Its versatility and effectiveness make it a significant tool for furthering scientific understanding and innovation (Bhargava *et al.*, 2024). Interestingly, it has been found that high gene-silencing effects in terms of strong RNAi phenotypes, including typical deformities and very high mortality rates at maggot, pupae, and adult developmental stages of *B. dorsalis*, can be achieved by ingestion of bacteria expressing dsRNA. This could be due to the absence of either dsRNases in the gut or the transfer of intact dsRNA through feeding of *E. coli* expressing dsRNA of two non-midgut target genes in *B. dorsalis* (Mohanpuria *et al.*, 2021)

18.7.3 Artificial intelligence and smart surveillance tools

The identification and damage symptoms of insect pests mainly rely on manual observations by experts and is a tedious and time-consuming process (Liu *et al.*, 2018). However, the advancement of new technology such as Artificial Intelligence (AI) has made it easier to detect pests in a much shorter time with higher precision (Zhu *et al.*, 2024). By using artificial intelligence, a study on fruit fly species and host preference, oviposition preference, number of pupae, weight of pupae, adult emergence, and sex ratio through an Artificial Neural Network ANN/machine learning (ML) algorithm revealed that *Z. tau* preferred pumpkin > cucumber > bitter gourd in order, *Z. cucurbitae* preferred bitter gourd > pumpkin > cucumber in order, and *Z. signata* also preferred pumpkin but followed by bitter gourd and cucumber for oviposition (Nazir *et al.*, 2024). Also, a study has been conducted on regurgitation behaviour using the YOLOv5 model, with accuracy found to be 99.8% (Zhou *et al.*, 2024). AI has been used in other insects too, for instance, a study conducted on identification of agriculturally important insects and damage symptoms of crucifers using a deep learning approach (YOLOv5-based single-stage object detection model) was found to be the best-performing model, acquiring an average accuracy, precision, recall, and F1-score of 99.5%, 92.0%, 83.0%, and 87.3%, respectively in the test images (Chakrabarthy *et al.*, 2024).

18.7.4 Climate resilient strategies for the management of fruit flies (Tephritidae)

Climate change, one of the most discussed concerns, poses serious challenges for farming on a global scale. Both agricultural output and insect pests are significantly impacted by shifts in precipitation patterns, rising CO_2 levels in the atmosphere, and rising temperatures (Ahmad *et al.*, 2025). Hence, climate-resilient strategies need to be implemented for sustainable food production.

18.7.4.1 Biopesticides and reduced-risk insecticides

A study conducted on the use of biopesticides and reduced-risk pesticides, such as spinosad at 0.002%, resulted in 14.44 and 8.89% infestation in 2016 and 2017, respectively, and was considered as most effective treatment compared with the control. It was followed by 0.01% azadirachtin, which was more effective than the recommended 0.1% malathion against *Bactrocera* spp. in mango orchards (Sharma and Gupta, 2018). Also, it has been reported that there was less impact of spinosad and neem-based products on non-target organisms (Dhillon *et al.*, 2005). In addition to this, a combination of malathion at 2 ml + jaggery 3 g/L of water was considered to be the best in controlling melon fruit fly (*Z. cucurbitae*) in bitter gourd crop (Gaddanakeri *et al.*, 2024).

18.7.4.2 Climate-smart agricultural practices

Climate-smart pest management (CSPM) encompasses a set of interdisciplinary approaches and strategies needed for primary production in adapting to a changing climatic environment. To attain global food security and safety, CSPM is regarded as a cutting-edge strategy that can assist in redefining and transforming sustainable agricultural systems. Through rigorous surveillance and forecasting systems, CSPM techniques seek to anticipate changes in pest scenarios brought on by climate change to provide early diagnosis and effective risk monitoring actions. Environmentally friendly management practices are the cornerstone of successful CSPM interventions, which aid better agricultural outputs for both people and ecosystems (Bouri *et al.*, 2023). For instance, the use of climate-smart models, such as weather-based population prediction in fruit flies (*B. dorsalis*), aided in estimation and prediction of possible variation in the population of the insect pest at least 1–2 weeks in advance in the mid-hill locations to an appreciable accuracy (Chandra *et al.*, 2022).

18.7.5 Policy needs and regional collaboration

India has implemented several policies and regulatory frameworks to manage and control fruit flies, particularly species such as *B. dorsalis*, which pose significant threats to horticultural crops. These measures encompass quarantine regulations, pest surveillance and integrated pest management (IPM), and village-level community-based pest management.

18.7.5.1 Plant quarantine (Regulation of import into India) order, 2003

This Order is issued by the Ministry of Agriculture and Farmers Welfare, outlining the phytosanitary requirements for importing plants and plant products into India. For instance, phytosanitary measures to enhance mango exports in terms of pest-free areas (PFA) have been undertaken at IIHR, Bangalore for *B. caryeae* (Kapoor) and commodity-based PFA for *Bactrocera* spp. including *B. dorsalis* (Verghese and Rashmi, 2016).

18.7.5.2 National standards for phytosanitary measures (NSPM) No. 14

This standard on 'Requirements for Establishment of Pest Free Areas for Tephritid Fruit Flies' provides necessary guidance and requirements for establishment, maintenance, and verification of pest-free areas for fruit flies, for use as a risk management option for undertaking phytosanitary certification of export of fresh fruits of mango from pest-free areas or to provide scientific justification for phytosanitary measures for protection of an endangered pest-free area. This standard aims to enable the recognition of pest-free areas in line with provisions of international agreements and thus facilitate trade.

18.7.5.3 Village-level community-based pest management

A study analysing the effectiveness of control strategies at the county and village levels found that village-level treatments, such as the Bait Application Technique (BAT) and Male Annihilation Technique (MAT), work noticeably better than isolated farm initiatives. Even though individual farmers can benefit from using pest controls, the research indicates that cooperative action at the village level has more observable impacts since treatments interact over larger areas. These observations highlight the importance of cooperative pest management strategies and address the "free rider" problem, where some farmers benefit from control measures without assisting in their implementation. Planning community-based pest management programmes and promoting participation can significantly boost agricultural output, enhancing food security in vulnerable areas (Stonehouse *et al.*, 2007).

18.8 Conclusion

The rapid increase in invasive insect pests, particularly fruit flies of the family Tephritidae, poses a mounting global challenge. Accelerated by international trade and exacerbated by climate change, these pests now threaten natural ecosystems, agricultural sustainability, and food security especially in tropical and subtropical regions. Known for their ecological plasticity, prolific reproduction, and broad host range, fruit flies such as *B. dorsalis*, *B. zonata*, and *C. capitata* have become highly invasive and difficult to control using conventional pest management tools. In tropical environments, year-round favourable conditions and continuous host availability enable these fruit flies to complete multiple generations annually, intensifying pressure on farming systems. This cycle severely impacts smallholder farmers, amplifies economic disparities, and disrupts ecological balance. The spread of invasive species further enforces strict quarantine measures and increases chemical usage, affecting export markets and environmental health. Traditional pest control methods that rely on broad-spectrum insecticides have proven inadequate, leading to pesticide resistance, ecological harm, and non-target effects. This necessitates a strategic shift toward Integrated Pest Management (IPM) frameworks that are science-driven, environmentally sound, and climate-resilient. Such approaches emphasize robust quarantine protocols, early detection, continuous surveillance, and the use of sustainable control methods including biological control agents, the sterile insect technique (SIT), male annihilation technique (MAT), and reduced-risk biopesticides. The dynamics of fruit fly invasions are influenced by a mix of biological, climatic, and anthropogenic factors. Climate change, in particular, expands suitable habitats and alters pest–host interactions, enhancing the colonization potential of genera such as *Bactrocera*, *Ceratitis*, and *Zeugodacus*. Understanding their genetic diversity and biogeography is critical for region-specific management. Southeast Asia and Africa, as biodiversity hotspots, demand international cooperation in research, surveillance, and policy enforcement. Modern tools such as molecular diagnostics, remote sensing, and AI-powered surveillance (e.g. SmartTrap systems) are transforming fruit fly detection and response strategies. When combined with traditional control practices, they enhance the precision and sustainability of pest suppression. Successful area-wide programmes in countries like Mexico, Thailand, and Guatemala demonstrate the effectiveness of multi-tactic, integrated approaches in reducing populations, restoring market access, and minimizing chemical dependence. Challenges remain, including high implementation costs, logistical complexity, and the need for coordinated action across farming communities. In India and other tropical nations, the strengthening of quarantine laws, farmer education, and regional collaboration have begun to show promise in mitigating pest impacts. In conclusion, addressing the rising threat of invasive fruit flies requires a multidisciplinary, globally coordinated response. By integrating scientific innovation, smart technologies, policy frameworks, and community engagement, we can develop resilient agricultural systems capable of withstanding biological invasions. This proactive approach is essential to preserve tropical agroecosystems, protect biodiversity, and ensure long-term food security in an increasingly interconnected and climate-impacted world.

Acknowledgements

We thank Dr V.V. Ramamurthy, President, ESI for critical reading and suggestions for improvement of the manuscript. Dr Shashank acknowledges support from the project titled "Development of pest and disease management strategies for major crops (CRSCIARISIL20140333265)".

References

Aarthi, N., Dubey, V.K., Shylesha, A.N., Kukreti, A., Patil, J. *et al.* (2024) Insights into the whole genome sequence of *Bacillus thuringiensis* NBAIR BtPI, a strain toxic to the melon fruit fly, *Zeugodacus cucurbitae*. *Current Genetics* 70(1), 13.

Afriansyah, A. and Ardiastuti, E. (2017) Implementation of the International Plant Protection Convention (IPPC) in the regulation of the agreement on the application of sanitary and phytosanitary measures (SPS agreement) in the World Trade Organization (WTO): Notes to Indonesia. *Yustisia* 6(1), 51–80.

Afroosheh, S. and Askari, M. (2024) Fusion of deep learning and GIS for advanced remote sensing image analysis. *arXiv Preprint* 4(2), 19–856.

Ahmad, M.A., Mahanta, D.K., Sattar, A. and Jha, R.K. (2025) Integrated pest management: Need for climate-resilient technologies. In: *Sustainable Production and Food Security: An Overview through Climate Smart Agricultural Interventions*. pp. 61–109.

Al Baki, M.A., Hossain, M.A., Khan, M.A.M. and Islam, M.S. (2017) Sexual maturation, mating behavior and daily mating rhythm of the oriental fruit fly *Bactrocera dorsalis* (Hendel) (Diptera: Tephritidae). *Journal of Asia-Pacific Entomology*, 20(4), 1237–1242.

Aketarawong, N., Isasawin, S. and Thanaphum, S. (2007) Evidence of cryptic species in *Bactrocera dorsalis* (Hendel) inferred from mitochondrial DNA sequences. *Molecular Phylogenetics and Evolution* 45(2), 450–460.

Aketarawong, N., Isasawin, S. and Thanaphum, S. (2014) Population genetic structure of *Bactrocera dorsalis* s.l. populations from the Greater Mekong Subregion inferred from microsatellite DNA. *BMC Genetics* 15, 71.

Anderson, D.T. (1963) The larval development of *Dacus tryoni* (Frogg) (Diptera: Trypetidae) 1. larval instars, Imaginal discs, and haemocytes. *Australian Journal of Zoology* 11(2), 202–218.

Andongma, A.A., Wan, L., Dong, Y.C., Wang, Y.L., He, J. *et al.* (2019) Assessment of the bacteria community structure across life stages of the Chinese citrus fly, *Bactrocera minax* (Diptera: Tephritidae). *BMC Microbiology* 19, 1–9.

Arias, M., Hernández-Ortiz, V. and Frías, D. (2014) Distribution, host plants and economic importance of *Anastrepha grandis* (Diptera: Tephritidae) in South America. *Florida Entomologist* 97(4), 1810–1817.

Aziz, D., Rafiq, S., Saini, P., Ahad, I., Gonal, B. *et al.* (2025) Remote sensing and artificial intelligence: Revolutionizing pest management in agriculture. *Frontiers in Sustainable Food Systems* 9. DOI: 10.3389/fsufs.2025.1551460.

Baronio, C.A., Machota, R., Nunes, M.Z., Paranhos, B.A., Bernardi, D. *et al.* (2023) The use of toxic baits for the suppression of mediterranean fruit fly in mango orchards. *Anais Da Academia Brasileira de Ciencias* 95(4), e20191046.

Bateman, M.A. (1972) The ecology of fruit flies. *Annual Review of Entomology* 17(1), 493–518.

Bhargava, C.N., Ashok, K., Asokan, R., Prasad Babu, K., Parvathy, M.S. *et al.* (2024) CRISPR/Cas9 Mediated Editing of the white (wh) locus affects body size and reproduction of the Oriental fruit fly, *Bactrocera dorsalis* (Hendel). *Agricultural Research* 1–8.

Biondi, A., Guedes, R., Wan, F. and Desneux, N. (2018) Ecology, worldwide spread, and management of invasive south tomato pinword, *Tuta absoluta*: Past, present, future. *Annual Review of Entomology* 63, 239–258.

Blackburn, T.M., Pyšek, P., Bacher, S., Carlton, J.T., Duncan, R.P. *et al.* (2011) A proposed unified framework for biological invasions. *Trends in Ecology & Evolution* 26(7), 333–339.

Bouri, M., Arslan, K.S. and Şahin, F. (2023) Climate-smart pest management in sustainable agriculture: Promises and challenges. *Sustainability* 15(5), 45–92.

Bourtzis, K. and Vreysen, M.J. (2021) Sterile insect technique (SIT) and its applications. *Insects* 12(7), 638.

Broennimann, O., Treier, U.A., Müller-Schärer, H., Thuiller, W., Peterson, A.T. *et al*. (2007) Evidence of climatic niche shift during biological invasion. *Ecology Letters* 10(8), 701–709.

Burns, R.E., Harris, D.L., Moreno, D.S. and Eger, J.E. (2001) Efficacy of spinosad bait sprays to control Mediterranean and Caribbean fruit flies (Diptera: Tephritidae) in commercial citrus in Florida. *Florida Entomologist* 84, 672–678.

Cancino, J., Montoya, P., Gálvez, F.O., Gálvez, C. and Liedo, P. (2023) Effect of the sterile insect technique and augmentative parasitoid releases in a fruit fly suppression program in mango-producing areas of southeast Mexico. *Insects* 14(9), 719.

Carey, J.R., Yang, P. and Foote, D. (1988) Demographic analysis of insect reproductive levels, patterns and heterogeneity: Case study of laboratory strains of three Hawaiian tephritids. *Entomologia Experimentalis et Applicata* 46(1), 85–91.

Chakrabarthy, S., Ramesh, K., Singh, V., Sharma, S. and Kumar, R. (2024) Deep learning–based identification of agriculturally important insects and damage symptoms of crucifer crops using a YOLOv5 object detection model. *Computers and Electronics in Agriculture*, 216, 108–494.

Chandra, A., Bhagawati, K., Kalita, H. and Angami, T. (2022) Weather-based fruit fly population dynamics prediction model for the mid-hills of Eastern Himalayan region of India. *Current World Environment* 17(3), 690.

Chaneiko, S.M., Brida, A.L., Bassa, P.G., Telles, M.H., Santos, L.A. *et al*. (2019) Pathogenicity of *Beauveria bassiana* and *Metarhizium anisopliae* to *Anastrepha fraterculus* (Diptera: Tephritidae) and effects on adult longevity. *Journal of Agricultural Science* 11(16), 132.

Chen, P., Ye, H. and Mu, Q.A. (2007) Migration and dispersal of the oriental fruit fly, *Bactrocera dorsalis* in regions of Nujiang River based on fluorescence mark. *Acta Ecologica Sinica* 27(6), 2468–2476.

Chen, P., Li, Z., Xu, L. and Zeng, L. (2011) Invasion and spread of *Bactrocera dorsalis* (Diptera: Tephritidae) in mainland China. *Journal of Economic Entomology* 104(6), 1992–1998.

Christenson, L.D. and Foote, R.H. (1960) Biology of fruit flies. *Annual Review of Entomology* 5(1), 171–192.

Clarke, A.R. and Measham, P.F. (2022) Competition: A Missing component of fruit fly (Diptera: Tephritidae) risk assessment and planning. *Insects* 13(11), 1065.

Clarke, A.R., Armstrong, K.F., Carmichael, A.E., Milne, J.R., Raghu, S. *et al*. (2005) Invasive phytophagous pests arising through a recent tropical evolutionary radiation: The *Bactrocera dorsalis* complex of fruit flies. *Annual Review of Entomology* 50, 293–319.

Clarke, A.R., Allwood, A.J., Chinajariyawong, A., Drew, R.A.I., Hengsawad, C. *et al*. (2001) Seasonal abundance and host use patterns of seven *Bactrocera* Macquart species (Diptera: Tephritidae) in an orchard in tropical Queensland. *Australian Journal of Entomology* 40(1), 1–6.

Cook, D.C., Fraser, R.W. and Mumford, J.D. (2007) Managing biosecurity risk: System integrity and the effective management of risk. *Risk Analysis* 27(6), 1495–1507.

Cornelissen, T., Wilson Fernandes, G., Vasconcellos-Neto, J. and Carneiro, M.A.A. (2019) Climate change and insect herbivory. In: *Insect-Plant Interactions in a Crop Protection Perspective*. Springer, pp. 129–149.

Courtney, G.W., Pape, T., Skevington, J.H. and Sinclair, B.J. (2009) Biodiversity of diptera. In: Foottit, R.G. and Adler, P.H. (eds) *Science and Society. Insect Biodiversity*. Wiley-Blackwell, pp. 185–222.

Cugala, D., Massimiliano, V., Maulid, M., De Meyer, M. and Canhanga, L. (2020) Economic injury level of the Oriental fruit fly, *Bactrocera dorsalis* (Diptera: Tephritidae), on commercial mango farms in Manica Province, Mozambique. *African Entomology* 28(2), 278–289.

David, P., Thébault, E., Anneville, O., Duyck, P.F., Chapuis, E. *et al*. (2017) Impacts of invasive species on food webs: A review of empirical data. *Advances in Ecological Research* 56, 1–60.

de Deus, E., Passos, J., Sauers-Muller, A., Jesus, C., Silva, J.G. *et al*. (2024) Phylogeography of the invasive fruit fly species *Bactrocera carambolae* Drew & Hancock (Diptera: Tephritidae) in South America. *Insects* 15(12), 949.

Deng, J.Q., Zhu, X.M., Han, P., Yang, G.P. and Li, Z.C. (2021) Research progress of the melon fly, *Bactrocera cucurbitae* (Diptera: Tephritidae) in China. *Plant Quarantine* 35, 1–7.

Dhillon, M.K., Singh, R., Naresh, J.S. and Sharma, H.C. (2005) The melon fruit fly, *Bactrocera cucurbitae*: A review of its biology and management. *Journal of Insect Science* 5(40), 1–16.

Dias, N.P., Zotti, M.J., Montoya, P., Carvalho, I.R. and Nava, D.E. (2018) Fruit fly management research: A systematic review of monitoring and control tactics in the world. *Crop Protection* 112, 187–200.

Didham, R.K., Tylianakis, J.M., Hutchison, M.A., Ewers, R.M. and Gemmell, N.J. (2007) Are invasive species the drivers of ecological change. *Trends in Ecology & Evolution* 22(9), 470–476.

Diller, E., Meyer, M., Steinbauer, M. J. and Kriticos, D. J. (2023). Artificial intelligence and machine learning improve surveillance and prediction of invasive insect pests. *Biological Invasions*, 25, 2851–2866.

Dimopoulou, A., Theologidis, I., Benaki, D., Koukounia, M., Zervakou, A. *et al*. (2021) Direct antibiotic activity of bacillibactin broadens the biocontrol range of *Bacillus amyloliquefaciens* MBI600. *Msphere* 6(4).

Drew, R.A.I. (2004) Biogeography and speciation in the Dacini (Diptera: Tephritidae: Dacinae). *The Raffles Bulletin of Zoology* Supplement 12, 19–27.

Drew, R.A.I. and Hancock, D.L. (1994) The *Bactrocera dorsalis* complex of fruit flies (Diptera: Tephritidae: Dacinae) in Asia. *Bulletin of Entomological Research* Supplement 2, 1–68.

Drew, R.A.I., Tsuruta, K. and White, I.M. (2005) A new species of pest fruit fly (Diptera: Tephritidae: Dacinae) from Sri Lanka and Africa. *African Entomology* 13(1), 149–154.

Duyck, P.F., David, P. and Quilici, S. (2004) A review of relationships between interspecific competition and invasions in fruit flies (Diptera: Tephritidae). *Ecological Entomology* 29(5), 511–520.

Dyck, V.A., Hendrichs, J. and Robinson, A.S. (eds) (2021) *Sterile Insect Technique: Principles and Practice in Area-Wide Integrated Pest Management*. Taylor & Francis, pp. 12–16.

Early, R., Bradley, B.A., Dukes, J.S., Lawler, J.J., Olden, J.D. *et al*. (2016) Global threats from invasive alien species in the twenty-first century and national response capacities. *Nature Communications* 7, 12485.

EFSA Panel on Plant Health (PLH), Bragard, C., Dehnen-Schmutz, K., Di Serio, F., Gonthier, P., Jacques, M.A. *et al*. (2020) Pest categorisation of non-EU tephritidae. *EFSA Journal* 18(1).

Ekesi, S. and Billah, M.K. (2007) *A Field Guide to the Management of Economically Important Tephritid Fruit Flies in Africa*. ICIPE Science Press, Nairobi, pp. 90–97.

Ekesi, S., Nderitu, P.W. and Rwomushana, I. (2006) Field infestation, life history and demographic parameters of the fruit fly *Bactrocera invadens* (Diptera: Tephritidae) in Africa. *Bulletin of Entomological Research* 96(4), 379–386.

Ekesi, S., Meyer, M., Mohamed, S.A., Virgilio, M. and Borgemeister, C. (2016) Taxonomy, ecology, and management of native and exotic fruit fly species in Africa. *Annual Review of Entomology* 61, 219–238.

Elqdhy, M.B., Ait Hamza, M., Askarne, L., Fossati-Gaschignard, O., Lakhtar, H. *et al*. (2024) Biology, ecology and control of the Mediterranean fruit fly, *Ceratitis capitata* (Diptera: Tephritidae), with special reference to biological control using entomopathogenic nematode (EPN): A review. *Journal of Plant Diseases and Protection* 131(2), 365–402.

Enkerlin, W., Arevalo, E., Caballero, J.E., Fezza, T., Garavelli, E. *et al*. (2025) Searching for more effective food baits for tephritid fruit flies (Diptera: Tephritidae): Performance of newly developed vial-lures relative to torula yeast borax. *Insects* 16(1), 53.

Enkerlin, W.R. (2021) Impact of fruit fly control programmes using the sterile insect technique. In: *Sterile Insect Technique*. CRC Press, pp. 979–1006.

Essl, F., Dullinger, S., Genovesi, P., Hulme, P.E. and Jeschke, J.M. (2019) A conceptual framework for range expanding species that track human-induced environmental change. *Bioscience* 69, 908–919.

European Food Safety Authority (EFSA), Tramontini, S., Gilioli, G., de la Peña, E., Rzepecka, D., Scala, M. *et al*. (2024) *Bactrocera dorsalis* pest report to support the ranking of EU candidate priority pests. *EFSA Supporting Publications* 22(3), 93–126.

Fay, H.A.C. and Meats, A. (1983) The influence of age, ambient temperature, thermal history and mating history on mating frequency in males of the queensland fruit fly, *Dacus tryoni*. *Entomologia Experimentalis et Applicata* 34(3), 273–276.

Fitt, G.P. (1981a) The ecology of Northern Australian Dacinae (Diptera: Tephritidae) I. Host phenology and utilization of *Opilia amentacea* Roxb (Opiliaceae) by *Dacus* (Bactrocera) *opiliae* Drew & Hardy, with notes on some other species. *Australian Journal of Zoology* 29(5), 691–705.

Fitt, G.P. (1981b) Ecology of Northern Australian Dacinae (Diptera: Tephritidae). Seasonal fluctuations in trap catches of *Dacus opiliae* and *D. tenuifascia*, and their relationship to host phenology and climatic factors. *Australian Journal of Zoology* 29(6), 885–894.

Fitt, G.P. (1983) Factors limiting the host range of tephritid fruit flies: With particular emphasis on the influence of *Dacus tryoni* on the distribution and abundance of *Dacus jarvisi*. Doctoral dissertation.

Gaddanakeri, S., Rolania, K., Maradi, R.M. and Duhan, D.S. (2024) Effect of various treatments against melon fruit fly, *Bactrocera cucurbitae* (Tephritidae: Diptera) in bitter gourd (*Momordica charantia* L.). *International Journal of Advanced Biochemistry Research* 8(8), 781–784.

Geurts, K., Mwatawala, M.W. and Meyer, M. (2012) Indigenous and invasive fruit fly diversity along an altitudinal transect in eastern Central Tanzania. *Journal of Insect Science* 12(1), 12.

Grimm, N.B., Faeth, S.H., Golubiewski, N.E., Redman, C.L., Wu, J. *et al.* (2008) Global change and the ecology of cities. *Science* 319(5864), 756–760.

Gujjar, N.R., Govindan, S., Verghese, A., Subramaniam, S. and More, R. (2017) Diversity of the cultivable gut bacterial communities associated with the fruit flies *Bactrocera dorsalis* and *Bactrocera cucurbitae* (Diptera: Tephritidae). *Phytoparasitica* 45, 453–460.

Guo, S., Ge, X., Zou, Y., Zhou, Y., Wang, T. *et al.* (2019) Projecting the potential global distribution of *Carpomya vesuviana* (Diptera: Tephritidae), considering climate change and irrigation patterns. *Forests* 10(4), 355.

Hallman, G.J. and Thomas, D.B. (2010) Ionizing radiation as a phytosanitary treatment against fruit flies (Diptera: Tephritidae): Efficacy in naturally versus artificially infested fruit. *Journal of Economic Entomology* 103(4), 1129–1134.

Hancock, D.L. (1985) New species and records of African Dacinae (Diptera: Tephritidae). *Arnoldia* 9(22), 299–314.

Hardy, D.E. (1973) The fruit flies of the subtribe Dacina (trypetidae, tephritidae). *Pacific Insects Monograph* 31, 1–370.

He WanZhong, H.W., Sun BingZhao, S.B., Li CuiJu, L.C. and Long ZhongBao, L.Z. (2002) Bionomics of *Bactrocera dorsalis* and its control in Hekou County of Yunnan Province 6(3), 26–67.

He, X., Wang, Q., Liu, W. and Hu, F. (2011). Effects of adult diet on sexual maturation and mating behavior of the oriental fruit fly *Bactrocera dorsalis* (Hendel) (Diptera: Tephritidae). *Journal of Insect Physiology*, 57(8), 1113–1118.

He, Y., Xu, Y. and Chen, X. (2023). Biology, ecology and management of tephritid fruit flies in China: A Review. *Insects*, 14(2), 196. DOI: 10.3390/insects14020196.

Hoskins, J.L., Hester, S.M., Taylor, S.C. and Cook, D.C. (2023a) The need for updated surveillance strategies for exotic fruit flies in Australia. *Austral Entomology* 62(2), 276–289.

Hoskins, J.L., Rempoulakis, P., Stevens, M.M. and Dominiak, B.C. (2023b) Biosecurity and management strategies for economically important exotic tephritid fruit fly species in Australia. *Insects* 14(10), 801.

Headrick, D.H. and Goeden, R.D. (1996) Problems with generalists: Historical perspective on biological control of the Mediterranean fruit fly (*Ceratitis capitata*). *Biological Control* 6(1), 20–27.

Hulme, P.E. (2006) Beyond control: Wider implications for the management of biological invasions. *Journal of Applied Ecology* 43(5), 835–847.

Hulme, P.E. (2009) Trade, transport and trouble: Managing invasive species pathways in an era of globalization. *Journal of Applied Ecology* 46(1), 10–18.

Hulme, P.E. (2017) Climate change and biological invasions: Evidence, expectations, and response options. *Biological Reviews* 92(3), 1297–1313.

IAEA (International Atomic Energy Agency) (2003) Trapping guidelines for area-wide fruit fly programmes. *International Atomic Energy Agency, IAEAS-TECDOC* 47–1336.

IAEA/FAO (International Atomic Energy Agency and Food and Agriculture Organization) (2013) Trapping manual for area-wide fruit fly programmes. *International Atomic Energy Agency* 128.

Jaffar, S., Rizvi, S.A.H. and Lu, Y. (2023) Understanding the invasion, ecological adaptations, and management strategies of *Bactrocera dorsalis* in China: A review. *Horticulturae* 9(9), 1004.

Jaleel, W., Saeed, S., Naqqash, M.N. and Zaka, S.M. (2018) Distribution and host range of *Zeugodacus tau* (Walker) in Asia. *Phytoparasitica* 46(4), 555–567.

Karsten, M., Vuuren, B., Addison, P. and Terblanche, J.S. (2015) Deconstructing intercontinental invasion pathway hypotheses of the Mediterranean fruit fly (*Ceratitis capitata*) using a Bayesian inference approach: Are port interceptions and quarantine protocols successfully preventing new invasions?. *Diversity and Distributions* 21(7), 813–825.

Karsten, M., Barton, M., Addison, P., Addison, M.F. and Terblanche, J.S. (2018) A synthesis for managing invasions and pest risks simultaneously for tephritid fruit flies in South Africa. *Entomologia Experimentalis et Applicata* 166(5), 344–356.

Khan, M.A., Ashfaq, M., Akram, W. and Lee, J.J. (2005) Management of fruit flies (Diptera: Tephritidae) of the most perishable fruits. *Entomological Research* 35(2), 79–84.

Kiritani, K. and Yamamura, K. (2003) Exotic insects and their pathways for invasion. In: G.M., R. and J.T, C. (eds) *Invasive Species: Vectors and Management Strategies*. Washington, DC: Island Press, pp. 44–67.

Kriticos, D.J., Ota, N., Hutchison, W.D., Beddow, J., Walsh, T. *et al.* (2013) The potential distribution of *Bactrocera dorsalis* (Diptera: Tephritidae): Evaluating against climate, host plants and surveillance data. *Bulletin of Entomological Research* 103(1), 1–13.

Lasa, R. and Williams, T. (2022) Alkaline hydrolyzed torula yeast as an attractant for *Anastrepha obliqua* (Diptera: Tephritidae). *Journal of Applied Entomology* 146(6), 676–686.

Lengai, G.M.W., Ndzi, D.L. and Ekesi, S. (2022). Economic and livelihood impacts of fruit fly (*Bactrocera dorsalis*) invasions on smallholder horticulture in sub-Saharan Africa. *CABI Agriculture and Food Security*, 11, 28.

Li, Z., Jiang, F., Ma, X., Fang, Y., Sun, Z. *et al.* (2013) Review on prevention and control techniques of Tephritidae invasion. *Plant Quarantine* 27(2), 1–10.

Li, Z., Pan, H., Lu, X. and Liu, Y. (2020) Distribution and host plants of *Zeugodacus tau* in southern China. *Plant Protection* 46(1), 42–46.

Li, Z.H., Jiang, X.F., Zeng, L. and Zhang, H.Y. (2012) Risk analysis of the potential invasion of *Bactrocera dorsalis* in China. *Journal of Integrative Agriculture* 11(3), 451–458.

Liang, G., Zheng, Y. and Chen, B. (2003) Occurrence and spread of *Bactrocera dorsalis* in China. *Entomotaxonomia* 25(4), 287–290.

Liebhold, A.M., Work, T.T., McCullough, D.G. and Cavey, J.F. (2006) Airline baggage as a pathway for alien insect species invading the United States. *American Entomologist* 52(1), 48–54.

Liquido, N.J. and Griffin, R.L. (2015) Quarantine mitigation for fruit flies in citrus. *Acta Horticultare* 1065, 1063–1070.

Liquido, N.J., Shinoda, L.A. and Cunningham, R.T. (1991) Host plants of the Mediterranean fruit fly (*Ceratitis capitata*) on the island of Hawaii (1949–1985 survey). *Journal of Economic Entomology* 84(6), 1863–1878.

Liu, Z.Y., Qi, J.G., Wang, N.N., Zhu, Z.R. and Luo, J. (2018) Hyperspectral discrimination of foliar biotic damages in rice using principal component analysis and probabilistic neural network. *Precision Agriculture* 19, 973–991.

Lux, S.A., Copeland, R.S., White, I.M., Manrakhan, A. and Billah, M.K. (2003) A new invasive fruit fly species from the *Bactrocera dorsalis* (Hendel) group detected in East Africa. *Insect Science and Its Application* 23(4), 355–361.

Ma, W., Liu, X., Zhang, Q., Zhao, Z. and Chen, H. (2013). Effects of adult nutrition on mating behavior and reproductive fitness of *Bactrocera dorsalis* (Diptera: Tephritidae). *Journal of Insect Physiology*, 59(10), 1004–1011.

Magagula, C.N., Manrakhan, A. and Addison, P. (2015) Establishment and spread of *Bactrocera dorsalis* in Swaziland. *Journal of Applied Entomology* 139(9), 749–757.

Malacrida, A.R., Gomulski, L.M., Bonizzoni, M., Bertin, S., Gasperi, G. *et al.* (2007) Globalization and fruit fly invasion and expansion: The medfly paradigm. *Genetica* 131(1), 1–9.

Manrakhan, A., Hattingh, V., Venter, N. and Holtzhausen, M. (2015) Status of *Bactrocera dorsalis* in South Africa. *Insects* 6(1), 171–179.

McQuate, G.T., Liquido, N.J. and Nakamichi, K.A. (2017) Annotated world bibliography of host plants of the melon fly, *Bactrocera cucurbitae* (Coquillett) (Diptera: Tephritidae). *Insecta Mundi*.

Meyerson, L.A. and Reaser, J.K. (2002) Biosecurity: Moving toward a comprehensive approach. *BioScience*, 52(7), 593–600.

Mohanpuria, P., Govindaswamy, M., Sidhu, G.S., Singh, S., Kaur, S. *et al.* (2021) Ingestion of bacteria expressing dsRNA to maggots produces severe mortality and deformities in fruit fly, *Bactrocera dorsalis* (Hendel) (Diptera: Tephritidae). *Egyptian Journal of Biological Pest Control* 31, 1–11.

Molina-Rotger, M., Moran, A., Miranda, M.A. and Alorda-Ladaria, B. (2023) Remote fruit fly detection using computer vision and machine learning-based electronic trap. *Frontiers in Plant Science* 14, 1241576.

Mumford, J.D. (2002) Economic issues related to quarantine in international trade. *European Review of Agricultural Economics* 29(3), 329–348.

Muriithi, B.W., Affognon, H.D., Diiro, G.M., Kingori, S.W. and Tanga, C.M. (2016) Impact assessment of Integrated Pest Management (IPM) strategy for suppression of mango-infesting fruit flies in Kenya. *Crop Protection* 81, 20–29.

Mwatawala, M.W., Meyer, M., Makundi, R.H. and Maerere, A.P. (2006) Seasonality and host utilization of *Bactrocera invadens* (Diptera: Tephritidae) in central Tanzania. *Journal of Applied Entomology* 130(9–10), 530–537.

Mwatawala, M.W., Meyer, M., Makundi, R.H. and Maerere, A.P. (2009) Host range and distribution of fruit-infesting pestiferous fruit flies (Diptera: Tephritidae) in selected areas of Central Tanzania. *Bulletin of Entomological Research* 99(6), 629–641.

Naaz, N., Choudhary, J.S., Choudhary, A., Dutta, A. and Das, B. (2020) Developmental stage-associated microbiota profile of the peach fruit fly, *Bactrocera zonata* (Diptera: Tephritidae) and their functional prediction using 16S rRNA gene metabarcoding sequencing. *Biotech* 10(9), 390.

Nair, S., Rao, N.S. and Josephrajkumar, A. (2017) Distribution and host range of *Zeugodacus tau* in India. *Indian Journal of Entomology* 79(3), 274–277.

Nazir, N., Fatima, S., Aasim, M., Yaqoob, F., Mahmood, K. *et al.* (2024) Zeugodacus fruit flies (Diptera: Tephritidae) host preference analysis by machine learning-based approaches. *Computers and Electronics in Agriculture* 222, 109095.

Norrbom, A.L., Carroll, L.E., Thompson, F.C., White, I.M. and Freidberg, A. (1999) Systematic database of names. *Fruit Fly Expert Identification System and Systematic Information Database* Myia 9, 65–251.

Opoku, E., Haseeb, M., Rodriguez, E.J., Steck, G.J. and Cabral, M.J. (2025) Economically important fruit flies (Diptera: Tephritidae) in Ghana and their regulatory pest management. *Insects* 16(3), 285.

Papadopoulos, N.T. (2014) Fruit fly invasion: Historical, biological, economic aspects and management. In: T., S. (ed.) *Trapping and the Detection, Control and Regulation of Tephritid Fruit Flies*. Dordrecht: Springer, pp. 219–252.

Papadopoulos, N.T., Meyer, M., Terblanche, J.S. and Kriticos, D.J. (2024) Fruit flies: Challenges and opportunities to stem the tide of global invasions. *Annual Review of Entomology* 69(1), 355–373.

Parker, A. and Mehta, K. (2007) Sterile insect technique: A model for dose optimization for improved sterile insect quality. *Florida Entomologist* 90(1), 88–95.

Pemberton, C.E. (1946) The oriental fruit fly (*Dacus dorsalis* Hendel) in Hawaii. *Proceedings of the Hawaiian Entomological Society* 12(1), 55–63.

Pereira, R., Yuval, B., Liedo, P., Teal, P.E.A., Shelly, T.E. *et al.* (2021) Improving post-factory performance of sterile male fruit flies in support of the sterile insect technique. In: *Principles and Practice in Area-Wide Integrated Pest Management*. pp. 631–656.

Phillips, C. (2013) Living without fruit flies: Biosecuring horticulture and its markets. *Environment and Planning A* 45(7), 1679–1694.

Pla, I., Caceres, C., Rendón, P. and Vreysen, M.J.B. (2021) The sterile insect technique: Principles and practice in area-wide integrated pest management. *Insects*, 12(9), 783.

Prabhakar, C.S. (2011) Diversity and host range of fruit flies (Diptera: Tephritidae) in north India. *Indian Journal of Entomology* 73(4), 343–348.

Prabhakar, C.S. (2012) Urban environments and the persistence of fruit flies: A case study. *Entomologia Experimentalis et Applicata* 145(2), 232–241.

Prokopy, R.J., Miller, N.W., Piñero, J.C., Barry, J.D., Tran, L.C. *et al.* (2003) Effectiveness of GF-120 fruit fly bait spray applied to border area plants for control of melon flies (Diptera: Tephritidae). *Journal of Economic Entomology* 96(5), 1485–1493.

Qin, Y., Paini, D.R., Wang, C., Fang, Y. and Li, Z. (2015) Global establishment risk of economically important fruit fly species (Tephritidae). *PLoS ONE* 10(1), 18.

Qin, Y., Wang, C., Zhao, Z., Pan, X. and Li, Z. (2019) Climate change impacts on the global potential geographical distribution of the agricultural invasive pest, *Bactrocera dorsalis* (Hendel) (Diptera: Tephritidae). *Climatic Change* 155, 145–156.

Quilici, S. and Donner, P. (2012) Analysis of exotic fruit fly trapping networks. *EPPO Bulletin* 42(1), 102–108.

Rajput, V.S., Jhala, J. and Acharya, V.S. (2020) Biopesticides and their mode of action against insect pests: A review. *International Journal of Chemical Studies* 8, 2856–2862.

Raza, M.F., Yao, Z., Bai, S., Cai, Z. and Zhang, H. (2020) Tephritidae fruit fly gut microbiome diversity, function and potential for applications. *Bulletin of Entomological Research* 110(4), 423–437.

Reddy, K.V., Devi, Y.K. and Komala, G. (2020) Management strategies for fruit flies in fruitcrops – a review. *Journal of Emerging Technologies and Innovative Research* 7(12), 1472–1480.

Reynolds, O.L., Osborne, T.J. and Barchia, I. (2017) Efficacy of chemicals for the potential management of the Queensland fruit fly *Bactrocera tryoni* (Froggatt) (Diptera: Tephritidae). *Insects* 8(2), 49.

Roy, H.E., Pauchard, A., Stoett, P.J., Renard Truong, T., Meyerson, L.A. *et al.* (2024) Curbing the major and growing threats from invasive alien species is urgent and achievable. *Nature Ecology & Eolution* 8(7), 1216–1223.

Rull, J., Encarnación, N. and Birke, A. (2012) Mass rearing history and irradiation affect mating performance of the male fruit fly, *Anastrepha obliqua*. *Journal of Insect Science* 12(1), 45.

Russell, J., Meyer, J., Holmes, N. and Pagad, S. (2017) Invasive alien species on islands: Impacts, distribution, interactions and management. *Environmental Conservation* 44(4), 359–370.

Sallam, R.F., Shalaby, F.F., Hafez, A.A., Abdel-Hameid, N.F. and Shairra, S.A. (2024) Efficacy of different entomopathogenic nematode isolates, against the peach fruit fly, *Bactrocera zonata* (Saund.) (Diptera: Tephritidae). *Egyptian Journal of Biological Pest Control* 34(1), 12.

Santana, P.A., Kumar, L., Silva, R.S. and Picanço, M.C. (2019) Global geographic distribution of *Tuta absoluta* as affected by climate change. *Journal of Pest Science* 92(4), 1373–1385.

Savaris, M., Marinoni, L. and Norrbom, A.L. (2016) Fruit flies (Diptera, Tephritidae) from Brazil. *Zootaxa* 4072(1), 1–264.

Serrão, J.E., Plata-Rueda, A., Martínez, L.C. and Zanuncio, J.C. (2022) Side-effects of pesticides on non-target insects in agriculture: A mini-review. *The Science of Nature* 109(2), 17.

Sharma, I. and Gupta, D. (2018) Management of fruit flies (*Bactrocera* spp.) in mango using biopesticides and clay. *Indian Journal of Entomology* 80(4), 1478–1481.

Sharma, K., Sharma, R.K., Chander, S. and Jilu, V. (2015) Effects of weather parameters on Guava fruit fly (*Bactrocera zonata*) population at IARI, New Delhi. *Journal of Agrometeorology* 17(2), 227–229.

Shelly, T., Fezza, T. and Kurashima, R. (2022) Captures of Oriental fruit flies and melon flies (Diptera: Tephritidae) in traps baited with torula yeast borax solution or 2- or 3-component synthetic food cones in Hawaii. *Florida Entomologist* 105(1), 37–43.

Shelly, T.E., Kurashima, R., Mesa Martin, R., Fezza, T. and Bazelet, C. (2023) Weathering of 3-component synthetic food cones: Effects on residual amount, release rate, and field capture of 3 pest species of fruit flies (Diptera: Tephritidae). *Environmental Entomology* 52(3), 408–415.

Shi, W., Kerdelhué, C. and Ye, H. (2010) Population genetic structure of the Oriental fruit fly, *Bactrocera dorsalis* (Hendel) (Diptera: Tephritidae) from Yunnan province (China) and nearby sites across the border. *Genetica* 138(3), 377–385.

Siebert, J. and Cooper, T. (1995) If medfly infestation triggered a trade ban: Embargo on California produce would cause revenue, job loss. *California Agriculture* 49(4), 7–12.

Silva, J.G. and Malavasi, A. (1993) The South American cucurbit fruit fly, *Anastrepha grandis* (Diptera: Tephritidae): An important pest of cucurbits. *Florida Entomologist* 76(1), 169–174.

Singh, M., Gupta, D. and Gupta, P.R. (2013) Population suppression of fruit flies (*Bactrocera* spp.) in mango (*Mangifera indica*) orchards. *Indian Journal of Agricultural Sciences* 83(10), 1064–1068.

Stokstad, E. (2018). Study reports increases in invasive species and broader ecosystem impacts. *Science*, 360(6386), 724–725.

Stonehouse, J.M., Mumford, J.D., Verghese, A., Shukla, R.P., Satpathy, S. *et al.* (2007) Village-level area-wide fruit fly suppression in India: Bait application and male annihilation at village level and farm level. *Crop Protection* 26(5), 788–793.

Suckling, D.M., Kean, J.M., Stringer, L.D., Cáceres-Barrios, C., Hendrichs, J. *et al.* (2016) Eradication of tephritid fruit fly pest populations: Outcomes and prospects. *Pest Management Science* 72(3), 456–465.

Sumathi, E., Manimaran, R., Nirmala, D., Ilamaran, M. and Agila, R. (2019) Population dynamics and management of mango fruit fly *Bactrocera dorsalis* (Hendel) (Diptera: Tephritidae). *International Journal of Current Microbiology and Applied Sciences* 8(1), 2705–2710.

Tortorici, S., Zeni, V., Perinelli, D.R., Ferrati, M., Spinozzi, E. *et al.* (2025) Toxicity and repellent activity of a carlina oxide nanoemulsion toward the South American tomato pinworm, *Tuta absoluta*. *Journal of Pest Science* 98(1), 309–320.

Touré, Y., Ongena, M.A.R.C., Jacques, P., Guiro, A. and Thonart, P. (2004) Role of lipopeptides produced by *Bacillus subtilis* GA1 in the reduction of grey mould disease caused by *Botrytis cinerea* on apple. *Journal of Applied Microbiology* 96(5), 1151–1160.

Trombik, J., Pergl, J., Skuhrovec, J., Barták, M. and Pyšek, P. (2023) Global patterns of alien Tephritidae invasions. *NeoBiota* 84, 181–207.

Tzanakakis, M.E., Tsitsipis, J.A. and Economopoulos, A.P. (1968) Frequency of mating in females of the olive fruit fly under laboratory conditions. *Journal of Economic Entomology* 61(5), 1309–1312.

Ullah, F., Zhang, Y., Gul, H., Hafeez, M., Desneux, N. *et al.* (2023) Estimation of the potential geographical distribution of invasive peach fruit fly under climate change by integrated ecological niche models. *CABI Agriculture and Bioscience* 4(1), 46.

Vadivelu, K. (2014) Biology and management of ber fruit fly, *Carpomyia vesuviana* Costa (Diptera: Tephritidae): A review. *African Journal of Agricultural Research* 9(16), 1310–1317.

Vargas, R.I., Miyashita, D. and Nishida, T. (1984) Life history and demographic parameters of three laboratory-reared tephritids (Diptera: Tephritidae). *Annals of the Entomological Society of America* 77(6), 651–656.

Vargas, R.I., Walsh, W.A., Kanehisa, D., Stark, J.D. and Nishida, T. (2000) Comparative demography of three Hawaiian fruit flies (Diptera: Tephritidae) at alternating temperatures. *Annals of the Entomological Society of America* 93(1), 75–81.

Vargas, R.I., Stark, J.D., Banks, J., Leblanc, L., Manoukis, N.C. *et al.* (2013) Spatial dynamics of two Oriental fruit fly (Diptera: Tephritidae) parasitoids, *Fopius arisanus* and *Diachasmimorpha longicaudata* (Hymenoptera: Braconidae), in a guava orchard in Hawaii. *Environmental Entomology* 42(5), 888–901.

Vayssières, J.F., Korie, S., Coulibaly, O., Temple, L. and Boueyi, S.P. (2008) The mango tree in northern and central Benin: Cultivars, production systems and constraints. *Fruits* 63(2), 111–120.

Vayssières, J.F., Carel, Y., Coubes, M. and Duyck, P.F. (2009) Development of immature stages and comparative demography of two invasive fruit fly species (*Ceratitis rosa* and *Bactrocera zonata*) in Réunion Island. *Environmental Entomology* 37(2), 307–314.

Vazquez, A., Cloonan, K.R., Rohde, B.B., Gill, M.A. and Mosser, L.K. (2022) Attraction and longevity of 2- and 3-component food cone lures for the Caribbean fruit fly, *Anastrepha suspensa* (Diptera: Tephritidae). *Journal of Economic Entomology* 115(4), 1231–1239.

Verghese, A. and Rashmi, M.A. (2016) *Management of Pre-harvest Insect Infestation for Quality Assurance of Fruit Crops in Relation to Quarantine and Export*. National Bureau of Agriculturally Important Insects, pp. 428–450.

Verghese, A., Madhura, H.S., Kamala Jayanthi, P.D. and Stonehouse, J.M. (2002) Fruit flies of economic significance in India, with special reference to *Bactrocera dorsalis* (Hendel). In: *Proceedings of 6th International Fruit fly Symposium*, pp. 6–10.

Vreysen, M.J., Robinson, A.S. and Hendrichs, J. (2007) *Area-wide Control of Insect Pests: From Research to Field Implementation*. Dordrecht: Springer.

Wakil, W., Usman, M., Piñero, J.C., Wu, S., Toews, M.D. *et al.* (2022) Combined application of entomopathogenic nematodes and fungi against fruit flies, *Bactrocera zonata* and *B. dorsalis* (Diptera: Tephritidae): Laboratory cups to field study. *Pest Management Science* 78(7), 2779–2791.

Walther, G.R., Roques, A., Hulme, P.E., Sykes, M.T., Pyšek, P. *et al.* (2009) Alien species in a warmer world: Risks and opportunities. *Trends in Ecology & Evolution* 24(12), 686–693.

Wan, X., Liu, Y., Zhang, B. and Wang, B. (2012) Invasion and management of *Bactrocera dorsalis* in China. *Journal of Integrative Agriculture* 11(3), 457–463.

White, I.M. (2006) Taxonomy of the *Dacina* (Diptera: Tephritidae) of Africa and the Middle East. *African Entomology Memoir* 2, 1–156.

White, I.M. and Elsonharris, M.M. (1992) Fruit flies of economic significance: Their identification and bionomics. *Environmental Entomology* 22, 1408.

Xie, Q. and Zhang, R.J. (2005) Study advance on biology and ecology of *Bactrocera dorsalis* (Hendel) and its control. *Ecology and Environmental Science* 24, 52–56.

Yan, Z.H., Zhang, J.L., Fang, X.J., Zhu, W.L., Zhang, W.H. *et al.* (2016) Host species of the insect pest *Bactrocera dorsalis* in Mengzi, Yunnan, and population dynamics of *Bactrocera dorsalis* and its related environmental factors in loquat garden. *Chinese Journal of Ecology* 35(11), 3037.

Yuan, S.Y., Kong, Q., Li, Z.Y., Xiao, C., Chen, B. *et al.* (2005) Study on biology of *Bactrocera* (Zeugodacus) *cucurbitae*. *Acta Agriculturae Boreali-Occidentalis Sinica* 14, 43–45.

Yuan RuiLing, Y.R., Yang Shan, Y.S., Wang XiaoWei, W.X. and Chen Peng, C.P. (2014) Test on flight ability of *Bactrocera dorsalis*. *Journal of Insect Science* 15(1), 141.

Zeng, Y., Reddy, G.V., Li, Z., Qin, Y., Wang, Y. *et al.* (2019) Global distribution and invasion pattern of Oriental fruit fly, *Bactrocera dorsalis* (Diptera: Tephritidae). *Journal of Applied Entomology* 143(3), 165–176.

Zhai, Q.H., Ni, M. and Xu, L.J. (2018) Study on the occurrence pattern and influencing factors of *Bactrocera dorsalis* (Hendel) in Wuxi area. *Xiandai Horticulture* 150–151.

Zhang, Q.Y., Lin, Z.J., Liu, J.Y. and Lin, X.Y. (1991) Biological characteristics of *Bactrocera tau* (Walker). *Plant Quarantine* 5, 164–166.

Zhang, Q.Y., Lin, Z.J., Liu, J.Y., Chen, H.Z., Gao, Q.Z. *et al.* (1998) Study on the biology of *Bactrocera dorsalis*. *Entomological Journal of East China* 7, 68–71.

Zhao, S., Xing, Z., Liu, Z., Liu, Y., Liu, X. *et al.* (2019) Efficient somatic and germline genome engineering of *Bactrocera dorsalis* by the CRISPR/Cas9 system. *Pest Management Science* 75(7), 1921–1932.

Zhou, T., Zhan, W. and Xiong, M. (2024) A series of methods incorporating deep learning and computer vision techniques in the study of fruit fly (Diptera: Tephritidae) regurgitation. *Frontiers in Plant Science* 14, 1337467.

Zhu, L., Li, X., Sun, H. and Han, Y. (2024) Research on CBF-YOLO detection model for common soybean pests in complex environment. *Computers and Electronics in Agriculture* 216, 108515.

Zida, I., Nébié, K., Sawadogo, A., Tassembédo, B., Kiénou, T. *et al.* (2023) Effectiveness of four integrated pest management approaches in the control of fruit flies (Diptera: Tephritidae) in mango agro-ecosystems in the South-Sudanian zone of Burkina Faso. *Advances in Entomology* 11(3), 124–142.

Zingore, M.J., Mohamed, S.A., Tanga, C.M. and Ekesi, S. (2020) Biology and rearing techniques for *Bactrocera zonata* (Saunders) (Diptera: Tephritidae). *Journal of Economic Entomology* 113(2), 855–864.

Index

Printed and bound by CPI Group (UK) Ltd, Croydon, CR0 4YY
16/09/2026
14963051-0001